MW01516697

Toshiaki Kabe, Atsushi Ishihara, Weihua Qian

Hydrodesulfurization and Hydrodenitrogenation

Chemistry and Engineering

KODANSHA

WILEY-VCH

Toshiaki Kabe, Atsushi Ishihara, Weihua Qian

Hydrodesulfurization and Hydrodenitrogenation

Chemistry and Engineering

KODANSHA

WILEY-VCH

Weinheim · Berlin · New York · Chichester·
Brisbane · Singapore · Toronto

Prof. T. Kabe, Prof. A. Ishihara, Prof. W. Qian
Department of Applied Chemistry
Tokyo University of Agriculture and Technology
Tokyo
Japan

This book was carefully produced. Nevertheless, authors and publisher do not warrant the information contained therein to be free of errors. Readers are advised to keep in mind that statements, data, illustrations, procedural detailes or other items may inadvertently be inaccurate.

Published jointly by
Kodansha Ltd., Tokyo (Japan),
WILEY-VCH Verlag GmbH, Weinheim (Federal Republic of Germany)

Library of Congress Card No. applied for.

A catalogue record for this book is available from the British Library.

Deutsche Bibliothek Cataloguing-in-Publication Data:

Kabe, Toshiaki:
Hydrodesulfurization and Hydrodenitrogenation / T. Kabe; A. Ishihara; W. Qian-Weinheim; New York; Chichester; Brisbane; Singapore: Toronto; Wiley-VCH. 1999
ISBN 3-527-30116-X (WILEY-VCH)
ISBN 4-06-209596-3 (KODANSHA)

Copyright © Kodansha Ltd., Tokyo, 1999
All rights reserved. No part of this book may be reproduced in any form, by photostat, microfilm, retrieval system, or any other means, without the written permission of Kodansha Ltd.(except in the case of brief quotation for criticism or review).

Printed in Japan

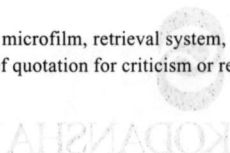

Preface

In recent years, air pollution in large city areas has become a serious problem, making it important to reduce heteroatoms such as sulfur and nitrogen in liquid fuels. Therefore hydrotreatment is key to the production of clean fuels. Hydrotreating processes in industry include various important catalytic reactions. Hydrodesulfurization (HDS) and hydrodenitrogenation (HDN) are among the most important. To improve such processes, it is necessary to understand the properties of catalytic reactions which are closely related to the study of applied chemistry and chemical reaction engineering. These catalytic reactions are often complicated and consist of wide-ranging fields such as reaction mechanisms, structure of catalysts and process engineering. Thus it has been difficult to study these subjects sysytematically and in detail. Nevertheless, recent progress in various analytical technologies has led to mutual understanding. A comprehensive monograph covering both the scientific and practical aspects of these catalytic reactions has been in demand. In this volume, we have tried to describe the basic and practical aspects of the science and technology of catalytic reactions related to hydrotreating reactions, especially HDS and HDN, which have been developed over the past two decades. These are impotant measures in coping with environmental pollution and the subject of strong interest in not only in Japan but throughout the world.

We would like to express our gratitude to Ms. Cecilia, M. Hamagami and Mr. Ippei Ohta of Kodansha Scientific Ltd. for their invaluable assistance in the preparation of the English manuscript.

The publication of this volume is supported in part by a Grant-in-Aid for Publication of Scientific Research Result, Grant-in-Aid for Scientific Research, the Ministry of Education, Science, Sports and Culture in 1998.

July 1998

Toshiaki Kabe
Atsushi Ishihara
Weihua Qian

Contents

vii

3   Structure of Hydrodesulfurization and Hydrodenitrogenation Catalysts   133

1

Introduction to Hydrodesulfurization and Hydrodenitrogenation

Hydrodesulfurization (HDS) and hydrodenitrogenation (HDN) are catalytic hydrogenation processes which remove sulfur and nitrogen in petroleum, and primarily constitute hydroprocessing or hydrotreating. In the 1960's, the process of HDS developed remarkably to remove high concentrations of sulfur in fuels. In recent years, the air pollution by nitrogen oxides and particulate matters included in diesel exhaust gas in large city area has become a serious problem, and it is important to reduce the sulfur content in light gas oil.[1,2] The problem has been caused by the rapid increase in petroleum product demand throughout the world, especially the Asia-Pacific area. The demand for petroleum in this area up to the early part of the 21st century is increasing approximately parallel to the increase in the demand worldwide, indicating that most of the increase in demand can be attributed to this area.[3] According to a survey of the increase in demand for each fraction of fuel, although demand for fuel oil will hardly change early in the 21st century, demand for gasoline and middle distillates is expected to increase twofold.[3-5] Most of advanced countries have already started regulating sulfur content, cetane index and aromatics content in distillate. The sulfur content of distillate in Japan was also regulated to 0.05 wt% in 1997. In 0.05 wt% of sulfur content most products must be hydrotreated, thus the hydrotreating process plays a very important role under such regulations.

On the other hand, hydrotreating catalysts have been developed for over several decades. Their activities increased nearly twofold over the past 30 years.[6] Although catalytic activity generally increases with respect to increase in the surface area of a catalyst, there is a limit as to has much the surface area of support can be extended. To cope with not only deep hydrotreatment but also the regulations of fuel oils which are growing stricter, it is essential to develop highly active deep hydrotreating catalysts, and elucidation of the reaction mechanism under deep hydrotreating conditions is required for the development of such catalyst. A number of attempts have been made to elucidate the mechanism of HDS and HDN by studying the kinetics. However, it is not well known what kind of sulfur compounds exists in liquid fuels and which among them is still not desulfurized at the final stage of the reaction. It has not been clarified either why these compounds remain until the final stage, that is, what is different in the mechanisms. The effects of the components in liquid fuels become very important in controlling the catalysis of HDS and HDN under deep hydrotreating conditions. Further, there are few examples clarifying the behavior of heteroatoms on the hydrotreating catalyst. In order to develop new catalysts on a continuous basis, it is essential to understand in more detail the HDS and HDN mechanisms. A deep understanding of these mechanisms will be the foundation for making rapid progress in the development of the necessary catalysts.

Here, we discuss in detail the following four aspects which are important for catalyst development and reaction mechanism study in HDS and HDN: 1) reaction profiles in HDS and HDN; 2) structure of HDS and HDN catalysts; 3) development of novel HDS and HDN catalysts; 4) process engineering. This chapter provides the fundamentals for understanding the processes of HDS and HDN and summarizes the contents of each chapter.

1.1 Components in Raw Oil Materials

The sulfur compounds found in petroleum or synthetic oils are generally classified into one of two types: heterocycles or nonheterocycles. The latter comprises thiols, sulfides and disulfides. Heterocycles are mainly composed of thiophenes with one to several rings and their alkyl or aryl substituents. Examples of sulfur compounds are shown in Fig. 1.1.

Although the determination of each component in fuels was difficult because the raw fuels were complex mixtures of various compounds, recent advances in analytical methods have

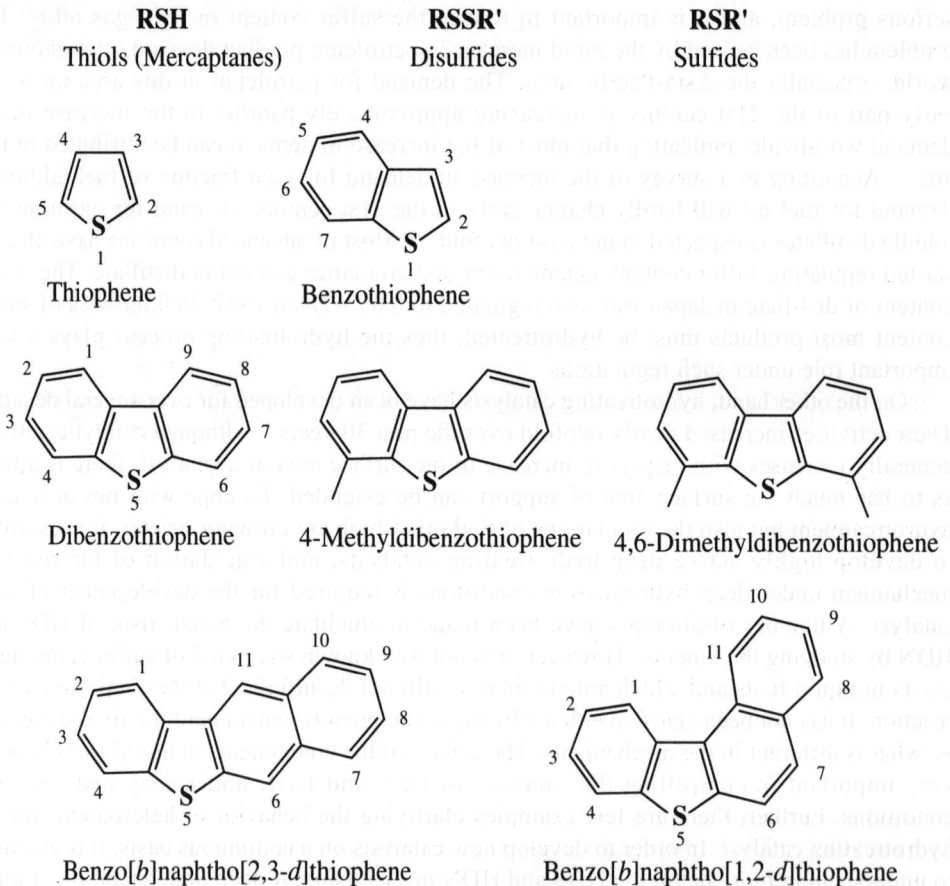

Fig. 1.1 Sulfur-containing compounds in petroleum.

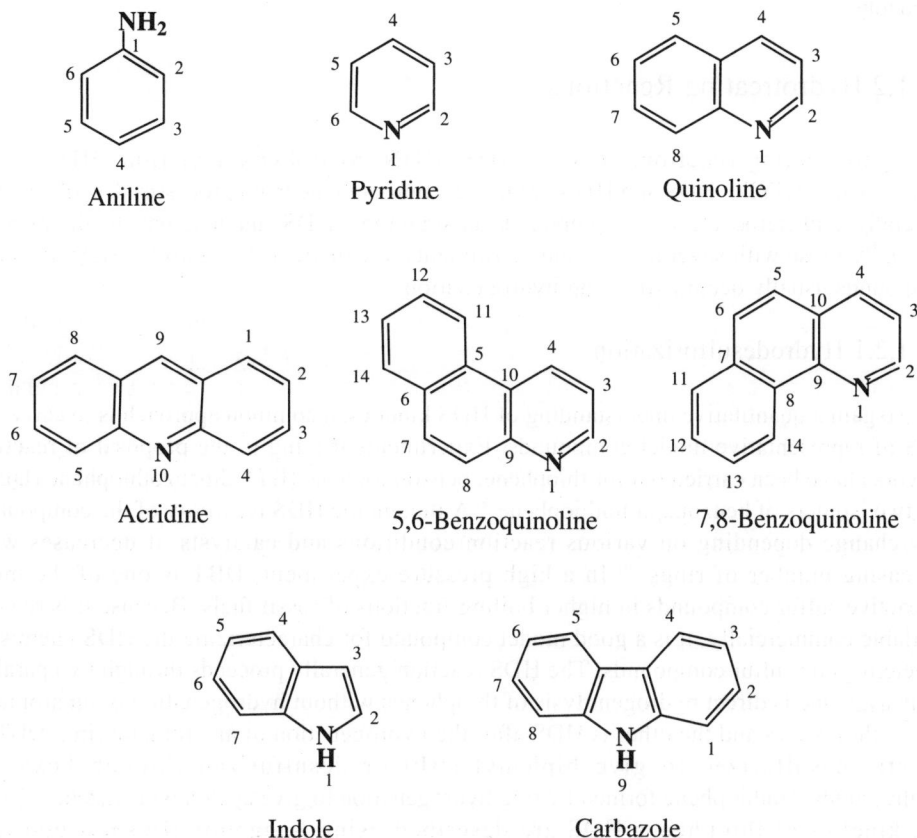

Fig. 1.2 Nitrogen-containing compound in petroleum.

enabled us to determine the sulfur components in raw oil materials. Sulfur containing polyaromatic compounds in straight run gas oil from Arabian Light were analyzed and determined by a gas chromatography-atomic emission detector (GC-AED) and a gas chromatography-mass spectroscopy (GC-MS).[7] It was found that 42 kinds of alkylbenzothiophene and 29 kinds of alkyldibenzothiophene were included in the oil. When this oil was desulfurized using Co-Mo/Al$_2$O$_3$ catalysts at 300–410°C, 4-methyldibenzothiophene (4-MDBT) and 4,6-dimethyldibenzothiophene (4,6-DMDBT) were most difficult to desulfurize. This result suggested that HDS of DBTs substituted at the 4,6-positions is the key reaction to achieve deep desulfurization.

Nitrogen compounds included in the feedstocks are divided to two types: heterocycles and nonheterocycles. Some of these are shown in Fig. 1.2. Among nonheterocyclic compounds, aniline derivatives which necessarily appear in the HDN network of heterocycles are important in HDN. Heterocyclic nitrogen compounds are most important in HDN because these are included in the feedstocks in larger amounts and are the most difficult to remove. Heterocyclic nitrogen compounds can be divided into basic compounds and nonbasic compounds. Basic compounds include six-membered ring heterocycles such as pyridine, quinoline and acridine. Nonbasic compounds include five-membered ring heterocycles such as pyrrole, indole and carbazole.

1.2 Hydrotreating Reactions

Hydrotreating reactions include HDS, HDN, hydrodeoxygenation (HDO), and hydrogenation (HYD). HDS and HDN are the reactions and the main processes described here. Generally nonheterocycle sulfur compounds undergo rapid HDS, and heterocyclic thiophenes, especially those with several rings and substituents, are difficult to remove.[8] HDN of these compounds usually occurs after ring hydrogenation.

1.2.1 Hydrodesulfurization

To gain a quantitative understanding of HDS kinetics, a common approach is to carry out HDS of representative model compounds. Experiments leading to the proposal of reaction networks have been carried out for thiophene, benzothiophene (BT), dibenzothiophene (DBT) and two isomers of benzonaphthothiophene.[9] Although the HDS reactivity of the compounds may change depending on various reaction conditions and catalysts, it decreases with increasing number of rings.[10] In a high pressure experiment, DBT is one of the most unreactive sulfur compounds in higher boiling fractions of fossil fuels. Because it is readily available commercially, it is a good model compound for characterizing the HDS chemistry of heterocyclic sulfur compounds. The HDS reaction generally proceeds through two parallel pathways: One is direct hydrogenolysis of thiophenes without hydrogenation of an aromatic ring in thiophenes and the other is HDS after the hydrogenation of an aromatic ring. DBT is directly desulfurized to give biphenyl (BP) or desulfurized through hexa- or tetrahydrodibenzothiophene formed by ring hydrogenation to give cyclohexylbenzene (CHB). The kinetics of thiophenes HDS are described using Langmuir-Hinshelwood rate equations.[8,9,11] The rate equations indicate that thiophenes inhibit their own HDS and that hydrogen sulfide inhibits HDS. Hydrogen sulfide (H_2S) is a strong inhibitor of hydrogenolysis but is not an inhibitor of hydrogenation in many cases. Further, there is a difference in form between the thiophenes hydrogenolysis rate equation and that for its hydrogenation. These results indicates that the two types of reactions take place on separate kinds of catalytic sites.

1.2.2 Effects of Methyl Substituents on Deep Hydrodesulfurization

It is well known that methyl substituents affect the reactivity of DBT.[12,13] Methyl substituents in the 2 and 8 positions of DBT hardly changed the reactivity and methyl groups in the 3 and 7 positions also had only a small effect on the reactivity.[12] In contrast, a methyl substituent at the 4 and 6 positions reduced the reactivity by one order of magnitude.[12,13]

In order to know why it is difficult to desulfurize methyl-substituted DBTs such as 4-MDBT and 4,6-DMDBT, HDS of DBT, 4-MDBT and 4,6-DMDBT was kinetically investigated in the range 190–340°C with the use of Co-Mo/Al_2O_3 and Ni-Mo/Al_2O_3.[13] Products were biphenyls (BPs) and cyclohexylbenzenes (CHBs). The total conversions of DBTs and the conversions of DBTs into BPs decreased in the order DBT > 4-MDBT > 4,6-DMDBT. In contrast, the conversion of DBTs into CHBs was almost the same at every temperature examined. It is indicated that, when an aromatic ring in DBTs is hydrogenated prior to desulfurization, the steric hindrance of a methyl group is weakened significantly. In order to elucidate the differences in reactivities of DBTs, activation energies of HDS and heats of adsorption of DBTs were estimated using the Langmuir-Hinshelwood rate equation. The results showed that the retarding effect of methyl substituents on HDS was not due to the inhibition of adsorption of DBTs onto the catalyst.

1.2.3 Retarding Effects of Components on Deep Hydrodesulfurization

Because a liquid fuel contains various components such as aromatics and N-containing hydrocarbons, the effects of components in light gas oil on catalytic activity become large under deep HDS conditions. In order to understand these effects, the effects of solvents on HDS of BT and DBT catalyzed by Co-Mo/Al$_2$O$_3$ were investigated.[10] The order of HDS activity was toluene > decalin > n-pentadecane > 1-methylnaphthalene in HDS of BT and n-heptane > xylene > decalin > tetralin in HDS of DBT. In order to estimate the retarding effects of solvents quantitatively, a retarding term by a solvent was introduced into the Langmuir-Hinshelwood equation. Van't Hoff plots of the adsorption equilibrium constant of a solvent revealed the linear relationship and heats of adsorption of solvents were calculated from the slopes. The results indicate that a solvent such as 1-methylnaphthalene, which has high aromaticity, is adsorbed competitively with thiophenes such as BT and DBT on Co-Mo/Al$_2$O$_3$ by more than a 1:1 ratio.

In HDS of light gas oil, about 3% of H$_2$S is produced. It is reasonable to assume that H$_2$S will markedly retard the HDS of light gas oil. However, the effect of H$_2$S on HDS of 4,6-DMDBT is not well known. To achieve deep desulfurization of liquid fuels, it becomes important to elucidate the reaction mechanism and the effect of H$_2$S on the HDS of 4,6-DMDBT.[14] Thus, the effects of H$_2$S on catalytic activity and selectivity in HDS of DBT and 4,6-DMDBT were investigated under deep desulfurization conditions (sulfur concentration < 0.05 wt%) using a commercial Co-Mo/Al$_2$O$_3$ catalyst. The conversion of DBTs decreased with increasing partial pressure of H$_2$S. The formation of both BPs and CHBs was inhibited by H$_2$S, but the former was inhibited more significantly. The HDS reactions of DBTs could be described using the Langmuir-Hinshelwood rate equation. The results showed that H$_2$S was adsorbed more strongly on the catalyst than DBT and 4,6-DMDBT, and inhibited the HDS of DBT and 4,6-DMDBT.

1.2.4 Model Studies of HDS Mechanisms

A number of HDS mechanisms have been suggested, and it has been believed that sulfur anion vacancies are the active sites for HDS because the coordinatively unsaturated sites are needed for the adsorption of thiophenes to occur. There are two adsorption modes of thiophenes: one-point (end-on) mode with sulfur atom or multipoint (side-on) mode. The latter may involve the adsorption with all or some member of the thiophene ring. It has been proposed that HDS proceeds through the adsorption of thiophenes on anion vacancies, hydrogenolysis and hydrogenation with metal hydride or SH species, desorption of products and H$_2$S, and the regeneration of anion vacancies. In HDS of DBT, 4-MDBT and 4,6-DMDBT, these DBTs are adsorbed on the surface with π-bond through aromatic rings.[13] It was proposed that DBTs rotated around a sulfur atom from parallel to perpendicular configuration when the C-S bond scission occurred. It was concluded that HDS of 4,6-

DMDBT and 4-MDBT was retarded in comparison with that of DBT because the rotation of 4,6-DMDBT and 4-MDBT during the C-S bond scission was inhibited due to the steric hindrance of the methyl group.

To understand the mechanism of thiophene HDS, the structures and the reactivities of thiophenes on organometallic complexes have been investigated.[15,16] Organometallic chemistry provides not only visual pictures (see Section 2.1.5.B) of the coordination and reaction of thiophenes and thiophene derivatives but also the possibility that such similar reactions as seen in those pictures may occur actually on a catalysts surface. It seems difficult to cleave the C-S bond compared with thiols or sulfides because the thiophene ring is conjugated with π-electrons and its C-S bonds have the double bond character. However, the C-S bonds in thiophenes can easily be cleaved on organometallic complexes. Thiophenes may coordinate to metal centers in organometallic complexes before the C-S bond scission occurs. There are several coordination modes of thiophenes on complexes, which could be models of adsorption modes of thiophenes on heterogeneous surface. The coordinated thiophenes reacted with various nucleophiles to give products in which the C-S bond of thiophenes was cleaved. In another type of C-S bond scission of thiophenes, transition metal centers insert into C-S bond of the thiophenes to give metallathiabenzene derivatives. Metal hydrides also cleave the C-S bond of thiophenes. Since molecular hydrogen is readily adsorbed dissociatively on surfaces to form surface hydride species, a metal hydride complex may be one of the model compounds of surface hydride species and the reactions of the complex with thiophenes can provide interesting models for the heterogeneous surface reactions. Reactions of thiophenes with metal clusters also represent a homogeneous analogue for heterogeneous HDS catalysis.

Surface science studies of the structures and reactivities of thiophenes on metal single-crystal surfaces also improve our understanding of the mechanism of thiophenes HDS.[17,18] Reactions of thiophene on various metal surfaces were carried out and analyzed using temperature-programmed desorption (TDS), edge X-ray adsorption fine structure (EXAFS), high-resolution electron-energy-loss spectroscopy (HREELS), and other methods. The coverage of thiophene and coadsorbed hydrocarbons and sulfur affects the orientation of thiophene with respect to metal surfaces. The metal surface acts as not only an electron donor but also an electrophile in the reaction of sulfur compounds. Some results from the reactions on the single-crystal surface are very similar to those on supported MoS_2, indicating that the single-crystal surface studies can be regarded as a good model reaction of actual heterogeneous HDS.

1.2.5 Hydrodenitrogenation

The concentration of the nitrogen-containing compounds in the heavier feedstocks is generally much higher than that in straight run distillates. Further, the acidic catalysts used for catalytic cracking of the heavier feedstocks are poisoned by nitrogen compounds and polyaromatics. Therefore, HDN becomes more important in the hydroprocessing of heavier feedstocks.[9,19-21] Hydrogenation of the N-containing ring occurs prior to C-N bond scission over conventional catalysts in order to reduce the strong bond energy of the C-N bond in N heterocycles (C = N, 615 kJ/mol, C-N, 305 kJ/mol). HDN is virtually irreversible under practical reaction conditions and is generally controlled by kinetics rather than by thermodynamics. The reactivities for N-ring hydrogenation decreased in the order quinoline > pyridine > isoquinoline > indole > pyrrole, and the reactivities for ring hydrogenation decrease in the order N-ring > aniline-like compounds > comparable aromatic.[22)] In general, the steric effect is not observed in HDN reaction so much as HDS probably because C-N bond scission and successive N removal occur only after hydrogenation of N-ring and the steric effect is extremely weakened.

Reaction networks and kinetics for HDN of pyridine, quinoline, acridine, indole, several anilines and their derivatives have been investigated. Major common pathways of N heterocycles are as follows: 1) hydrogenation of the N-ring, 2) C-N bond scission to an amine, and 3) hydrogenolysis of the amine to hydrocarbons and ammonia. For example, pyridine is hydrogenated to piperidine of which one C-N bond is cleaved to form n-pentylamine. n-Pentane and ammonia are formed by hydrogenolysis of the amine. The HDN rate of n-pentylamine is an order of magnitude faster than that of pyridine.[23)] There are few examples for quantitative analysis of HDN of nonbasic five-membered nitrogen heterocycles. Aniline intermediates appear in almost all the reaction networks of HDN of N heterocycles such as quinoline, acridine and indole, and thus nitrogen removal from aniline intermediates has often been one of the most important reactions in a network. Hydrogenation of aromatic ring is generally needed before C-N bond scission.

It is well known that H_2S promotes hydrogenolysis in HDN.[23,24)] The effect of several organic sulfur compounds such as thiophenes on HDN has been investigated and may be regarded as the effect of H_2S formed. Some positive effects as well as negative effects of these compounds on hydrogenolysis and hydrogenation included in HDN have been observed and these effects changed depending on the factors of kinds of sulfur compounds and catalysts, reaction conditions, etc. The available results indicate that the presence of sulfur plays an important role in bringing out the high activity of HDN catalysts.

Nitrogen compounds have a retarding effect on HDN. Basic nitrogen compounds such as quinoline are more strongly adsorbed on the catalyst surface than nonbasic nitrogen compounds such as indole. Although alkylanilines are known to be relatively reactive when they are pure, alkylanilines HDN become significantly unreactive in the presence of nitrogen heterocycles.[25,26)] The effect is attributed to the competitive adsorption onto the catalyst surface between anilines and nitrogen heterocycles.

The mechanisms of C-N bond scission have been explained based on classical Hofmann-type elimination and nucleophilic substitution. These two routes can be catalyzed by Brønsted acid, which also increases with increase in H_2S concentration. The above explains the fact that H_2S promotes the C-N bond scission.

1.3 Structure and Active Sites of Hydrotreating Catalysts

1.3.1 Conventional Molybdenum- and Tungsten-based Catalyst

HDS catalysts are normally prepared by pore volume impregnation of alumina with aqueous solutions of $(NH_4)_6Mo_7O_{24}$, $Co(NO_3)_2$, and $Ni(NO_3)_2$ with intermediate drying and calcination steps. The resulting oxide precursor is presulfided prior to HDS reaction according to a sulfiding procedure that may consist of reacting in a mixture of H_2S and H_2 or thiophene and H_2, or in a liquid feed of sulfur-containing molecules and H_2. During sulfiding as well as during actual HDS reaction, because the catalysts are highly reduced with H_2S, which is always present, thermodynamics predict that molybdenum should be in the MoS_2 form, cobalt in the Co_9S_8, and nickel in the Ni_3S_2 or NiS form.

MoS$_2$ belongs to a group of materials that crystallize with the layered structure, and it is found that each layer is composed of sheets of Mo atoms sandwiched between sheets of sulfur atoms. Within a given layer the bonding is mainly covalent, whereas between layers the bonding is mainly of the van der Waals type.[27] Topsøe et al. claim that MoS$_2$ can be present on industrial supports as very large patches of a wrinkled, one-slab-thick MoS$_2$ layer.[28] Therefore catalysis most likely occurs at edges and corners, and not at basal planes.[29]

The structure of the catalyst containing a promoter such as cobalt and nicke is more complicated than Mo or W/Al$_2$O$_3$ catalyst. Many spectroscopic techniques are capable of detecting the presence of cobalt in one structure or another. Based on these advances, a variety of structure models were proposed. The four main types of structural models are the monolayer model,[30] pseudo-intercalation model,[31,32] contact synergy model[33] and edge decoration model or "CoMoS" phase model.[28,34] At present the "CoMoS" model has gained the greatest recognition.

1.3.2 Behavior of Sulfur on Sulfided Catalyst in Hydrotreating Reaction

In recent years, the radioisotope tracer method using radioactive ^{35}S has been developed to clarify the behavior of sulfur because this method makes it possible to understand more precisely how sulfur in sulfur compounds is translated to H$_2$S and how sulfur in the sulfided catalyst participates in the actual HDS reaction. Sulfur has a radioisotope ^{35}S with a half-life of 87.5 days emitting soft β-radiation (167 kev). Lukens et al.[35] have measured the accessible surface area of supported transition metal sulfides by isotope exchange with a labeled H$_2$S in liquid scintillation solution. Kalechits and co-workers[36] have shown that in the hydrogenation of a mixture of benzene and ^{35}S-labeled CS$_2$ ([^{35}S]CS$_2$) on WS$_2$ catalyst, the sulfur in the catalyst was exchanged with radioactive sulfur of the feedstocks. This labile sulfur would be a part of the non-stoichiometric sulfur of the catalyst that would be responsible for the acceleration of acid catalyzed reaction (isomerization and cracking). Gachet et al.[37] presulfided a commercial Co-Mo/Al$_2$O$_3$ catalyst with the gas mixture of ^{35}S-labeled H$_2$S ([^{35}S]H$_2$S) and H$_2$, then carried out the HDS of DBT at atmospheric pressure. They postulated that two types of sulfur appeared over the sulfided catalyst: labile sulfur and relatively fixed one. Gellman et al.[38] used radiotracer ([^{35}S]CS$_2$) labeling technique to measure removal rate of sulfur adsorbed on the Mo (100) surface in HDS of thiophene and postulated that the rate was fit with a first-order kinetic equation. Isagulyants et al.[39] made an attempt to estimate quantitatively the amount of sulfur held on the catalysts in a study where HDS of thiophene was conducted on a series of Co-Mo catalysts sulfided by ^{35}S elements and ^{35}S-thiophene, respectively. Their study pointed out that about 20–30% of sulfur held on vacancy sites was exchangeable. Paal and his group[40] carried out hydrogenation reaction of cyclohexanol and HDS reaction of thiophene in a pulse microreactor over the MoO$_3$, CoMoO$_4$, Mo/Al$_2$O$_3$ and Co-Mo/Al$_2$O$_3$ catalysts, promoted by Pd, Ir and Ru. In their study, the catalysts were sulfided by a gas mixture of [^{35}S]H$_2$S and H$_2$. In these researches, however, all reactions were carried out at low pressure and over ^{35}S-labeled catalysts.

In recent years, the HDS of ^{35}S-labeled dibenzothiophene ([^{35}S]DBT) catalyzed by Mo/Al$_2$O$_3$, Co/Al$_2$O$_3$, Co-Mo/Al$_2$O$_3$, Ni/Al$_2$O$_3$ and Ni-Mo/Al$_2$O$_3$ has been carried out.[41–46] After the reaction of [^{32}S]DBT reached the steady state, [^{35}S]DBT was substituted for [^{32}S]DBT and the radioactivities of unreacted [^{35}S]DBT and formed [^{35}S]H$_2$S were traced at the outlet of a reactor. The radioactivities of unreacted [^{35}S]DBT in liquid products reached a steady state immediately. On the contrary, time delay in the case of [^{35}S]H$_2$S was detected to achieve the steady state of the radioactivity. Then, the decalin solution of [^{32}S]DBT was again substituted for [^{35}S]DBT in a similar way, and the radioactivities of unreacted [^{35}S]DBT decreased immediately from the steady state to normal state while a time delay in the case of [^{35}S]H$_2$S was also observed. The result indicates that the sulfur in DBT is not directly released as H$_2$S, but accommodated on the catalyst. From the amount of ^{35}S accommodated on the catalyst, the amount of labile sulfur could be calculated. It was suggested from this calculation that in Co-Mo/Al$_2$O$_3$ and Ni-Mo/Al$_2$O$_3$, the sulfur located between Mo and Co or Ni would be the labile sulfur.

The ^{35}S tracer method was also used to investigate HDO and HDN. It was found that the sulfur exchange rate only depends upon the rate of sulfur incorporation from HDS reactions, and is irrelevant with kinds of sulfur-containing compounds. It is likely that HDO and HDN reactions occur mainly on active sites different from that of HDS reaction on sulfided Mo/Al$_2$O$_3$.

1.3.3 Correlation between Structure and Catalytic Activity

Scheffer et al.[47] found that the reduction behavior of sulfided Mo/Al$_2$O$_3$ catalysts correlates with HDS activity. The most effective factor for the catalytic activity is the loading of active metals as well as supports. For low Mo-loading catalysts, increasing the Mo loading will increase the total concentration of MoS$_2$ edge sites.[48] To correlate the sulfided state of the catalyst with the behavior of catalyst during the HDS reaction, several groups have developed a radioisotope ^{35}S tracer method to trace the behavior of sulfur in HDS.[37,38,40,49]

Concerning the effect of molybdenum content on the structure of sulfided catalysts, the sulfur exchange rates were approximately same at the same temperature and apparent activation energies of HDS reactions were 20 ± 2 kcal/mol for all catalysts.[43] These results suggest that the mechanism of HDS and the nature of active sites did not vary with molybdenum content. On the other hand, the amount of labile sulfur increased linearly with the molybdenum content up to 2.89 atom/nm^2 but then leveled off over this loading of molybdenum. Thus, it was suggested that monolayer dispersion of molybdenum sulfide on alumina was maintained up to 2.89 atom/nm^2 but some of crystallite of molybdenum sulfide would be formed when molybdenum was added beyond 2.89 atom/nm^2. Moreover, since the sulfur exchange rate hardly varied with the molybdenum content, the HDS rates of DBT on the catalysts paralleled the amount of labile sulfur.

^{35}S radioisotope pulse tracer method (RPTM) was also used to investigate the promotion of Co to Mo/Al$_2$O$_3$.[44] The increase in amount of labile sulfur increased remarkably with the addition of cobalt and then increased linearly with increasing ratio of cobalt-to-molybdenum up to about 0.5, but decreased slightly above 0.5 of Co/Mo molar ratio. In contrast to this, the rate constant of sulfur exchange remarkably increased once with the addition of cobalt whereas it no longer varied with increase in ratio of cobalt to molybdenum.

In most studies on hydrotreatment, the relationships between catalyst structure and HDS reactions of thiophenes have not been clarified and HDS of thiophenes was often carried out at atmospheric pressure. Further, there are few studies dealing with the promoting effect of Co on Mo/Al$_2$O$_3$ catalysts for HDS of DBT and 4,6-DMDBT. HDS of DBT and 4,6-DMDBT are described here using a series of Mo/Al$_2$O$_3$ and Co-Mo/Al$_2$O$_3$ catalysts.[50] The activities of

DBT and 4,6-DMDBT for HDS increased with increasing amount of cobalt added at lower Co/Mo molar ratios (below 0.5). At higher Co/Mo molar ratios (above *ca.* 0.5), however, the promoting effect of cobalt for DBT increased only slightly. Compared with the case of DBT, the promoting effect of cobalt for HDS of 4,6-DMDBT decreased with increasing Co/Mo molar ratio; the maximum effect was attained when the ratio was 0.5. At the same molar ratio, the rate constants of the formation of BPs were approximately 20 times that of unpromoted Mo/Al_2O_3, while the rate constants of the formation of CHBs were approximately 4 times that of unpromised Mo/Al_2O_3. Cobalt enhanced the activity of HDS more than that of hydrogenation. Further, the mode of formation of decalin and CHBs was nearly the same and it was suggested that hydrogenation of tetralin and DBTs occurred on the same active sites.

1.3.4 Genesis of Active Sites and Reaction Mechanism of HDS

A major part of hydrotreatment research has focused on obtaining insights on the genesis of the active sites and the reaction mechanism for the development of novel hydrotreating catalysts.

The inhibition of H_2S in HDS was compared between Mo/Al_2O_3 and $Co-Mo/Al_2O_3$ using a kinetic analysis method.[45] The heats of adsorption of DBT, 4,6-DMDBT and H_2S over each catalyst increased in the order DBT < 4,6-DMDBT < H_2S. Compared with the Mo/Al_2O_3 catalyst, all the adsorption equilibrium constants of the $Co-Mo/Al_2O_3$ catalyst were less than those of the Mo/Al_2O_3 catalyst because of the addition of Co. On the other hand, it was found that sulfur on the catalyst can be exchanged by the sulfur in H_2S. This portion of sulfur represented the total amount of sulfur exchanged on the catalyst under this reaction condition.[46,51] Sulfur exchange with H_2S was very rapid at each temperature. Comparison of the amounts of labile sulfur and the rate constants of sulfur exchange in HDS reactions with those in sulfur exchange with H_2S suggests that the transformation between the labile sulfur and the vacancies on the catalyst surface proceed predominantly through the sulfur exchange with H_2S even in the HDS reactions. It was also found that the addition of H_2S inhibited the HDS reaction of DBT while it increased the sulfur exchange rate.

The generation mechanism of active site and the HDS mechanism were discussed.[46] There are two routes where labile sulfur present in the form of bimetallic sulfur species, "CoMoS phases," desorbed as H_2S from the catalyst and formed a vacancy. In the sulfur exchange with H_2S, when one vacancy (active site) is occupied by sulfur in H_2S formed in the HDS reaction, a labile sulfur in another site is released as H_2S to form another new vacancy. In HDS, when a sulfur compound is adsorbed on a vacancy, the C-S bond is subsequently cleaved, and the sulfur remains on the catalyst. Simultaneously, another labile sulfur is released as H_2S and a new active site is formed. In the two routes, the migration of vacancies on the catalyst always occurs due to the transformation between labile sulfur and vacancies on the catalyst surface. Therefore, it can be assumed that a rapid adsorption/desorption of H_2S will always take place in the presence of H_2S under typical hydrotreating conditions, leading to a rapid interconversion of the active site and labile sulfur. Thus, the vacancies under reaction conditions will not be fixed, but be mobile.

1.4 Development of Novel Catalysts

Alumina-supported Co-Mo and Ni-Mo have been conventional catalysts of petroleum hydrotreatment for a long time. However, the development of novel catalysts with high catalytic activity is needed in order to achieve clean fuel production. Novel preparation methods and the reactivities of catalysts using molybdenum as a major active element and cobalt and nickel as promoters are described first, followed by approaches using noble metals.

1.4.1 Conventional Approaches

Molybdenum is a major active element for hydrotreatment, and cobalt and nickel are the most important promoters. Since supports directly act on such active elements, the use of different kinds of supports is one of the most simple methods for preparing novel catalysts. Alumina has many advantages and has been extensively studied. However, in order to develop novel hydrotreating catalysts with excellent capabilities, it is very important to study supports other than alumina.

Carbon-supported Co-Mo and Ni-Mo catalysts have higher activity in HDS than the conventional alumina-supported ones.[52] HDN catalyzed by carbon-supported catalysts has also been reported.[53] It has been reported that a fully sulfided CoMoS phase which is more active than another less sulfided CoMoS phase is formed on carbon support.[54] When nitrilotriacetic acid was used with cobalt nitrate and heptamolybdate, this highly active fully sulfided CoMoS was formed selectively on carbon.[55]

Other kinds of oxide supports such as TiO_2, ZrO_2, SiO_2-Al_2O_3 and SiO_2 have also been studied. Up to a molybdenum loading corresponding to 2.8 atom/nm^2 of molybdenum, Mo/TiO_2 catalysts were about five times more active per molybdenum atom than the corresponding Mo/Al_2O_3 samples in the hydrogenation of BP and the HDS of thiophene.[56] In an electron microscopic study, the length of MoS_2 crystallites for Mo/TiO_2 distributed toward the smaller particle range in comparison with Mo/Al_2O_3.[57] The differences in the catalytic activity between these two catalysts was explained in terms of different activities between the smaller and larger particles. However, the synergetic effect of cobalt was lower on TiO_2 than on Al_2O_3.[57,58]

Titania is unsuitable for industrial application due to its relatively low surface area and the low stability of the active anatase structure at high temperature. Therefore, some binary and ternary oxides such as TiO_2-Al_2O_3, TiO_2-ZrO_2 and TiO_2-ZrO_2-V_2O_5 have been used as supports in a catalyst for hydrotreating reactions.[59,60] It should be noted that the properties of such mixed oxides depend on the preparation procedure to a great extent.

The use of additives such as phosphorus, boron and fluorine to molybdenum-based conventional catalysts is also one of the simple methods for modifying the HDS and HDN reactivities of these catalysts. It is well known that not only nickel and molybdenum but phosphorus is also contained in commercial HDN catalysts for the treatment of heavy feedstocks.[61] The positive effects of phosphorus on HDS, HDN, hydrogenation, hydrocracking and hydrodemetallation catalyzed by molybdenum-based catalysts have been reported.

A temperature-programmed reaction of MoO_3 with NH_3 or CH_4 and H_2[62] has enabled the use of these as catalysts for several hydrogen transfer reactions. The behavior of nitrides and carbides in most of the reactions is similar to that of the group 8 metals. These properties of nitrides and carbides have promoted the development of novel hydrotreating catalysts using these materials.[63] HDN activities of several unsupported and supported molybdenum nitrides and carbides are superior to those of commercial catalysts. These results show that the products are aromatic compounds rather than completely saturated ones, indicating that the ability for hydrogenation is rather low. Supported and unsupported molybdenum nitrides are also active for HDS.

The addition of nitrilotriacetic acid (NTA), ethylenediamine (EDA) or citric acid to Co-Mo catalysts increases the catalytic activity of thiophene HDS.[55,64,65] Effects are observed for Al_2O_3, SiO_2 and carbon supports. The roles played by these reagents may be attributed to an increase in concentration or dispersion of active species on the catalyst, or a decrease in the lateral size of MoS_2-like crystallites.

Dispersed supported molybdena/alumina catalysts can be prepared from simple physical mixtures by the spreading of MoO_3 on the surface of an Al_2O_3 support (solid-solid wetting).[66] The spreading of MoS_2 on an Al_2O_3 support can be applied in the preparation of molybdenum sulfide/alumina catalysts.[67] The dispersed phase obtained exhibits activity for thiophene HDS comparable with a conventional catalyst.

1.4.2 Approaches Using Molybdenum Carbonyls

Generally molybdena-alumina catalysts have been prepared by using ammonium heptamolybdate. An alternative method is to use supported metal complexes. However, their reactivities in HDS of thiophenes, especially in a pressurized flow system, has been investigated only very slightly.

In this section, HDS of DBT catalyzed by supported anionic molybdenum carbonyl complexes is described, focusing on catalytic activity and product selectivity. The vaporization of the metal species in activation processes can be inhibited with the use of these salts in heterogeneous catalytic reactions,[68–70] because the vapor pressure of these anionic salts is much smaller than that of neutral metal carbonyls. Further, the hydrotreatment of metal carbonyls with metal sulfur bonds forms metal sulfides.

When a $Mo(CO)_6$-NEt_3-$EtSH/Al_2O_3$ system is activated by H_2 or H_2S, the catalytic activity is the highest among the catalysts derived from supported-metal carbonyls, and the conversion of DBT at 300°C is 43%, which is higher than that over a conventional molybdena-alumina.[68] When the effects of supports on catalysts prepared from supported anionic molybdenum carbonyls [69] were investigated, the yields of HDS products such as BP and CHB decreased in the order, SiO_2-Al_2O_3 > Al_2O_3 > TiO_2 = active carbon > SiO_2 > NaY zeolite = HZSM-5 > HY zeolite.

Further, HDS catalysts are prepared from silica-alumina and silica-supported molybdenum and cobalt carbonyls. The silica-alumina-supported catalyst shows higher catalytic activity in HDS of DBT and 4-MDBT than the alumina-supported catalyst, which shows activity similar to that of the conventional sulfided Co-Mo/Al_2O_3.[70]

1.4.3 Approaches Using Noble Metals

Since it has been found that among the transition metal sulfides, unsupported ruthenium sulfide is most active for HDS of DBT.[71] HDS, hydrogenation (HYD), and HDN catalyzed by noble metal sulfides, especially ruthenium sulfide, have been extensively investigated to develop a new generation of catalysts with different properties from the present Co-, Ni-, Mo-

and W-based catalysts. The deposition of ruthenium sulfide onto supports with large surface areas, e.g., alumina, silica, silica-alumina, zeolite, etc., increases catalytic activity per ruthenium atom. Although some catalysts show activity comparable to that of commercial catalysts in HDS or HDN, the HDS activities of other catalysts are rather low probably because sulfidation of ruthenium species is incomplete and RuS_2 on alumina is unstable in hydrogen atmosphere.[72]

Noble metals other than ruthenium have also been found to be active for HDS of thiophene. The volcano curves between the HDS activity and the periodic position are observed for the second and third row transition metals. At very low metal loading, carbon-supported rhodium and iridium sulfides show the highest activities instead of ruthenium in the second row and osmium in the third row, respectively.[73] Supported rhenium, osmium, rhodium, iridium, paradium and platinum catalysts as well as a supported ruthenium catalyst are also effective for HDN.[74]

Bimetallic catalysts have been prepared and some of these exhibit a synergy effect or higher activity in HDS, HDN and HYD than commercial catalysts.[75,76] The synergy effect seems to be related to the change in the strength of the metal-sulfur bond according to the combination of metals. A too strong or too weak metal-sulfur bond is not favorable for HDS reactions. Thus, ruthenium sulfide may have the most favorable bond strength for HDS reactions.

1.4.4 Approaches Using Ruthenium Carbonyls

The use of catalysts derived from supported anionic molybdenum carbonyls is effective for the preparation of highly dispersed molybdenum catalysts. Therefore, these techniques can be applied for the preparation of supported anionic ruthenium carbonyls for HDS of DBT.[77]

HDS of DBT catalyzed by alumina-supported ruthenium complexes is investigated in a pressurized flow system. The catalysts prepared by activation of $Ru_3(CO)_{12}$-NEt_3-EtSH/Al_2O_3 systems where metal carbonyls are treated with NEt_3 and EtSH to give anionic complexes with metal-sulfur bonds have been found to be active for HDS of DBT.

In HDS of DBT catalyzed by sulfided alumina-supported ruthenium compounds, the effects of addition of alkali metal hydroxide on the catalytic activity and product selectivity are described.[78] When sodium hydroxide is added to catalysts derived from alumina-supported $Ru_3(CO)_{12}$, the conversion of DBT remarkably increases from 44% to 71%. It is essential for high catalytic activity that, after the reaction of $Ru_3(CO)_{12}$ with sodium hydroxide to give a ruthenium hydride $Na[HRu_3(CO)_{11}]$, the hydride is supported on alumina. Among alkali metals, cesium is most effective. The conversion of DBT reaches maximum at $Cs/Ru = 2$. Further addition of cesium decreases the activity. In this system, BP is produced selectively.

In HDS of [^{35}S]DBT catalyzed by alumina-supported ruthenium carbonyl-cesium hydroxide systems, the role of cesium is elucidated by tracing the behavior of ^{35}S on the working ruthenium catalysts.[79] The rate constant of [^{35}S]H_2S release is estimated from the first-order plots of the increasing and decreasing radioactivities of the product [^{35}S]H_2S. The results obtained indicate that the dispersion of ruthenium species can be significantly high. Further, it is suggested that cesium promotes the C-S bond scission of DBT and increases the activity by stabilizing Ru-S bonds for ruthenium sulfide.

1.5 Process and Operation

1.5.1 The Role of Hydrotreatment

The hydrotreating consists mainly of HDS, HDN and hydrogenation. All the reactions are exothermic, so the control of temperature in the reactor, especially the catalyst bed, is very important in the practical operation. Although equilibrium constants decrease at higher temperatures, the heteroatom removal reactions are favorable under practical operating conditions. Hydrogenation of aromatics, however, is limited by thermodynamics at high temperature and lower hydrogen pressure.

Generally speaking, petroleum crudes are a complex mixture of various organic compounds, and the major components are hydrocarbons. Large amounts of heteroatoms are also included and their concentrations change depending on their origin. While sulfur is generally the most abundant heteroatom, the concentration of sulfur is specifically higher in crudes from the Middle East where 50% of world petroleum deposits exist. The compounds with these heteroatoms are distributed over the entire boiling range, and the concentration of heteroatoms increases with increasing the boiling point. In lighter fractions, sulfur is present in the form of thiols, sulfides, disulfides and thiophenes, while various alkylbenzothiophenes and alkyldibenzothiophenes are included in the heavier gas oil fraction. The components in atmospheric and vacuum residues, which include large amounts of not only sulfur and nitrogen but also nickel and vanadium, have not yet been well characterized.

Crude oil is initially separated into naphtha, kerosene, gas oil and atmospheric residue by atmospheric distillation. The atmospheric residue is further separated by vacuum distillation into a vacuum gas oil and a vacuum residue. Feeds in hydrotreating are various fractions with different boiling ranges. The properties required for various products are different from each other and change with increase in the severity of air pollution control. From the viewpoint of environmental conservation, deep desulfurization and aromatics saturation of diesel fuels are now key reactions to reduce NO_x and particulate matter. These reactions are required for reducing the concentration of sulfur compounds to prevent corrosion when an exhaust gas recirculation (EGR) system is introduced into a diesel engine to reduce NO_x concentration in diesel exhaust.

Heavy crudes such as atmospheric and vacuum residues comprise from almost half to more than half of raw petroleum crude, and the upgrading of these materials is very important because of limited fossil fuel resources. Atmospheric and vacuum residues are commonly higher in viscosity and pour point and include concentrated matter such as sulfur, nitrogen, heavy metals and Conradson carbon residue (CCR), which become environmentally harmful. Among such matter, sulfur and nitrogen is distributed extensively from light to heavy fractions while heavy metals such as vanadium and nickel and CCR are concentrated only in the heavier fraction.

Atmospheric and vacuum residues are approximately 1.5–1.9 in H/C ratio and most of these are converted by two different routes, carbon rejection and hydrogen addition to increase the H/C ratio of products. The major carbon rejection route is the coking process, which consists of thermal cracking processes with higher temperature and without catalyst to produce lighter hydrocarbons and cokes. Visbreaking is also a liquid phase thermal cracking process used to reduce the viscosity and pour point of the resids. In deasphalting, resins and asphaltenes are removed from vacuum resid by solvent extraction. Another carbon rejection route is the residue fluid catalytic cracking (RFCC). This process has been developed originally from FCC process which cracks gas oil and vacuum gas oil to give gasoline. An alternative route to increase the H/C ratio of residues is hydrogen addition. The hydrogen addition for residues upgrading includes hydrotreating and hydrocracking. Hydrogen is consumed for not only HDS, HDN, and HDM but also aromatic hydrogenation and hydrocracking of heavier fractions. Crudes are becoming rich in heavier fractions while lighter fractions are favorable for products. Therefore, the higher ability of desulfurization, demetallation and cracking are required for hydrotreating and hydrocracking processes.

1.5.2 Hydrotreating Processes

In a typical hydrotreating process, a mixture of feedstock, recycle gas and make-up gas pressurized to the expected pressure is heated in a furnace to the reaction temperature and introduced into a fixed bed reactor containing the catalyst. In the processing of distillates, gas or liquid-phase oil is flowed downward through the solid catalyst particles in the reactor. To produce the deeply hydrotreated light gas oil (LGO), there are several methods. The process flow of a two-stage deep desulfurization process is one of the new processes.[80] In this process, deep desulfurization of less than 0.05 wt% of sulfur is achieved at higher temperatures, higher SV and lower pressure in the first reactor, then color is removed from highly colored gas oil at lower temperatures (200–300°C) in the second reactor. In the hydrocracking process,[81] high grade middle distillates such as kerosene, jet fuel, gas oil are effectively produced from heavy gas oil, vacuum gas oil, light cycle oil, deasphalted oil, etc.

The operating conditions of hydrotreating reactors change depending on the reactivity of the feed and the quality and amount of products desired. The reaction conditions are remarkably different between distillates and residue. In general, since the HDS rate constant of each compound is remarkably different, HDS rate of total sulfur compounds apparently behaves like a second-order reaction with respect to the total sulfur concentration. In contrast to HDS rate, HDN rate of total nitrogen compounds follows first-order reaction with respect to the total nitrogen concentration.[82]

Down-flow fixed-bed reactors are still used in the HDS of heavy petroleum feeds such as atmospheric residue. Since the HDS reactivity of heavy feeds is low, the reaction condition is more severe. Further, residues include a large amounts of heavy metals and asphaltenes which bring about coke and metal depositions and subsequently the deactivation of desulfurization catalysts during operation. To avoid this, a demetallation reactor is generally set at the upstream of an HDS reactor. When the vanadium content in feeds are higher, the fixed bed reactor needs the exchange of HDM catalysts which stop the operation of the reactor for a long time. To solve the problems in the fixed-bed reactor, the OCR (Onstream Catalyst Replacement)[83] and bunker flow processes[84] using a moving-bed reactor, and the H-Oil and L-C fining processes[85,86] employing an ebullated-bed reactor, which enable the exchange of catalysts during the hydrotreating operation, have been developed.

1.5.3 Hydrotreating Catalysts

Large amounts of hydrotreating catalysts are used extensively for HDS, hydrocracking, and hydrogenation reactions in refineries. In hydrotreating catalysts, metal components with the activities of hydrotreating reactions such as HDS and hydrogenation are dispersed on inorganic oxide supports with high porosity. Generally, cobalt and molybdenum are used for the metal components and alumina is used for the support. In practical catalysts, about 2–5 wt% of CoO and 8–20 wt% of MoO_3 are supported on alumina. Ni and W are used instead of Co and Mo, respectively.

In general, hydrotreating catalysts are sulfided to obtain the active phase. This sulfiding procedure appears to affect the catalytic activity and stability significantly. The presulfiding is typically performed at the beginning of the hydrotreating process by introducing the sulfur-containing feed into the catalyst.[87] Although H_2S is also used for presulfiding by adding it to the recirculating hydrogen stream, this method is being replaced by other improved methods. For example, carbon disulfide, dimethylsulfide (DMS) and dimethyldisulfide (DMDS) are added to the recycle gas instead of H_2S.

Catalyst deactivation is mainly caused by coking and metal sulfide deposition. Coking is caused by high molecular weight polynuclear aromatics (PNA).[88,89] When the amount of high molecular weight PNA in feeds is large, a large amount of coke is observed on the catalysts.[88] Further, it has been shown that coking depends on the acidic properties of the catalyst rather than on pore size distribution.[89] The deactivation of catalysts caused by coking is related to the plugging of catalyst pore-mouths and covering of active sites.[90]

There are two types of metal deposition: one is pore-mouth plugging and the other is poisoning of active sites on the inside of the catalyst pore. In general, metal deposition occurs near the exterior surface of the catalyst particle.[91] Subsequently, this prevents the diffusion of reactants to the interior of the catalyst.[92] Demetallation is affected by diffusion more than HDS. The activity and stability of the catalyst are significantly affected by pore size.[84,93]

The regeneration of spent catalyst is an attractive process to recover the original activity and stability and a number of attempts have been made to regenerate the catalyst.[94] Regeneration for most spent catalysts has been performed in the exclusive *ex-situ* process. *Ex-situ* regeneration can offer a better performance recovery because of better temperature control.[95] Spent catalyst to which oil still adheres is extracted from the reactor followed by regenerative calcination or metal recovery.

References

1. A. Ishihara and T. Kabe, *Chemistry and Education*, **46** (3), 148 (1998) [in Japanese].
2. P. T. Vasudevan, J. L. G. Fierro, *Catal. Rev.-Sci. Eng.* **38** (2),161 (1996).
3. I. Naka, "*Nippon Ketjen Seminar 1995*," G-3, Nippon Ketjen Co. Ltd. (1995) [in Japanese].
4. J. R. Dosher and J. T. Carney, *Oil & Gas J.*, May 23, 43 (1994).
5. Y. Mori and T. Takatsuka, *Petrotech*, **18** (4), 289 (1995) [in Japanese].
6. K. Nonaka, S. Morita, T. Ishizaki and Y. Inoue, "*Nippon Ketjen Seminar 1995*," P-6, Nippon Ketjen Co. Ltd. (1995).
7. T. Kabe, A. Ishihara and H. Tajima, *Ind. & Eng. Chem. Res.*, 31, 1577 (1992).
8. B. C. Gates, J. R. Katzer and G. C. A. Schuit, *Chemistry of Catalytic Processes*, Chap. 5, p. 390, McGraw-Hill, Inc., New York (1979).
9. M. J. Girgis and B. C. Gates, *Ind. Eng. Chem. Res.*, **30**, 2021 (1991).
10. A. Ishihara, T. Itoh, T. Hino, P. Qi, M. Nomura and T. Kabe, *J. Catal.*, **140**, 184 (1993).
11. M. L. Vrinat, *Appl. Catal.*, **6**, 137 (1983).
12. M. Houalla, D. H. Broderick, A. V. Sapre, N. K. Nag, V. H. J. de Beer, B. C. Gates and H. Kwart, *J. Catal.*, **61**, 523 (1980).
13. T. Kabe, A. Ishihara and Q. Zhang, *Appl. Catal.*, A: Gen 97, L1 (1993).
14. Q. Zhang, W. Qian, A. Ishihara and T. Kabe, *J. Jpn. Petrol. Inst.*, **40**, 3, 185 (1997).
15. R. J. Angelici, *Acc. Chem. Res.*, **21**, 387 (1988).

16. R. A. Sanchez-Delgado, *J. Mol. Cat.,* **86**, 287 (1994).
17. R. Prins, V. H. J. de Beer and G. A. Somorjai, *Catal. Rev.-Sci. Eng.*, **31** (1&2), 1 (1989).
18. B. C. Wiegand and C. M. Friend, *Chem. Rev.*, **92** (4), 491 (1992).
19. H. Topsøe, B. S. Clausen and F. E. Massoth, *Hydrotreating Catalysis*, p.111, Springer Verlag, Berlin (1996).
20. T. C. Ho, *Catal. Rev.-Sci. Eng.*, **30**, 117 (1988).
21. G. Perot, *Catal. Today*, **10**, 447 (1991).
22. H. Schulz, M. Schon and N. M. Rahman, *Studies in Surface Science and Catalysis, Vol.27, p.204, Catalytic Hydrogenation, a Modern Approach* (L. Cerveny ed.), Elsevier, Amsterdam (1986).
23. R. T. Hanlon, *Energy & Fuels*, **1**, 424 (1987).
24. C. N. Satterfield and S. H. Yang, *Ind. Eng. Chem., Process Des. Dev.*, **23**, 11 (1984).
25. C. N. Satterfield and J. F. Cocchetto, *Ind. Eng. Chem., Process Des. Dev.*, **20**, 53 (1981).
26. G. Perot, S. Brunet, C. Canaff and H. Toulhoat, *Bull. Soc. Chim. Belg.*, **96**, 865 (1987).
27. R.H. Williams and A. J. Mcevoy, *J. Phys., D: Appl. Phys.*, **4**, 456 (1971).
28. H. Topsøe and B. S. Clausen, *Catal. Rev.-Sci. Eng.*, **26**, 395 (1984).
29. R. J. H. Voorhoeve and J. C. M. Stuiver, *J. Catal.,* **23**, 234 (1971).
30. J. M. J. G., Lipsch and G. C. A., Schuit, *J. Catal.,* **15**, 179 (1969).
31. A. L. Farragher and P. Cossee, in: *Proceedings, 5th International Congress on Catalysis*, Palm Beach, 1972 (J. W. Hightower ed.), p.1301, North-Holland, Amsterdam (1973).
32. R. J. H. Voorhoeve and J. C. M. Stuiver, *J. Catal.,* **23**, 228 (1971).
33. B. Delmon, *ACS. Div. Pet. Chem. Prepr.*, **22**, 503 (1977).
34. P. Ratnasamy and S. Sivasanker, *Catal. Rev. Sci. Eng.*, **22**, 401 (1980).
35. H. R. Lukens, J. R. G. Meisenheimer and J. N. Wilson, *J. Phys. Chem.*, **66**, 469 (1962).
36. K. A. Pavlova, B. D. Panteleea, E. N. Deryagina and I. V. Kalechits, *Kinet. Katal.*, **6**, 3493 (1965).
37. C. G. Gachet, E. Dhainaut, L. de Mourgues, J. P. Candy and P. Fouilloux, *Bull. Soc. Chim. Belg.*, **90** (12), 1279 (1981).
38. A. J. Gellman, M. E. Bussell and G. A. Somorjai, *J. Catal.,* **107**, 103 (1987).
39. G. V. Isagulyants, A. A. Greish and V. M. Kogan, *"Symposium of International Catalyst Annual Conference in Canada"*, p.35 (1988).
40. M. Dobrovolszky, Z. Paal and P. Tetenyi, *Catal. Today*, **9**, 113 (1991).
41. T. Kabe, W. Qian, S. Ogawa and A. Ishihara, *J. Catal.,* **143**, 239 (1993).
42. W. Qian, A. Ishihara, S. Ogawa and T. Kabe, *J. Phys. Chem.*, **98** (3), 907 (1994).
43. W. Qian, Q. Zhang, Y. Okoshi, A. Ishihara and T. Kabe, *J. Chem. Soc., Faraday Trans.*, **93** (9), 1821 (1997).
44. W. Qian, A. Ishihara, Y. Okoshi, W. Nakagami, M. Godo and T. Kabe, *J. Chem. Soc., Faraday Trans.*, **93** (24), 4395 (1997).
45. T. Kabe, W. Qian and A. Ishihara, *Catal. Today*, **39**, 3 (1997).
46. W. Qian, A. Ishihara, G. Wang, T. Tsuzuki, M. Godo and T. Kabe, *J. Catal.,* **170**, 286 (1997).
47. B, Scheffer, N. J. J., Dekker, P. J., Mangnus and J. A., Moulijn, *J. Catal.,* **121**, 31 (1990).
48. J. Bachelier, J. C. Duchet and D. Cornet, *Bull. Soc. Chim. Belg*, **90**, 1301 (1981).
49. M. Dobrovolszky, Z. Paal and P. Tetenyi, *Appl. Catal. A*, **142**, 159 (1996).
50. Q. Zhang, W. Qian, S. Oshima, A. Ishihara and T. Kabe, *J. Jpn. Petrol. Inst.*, **40** (5), 408 (1997).
51. F. E. Massoth and P. Zeuthen, *J. Catal.,* **145**, 216 (1994).
52. M. Breysse, J. L. Portefaix and M. Vrinat, *Catal. Today*, **10**, 489 (1991).
53. S. Eijsbouts, C. Dudhakar, V. H. J. de Beer and R. Prins, *J. Catal.,* **127** (2), 605 (1991).
54. R. Candia, O. Sørensen, J. Villadsen, N. Topsøe, B. S. Clausen and H. Topsøe, *Bull. Chem. Chim. Belg.*, **93**, 763 (1984).
55. J. A. van Veen, E. Gerkema, A. M. van der Kraan and A. Knoester, *J. Chem. Soc., Chem. Commun*, 1684 (1987).
56. M. Vrinat, M. Breysse, S. Fuentes, M. Lacroix and J. Ramirez in: *Proc. IX Simposio, Iberoamericano de Catalisis*, (O. Bermudo, G. Del Angel and R. Gomez eds.), p.1029, F. Cossio (1988).
57. J. Ramírez, S. Fuentes, G. Diaz, M. Vrinat, M. Breysse and M. Lacroix, *Appl. Catal.*, **52**, 211 (1989).
58. G. Muralidhar, F. E. Massoth and J. Shabtai, *J. Catal.,* **85**, 44 (1984).
59. C. Moreau, L. Bekakra, P. Geneste, J. L. Olivé, J. C. Duchet, M. J. Tilliette and J. Grimblot, *Bull. Soc. Chim. Belg.*, **100**, 11–12, 841 (1991).
60. I. Wang and R. C. Chang, *J. Catal.,* **117**, 266 (1989).
61. S. Eijsbouts, J. N. M. van Gestel, J. A. R. van Veen, V. H. J. de Beer and R. Prins, *J. Catal.,* **131**, 412 (1991) and literature cited therein.
62. S. T. Oyama, J. C. Schlatter, J. E. Metcalfe III and J. M. Lambert Jr., *Ind. Eng. Chem. Res.*, **27**, 1639 (1988).
63. C. W. Colling and L. T. Thompson, *J. Catal.,* **146**, 193 (1994).
64. P. Blanchard, C. Mauchausse, E. Payen, J. Grimblot, O. Poulet, N. Boisdron and R. Loutaty, *Preparation of Catalysts VI* (G. Poncelet *et al.* eds.), p.1037, Elsevier Science B. V. (1995).
65. Y. Yoshimura, N. Matsubayashi, T. Sato, H. Shimada and A. Nishijima, *Appl. Catal. A: Gen.*, **79**, 145 (1991).

66. J. Leryer, M. I. Zaki and H. Knözinger, *J. Phys. Chem.*, **90**, 4775 (1986).
67. B. M. Reddy and B. Manohar, *J. Chem. Soc., Chem. Commun.*, 1435 (1991).
68. A. Ishihara, M. Azuma, M. Matsushita and T. Kabe, *J. Jpn. Petrol. Inst.*, **36**, 5, 360 (1993).
69. A. Ishihara, K. Shirouchi and T. Kabe, *Chem. Lett.*, 589 (1993).
70. A. Ishihara, M. Matsushita, K. Shirouchi, Q. Zhang and T. Kabe, *J. Jpn. Petrol. Inst.*, **39** (1), 26 (1996).
71. T. A. Pecoraro and R. R. Chianelli, *J. Catal.*, **67**, 430 (1981).
72. T. G. Harvey and T. W. Matheson, *J. Catal.*, **101**, 253 (1986).
73. M. J. Ledoux, O. Michaux, G. Agostini and P. Panissod, *J. Catal.*, **102**, 275 (1986).
74. S. Eijsbouts, V. H. J. de Beer and R. Prins, *J. Catal.*, **127** (2), 619 (1991).
75. B. Delmon, *Catal. Lett.*, **22**, 1 (1993).
76. C.-A. Jan, T.-B. Lin and J.-R. Chang, *Ind. Eng. Chem. Res.* **35**, 3893 (1996).
77. A. Ishihara, M. Nomura and T. Kabe, *J. Catal.*, **150**, 212 (1994).
78. A. Ishihara, M. Nomura and T. Kabe, *Chem. Lett.*, 2285 (1992).
79. A. Ishihara, M. Yamaguchi, H. Godo, W. Qian, M. Godo and T. Kabe, *Chem. Lett.*, 743 (1996).
80. M. Ushio and M. Hatayama, *Petrotech*, **17** (8), 701 (1994) [in Japanese].
81. M. Ishii, *Petrotech*, **17** (2), 174 (1994) [in Japanese].
82. A. G. Bridge and E. M. Blue, *Kagaku Hannou To Hannouki Sekkei*, (H. Tominaga and M. Tamaoki eds.), p.265, Maruzen (1996) [in Japanese].
83. B. E. Reynolds and R. W. Bachtel, *Petrotech*, **17** (8), 659 (1994) [in Japanese].
84. P. B. Kwant, *Petrotech*, **8** (6), 543 (1985) [in Japanese].
85. T. Takahashi, *Petrotech*, **14** (9), 863 (1991) [in Japanese].
86. R. E. Boening, N. K. McDaniel, R. D. Petersen and R. P. Van Driesen, *Hydrocarbon Proc.*, 59, September (1987).
87. Y. Morimura, S. Nakata, T. Takatsuka and M. Nakamura, *Sekiyu Gakkaishi*, **38** (3), 192 (1995) [in Japanese].
88. P. Wiwel, P. Zeuthen and A. C. Jacobsen, *Stud. Surf. Sci. Catal.*, **68**, 257 (1991).
89. L. W. Brunn, A. A. Montagna and J. A. Paraskos, *Am. Chem. Soc., Div. Petrol. Chem. Prepr.*, **21**, 173 (1976).
90. P. N. Hannerup and A. C. Jacobsen, *Am. Chem. Soc., Pet. Div. Prepr.*, **28**, 576 (1983).
91. Z. Sarbak and S. L. T. Andersson, *Appl. Catal., A: Gen.*, **79**, 191 (1991).
92. B. G. Johnson, F. E. Massoth and J. Bartholdy, *AIChE. J.* **32**, 1980 (1986).
93. R. J. Quann, R. A. Ware, C.-W. Hung and J. Wei, *Adv. Chem. Eng.*, **14**, 95 (1988).
94. Y. Yoshimura and E. Furimsky, *Appl. Catal.*, **23**, 157 (1986).
95. T. Suzuki and P. Dufresne, *Stud. Surf. Sci. Catal.*, *92, Science and Technology in Catalysis 1994*, 215 (1995).

2

Reaction Profile in Hydrodesulfurization and Hydrodenitrogenation

2.1 Hydrodesulfurization

2.1.1 Analysis and Desulfurization of Sulfur-containing Compounds

A. Separation and Analysis of Polycyclic Aromatic Sulfur Heterocycles

Polycyclic aromatic sulfur heterocycles (PASH) are known to be major components in petroleum- and coal-derived products. The extensive use of these products causes pollution in big city areas and detailed studies on the analysis and removal of these compounds in fossil fuels are of great importance.[1] The analysis of sulfur-containing compounds in petroleum fractions is composed of separation, concentration and determination of those compounds. In the 1960's, most analytical study for determining the amount of sulfur in an original oil consisted of measuring total sulfur by coulometric titration without quantitative and qualitative

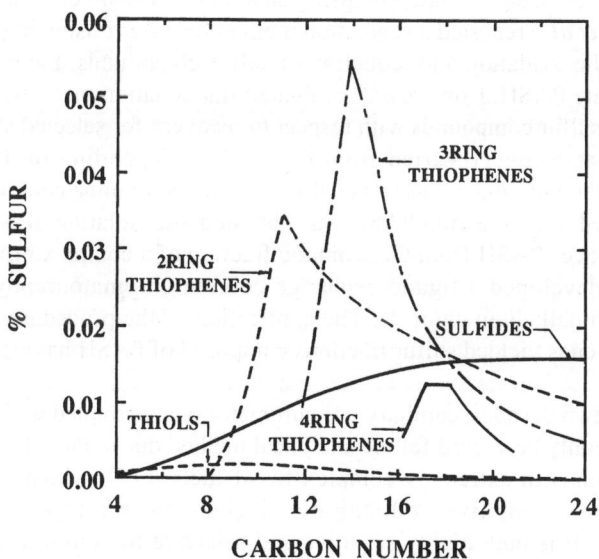

Fig. 2.1 Sulfur types by carbon number in Middle-East crude.
(From R. L. Martin and J. A. Grant, *Anal. Chem.*, **37**, 648 (1965))

analyses for various components of PASH such as benzothiophenes (BTs) and dibenzothiophenes (DBTs) and other sulfur-containing compounds, e.g., alkyl and aryl sulfides and mercaptanes, in the oil. Martin and Grant[2,3] separated Middle-East crude to some fractions by distillation and determined the distributions of sulfur compounds in a variety of petroleum samples by a combination of gas chromatography and microcoulometric sulfur detection. As shown in Fig. 2.1, sulfides gradually increased with an increase in the carbon number while 2-ring, 3-ring and 4-ring thiophenes revealed maxima at about C_{11}, C_{14} and C_{18}, respectively. Although the sulfur concentration of thiols was very small, 2-ring and 3-ring thiophenes concentrations were significantly high. Further, they also showed that the concentrations of 2-ring and 3-ring thiophenes among sulfur-containing compounds in vacuum gas oil and light catalytic cycle oil were predominant, more than 70% and 90%, respectively.

In order to identify the components of sulfur-containing compounds in petroleum or solvent refined coal in detail, however, it was necessary to synthesize the standard compounds and develop separation and analytical techniques using these compounds. Since then, separation and analytical technologies of sulfur-containing compounds using adsorption column chromatography,[4–9] ligand-exchange column chromatography,[10–19] supercritical extraction,[20] size exclusion chromatography with inductively coupled plasma optical emission detector (SEC-ICP),[21–24] capillary gas chromatography-flame photometric detector (GC-FPD),[8,16,17,19,25–29] GC-sulfur chemiluminescence detector (GC-SCD),[30] GC-radio frequency plasma detector (GC-RFPD),[31] GC-atomic emission detector (GC-AED) [32–37] and GC-mass spectroscopy (GC-MS)[6–8,17,29,32–35,37,38] have been developed, enabling the detail analysis of sulfur compounds.

Drushel et al.[4] separated sulfur compounds in a distillate fraction of vacuum gas oil (boiling range between 425°C and 455°C) using adsorption column chromatography (silicagel-benzene). Sulfur compounds in the benzene eluent was oxidized by hydrogen peroxide to sulfones, which were also separated by using adsorption column chromatography (silicagel-benzene). Jewell et al.[5] reported a separation method for PASH using hydrated basic alumina which included the oxidation and reduction of sulfur compounds. Later et al.[6] used neutral alumina to separate PASH. Kong et al.[8] evaluated fractionation methods including oxidation and reduction of sulfur compounds with respect to recovery for selected standard compounds. They showed that recovery varied from 0% to 70% depending on the structure of the compounds. Poirier and Smiley[9] used a dual-packed silica-alumina column method. Although a fraction enriched in 1- to 3-ring PASH was obtained, the isolation of 3-ring PASH was not complete. To recover PASH from the aromatic fraction of a complex mixture quantitatively, Nishioka et al. developed a ligand-exchange column chromatography[17] using silica gel impregnated with palladium chloride. The application of the procedure to a coal liquid and petroleum heavy ends yielded sulfur fractions composed of PASH having two to six aromatic rings.

For the past two decades, capillary gas chromatography coupled with a sensitive detector to sulfur has generally been used for the analytical method due to the relatively low abundance of sulfur compounds in extremely complex fossil fuels. FPD, which has existed for three decades,[39] is more sensitive and stable and gives more reproducible results than a microcoulometer. It is both highly sensitive and selective for sulfur and is one of the most suitable GC detectors available for determining sulfur compounds in complex mixtures such as petroleum distillates. However, there are disadvantages in that its response varies with sulfur compound type and is nonlinear for a given type of sulfur compound.[40] Further, the detector response is also quenched by coeluting water or hydrocarbons.[41,42] This means that a calibration curve is needed for each sulfur compound. When this detector is used for complex mixtures such as petroleum samples, many standards for sulfur-containing compounds included in petroleum are required to make the calibration curves. It is difficult to obtain a complete calibration. To overcome this situation, Bradley and Schiller[28] used FPD after converting all the sulfur-containing compounds to SO_2. In this method, a single calibration equation was formulated expressing a uniform and linear relationship between detector response and sulfur concentration for the various types of sulfur compounds in petroleum. An alternative approach is to use another detector which can be coupled with GC. Skelton et al.[31] showed that the element-selective radio frequency plasma detector (RFPD) possesses low limits of detection of sulfur (0.5 pg/s) and good linear response. It was also reported that various numbers of phenyl groups attached to the base thiophenic ring had little effect on the intensity of the sulfur emission signal and that coelution of hydrocarbons was found to affect sulfur response only at high concentrations. In more recent years, Ishihara et al.,[32] Kabe et al.,[33,34] Tajima et al.[35] and Amorelli et al.[37] used a gas chromatography-helium microwave-induced plasma atomic emission detector (GC-AED) and GC-mass spectroscopy (GC-MS) for the characterization of sulfur compounds included in middle distillate. AED is a new detector for GC and selectively detects the elements in a compound. Because AED is the same as RFPD in principle, it has characteristics similar to those of RFPD mentioned above: it provides high element selectivity independent of the structure of a compound, low limits of detection of sulfur, good linear response, etc. Further, although a flame ionization detector (FID) must be used for the detection of various hydrocarbons when FPD is used to detect sulfur compounds, it is possible to detect carbon and sulfur simultaneously employing AED. With GC-AED it is possible to detect selectively sulfur compounds in fractions from adsorption chromatography or ligand exchange chromatography. Further, simultaneous use of GC-AED and GC-MS remarkably increases the precision for the structure analysis of sulfur compounds.

An example of successful separation and analysis of PASH in light gas oil (LGO) is shown here.[32-35] The LGO used was the middle distillate of Arabian Light and the results of distillation are shown in Table 2.1. Fig. 2.2 shows the flow chart for separation of PASH. Separation of PASH from the samples and qualitative analysis of components were carried out

Table 2.1 Distillation Properties of Raw Light Gas Oil (Arabian Light)

Amount of Distillate (%)	Boiling Point (°C)
Initial Boiling Point	245
10	274
20	283
30	289
40	296
50	304
60	313
70	324
80	337
90	354
95	369
End Point	374

(From T. Kabe, A. Ishihara et al., Ind. Eng. Chem. Res., 31, 1577 (1992))

OIL Original

Separation by LC-1 (Silicagel)

Paraffins
(OIL 1)

Aromatics with
Single Ring
(OIL 2)

Aromatics with
Double and Triple Rings
+   (OIL 3)
PASC

Separation by LC-2(PdCl₂/Silicagel
Ligand Exchange Chromatography)

Aromatics
(OIL 3-1)

Thiophenes and
Other Sulfur-containing
Compounds (OIL 3-2)

Thiophenes with
Double and Triple
Rings (OIL 3-3)

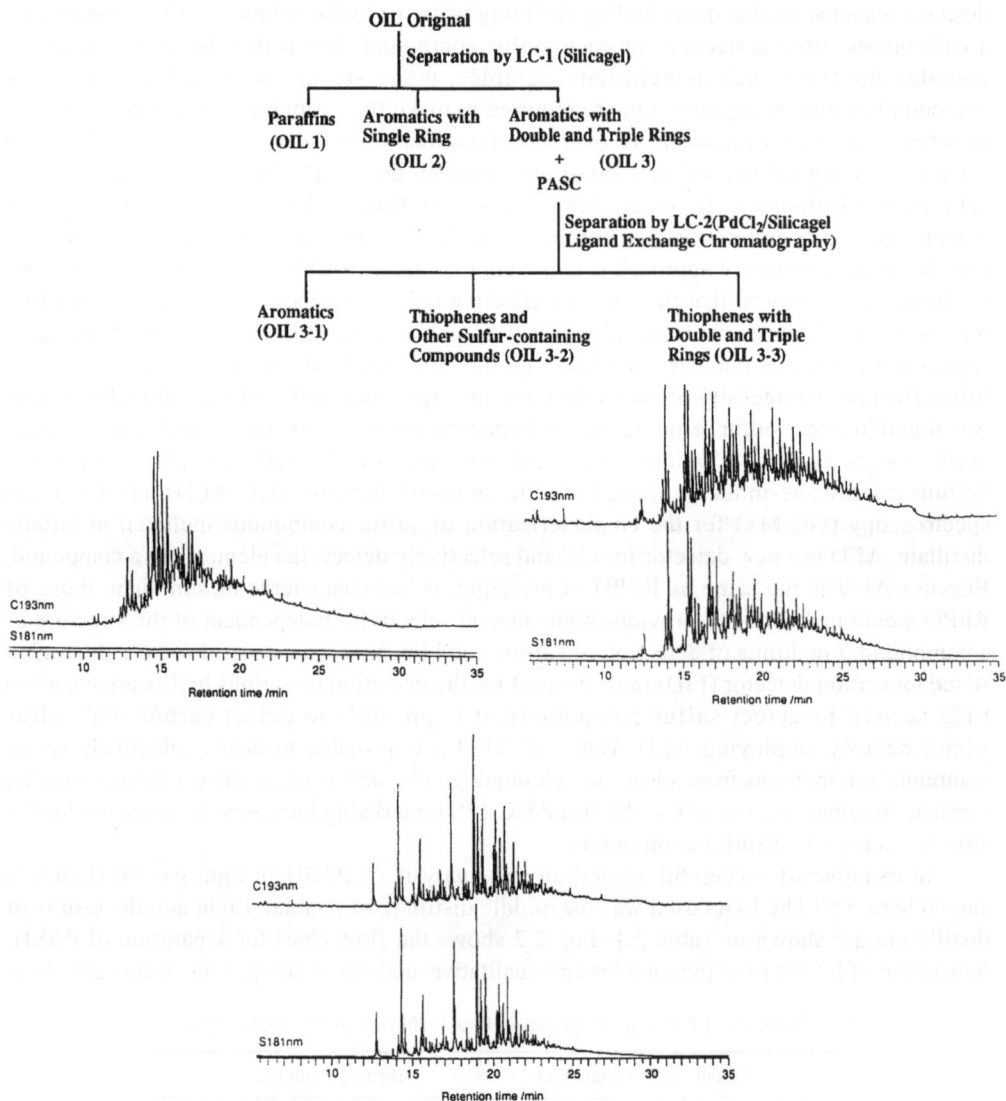

Fig. 2.2   Separation and analysis of polyaromatic sulfur-containing compounds. Analysis of OIL 3-1, 3-2 and 3-3 by GC-AED and GC-MS.
(From T. Kabe, A. Ishihara *et al.*, *J. Jpn. Petrol. Inst.*, **36**, 46 (1993))

as follows. Each sample was diluted to 10% with *n*-hexane and separated into three fractions: aliphatic compounds (oil 1), aromatic compounds with a single ring (oil 2), and aromatic compounds with double or triple rings (oil 3) by high performance liquid chromatography (HPLC) with a column packed with silica gel. Each fraction was analyzed by GC-AED, and the presence of PASH was confirmed in oil 3. The oil 3 fraction was concentrated and further separated into three fractions, including aromatics (oil 3-1), thiophenes and other sulfur-containing compounds (oil 3-2), and thiophenes with double or triple rings (oil 3-3), by ligand exchange chromatography (HPLC) using a column packed with 10% PdCl$_2$-silica gel. Each fraction was monitored by a UV detector. The PASH components in oil 3-2 and oil 3-3 fractions were analyzed using GC-AED and GC-MS and identified by comparing their retention time and mass spectra with those of the standard compounds. 2 μl of each sample was analyzed by GC-AED using a Ultra-1 column (i.d.: 0.32 mm, film thickness: 0.17 μm, length: 25 m) programmed from 50°C to 300°C (heating rate: 8°C/min, inj. temp.: 350°C, transfer line: 320°C, carrier gas: He, pressure of column head: 80 kPa, elements and wavelength: carbon-193 nm and sulfur-181 nm) and by GC-MS using DB-1 column (i.d.: 0.25 mm, film thickness: 0.25 μm, length 60 m) programmed from 50°C to 320°C (heating rate: 8°C/min, inj. temp.: 350°C, transfer line: 350°C, carrier gas: He, pressure of column head: 150 kPa, scan range: 50–400 m/z). GC-AED charts of oil 3-1, oil 3-2, and oil 3-3 are also shown in Fig. 2.2. In these charts, the upper portion represents a chromatogram of carbon (193 nm), the lower portion that of sulfur (181 nm), and the horizontal line retention time. The shapes of signals in carbon and sulfur chromatograms for oil 3-2 or oil 3-3 were very similar, indicating that only sulfur-containing compounds in the original oil were selectively separated by this method. GC-AED charts other than these fractions and more detailed analytical methods appear in the literature.[35] The standard compounds, BT and DBT, were commercially available. Substituted BTs and substituted DBTs were synthesized by reported methods.[43–48] Other compounds were identified by comparing the ratio of composed elements (carbon:sulfur ratio) in GC-AED analysis and mass spectra and mass chromatogram in GC-MS analysis of oil 3-3. Every component was quantitatively analyzed by comparing its peak area with that of DBT using a chromatogram of sulfur (181 nm) in GC-AED. A GC-AED chart of the original oil is shown in Fig. 2.3. 42 kinds of alkylbenzothiophenes and 29 kinds of alkyldibenzothiophenes were detected. Sulfur-containing compounds in this LGO and sulfur contents are shown in Table 2.2. The amounts of sulfur in alkylbenzothiophenes and

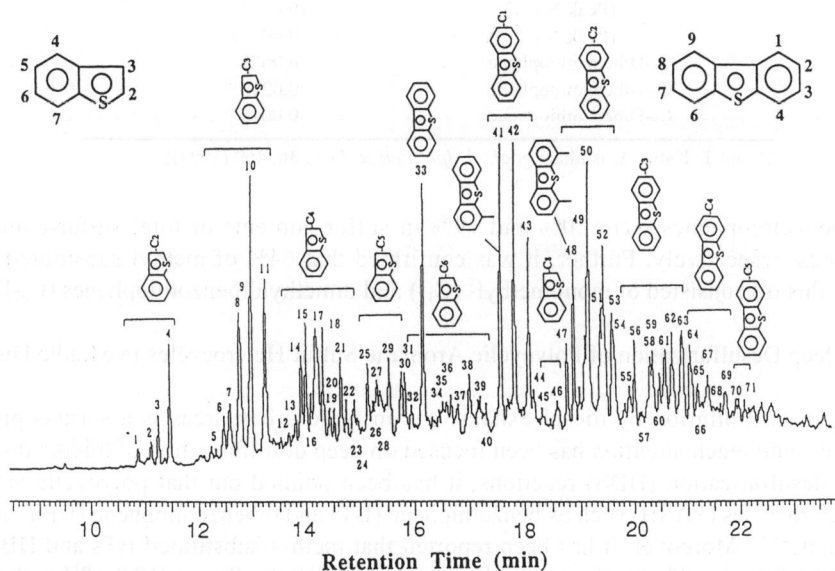

Fig. 2.3 GC-AED chart of polyaromatic sulfur-containing compounds in original oil.
(From T. Kabe, A. Ishihara et al., Ind. Eng. Chem. Res., 31, 1578 (1992))

Table 2.2 Sulfur Contents of Benzothiophenes and Dibenzothiophenes in Raw Light Gas Oil (Arabian Light)

Sulfur-containing Compounds	Sulfur Content (wt%)
Total Sulfur	1.4600
Benzothiophenes	0.2936
Dibenzothiophenes	0.2798
Others	0.8866
Benzothiophenes	
C_2-Benzothiophene	0.0165
C_3-Benzothiophene	0.0796
C_4-Benzothiophene	0.0891
C_5-Benzothiophene	0.0465
C_6-Benzothiophene	0.0499
C_7-Benzothiophene	0.0077
C_8-Benzothiophene	0.0042
Dibenzothiophenes	
Dibenzothiophene	0.0233
C_1-Dibenzothiophene	0.0693
(Peak No. 41)	(0.0270)
(Peak No. 42)	(0.0257)
(Peak No. 43)	(0.0167)
C_2-Dibenzothiophene	0.0960
(Peak No. 47)	(0.0060)
(Peak No. 48)	(0.0122)
(Peak No. 49)	(0.0071)
(Peak No. 50)	(0.0260)
(Peak No. 51)	(0.0089)
(Peak No. 52)	(0.0198)
(Peak No. 53)	(0.0083)
(Peak No. 54)	(0.0077)
C_3-Dibenzothiophene	0.0688
C_4-Dibenzothiophene	0.0210
C_5-Dibenzothiophene	0.0015

(From T. Kabe, A. Ishihara *et al.*, *J. Jpn. Petrol. Inst.*, **36**, 469 (1993))

alkyldibenzothiophenes were 20% and 19% in sulfur contents of total sulfur-containing compounds, respectively. Further, it was confirmed that 64% of methyl-substituted DBTs found in this oil consisted of monomethyl- (C_1-) and dimethyldibenzothiophenes (C_2-DBTs).

B. Deep Desulfurization of Polycyclic Aromatic Sulfur Heterocycles in Middle Distillate

Recent air pollution by diesel exhaust gas in large urban areas is a serious problem worldwide, and much attention has been focused on deep desulfurization of middle distillate. In hydrodesulfurization (HDS) reactions, it has been pointed out that polycyclic aromatic sulfur heterocycles (PASH) such as benzothiphene (BT) and dibenzothiophene (DBT) are key compounds.[49-59] Moreover, it has been reported that methyl-substituted BTs and DBTs are more difficult to desulfurize than nonsubstituted ones.[60,61] Studies of HDS of thiophenes in naphtha, white oil, diesel oil, light catalytic cycle oil (LCO), etc. have also been reported.[62-65] Frye and Mosby reported a kinetic study of HDS of LCO catalyzed by Co-Mo/Al$_2$O$_3$ using a trickle bed reactor. They used the same analytical method for sulfur compounds in LCO by a combination of GC and microcoulometric sulfur detection as Martin and Grant.[2,3)] Fig. 2.4 shows GC charts for feed and hydrotreated oils. In Fig. 2.4, it is revealed that the low molecular weight compounds are desulfurized more easily than high molecular weight compounds and that significant amounts of C_{13} and C_{14} compounds which may show alkyl-substituted DBTs remain even after the hydrotreatment at 330°C and 20 atm. They traced HDS of DBT and two kinds of trimethylbenzothiophene in Fig. 2.4; the reaction was shown to be pseudo-first-order with respect to the concentration of these thiophenes.[62)] However, there are few examples which determined each component separately in HDS of a fraction of petroleum under industrial conditions. Especially under deep desulfurization conditions where sulfur compounds must be desulfurized up to ppm orders of concentration, it is not well known what kind of sulfur compounds exist in middle distillate and which of them is still not desulfurized at the final stage of the reaction.

As shown in Section 2.1.1A, the latest developments in analytical techniques using GC-AED and GC-MS have enabled the determination of PASH such as alkylbenzothiophenes and alkyldibenzothiophenes, which are considered to be key compounds in HDS of middle distillate. The use of GC-AED has also enabled the determination of each component in hydrotreated oils and has clarified which compounds are not desulfurized even at the final stage of the reaction.[32,33)] The same oil as that shown in Fig. 2.3 was hydrodesulfurized by a Co-Mo/Al_2O_3 (CoO: 4.5 wt%, MoO_3: 17.0 wt%); the reactor was a 20 mm-i.d. stainless steel tube packed with 5.9 g of catalyst particles diluted with quartz sand to fill 12 cc. After the catalyst bed was heated for 24 h at 450°C in air, it was presulfided with a mixture of 3% H_2S in H_2 flowing at 30 l/h at atmospheric pressure and 400°C for 3 h. After the pretreatment, the reactor was pressurized with hydrogen. Then light gas oil (LGO) was supplied to a feed pump. The HDS reaction was carried out under the following conditions: temperature:

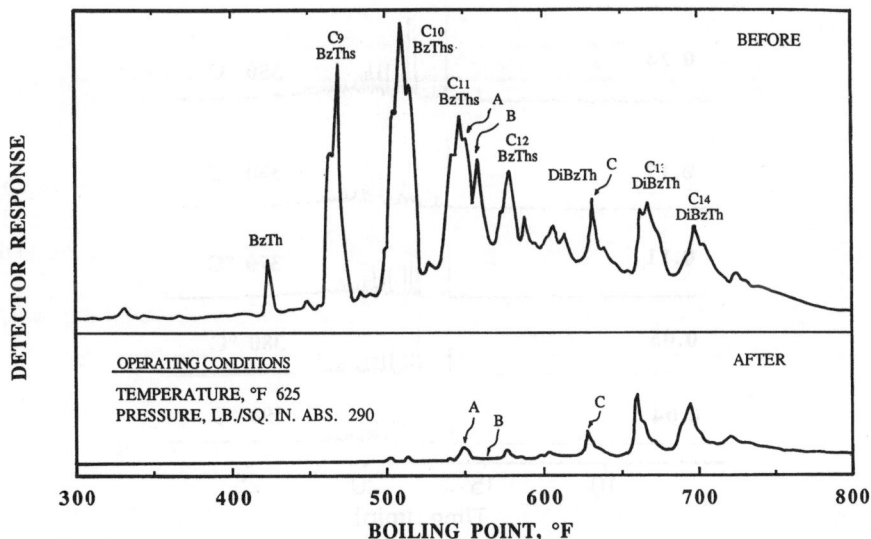

Fig. 2.4 Gas chromatographic analyses of light catalytic cycle oil before and after hydrodesulfurization. (From C. G. Frye and J. F. Mosby, *Chem. Eng. Prog.*, **63**, 68 (1967))

300–410°C, LHSV: 4 h⁻¹; Gas/Oil: 120 Nl/l. After HDS of LGO for 10 h at 350°C, samples were collected from the gas-liquid separator. Then the reaction temperature was changed and after 4–5 h, sampling was carried out in a similar manner.

GC-AED charts of desulfurized oil at each temperature are shown in Fig. 2.5. When the original oil was hydrodesulfurized at 390°C, 1.46 wt% sulfur content in original oil decreased to 0.04 wt%. Alkylbenzothiophenes were completely desulfurized even at 350°C independent of the presence of substituents. Nonsubstituted DBT was initially desulfurized, followed by a decrease in C_5- and C_4-substituted DBTs. In the case of nonsubstituted DBT, this result is due to the relative ease with which DBT desulfurizes. With respect to C_5- and C_4-DBTs, the result may be due to the paucity of the initial amounts rather than the relative ease with which they desulfurize. By contrast, C_1-, C_2-, and C_3-substituted DBTs remained until the final stage. Two percent of C_1-substituted DBT, 12% of C_2-substituted DBT, and 10% of C_3-substituted DBT did not convert even at 390°C. It was found that among PASH methyl-substituted dibenzothiophenes, 4-methyldibenzothiophene (4-MDBT) and 4,6-dimethyldibenzothiophene (4,6-DMDBT) in particular were most difficult to desulfurize in direct desulfurization of middle distillate by Co-Mo/Al₂O₃. 6% of 4-MDBT and 52% of 4,6-DMDBT were not converted even at 390°C.

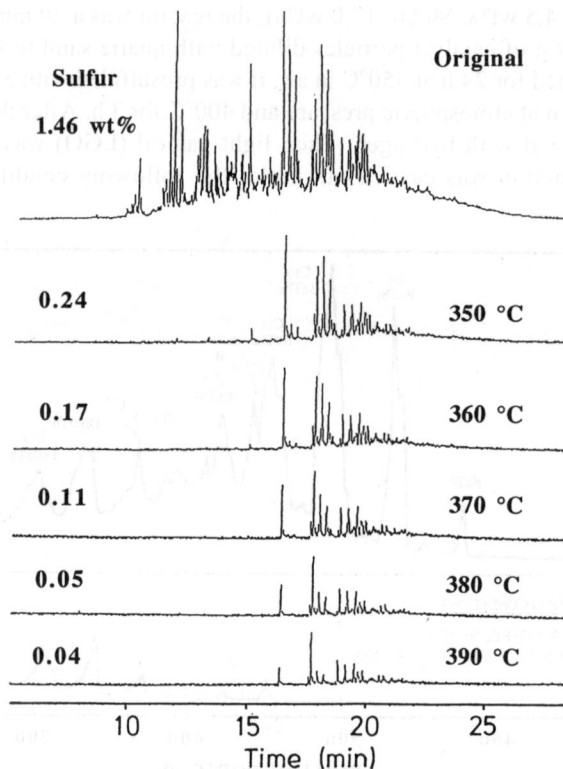

Fig. 2.5 GC-AED charts of desulfurized oil at each temperature.
(From T. Kabe, A. Ishihara et al., Ind. Eng. Chem. Res., 31, 1579 (1992))

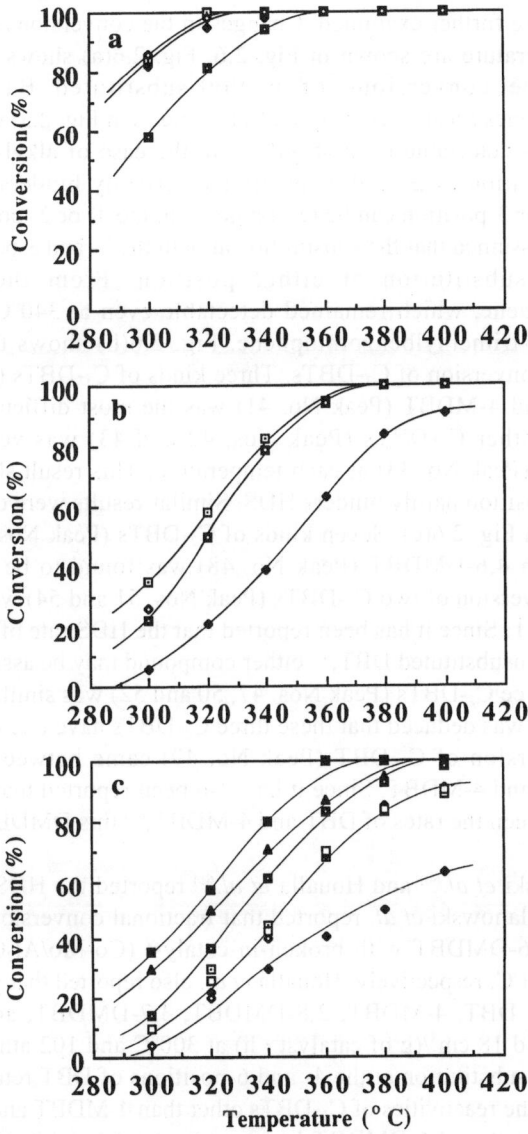

Fig. 2.6   Effect of temperature on conversions of methylsubstituted benzothiophenes, monomethyldibenzothiophenes
and dimethyldibenzothiophenes.
(a) methylsubstituted benzothiophenes.
    Peak No.: ●: 4; ◆: 8; ■: 10; ◇:11.
(b) monomethyldibenzothiophenes.
    Peak No.: ●: 33; ◆: 41; ■: 42; ◇:43.
(c) dimethyldibenzothiophenes.
    Peak No.: ●: 47; ◆: 48; ■: 49; ◇:50; ■: 51; □: 52; △:54.
    Total pressure: 30 atm, LHSV: 4 h⁻¹, Gas/Oil: 120 NL/L, Cat B.
(From T. Kabe, A. Ishihara *et al.*, *J. Jpn. Petrol. Inst.*, **36**, 469–470 (1993))

The behaviors of some alkyl-substituted BTs and C_1- and C_2-DBTs other than 4-MDBT and 4,6-DMDBT were further examined. Changes in the conversions of BTs, C_1-DBTs and C_2-DBTs with temperature are shown in Fig. 2.6. Fig. 2.6(a) shows the effect of reaction temperature on the conversions of methyl-substituted BTs. Three kinds of trimethylbenzothiophenes (Peak Nos. 8, 10 and 11, same as in Fig. 2.3) were detected at 300°C and one of them was detectable even at 340°C. In the case of alkyl-substituted BTs, it is reported that substitution of 2- and 7- positions sterically hinders HDS. However, the substituent at the 2- or 3-position can be rearranged into the 3- or 2- position, respectively.[66] Therefore, it can be assumed that the substitution at both the 2- and 3-positions hindered HDS much more than substitution at either position. From these considerations, trimethylbenzothiophene, which remained detectable even at 340°C (Peak No. 10), was deduced to be 2,3,7-trimethylbenzothiophene. Fig. 2.6(b) shows the effect of reaction temperature on the conversion of C_1-DBTs. Three kinds of C_1-DBTs (Peak Nos. 41, 42 and 43) were detected and 4-MDBT (Peak No. 41) was the most difficult to desulfurize. The conversion of two other C_1-DBTs (Peak Nos. 42 and 43) was very similar to that of nonsubstituted DBT (Peak No. 33) at each temperature. This result shows that substitution other than at the 4-position hardly hinders HDS. Similar results were obtained in the case of C_2-DBT, as shown in Fig. 2.6(c). Seven kinds of C_2-DBTs (Peak Nos. 47–52 and 54) were traced, among which 4,6-DMDBT (Peak No. 48) was found to be the most difficult to desulfurize. The conversion of two C_2-DBTs (Peak Nos. 51 and 54) was very similar to that of nonsubstituted DBT. Since it has been reported that the HDS rate of 2,8-DMDBT is about the same as that of nonsubstituted DBT,[61] either compound may be assigned to 2,8-DMDBT. The conversion of three C_2-DBTs (Peak Nos. 47, 50 and 52) was similar to that of 4-MDBT. From these results, it was deduced that these three C_2-DBTs have one methyl group at the 4-position. The conversion of C_2-DBT (Peak No. 49) came between the conversions of nonsubstituted DBT and 4-MDBT. Since it has also been reported that the HDS rate of 3,7-DMDBT comes between the rates of DBT and 4-MDBT,[61] this DMDBT may be assigned to 3,7-DMDBT.

So far, Kilanowski et al.[60] and Houalla et al.[61] reported the HDS reactions of methyl-substituted DBTs. Kilanowski et al. reported that fractional conversions of DBT, 4-MDBT, 2,8-DMDBT, and 4,6-DMDBT with broken-in catalyst (Co-Mo/Al_2O_3) were 0.12, 0.047, 0.24, and 0.023 at 450°C, respectively. Houalla et al. also reported that pseudo-first-order rate constants of HDS of DBT, 4-MDBT, 2,8-DMDBT, 3,7-DMDBT, and 4,6-DMDBT were 258, 24, 242, 127, and 18 cm^3/(g of catalyst · h) at 300°C and 102 atm, respectively. It was evident that only the substitution at the 4- and 6-positions of DBT retarded the rate of HDS markedly. However, the reactivities of C_1-DBTs other than 4-MDBT and C_2-DBTs substituted at the 4-position other than 4,6-DMDBT have not been reported in HDS catalyzed by Co-Mo/Al_2O_3. Although all C_1- and C_2-DBTs were not identified in the study, the reactivities of C_1- and C_2-DBTs other than 4-MDBT and 4,6-DMDBT could be estimated in hydrodesulfurization of middle distillate. From the above results, PASH in middle distillate may be classified with respect to the HDS reactivities as follows: 1) alkyl-substituted BTs, 2) nonsubstituted DBT and alkyl-substituted DBTs with no substituents at 4- and 6-positions, 3) alkyl-substituted DBTs with a substituent at the 4-position, and 4) alkyl-substituted DBTs with substituents at both the 4- and 6-positions. The results suggest that development of catalysts to enable selective desulfurization of alkylbenzothiophenes with substituents at the 4- and 6-positions are needed to obtain a highly clean oil.

A number of efforts have been made to prepare catalysts having high catalytic activity. The reactivities for some of these are introduced here. To investigate the effect of loading and composition of catalysts on the HDS reactions of LGO, the behavior of alkyl-substituted DBTs in desulfurization of LGO has been traced using three kinds of Co-Mo/Al$_2$O$_3$ catalysts.[67] The catalysts were commercial Co-Mo/Al$_2$O$_3$ catalysts (Cat. A: CoO 3.8 wt%, MoO$_3$ 12.3 wt%; Cat. B: CoO 4.5 wt%, MoO$_3$ 17.0 wt%; Cat. C: for deep desulfurization). Cat. B was the same one used in the preceding paragraphs. Figs. 2.7(a), (b) and (c) show the effects of temperature on the conversions of DBT, 4-MDBT and 4,6-DMDBT by the use of Cat. A, Cat. B and Cat. C, respectively. Comparing Cat. A with Cat B, 15°C, 10°C and 20°C higher temperatures were required to obtain the same conversions of DBT, 4-MDBT and 4,6-DMDBT, respectively, using Cat. A. This can be explained by the difference in the amounts of molybdenum and cobalt involved in the catalysts. Comparing Cat. B with Cat. C, although there were no differences between conversions of DBT or 4-MDBT, there was a slight difference in the conversion of 4,6-DMDBT above *ca.* 350°C. Above 350°C, sulfur concentration approached deep desulfurization conditions (less than 0.05 wt% sulfur). The result indicates that Cat. C favored the conversion of 4,6-DMDBT more than Cat. B under deep desulfurization conditions probably because of the relative ease of adsorption of 4,6-DMDBT and hydrogenation of an aromatic ring.

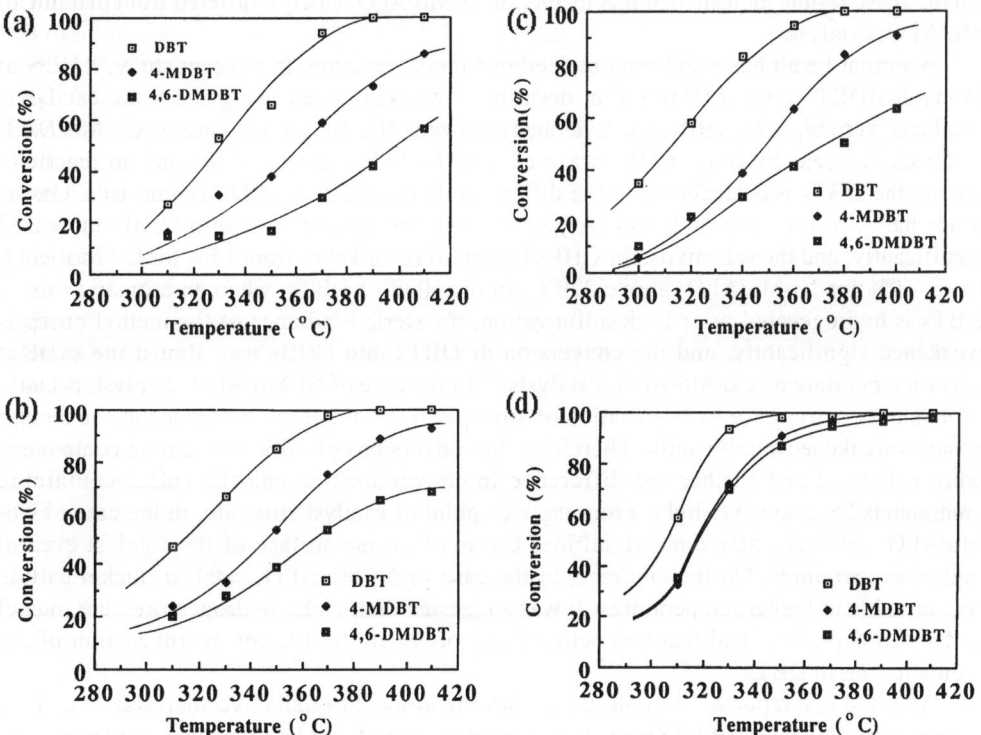

Fig. 2.7 Effect of temperature on conversions of DBT, 4-MDBT and 4,6-DMDBT in light gas oil. Total pressure: 30 atm, LHSV: 4 h^{-1}, Gas/Oil: 120 NL/L. (a) Cat A; (b) Cat B; (c) Cat C; (d) Cat D.
(From T. Kabe, A. Ishihara *et al.*, *J. Jpn. Petrol. Inst.*, **40**, 32 (1997))

It is important to compare the results from deep HDS of PASH in middle distillate using the Co-Mo/Al$_2$O$_3$ catalysts with those using a commercial Ni-Mo/Al$_2$O$_3$ catalyst (Cat. D: NiO 2.9%, MoO$_3$ 15.8%), which has higher activity for hydrogenation of aromatic rings.[67] Similar to Co-Mo/Al$_2$O$_3$ catalysts, one kind of trimethylbenzothiophene (Peak No.10 same as in Fig. 2.3) was most difficult to desulfurize. Compared with Co-Mo/Al$_2$O$_3$ catalysts, all the conversions of methyl-substituted BTs increased, and the difference in conversion among the four compounds (Peak Nos. 4, 8, 10 and 11) became smaller for the Ni-Mo/Al$_2$O$_3$ catalyst. Similar to the Co-Mo/Al$_2$O$_3$ catalysts, 4-MDBT (Peak No. 41) was most difficult to desulfurize. Compared with Co-Mo/Al$_2$O$_3$ catalysts, the Ni-Mo/Al$_2$O$_3$ catalyst enhanced all the conversions of C$_1$-DBTs at high temperatures especially (above 330°C), and the difference in conversion among the four compounds (Peak Nos. 33, 41, 42 and 43) was smaller than that of the Co-Mo/Al$_2$O$_3$ catalysts. These trend varied more significantly in the case of C$_2$-DBTs. In the case of Co-Mo/Al$_2$O$_3$ catalysts, the conversions decreased in the order No. 51 > 54 > 49 > 52 = 50 > 47 > 48, and the difference of reaction activity among seven compounds was very obvious. In the case of Ni-Mo/Al$_2$O$_3$ catalyst, however, the difference in reaction activity among seven compounds was rather small. The conversions of DBT, 4-MDBT and 4,6-DMDBT on Ni-Mo/Al$_2$O$_3$ catalyst are plotted against temperature in Fig. 2.7(d). Compared with Co-Mo/Al$_2$O$_3$ catalysts, the difference in reaction activity among DBTs was rather small. These results indicate that this feature of Ni-Mo/Al$_2$O$_3$ catalyst differed from that of Co-Mo/Al$_2$O$_3$ catalyst.

A similar result has also been obtained in a model reaction. In a recent study,[68] HDS of DBT, 4-MDBT and 4,6-DMDBT in decalin were investigated using the same catalysts. Products were biphenyl (BPs) and cyclohexylbenzene (CHBs). In the case of Co-Mo/Al$_2$O$_3$ catalysts, the selectivity for CHBs was very low (< 15%), and the difference in reactivity among the DBTs was attributed to the difference in conversions of DBTs into BPs. On the other hand, in the case of Ni-Mo/Al$_2$O$_3$ catalyst, the conversions into CHBs increased significantly, and the selectivity for CHBs increased remarkably from 15% for Co-Mo/Al$_2$O$_3$ to $ca.$ 60% for Ni-Mo/Al$_2$O$_3$ above 320°C. In the HDS of DBTs, when an aromatic ring in DBTs is hydrogenated prior to desulfurization, the steric hindrance of the methyl group is weakened significantly, and the conversion of DBTs into CHBs was almost the same at every temperature on Co-Mo/Al$_2$O$_3$ catalysts.[68] In the case of Ni-Mo/Al$_2$O$_3$ catalyst, because of its higher activity for hydrogenation of aromatic rings, the steric hindrance of the methyl group is weakened significantly. Therefore, the conversions of sulfur-containing compounds were enhanced and further the difference in conversions among the sulfur-containing compounds became very slight. From the viewpoint of catalyst structure, in the case of Co-Mo/Al$_2$O$_3$ catalysts, active metal sulfides can exist on the surface of the catalyst even at higher temperatures. On the contrary, in the case of Ni-Mo/Al$_2$O$_3$ catalyst, nickel sulfide was unstable at higher temperatures. It was suggested that, under hydrogen pressure, nickel metal was deposited, and that this active metal provided the site for hydrogenation of the aromatic ring in DBTs.

After the first reports,[32,33] some groups have reported the qualitative analysis[69–72] and the determination[34,37,67,73,74] of PASH in deep desulfurization of middle distillates. For analysis of PASH, GC-AED,[32–34,37] GC-FPD,[67–73] GC-SCD (sulfur chemiluminescence detector),[74] etc. were used, and for the activity test a fixed bed reactor[32–34,37,67–69,71,74] or an autoclave[70,72,73] and Co-Mo/Al$_2$O$_3$[32–34,37,67–74] or Ni-Mo/Al$_2$O$_3$ [67,68,70,72–74] were used. In most of these reports, it was pointed out that both 4-MDBT and 4,6-DMDBT remained even under deep desulfurization conditions and that the substitution only at the 4- and 6-positions of DBT remarkably retarded the HDS rate. However, it has not been clarified why sulfur compounds with substituents at the 4- and 6-positions of DBT such as 4-MDBT and 4,6-DMDBT are difficult to desulfurize even under deep desulfurization conditions. In order to answer this question, the HDS mechanisms of 4-MDBT and 4,6-DMDBT must be elucidated by examining the reaction kinetics of these compounds in detail. Before describing the effects of the substituents on HDS reactions, the basic HDS reaction kinetics of thiophene, benzothiophene and dibenzothiophene will be discussed.

2.1.2 Kinetics of Catalytic Hydrodesulfurization Reactions

Hydrodesulfurization (HDS) has long been one of the most important processes in the petroleum industry. A number of attempts have been carried out to develop the process and elucidate the basic scientific aspects over the last several decades. The conversion of sulfur-containing compounds in petroleum and other fossil fuels to hydrocarbons and hydrogen sulfide are included in the reaction. Classic objectives in this process include prevention of sulfur poisoning of platinum catalysts in the reforming process and avoiding SO_2 pollution caused by the combustion of sulfur-containing fuel oils. In recent years, the increase in the demand for white oil worldwide as well as the concern for environmental regulations has required the hydroprocessing of heavy and residual oils. In relation to this, much attention has been focused on HDS of the least reactive sulfur heterocycles such as thiophenic compounds. Further, in the 1990's deep desulfurization of middle distillate led to renewed interest in the development of new catalysts and the elucidation of deep desulfurization mechanisms. Among the large number of studies, kinetics can give us much information related to the vital behavior of reacting species on an actually working catalyst in a reactor.

Comprehensive review papers on this subject have been published by many authors: Gates et al.,[49] Weisser and Landa,[75] and Schuit and Gates,[76] in the 1970's, Vrinat[50,51] and Massoth and Muralidhar[52] in the 1980's, and Girgis and Gates,[53] Topsøe et al.,[77] A. N. Startsev,[78] and B. Delmon,[79] in the 1990's.

A. Thermodynamics and Reactivities

HDS of mercaptanes, sulfides, disulfides and thiophenic compounds is exothermic and there is no thermodynamic limitation under industrial reaction conditions where the equilibrium constants all decrease with increase in temperature and have values more than 1.[50,53,80] It has also been reported that HDS of dibenzothiophene (DBT) to biphenyl (BP) is favored thermodynamically under practical HDS conditions.[50] In contrast, the equilibrium constant for hydrogenation of thiophene to tetrahydrothiophene is less than 1 above 350°C, indicating that a HDS pathway via hydrogenation of thiophene ring may be inhibited at higher temperatures because of the low equilibrium concentration of tetrahydrothiophene.

It has long been recognized that thiophenic compounds are less reactive in HDS than mercaptanes, sulfides and disulfides.[81–83] Therefore, recent researches in HDS have been directed at the conversion of these thiophenic compounds. There is a large number of reports on HDS of thiophene (T), benzothiophene (BT) and DBT because they are among the simplest compounds in model reactions for petroleum refineries. Nag et al. reported the reactivities of typical thiophenic compounds as shown in Table 2.3.[84] The rate constants decreased in the

Table 2.3 Reactivities of Several Heterocyclic Sulfur Compounds [84]

Reactant	Structure	Pseudo-first-order rate constants (L/s · g-cat)
Thiophene		1.38×10^{-3}
Benzothiophene		8.11×10^{-4}
Dibenzothiophene		6.11×10^{-5}
Benzo[*b*]naphtho-[2,3-*d*]thiophene		1.61×10^{-4}
7,8,9,10-Tetrahydro-benzo[*b*]naphtho-[2,3-*d*]thiophene		7.78×10^{-5}

Batch reactor, React, Temp. 300°C, 71 atm, *n*-hexadecane solvent, concentration of reactant: 0.25 mol%, Co-Mo/Al$_2$O$_3$.
(From N. K. Nag, A. V. Sapre, D. H. Broderick and B. C. Gate, *J. Catal.*, **57**, 510 (1979))

order T (1) > BT (0.59 of T) > DBT (0.04 of T). DBT was one order of magnitude less reactive than BT. Benzonaphthothiophene (BNT) and its hydrogenated derivative has similar or rather higher reactivities than DBT. The result suggests that HDS of three-ring compounds may be a key reaction in making deeply desulfurized oil from heavier fractions. Hydrogenated derivatives are more easily desulfurized than thiophenic compounds.[60,75]

It seems that HDS reactivity depends on the molecular size and the structure of the sulfur-containing compound. However, there are some discrepancies in the literature. For the reactions of T and BT, there are three types of reports, that is, T > BT,[84,85] T = BT[60,86] and T < BT.[87-90] For the reactions of BT and DBT, there are two types of reports, that is, BT > DBT,[54,55,84,85,91] and BT = DBT.[60,86,92] The results suggest that the reactivities of these compounds change depending on the reaction conditions and that it is difficult to compare the results obtained under different conditions.

B. Reaction Networks and Kinetics

There are few examples of kinetic studies for HDS of thiophenes where the effects of all the reactants and products have been considered under extensive reaction conditions. Therefore reaction networks for HDS of thiophenes are not completely understood. However, much effort has been made to elucidate the reaction pathways and mechanisms for HDS of thiophene, benzothiophene, dibenzothiophene, etc. Extensive work on HDS of thiophenes has been reviewed by Gates *et al.*,[49] Vrinat,[50] Girgis and Gates,[53] Topsøe *et al.*[77] and Startsev.[78]

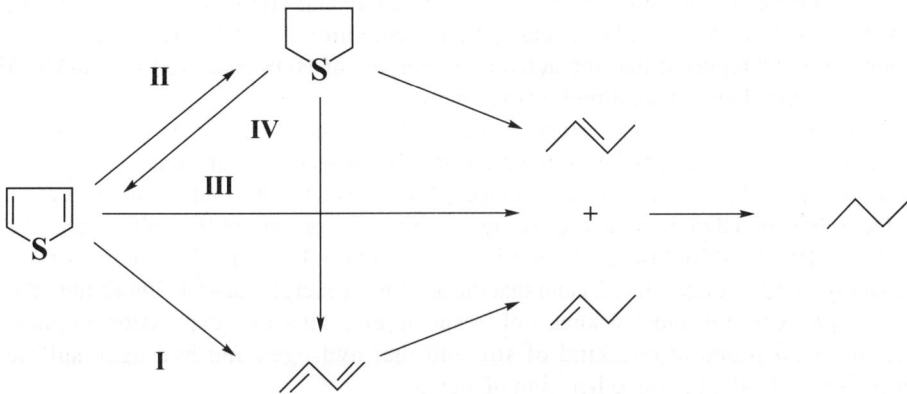

Fig. 2.8 Proposed reaction pathways for thiophene HDS.

1. Thiophene

Figure 2.8 shows the possible reaction pathways for thiophene HDS. Two major reaction pathways have been proposed in the literature: 1) Thiophene is desulfurized prior to hydrogenation to give butadiene which is hydrogenated to butene; 2) Thiophene is hydrogenated prior to desulfurization to give tetrahydrothiophene (THT) which is desulfurized to butene. The former was proposed originally by Owens and Amberg[93] and later by other groups.[94–97] Other possible routes have been proposed: 3) THT is desulfurized to butadiene which is hydrogenated to butene;[98–100] 4) Thiophene is directly desulfurized to butene.[101] In the fourth route, a concerted mechanism was proposed in which saturation of the thiophene ring and hydrogenolysis of the C-S bond occurred on one active center without intermediates (butadiene and THT) evolving into the gas phase.

Satterfield and Roberts reported the first example of the determination of the kinetics of thiophene HDS.[102] The reaction was performed in a steady-state recirculation flow reactor using Co-Mo/Al$_2$O$_3$ (Co 3%, Mo 7%) in the absence of mass transfer influence (235–265°C). The thiophene disappearance was represented by the following Langmuir-Hinshelwood (L-H) rate equation:

$$r_{HDS} = k_{HDS}K_TP_TK_HP_H/(1 + K_TP_T + K_{H2S}P_{H2S})^2 \tag{2.1}$$

where r_{HDS} = rate of HDS; k_{HDS} = rate constant of HDS; K_T, K_{H2S} and K_H=adsorption equilibrium constants of thiophene, hydrogen sulfide and hydrogen; P_T, P_{H2S} and P_H = partial pressure of thiophene, hydrogen sulfide and hydrogen. The rate expression implies that the surface reaction between thiophene and hydrogen is the rate-determining step and that thiophene and hydrogen sulfide which retards the HDS reaction are competitively adsorbed at one type of catalytic site. It cannot be clarified whether hydrogen is adsorbed at the same catalytic site as thiophene and hydrogen sulfide because the hydrogen partial pressure was varied to only a small extent. Further, the activation energy of HDS was quite small 3.7 kcal/mol. The rate of butene hydrogenation was also estimated using a L-H equation and it has been concluded that hydrogenolysis of thiophene and hydrogenation of butene proceed on separate catalytic sites since hydrogen sulfide affects these reactions differently. Similarly some other rate equations have been applied for thiophene HDS using Co-Mo/Al$_2$O$_3$, Ni-Mo/Al$_2$O$_3$ and Ni-W/Al$_2$O$_3$ under several different conditions.[63,88,103–113] Ozimek et al.[103] and Morooka et al.[105] reported that the activation energies of HDS catalyzed by Co-Mo/Al$_2$O$_3$ were 21.6 kcal/mol and 20 kcal/mol, respectively.

In most studies of thiophene HDS the range of hydrogen pressure was very narrow and near atmospheric pressure, and a large variation of hydrogen pressure was needed to obtain the hydrogen partial pressure term in the rate expression exactly. Recently Radomysky et al.[111] reported thiophene HDS under a wider range of hydrogen pressure (469–509 K, 583–623 K, 0.10–1.81 MPa, Co-Mo/Al$_2$O$_3$; MoO$_3$ 15.8 wt%; CoO 5.3 wt%). They showed that data successfully fitted through Eq.(2.2) and that the activation energies at 469–509 K and 583–623 K were 10.8 kcal/mol and 5.6 kcal/mol, respectively. This rate expression implies that thiophene is adsorbed at one kind of site and that hydrogen and hydrogen sulfide are competitively adsorbed at the other kind of site.

$$r_{HDS} = k_{HDS}K_T P_T/(1 + K_T P_T) \times K_H P_H/(1 + K_H P_H + K_{H2S} P_{H2S}) \tag{2.2}$$

Van Parijs and Froment[88] also performed thiophene HDS under conditions varied over a wide range (260–350°C, 2–30 atm, H$_2$/thiophene 4–9 (mol/mol), Co-Mo/Al$_2$O$_3$). Products were 1-butene, cis- and trans-2-butene and butane and neither tetrahydrothiophene nor butadiene was detected. The data were correlated by L-H equation (2.3):

$$r_{HDS} = k_{HDS}K_T P_T\, K_H P_H\,/[1 + (K_H P_H)^{1/2} + K_T P_T + K_{H2S} P_{H2S}/P_H]^3 \tag{2.3}$$

where it was assumed that hydrogen was dissociatively adsorbed. This rate expression means that thiophene, hydrogen and hydrogen sulfide are competitively adsorbed at one kind of site. Activation energies of thiophene HDS and heat of adsorption of thiophene were calculated to be 29.9 and 10.7 kcal/mol, respectively. Equilibrium constants of thiophene, hydrogen sulfide and hydrogen were calculated to be 13.7, 91.2 and 0.536 atm^{-1} at 260°C. The rate of butene hydrogenation was also estimated using a L-H equation, and an equilibrium constant of hydrogen calculated for butene hydrogenation (6.02 × 10^{-2} atm^{-1}) was one order of magnitude less than that for thiophene HDS. In accordance with results of low-pressure experiments the result suggested that hydrogenolysis of thiophene and hydrogenation of butene proceeded separately on different kinds of catalytic sites.

2. Benzothiophene

Figure 2.9 shows the possible reaction pathways for benzothiophene (BT) HDS. Two major reaction pathways have been proposed in the literature[49–53]: 1) BT is desulfurized prior to partial hydrogenation to dihydrobenzothiophene (DHBT) to give styrene, which is hydrogenated to ethylbenzene (EB); 2) BT is hydrogenated prior to desulfurization to give DHBT, which is desulfurized to EB. As noted elsewhere, the differences in the reaction conditions or experimental procedure lead to contradictory conclusions regarding the network.[50–53] For example, Daly reported that BT HDS gave EB by either of two parallel pathways.[114] Geneste et al. reported that ethylbenzene was formed through DHBT which was irreversibly produced from BT.[115] These experiments were performed with liquid-phase reactants under high hydrogen pressure in a batch system and only rate constants for BT disappearance were reported. Some groups suggested that the path through styrene was the

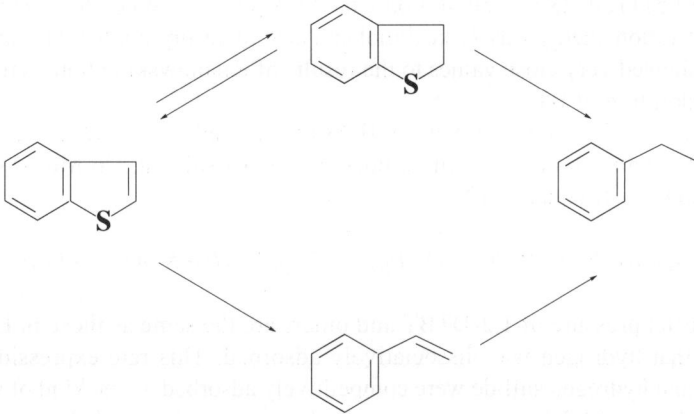

Fig. 2.9 Proposed reaction pathways for benzothiophene HDS.

major route since styrene was detected in the reaction products.[60,116] The two pathways were also described for vapor-phase or liquid-phase reaction of BT at high pressure using a flow reactor.[89,90,117,118] Although styrene was not detected, it was assumed that it was hydrogenated rapidly to give EB after being formed from BT. It was also observed that a hydrogenation-dehydrogenation equilibrium between BT and DHBT was established at a rate which was high compared with the rate of desulfurization.[49,66] Consequently, it was pointed out that it was impossible to establish whether one was an intermediate in the desulfurization of the other because these two compounds were desulfurized at approximately equal rates.[49,66]

Although BT is one of PASH compounds found in heavier feed stocks,[62] there have been few kinetic studies for BT HDS over a wide range of temperature and pressure.[50–53,89,90,118,119] The kinetic study of BT HDS over a wide range of temperature was reported by Kilanowski and Gates.[119] They performed the reaction of BT with hydrogen at temperatures 525, 575 and 605 K below 2 atm on Co-Mo/Al$_2$O$_3$. They showed that Eq.(2.4), where BT, hydrogen sulfide and hydrogen were competitively adsorbed at the same adsorption sites and the retarding term, $K_{H_2}P_{H_2}$, in the denominator was neglected, roughly fitted the data at all three temperatures and that Eq.(2.5), where BT and hydrogen were adsorbed at separate adsorption sites and BT and hydrogen sulfide were competitively adsorbed at the same adsorption sites, reasonably fitted the data at the two higher temperatures:

$$r_{HDS} = k_{HDS}K_{BT}P_{BT}K_HP_H /(1 + K_{BT}P_{BT} + K_{H_2S}P_{H_2S})^2 \tag{2.4}$$

$$r_{HDS} = k_{HDS}K_{BT}P_{BT}/(1 + K_{BT}P_{BT} + K_{H_2S}P_{H_2S}) \times K_HP_H/(1 + K_HP_H) \tag{2.5}$$

where K_{BT} = adsorption equilibrium constants of BT, P_{BT} = partial pressure of BT and others are the same as those in Eq.(2.1). When Eq.(2.4) was used in their no-solvent system, the activation energy was 20±3 kcal/mol and heats of adsorption for BT and H$_2$S were 15±10 and 6±6 kcal/mol, respectively. It was described that the uncertainty in these latter values reflects the poorness of fit of the data to Eq.(2.4) at 525°C. When Eq.(2.5) was used, the values for BT and H$_2$S, respectively, were 19 and 13 kcal/mol.

Ishihara *et al.* reported a kinetic study of BT HDS using a simplified rate expression similar to Eq.(2.5) (180–280°C, 50 atm on Co-Mo/Al$_2$O$_3$).[54,55] When toluene was used as a solvent, the activation energy was 21 kcal/mol and heat of adsorption for BT was 22 kcal/mol. These results showed very close values to the results of Kilanowski in both activation energy and heat of adsorption of BT.

Van Parijs *et al.*[89] also performed BT HDS under conditions varied over a wide range (240–300°C, 2–30 atm, H$_2$/BT 4–9 (mol/mol), *n*-heptane solvent, Co-Mo/Al$_2$O$_3$). The data were correlated by L-H equation (2.6):

$$r_{HDS} = k_{HDS}K_{BT}P_{BT}K_HP_H /[1 + (K_HP_H)^{1/2} + K_{H2S}P_{H2S}/P_H+K_{BT}(P_{BT} + P_D)]^3 \tag{2.6}$$

where P_D = partial pressure of 1,2-DTBT and others are the same as those in Eq.(2.4) and it was assumed that hydrogen was dissociatively adsorbed. This rate expression means that BT, hydrogen and hydrogen sulfide were competitively adsorbed at one kind of site. The rates of 1,2-DTBT HDS and BT hydrogenation were also estimated using L-H equations. Similar to the study of thiophene HDS,[88] it was assumed that hydrogenolysis of BT and 1,2-DTBT and hydrogenation of BT proceeded separately on different kinds of catalytic sites. Activation energies for BT and DTBT HDS were calculated to be 17.6 and 31.3 kcal/mol, respectively. Equilibrium constants of BT, hydrogen sulfide and hydrogen were calculated to be 19.3, 211 and 0.358 atm^{-1} at 260°C. Because these parameters have values similar to those for thiophene HDS, it is suggested that the parameters have a physical meaning in the context of Langmuir-Hinshelwood models.[53]

3. Dibenzothiophene

Figure 2.10 shows the possible reaction pathways for dibenzothiophene HDS. Houalla *et al.*[120] and Nagai and Kabe[121] have proposed two parallel pathways as follows: 1) Dibenzothiophene (DBT) is directly desulfurized to give biphenyl (BP); 2) DBT is hydrogenated prior to desulfurization to give 1,2,3,4-tetrahydrodibenzothiophene (TDBT) or 1,2,3,4,10,11-hexahydrodibenzothiophene (HDBT), which is desulfurized to cyclohexylbenzene (CHB). The pseudo-first-order rate constants in L/(g of catalyst · s) at 300°C determined by Houalla *et al.* are also shown in Fig. 2.10. The results shows that the desulfurization pathways of DBT to BP and of TDBT or HDBT to CHB are much faster than the hydrogenation pathways of DBT to TDBT or HDBT and of BP to CHB in HDS of DBT catalyzed by Co-Mo/Al$_2$O$_3$ under the elevated pressure. It was also reported that the selectivity for CHB did not exceed 15% in HDS of DBT catalyzed by various Co-Mo/Al$_2$O$_3$ catalysts even at 340°C and 50 atm.[68] Although a similar reaction pathway has been reported for Ni-Mo/Al$_2$O$_3$,[53] the selectivity for CHB remarkably increased from about 15% for Co-Mo/Al$_2$O$_3$ to about 60% for Ni-Mo/Al$_2$O$_3$ above 320°C.[68] The higher yield of CHB was due to its greater activity for hydrogenation. As shown in Fig. 2.10, the HDS rate of partially hydrogenated DBT (TDBT and HDBT) is higher than that of DBT. When HDS activities for Co-Mo/Al$_2$O$_3$ and Ni-Mo/Al$_2$O$_3$ with similar loadings of metals were compared under the same reaction conditions, the total rate of HDS for the latter was higher than that for the former, indicating that the hydrogenation pathway of DBT to TDBT or HDBT which is desulfurized to CHB as well as the direct S-extrusion pathway of DBT to BP is more important for Ni-Mo/Al$_2$O$_3$ compared with Co-Mo/Al$_2$O$_3$.[68] Since the higher hydrogenation activity for Ni-Mo/Al$_2$O$_3$ leads to excess consumption of expensive hydrogen, however, it is necessary to choose the way to use this catalyst.

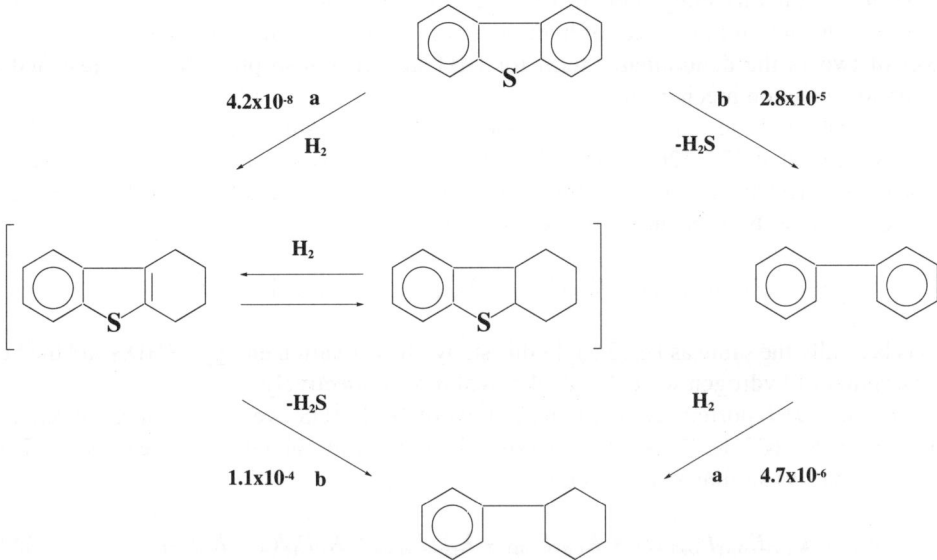

Fig. 2.10 Proposed reaction pathways for dibenzothiophene HDS.
(From M. Houalla, N. K. Nag, A. V. Sapre, D. H. Broderick and B. C. Gates, *AIChE J.*, **24**, 1019 (1978))

The kinetic studies of dibenzothiophene HDS have been reported by several investigators.[54,55,57–59,62,68,121–127] Experiments over a wide range of temperature and pressure was reported by Broderick and Gates.[57] They performed the reaction of DBT in *n*-hexadecane with hydrogen at the range of 548–598 K, and 34–160 atm on Co-Mo/Al$_2$O$_3$. For hydrogenolysis of DBT, they recommended the rate equation (2.7):

$$r_{HDS} = k_{HDS}K_{DBT}P_{DBT}/(1 + K_{DBT}P_{DBT} + K_{H2S}P_{H2S})^2 \times K_H P_H/(1 + K_H P_H) \qquad (2.7)$$

where K_{DBT} = adsorption equilibrium constants of dibenzothiophene, P_{DBT} = partial pressure of dibenzothiophene and others are the same as those in Eq.(2.1). The rate was estimated by the formation of biphenyl. The rate expression implies that the surface reaction between dibenzothiophene and hydrogen is the rate-determining step and that dibenzothiophene and hydrogen sulfide which retards the HDS reaction are adsorbed competitively at one type of catalytic site. Further, dibenzothiophene and hydrogen are adsorbed at separate adsorption sites. The rate of dibenzothiophene hydrogenation (HY) was also estimated by tracing the formation of CHB, TDBT and HDBT using a L-H equation. Hydrogenation of DBT is not inhibited by hydrogen sulfide, consistent with the results reported for thiophene and benzothiophene.[88,89] There was a difference in form between the rate equations for DBT hydrogenolysis and DBT hydrogenation. It is suggested that DBT hydrogenolysis and DBT hydrogenation may proceed separately on different kinds of catalytic sites, in good agreement with some of the results mentioned above for thiophene and benzothiophene. For DBT hydrogenolysis in this study, the activation energy, and the heats of adsorption of DBT, H$_2$S and hydrogen were 30, 4.5, 5.3 and −8.4 kcal/mol, respectively. The result revealed that hydrogen sulfide adsorbed on the catalysts surface more strongly than DBT. Although the important result in this study is the firmly established dependence of the hydrogenolysis rate on the H_2 concentration, the heat of hydrogen adsorption was a negative value. Further, the power of two in the denominator term for DBT and H_2S adsorption was not justified in relation to a reaction mechanism.

Vrinat and de Mourgues reported a vapor phase kinetic study of DBT on Co-Mo/Al_2O_3 industrial catalyst at 473–520 K and 5–50 atm.[59] The rate law is consistent with a Langmuir-Hinshelwood mechanism without competitive adsorptions between DBT and hydrogen, and with the inhibition by H_2S, and is best expressed as:

$$r_{HDS} = k_{HDS}K_{DBT}P_{DBT}/(1 + K_{DBT}P_{DBT} + K_{H2S}P_{H2S}) \times K_H P_H/(1 + K_H P_H). \tag{2.8}$$

This is basically the same as Eq.(2.5). In this study, the activation energy of HDS and the heat of adsorption of hydrogen were 23 and 4.5 kcal/mol, respectively.

Singhal et al. reported another kinetic study of DBT HDS over a wide range of reaction conditions (558–623 K, 7–26 atm, Co-Mo/Al_2O_3, tetralin solvent, flow reactor).[58] They proposed Eq.(2.9) similar to Eq.(2.8):

$$r_{HDS} = k_{HDS}K_{DBT}P_{DBT}/(1 + K_{DBT}P_{DBT} + K_{prod}P_{prod}) \times K_H P_H/(1 + K_H P_H) \tag{2.9}$$

where K_{prod} and P_{prod} mean the adsorption equilibrium constant and partial pressure of products (BP, CHB and H_2S). The equation shows that products other than hydrogen sulfide also contributed to the competitive adsorption with dibenzothiophene. However, this is not consistent with Broderick's result that biphenyl is not an inhibitor. In this study, the activation energy of HDS and heat of adsorption of hydrogen were 39 and 25 kcal/mol, respectively.

O'Brien et al. also reported another kinetic study of DBT HDS over a wide range of reaction conditions (540–695 K, 3.4–12.2 MPa, Co-Mo/Al_2O_3, continuous flow autoclave).[125] They proposed Eq.(2.10):

$$r_{HDS} = k_{HDS}K_{DBT}P_{DBT}K_H P_H/(1 + K_{DBT}P_{DBT} + K_H P_H + K_{H2S}P_{H2S})^2. \tag{2.10}$$

The equation shows that molecularly adsorbed hydrogen contributed to the competitive adsorption with dibenzothiophene. In this study, the activation energy of HDS and heats of adsorption of dibenzothiophene, hydrogen and hydrogen sulfide were 33, 12, 33 and 13 kcal/mol, respectively. The activation energy values were comparatively close to the results of Broderick et al. (30 kcal/mol) and Singhal et al. (39 kcal/mol). The heat of adsorption value of hydrogen was also close to the result of Singhal et al. (25 kcal/mol). Values between 20 and 60 kcal/mol for the heat of adsorption of hydrogen have been reported.[125] The relatively high value of the heat of adsorption of hydrogen may reveal that significant numbers of possible sites for adsorption are covered by hydrogen at high pressure. The heats of adsorption of dibenzothiophene and hydrogen sulfide were larger than those of Broderick et al. (4.5 and 5.3 kcal/mol, respectively). In each experiment, the values for dibenzothiophene and hydrogen sulfide were quite close. This shows that dibenzothiophene and hydrogen sulfide are competitively adsorbed at one catalytic site to the same extent.

Kabe and Ishihara reported a kinetic study of dibenzothiophene under deep desulfurization conditions (200–250°C, total pressure 12.5–75 atm, Co-Mo/Al_2O_3).[54,55,68,122,128] Under the experimental conditions, the amount of converted DBT was of zero order with

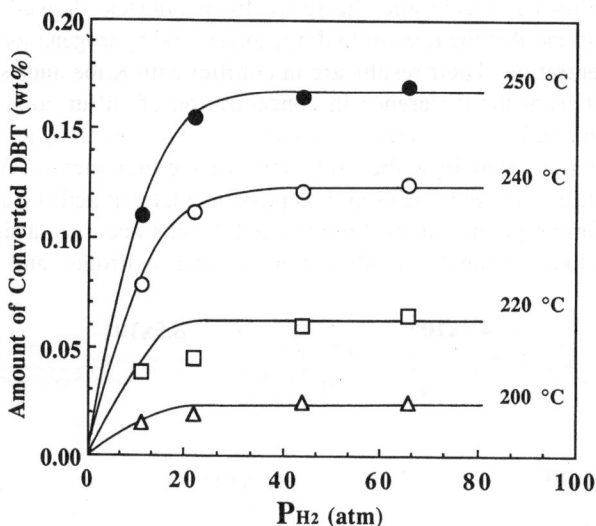

Fig. 2.11 Relation of amount of converted DBT vs. P_{H_2}.
Solvent: decalin; Total pressure: 50 atm; Catalyst: Co(4.0)-Mo(12.0).
(From T. Kabe, A. Ishihara *et al.*, *J. Jpn. Petrol. Inst.*, **40**, 186 (1997))

respect to partial pressure of hydrogen, as shown in Fig. 2.11. Using kinetic equation Eq.(2.8) or (2.9), the hydrogen pressure term can be included in the rate constant, and under deep desulfurization conditions (sulfur concentration 500 ppm), the inhibition due to H_2S is not significant because of the low concentration of H_2S in the reactor and the constant hydrogen sweeping. Further, it was assumed that K_{H_2S} was not much greater than K_{DBT}. Based on the above assumptions Eq.(2.8) or (2.9) could be simplified to

$$r_{HDS} = k_{HDS}K_{DBT}P_{DBT}/(1 + K_{DBT}P_{DBT}). \tag{2.11}$$

When decalin was used as the solvent, the activation energy was 22±2 kcal/mol and the heat of adsorption for DBT was 11±1 kcal/mol. This result for activation energy is closest to the result of Vrinat and de Mourgues. K_H and heat of adsorption of hydrogen were not obtained because hydrogen reached saturation of adsorption at over 25 atm in the range 200–250°C and the rate of HDS did not change with respect to partial pressure of hydrogen under low sulfur concentration. On the other hand, the rate of HDS increased with respect to partial pressure of DBT in the same temperature range. The result supports the mechanism that DBT and hydrogen are adsorbed at separate adsorption sites on the catalyst.

Differences in activation energy between the results described above are observed and may be due to the reaction temperature. When the range of temperature is relatively high, the activation energy of HDS tends to be high. Since Kabe and Ishihara's and Vrinat's experiments were performed at a similar lowest temperature range, similar results seemed to be obtained. Heat of adsorption of hydrogen appears to depend on the range of hydrogen pressure. When the range of hydrogen pressure is small and relatively low, heat of adsorption of hydrogen reveals high values as shown by Singhal's result. When the range of hydrogen pressure was extended to 50 atm by Vrinat or 160 atm by Broderick, heat of adsorption of hydrogen remarkably decreased and the result by Broderick showed a negative value. Nevertheless, they found that the rates of hydrogenolysis and hydrogenation of DBT depended on hydrogen concentration. Their results are in conflict with Kabe and Ishihara's data. The conflict may result from the difference in concentration of sulfur compounds. The molar ratio of sulfur compounds to hydrogen in Broderick's experiments is at least two orders of magnitude higher than that in Kabe and Ishihara's experiments. When such a high concentration of sulfur compound is used, it is possible that competitive adsorption between hydrogen and sulfur compounds at one kind of catalytic site occurs to a significant extent. If the mechanism includes competitive adsorption between hydrogen and sulfur compounds

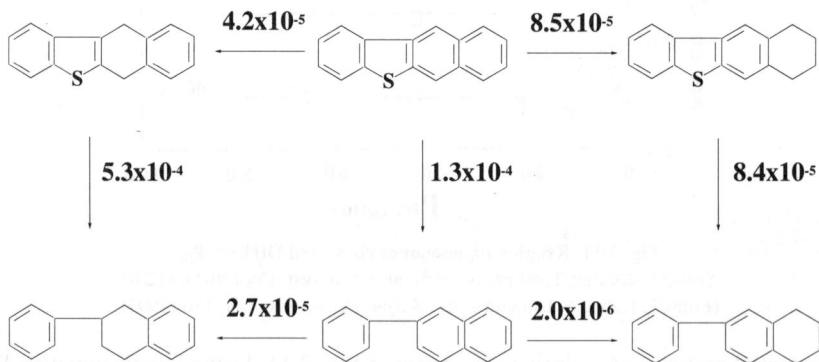

(A) The numbers next to the arrows denote pseudo-first-order rate constants
     in L/(g of catalyst·s) at 300 °C

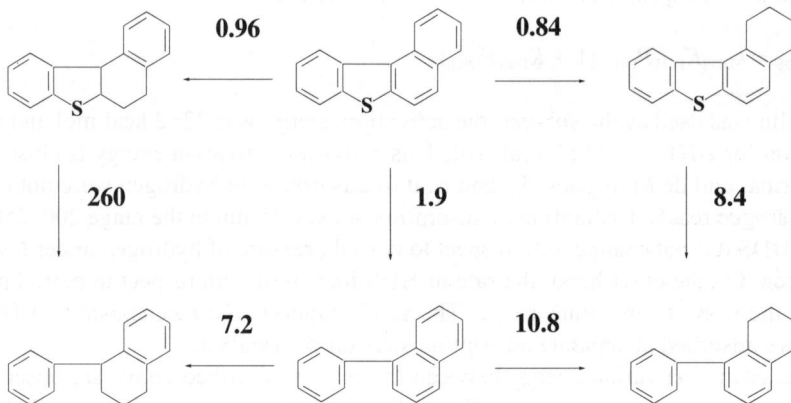

(B) The numbers next to the arrows are relative values of the pseudo-first-order
     rate constants at 250 °C

Fig. 2.12 Proposed reaction pathways for (A) benzo[b]naphtho[2,3-d]thiophene and (B) benzo[b]naphtho[1,2-d]thiophene.
(From M. J. Grigis and B. C. Cates, *Ind. Eng. Chem. Res.*, **30**, 2032 (1991))

under the condition of higher concentration of sulfur, the rate expression will be consistent not with Eq.(2.7) but with the equation for DBT similar to Eq.(2.2), (2.3) or (2.6), which was proposed for thiophene and benzothiophene HDS.

It is generally believed that it is impossible to predict industrial activity under high pressure by tracing the reaction of thiophene under atmospheric pressure. However, some attempts have been made to correlate lab scale experiment under normal pressure with that under practical high pressure.[129–131] It has been reported that when the kinetics of the reaction are carefully applied, thiophene HDS under atmospheric pressure of H_2 can correlate very accurately with dibenzothiophene HDS under industrial operating conditions.[129]

4. Benzonaphthothiophene

The reaction networks of two kinds of benzonaphthothiophene (BNT), which is the heaviest organosulfur compound investigated, have been reported: One is benzo[b]naphtho[2,3-d]thiophene[132] and the other is benzo[b]naphtho[1,2-d]thiophene.[50,53] As shown in Table 2.3, Nag et al. reported that the HDS rate for benzo[b]naphtho[2,3-d]thiophene is slightly higher than that for DBT (a batch reactor, Co-Mo/Al$_2$O$_3$, 573 K, 71 atm). In contrast, it has been reported that the HDS rate for benzo[b]naphtho[1,2-d]thiophene is lower than that for DBT (a batch reactor, Ni-Mo/Al$_2$O$_3$, 523 K, 40 atm). Although there is likely to be some reactivity differences between the networks of BNT, these may be explained by the differences in the catalyst or reaction conditions. In both networks (Fig. 2.12), the characteristic feature was that in the case of BNT hydrogenation prior to HDS occurred at a rate comparable with that of HDS, while in the case of DBT the rate of hydrogenation prior to HDS was two orders of magnitude less than that of HDS.[120] To explain the relatively rapid hydrogenation of BNT, it was suggested that these compounds were adsorbed flat on the catalyst to form π-bonded species to which hydrogen may be added.[53] No kinetic studies leading to rate expressions have been reported for benzonaphthothiophene HDS.

2.1.3 Effects of Substituents on Reaction

A. Reactivities of Alkyl-substituted Thiophene, Benzothiophene and Dibenzothiophene

The effects of substituents on hydrodesulfurization (HDS) of thiophene,[125,133–135] benzothiophene (BT)[60,66,115,136] and dibenzothiophene (DBT) [32–34,60,61,68,72,122,137–142] are well known and a comparison of the relative rate for thiophenes HDS discussed in the literature.[49–53,77] General tendencies are as follows: Substituents groups at the carbon adjacent to or near the sulfur atom retard the HDS rate significantly while substituent groups at the carbon far from the sulfur atom increase or do not affect the HDS activity. It is generally believed that the former effect is due to the inhibition of C-S bond cleavage or adsorption of substrates by the steric hindrance of substituents and that the latter effect is due to the increase in electron density of the sulfur atom, that is, inductive effect of alkyl substituents.

The substitution of thiophene at both the 2- and 5-positions retards the HDS rate significantly.[125,135] It seems that substitution only at the 3-position promotes the HDS while that only at the 2-position affects the rate slightly.[125,133–135] When a catalyst was reduced before HDS, the reactivities of thiophene and substituted ones showed the reverse order, indicating that the structure of the reduced catalyst is completely different from that of the sulfided one.[134] In BT HDS, it has been observed that a methyl group at the 2- and 3-position

Table 2.4 Effects of Substituents on HDS

Author	Sulfur compound	Relative reactivities [a]	Catalyst and reaction conditions [c]
O'Brien [125]	Thiophene	2,5-DM(0.51)<2-E(0.68) <2-M(0.74)<NS(1.0)	Co-Mo/Al$_2$O$_3$ (Katalco 477, MoO$_3$ 14.0 wt%, CoO 3.3 wt%), 575K, 4.1 Mpa, continuous flow autoclave, solvent toluene or benzene
Desikan [133]	Thiophene	2-M(0.9)<NS(1.0)<3-M(2.5) 498K	Co-Mo/Al$_2$O$_3$ (Girdler G35A) 1g, 343–498K, 1 atm, Feed rate: 55 cc/min, H$_2$ containing 1.67, 1.09, and 1.00 mole% of thiophene, 2-, and 3-methylthiophene, continuous flow
Zdrazil [134]	Thiophene	NS(1.0)<2-M(1.5)<3-M(1.9) <2,5-DM(2.0)	Co-Mo/Al$_2$O$_3$ reduced at 673K, 573–723K, 1 atm, thiophene 0.040 mol/h (0.1 atm), flow reactor
Miki [135]	Thiophene	2,5-DM(0.6)<2-E(0.8)<2-M(0.9)<NS(1.0)<3-M(1.2)<3-E(2.0) 275°C	Ni-Mo/Al$_2$O$_3$ (KF-153, MoO$_3$ 15 wt%, NiO 3 wt%), 225–300°C, initial pressure 12MPa, thiophene 10 wt%, solvent tetradecane, 40 mL autoclave
Givens [66]	Benzothiophene	2,3,7-TM(0.16)=2,3-DM (0.15)<3,7-DM(0.24)<3-M(0.32) <2,7-DM(0.43)<7-M(0.57) <2-M(0.66)<NS(1.0)	Co-Mo/Al$_2$O$_3$ (MoO$_3$ 10 wt%, CoO 3 wt%) hydrotreated at 550°C for 3 h before use, 400°C, 1 atm, LHSV0.30 h^{-1}, H$_2$/BT molar ratio=3–5, flow reactor, solvent hexane
Kilanowski [60]	Benzothiophene	3,7-DM(0.24)<3-M=(0.66) 2-M(0.71)=7-M(0.71) <NS(1.0)	Co-Mo/Al$_2$O$_3$ (HDS-16A, MoO$_3$ 11.2 wt%, CoO 5.6 wt%) (broken-in), 450°C, 1 atm, pulse microreactor, n-heptane:n-dodecane =1:1
Geneste [115]	Benzothiophene	2,3-DM(0.11)<3-M(0.18) <2-M(0.37)<NS(1.0)	Co-Mo/Al$_2$O$_3$ (HR-103, MoO$_3$ 13.5 wt%, CoO 2.8 wt%), 250°C, 50 atm, 0.5 L autoclave
Levaché [136]	Benzothiophene	3-M(0.5)<2-M(1.0)=NS(1.0)	Ni-Mo/Al$_2$O$_3$ (HR-346, MoO$_3$ 14 wt%, NiO 3 wt%), 250°C, 40 atm, 0.3 L autoclave, solvent dodecane
Kilanowski [60]	Dibenzothiophene	4,6-DM(0.19)<4-M(0.39) <NS(1.0)<2,8-DM(2.0)	Co-Mo/Al$_2$O$_3$ (HDS-16A, MoO$_3$ 11.2 wt%, CoO 5.6wt%) (broken-in), 450°C, 1 atm, pulse microreactor, n-heptane:n-dodecane=1:1
Houalla [61]	Dibenzothiophene	4,6-DM(0.07)<4-M(0.09) <3,7-DM(0.49) <2,8-DM(0.94)<NS(1.0)	Co-Mo/Al$_2$O$_3$ (HDS-16A, MoO$_3$ 11.2 wt%, CoO5.6 wt%), 300°C, 102 atm, LHSV 2–38 h^{-1}, flow reactor, solvent n-hexadecane

Author	Sulfur compound	Relative reactivities [a]	Catalyst and reaction conditions [c]
Katti [137]	Dibenzothiophene	4-M(0.31)<NS(1.0)<MDBT(1.8) [b]	Ni-Mo/Al$_2$O$_3$ (HDS-9A, MoO$_3$ 18.3 wt%, NiO 3.1 wt%), 355°C, 120 atm, WHSV 0.05–0.2 h^{-1}, flow reactor, neutral oils (coal-derived liquid) 0.62 wt% sulfur, 0.25 wt% solution in cyclohexane
Kabe [68,122] Ishihara	Dibenzothiophene	4,6-DM(0.01)<4-M(0.10)<NS(1.0) 240°C, Co-Mo/Al$_2$O$_3$ 4,6-DM(0.06)<4-M(0.19)<NS(1.0) 260°C, Ni-Mo/Al$_2$O$_3$	Co-Mo/Al$_2$O$_3$ (MoO$_3$ 17 wt%, CoO 4.5 wt%), 240–320°C, 50 atm, WHSV 70 h^{-1}, H$_2$ 18 L/h, fixed bed flow reactor, solvent decalin
Kabe [32–34,138] Ishihara	Dibenzothiophene	4,6-DM(0.13)<4-M(0.38)<NS(1.0) 350°C, Co-Mo/Al$_2$O$_3$ 4,6-DM(0.55)<4-M(0.60)<NS(1.0) 310°C, Ni-Mo/Al$_2$O$_3$	Co-Mo/Al$_2$O$_3$ (MoO$_3$ 17 wt%, CoO 4.5 wt%), Ni-Mo/Al$_2$O$_3$ (MoO$_3$ 15.8 wt%, NiO 2.9 wt%), 300–410°C, 30 atm, LHSV 4 h^{-1}, H$_2$/oil 120 L(NTP)/L, flow reactor, Arabian light 1.46 wt% sulfur, boiling range 245–374°C
Isoda [139]	Dibenzothiophene	4,6-DM(0.12)<4-M(0.29)<NS(1.0) 325°C, 2.5 MPa, Co-Mo/Al$_2$O$_3$ 4,6-DM(0.13)<4-M(0.39)<NS(1.0) 325°C, 2.9 MPa, Co-Mo/Al$_2$O$_3$	Co-Mo/Al$_2$O$_3$ (KF-742, MoO$_3$ 15 wt%, CoO 4wt%), Ni-Mo/Al$_2$O$_3$ (KF-842, MoO$_3$ 15 wt%, NiO 3 wt%), 270–340°C, 2.5–4.1 MPa, 50 mL autoclave, no solvent
Ma [72]	Dibenzothiophene	4,6-DM(0.10)<4-M(0.31)<NS(1.0) 360°C, 2.9 MPa, Co-Mo/Al$_2$O$_3$ 4,6-DM(0.14)<4-M(0.35)<NS(1.0) 360°C, 2.9 MPa, Ni-Mo/Al$_2$O$_3$	Co-Mo/Al$_2$O$_3$ (MoO$_3$ 14.9 wt%, CoO 4.4 wt%), Ni-Mo/Al$_2$O$_3$ (MoO$_3$ 14.9 wt%, NiO 3 wt%), 280–420°C, 2.9 MPa, diesel fuel 0.706 wt% sulfur, boiling range 232–365°C, 50 mL autoclave
Lamure-Meille [140]	Dibenzothiophene	4-M(0.21)<NS(1.0) Co-Mo/Al$_2$O$_3$ 4-M(0.20)<NS(1.0) Ni-Mo/Al$_2$O$_3$	Co-Mo/Al$_2$O$_3$ (Mo 8 wt%, Co 1.87 wt%), Ni-Mo/Al$_2$O$_3$ (Mo 9 wt%, Ni 2.4 wt%), 563K, 5.0 MPa, 200 mL autoclave, solvent dodecane
Lamure-Meille [141]	Dibenzothiophene	4,6-DM(0.02)4-M(0.15)<NS(1.0)<2,8-DM(2.37) Ni-Mo/Al$_2$O$_3$	Ni-Mo/Al$_2$O$_3$ (Mo 9 wt%, Ni 2.4 wt%), 573K, 5.0 MPa, 200 mL autoclave, solvent dodecane
Vanrysselberghe [142]	Dibenzothiophene	4,6-DM(0.08)<4-M(0.59)<NS(1.0) Co-Mo/Al$_2$O$_3$	Co-Mo/Al$_2$O$_3$ (Akzo, KF 742), 533–593K, 80 bar, DBT 2 wt%m 4-MDBT 0.27 wt%, 4,6-DMDBT 0.067 or 0.286 wt%, solvent a mixture of n-paraffins, Flow rate, DBT:1.71 × 10^{-6} – 3.92 × 10^{-6}, 4-MDBT:2.13 × 10^{-7} – 4.86 × 10^{-7}kmol/h, molar hydrogen-to-hydrocarbon ratio 1.1–4.2

[a] TM=trimethyl, DM=dimethyl, DE=diethyl, M=methyl, E=ethyl, NS=no substituent, Numbers mean substituted positions of sulfur compounds. The ratios of the HDS rate or the conversion of sulfur compounds are given in parentheses. [b] MDBT=methyldibenzothiophene, an undetermined isomer other than 4-MDBT. [c] Otherwise noted, catalysts have been presulfided.

easily transfers to the 3- and 2-position, respectively.[66] Therefore, the substitution at the 2-position slightly affects the HDS rate. In contrast to the result from thiophene, the substitution at the 3-position of BT retarded the HDS rate.[60,66,115,136] When the 2- and 3-positions of BT are substituted simultaneously, methyl groups cannot easily transfer to a different less-hindered position. Therefore, 2,3,7-trimethylbenzothiophene and 2,3-dimethylbenzothiophene are the most difficult to desulfurize among substituted BTs.[66] Although the substitution at the 7-position retards the rate, the effect seems to be relatively small. Geneste *et al.* reported the apparent activation energies estimated for BT HDS to be as follows: 25 kcal/mol (for NS); 32 (3-M); 37 (2,3-DM).[115] In DBT HDS, the substitution at the 2- and 8-positions seems to promote the reaction in a pulse microreactor [60] or a batch system [141] while the rates were inhibited slightly for 2,8-DMDBT and largely for 3,7-DMDBT in a flow system.[61] As shown in Section 2.1.1, the substituents at the 4- and 6-positions of DBT remarkably retard the HDS rate. This result is consistent with all the results from the model HDS reactions,[60,61,68,122,139–142] and direct HDS of middle distillate[32–34,72,138] and coal-derived liquid[137] in Table 2.4. When Co-Mo/Al_2O_3 is used, the ratios of HDS rate between DBT, 4-MDBT and 4,6-DMDBT in a flow system are larger than those with Ni-Mo/Al_2O_3. This may be due to the hydrogenation ability of Ni-Mo/Al_2O_3 as mentioned in Section 2.1.1B. At lower temperatures, the ratios of HDS rate between DBT, 4-MDBT and 4,6-DMDBT are also larger than those at higher temperatures. This may be due to the difference in activation energies of these compounds (*vide infra*). Further, in HDS of middle distillate which includes many components, the ratios of HDS rate between DBT, 4-MDBT and 4,6-DMDBT in a flow system are lower than those for the model reactions. This may be due to the effect of components where HDS of DBTs to BPs are inhibited (Section 2.1.1B). Some differences in HDS rates of DBTs are observed between a flow system and a batch system or a pulse microreactor. In the batch system, it is difficult to estimate the stationary state of catalysis because the formation of products such as hydrogen sulfide and organic hydrocarbons inhibits the reaction, and it takes time to obtain the expected reaction temperature and some substrates react during heating. In the pulse microreactor, it is impossible to estimate the stationary state of catalysis. The discrepancies seem to appear for these reasons.

B. Kinetics of 4-Methyldibenzothiophene and 4,6-Dimethyldibenzothiophene HDS

In Section 2.1.1, it has been clarified that 4-methyldibenzothiophene (4-MDBT) and 4,6-dimethyldibenzothiophene (4,6-DMDBT) are most difficult to desulfurize in deep desulfurization of middle distillate and that only substitution at the 4- and 6-positions retards HDS significantly. It has not been clarified, however, why those compounds were difficult to desulfurize and what their product selectivities, activation energies of HDS and heats of adsorption were. To elucidate mechanisms of desulfurization of 4-MDBT and 4,6-DMDBT, a kinetic approach should be a useful method. Although a number of attempts have been made to elucidate mechanisms using kinetic data, there were few examples dealing with the retarding effects of the methyl substitution on DBT, that is, the mechanisms of desulfurization of 4,6-DMDBT or 4-MDBT. Recent requirement for the elucidation of deep desulfurization mechanisms led the researcher to the kinetic study of HDS of 4,6-DMDBT and 4-MDBT. In this section, a limited example of a kinetic study of 4,6-DMDBT and 4-MDBT HDS using three kinds of commercially available Co-Mo/Al_2O_3 catalysts is presented: The mechanisms of HDS of DBTs over the catalysts were investigated by estimating product selectivity, activation energies of HDS and heats of adsorption of DBTs derived from the Langmuir-Hinshelwood (L-H) rate equation.[68,122] These results were compared among those using commercially available Ni-Mo/Al$_2$O$_3$ catalyst, and the difference in mechanisms among different Co-Mo/Al$_2$O$_3$ and of Ni-Mo/Al$_2$O$_3$ was investigated.[68]

Materials used in this study were as follows: DBT, 4-MDBT and 4,6-DMDBT were synthesized by methods reported.[143-145] The concentration of hydrogen sulfide in hydrogen was 3%. The catalysts were commercial grade Co-Mo/Al$_2$O$_3$ catalysts (Cat. A: CoO 3.8 wt%, MoO$_3$ 12.3 wt%; Cat. B: CoO 4.5 wt%, MoO$_3$ 17.0 wt%; Cat. C: for deep desulfurization) and Ni-Mo/Al$_2$O$_3$ catalyst (Cat. D: NiO 3.9%, MoO$_3$ 19.9%).

The reactor was an 8-mm-i.d. stainless steel tube packed with 0.2 g of catalyst particles, diluted with quartz sand to 2 cm^3. After being heated for 24 h at 450°C in air, the catalyst was presulfided in a stream of 3% H$_2$S in H$_2$ under the following conditions: 30 l/h, 1 atm, 400°C, 3 h. It was then pressurized with hydrogen and a reactant solution was fed in by a pump. The HDS reaction was carried out under the following conditions: temperature: 190–340°C; total pressure: 50 atm; WHSV: 70 h^{-1}; Gas/Oil: 1100 NL/L; initial concentrations of DBTs: 0.1–0.4 wt%. After the HDS reaction reached the steady state (3 h), samples were collected from a gas-liquid separator. Subsequently, the reaction temperature was changed and after 2 h, sampling was carried out again in a similar manner.

A sample was collected and analyzed by gas chromatography using a flame ionization detector and every component in the sample was identified by comparing its retention time with that of standard compounds. As standard compounds, DBT, 4-MDBT and 4,6-DMDBT were synthesized as noted above. Biphenyl (BP), cyclohexylbenzene (CHB), 3-methylbiphenyl (3-MBP) and 3,3′-dimethylbiphenyl (3,3′-DMBP) were of commercially available grade. The 3-methylcyclo-hexylbenzene (3-MCHB), 3,3′-dimethylcyclohexylbenzene (3,3′-DMCHB) and tetra- and hexahydro-DBTs were identified by a GC-MS for molecular weight, from reaction samples. 0.1 µl of each sample was analyzed using a G-column 100 (i.d.: 1.2 mm; film thickness: 1.0 µm; length: 40 m) programmed from 100°C to 230°C (heating rate: 8°C/min, injection temperature: 270°C, carrier gas: N$_2$).

To compare kinetic data for HDS of 4-MDBT and 4,6-DMDBT with that for DBT, Eq.(2.10) was used on the basis of the same assumptions. Eq.(2.12) can be obtained from taking the reciprocal of both sides of Eq.(2.11).

$$1/r_{HDS} = 1/k_{HDS}K_{DBT}P_{DBT} + 1/k_{HDS} \tag{2.12}$$

By plotting $1/r_{HDS}$ against $1/P_{DBT}$, k_{HDS} and K_{DBT} were calculated by slopes and intercepts. The rates of HDS reactions of DBTs were calculated using the following equation:

$$r_{HDS} = L \cdot C \cdot Conv./(M \cdot g\text{-}Cat.) \tag{2.13}$$

where r_{HDS} = rate of HDS (mol/g-cat·h); L = flow of reaction (g/h); C = initial concentration of DBT (wt%); $Conv.$ = conversion of DBT (%); M = molecular weight of DBT; $g\text{-}Cat.$ = weight of catalyst (g). The partial pressures of DBTs were calculated using the following equation:

$$P_{DBT} = P_{total} \cdot F_{DBT}/F_{total} \tag{2.14}$$

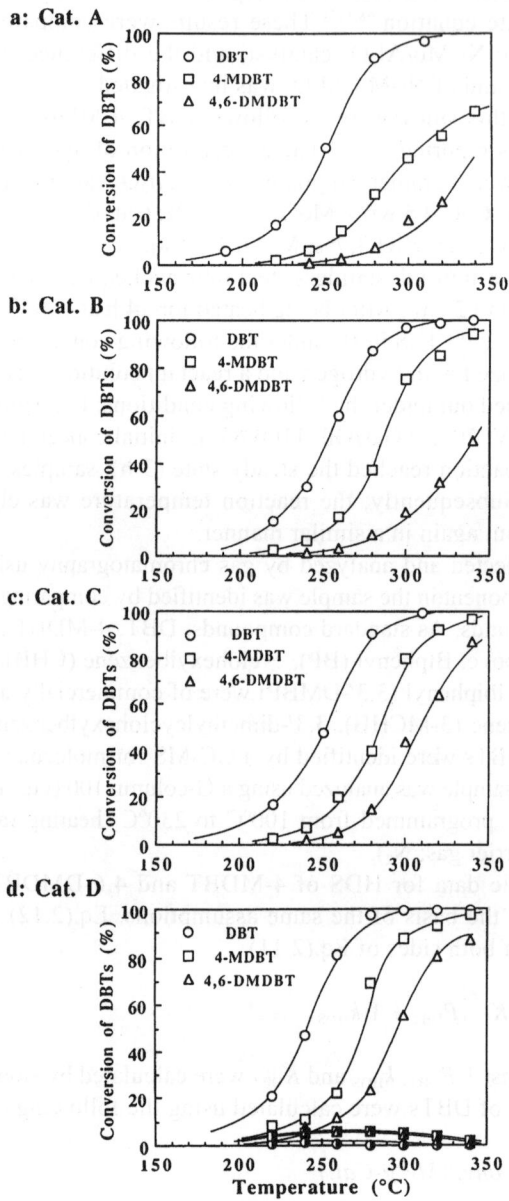

Fig. 2.13 Effects of temperature on conversions of DBTs.
DBTs: 0.1 wt% in decalin; total pressure: 50 atm; WHSV: 70 h⁻¹; Gas/Oil: 1100 NL/L. ◑, ◨, ▲ :
tetrahydro- and hexahydro-DBTs.
(From T. Kabe, A. Ishihara et al., J. Jpn. Petrol. Inst., **39**, 412 (1996))

where P_{DBT} = partial pressure of DBT; P_{total} = total pressure of reactants; F_{DBT} = flow of DBT (mol/h); F_{total} = total flow of reactants (mol/h).

1. Kinetics of Dibenzothiophenes HDS Catalyzed by Co-Mo/Al$_2$O$_3$ Catalysts

In the case of DBT, products were basically biphenyl (BP) and cyclohexylbenzene (CHB), while for 4-MDBT and 4,6-DMDBT, the corresponding methyl-substituted BPs and CHBs were produced. Shown in Figs. 2.13(a), (b) and (c) are the effect of temperature on the conversions of DBTs (DBT, 4-MDBT and 4,6-DMDBT) using three Co-Mo/Al$_2$O$_3$ catalysts, Cat. A, Cat. B and Cat. C, respectively (190–340°C; 50 atm, initial concentrations of DBTs, 0.1 wt%). The reactivities of DBTs decreased in the order DBT > 4-MDBT > 4,6-DMDBT. Compared with Cat. A, in the range of 190–340°C, the conversions of DBT catalyzed by Cat. B and Cat. C did not change while, in contrast, that of 4-MDBT clearly increased. Furthermore, Cat. C increased the conversion of 4,6-DMDBT although there was no substantial difference between that of 4-MDBT by Cat. B and Cat. C. The reaction temperatures needed to convert 40% of DBT, 4-MDBT and 4,6-DMDBT were 245°C, 295°C and 330°C on Cat. A, 245°C, 280°C and 330°C on Cat. B and 245°C, 280°C and 300°C on Cat. C, respectively. Evidently, to obtain the same conversions of 4-MDBT, in the cases of Cat. B and Cat. C, the temperatures were 15°C lower than in the case of Cat. A, and similarly, to obtain the same conversion of 4,6-DMDBT, Cat. C was 30°C lower than Cat. A. DBT conversion, however, remained the same on each catalyst.

Shown in Figs. 2.14(a), (b) and (c) are the effects of temperature on the conversions of DBTs to BPs on each catalyst. The shapes of curves in Figs. 2.14(a), (b) and (c) were very similar to those in Figs. 2.13(a), (b) and (c). Also, the conversions of DBTs to CHBs as shown in Figs. 2.15(a), (b) and (c) were almost the same (*ca.* 10%) above 280°C on each catalyst. These results show that the formation of BPs is markedly prevented by the methyl-substitution of aromatic rings. Furthermore, it is shown that, when an aromatic ring in DBTs is hydrogenated prior to desulfurization, the steric hindrance of the methyl group is weakened significantly. Under the same experimental conditions, BP or 3,3′-DMBP was added to DBT or 4,6-DMDBT solution at the steady state of the HDS reaction, but no significant increase in CHBs formation was observed. This revealed that CHBs were formed mainly through hydrogenation of DBTs prior to desulfurization. In other words, when an aromatic ring of DBT is hydrogenated, the cyclohexyl ring formed deviates from a plane of the remaining aromatic ring. Further, when a methyl group is located on an axial position (Fig. 2.16) in a cyclohexyl ring, desulfurization may be simplified and the steric hindrance by methyl-substitution decreased.[146]

In order to elucidate the differences in reactivities of DBTs, activation energies of HDS and heats of adsorption of DBTs were estimated using the Langmuir-Hinshelwood rate equation. In HDS of DBTs, the initial concentrations of DBTs were changed in the range of 0.1 to 0.4 wt%. The rate constants and adsorption equilibrium constants were calculated by Eq.(2.11). Rate constants of HDS of DBT, 4-MDBT, and 4,6-DMDBT for each catalyst at 240°C are listed in Table 2.5. In the case of Cat. A, the ratio of rate constants of DBT, 4-MDBT and 4,6-DMDBT was 100:6:1, while in the cases of Cat. B and Cat. C they were 100:10:1 and 100:10:2, respectively. Noticeably, there was no difference in the rates of HDS of DBT among the three catalysts.

Moreover, compared with Cat. A, both Cat. B and Cat. C increased the rates of 4-MDBT, but only Cat. C increased the rate of 4,6-DMDBT. Shown in Figs. 2.17, 2.18 and 2.19 are

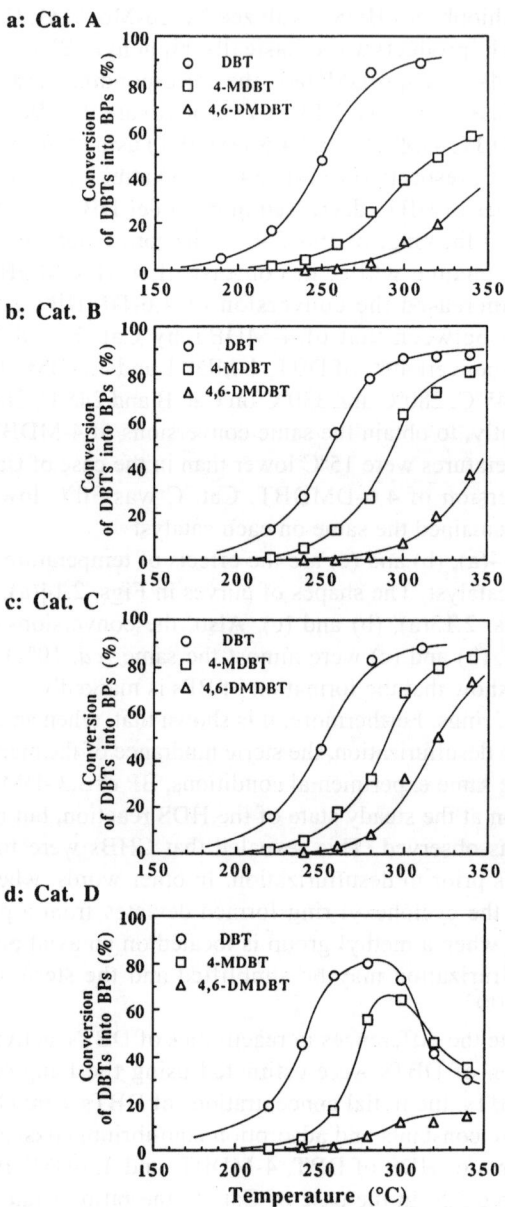

Fig. 2.14   Effects of temperature on conversions of DBTs to BPs.
DBTs: 0.1 wt% in decalin; total pressure: 50 atm; WHSV: 70 h^{-1}; Gas/Oil: 1100 NL/L.
(From T. Kabe, A. Ishihara *et al.*, *J. Jpn. Petrol. Inst.*, **39**, 413 (1996))

a: Cat. A

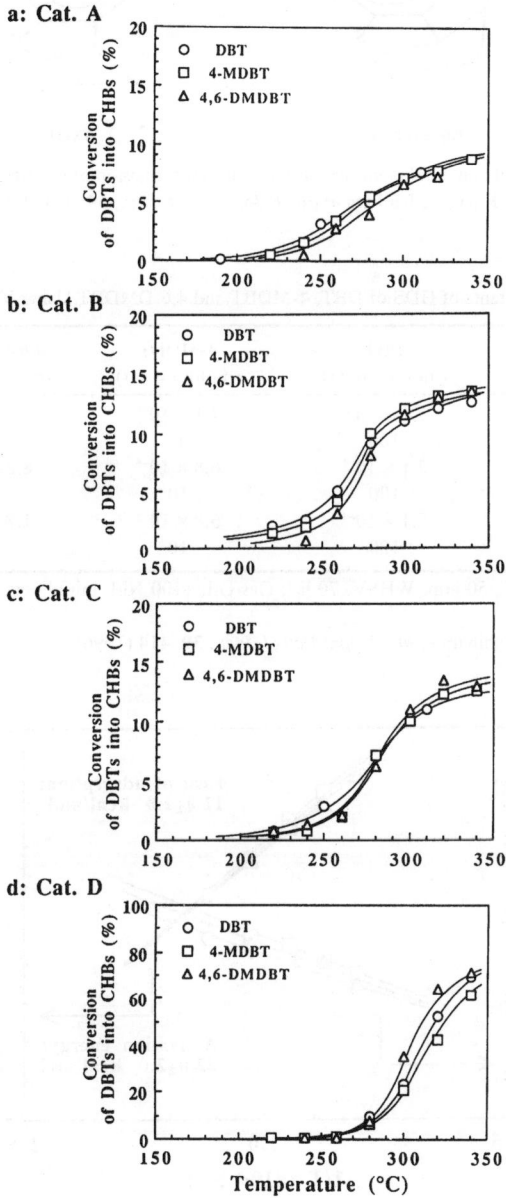

b: Cat. B

c: Cat. C

d: Cat. D

Fig. 2.15   Effects of temperature on conversions of DBTs to CHBs.
DBTs: 0.1 wt% in decalin; total pressure: 50 atm; WHSV: 70 h^{-1}; Gas/Oil: 1100 NL/L.
(From T. Kabe, A. Ishihara *et al.*, *J. Jpn. Petrol. Inst.*, **39**, 413 (1996))

a   b

Equatorial   Axial

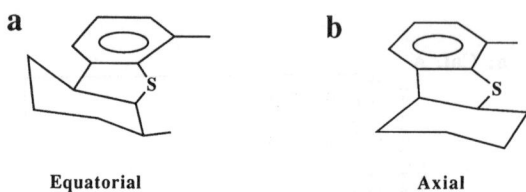

Fig. 2.16 Equatorial and axial positions on the chair conformation of a methylcyclohexane ring.
(From T. Kabe, A. Ishihara *et al.*, *J. Jpn. Petrol. Inst.*, **39**, 414 (1996))

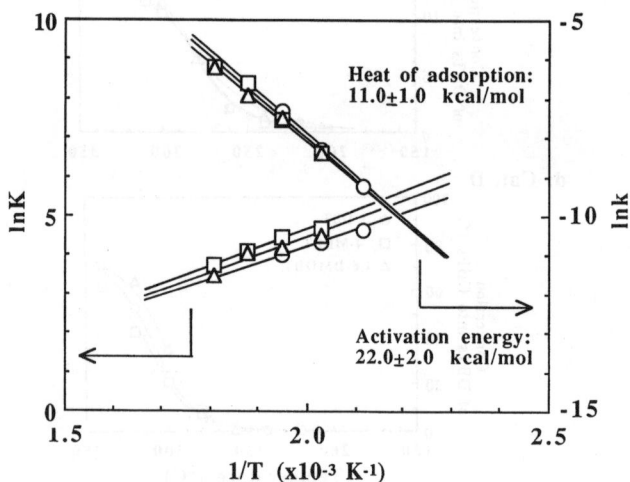

Table 2.5 Rate Constants of HDS of DBT, 4-MDBT and 4,6-DMDBT Using Various Co-Mo/Al$_2$O$_3$

Catalyst	DBT (mol/h · g-cat)	4-MDBT (mol/h · g-cat)	4,6-DMDBT (mol/h · g-cat)
Cat A	7.1×10^{-4}	4.1×10^{-5}	8.2×10^{-6}
Relative Ratio	100	6	1
Cat B	7.1×10^{-4}	6.8×10^{-5}	8.2×10^{-6}
Relative Ratio	100	10	1
Car C	7.1×10^{-4}	6.8×10^{-5}	1.7×10^{-5}
Relative Ratio	100	10	2

React. Temp. 240°C, 50 atm, WHSV: 70 h^{-1}; Gas/Oil: 1100 Nl/l, initial concentration of DBTs:
0.1–0.4 wt%.
(From T. Kabe, A. Ishihara *et al.*, *J. Jpn. Petrol. Inst.*, **39**, 414 (1996))

Heat of adsorption:
11.0±1.0 kcal/mol

Activation energy:
22.0±2.0 kcal/mol

1/T (x10^{-3} K^{-1})

Fig. 2.17 Arrhenius and Van't Hoff plots in HDS reaction of DBT.
Catalyst: ○: Cat A; □: Cat B; △: Cat C.
(From T. Kabe, A. Ishihara *et al.*, *J. Jpn. Petrol. Inst.*, **39**, 414 (1996))

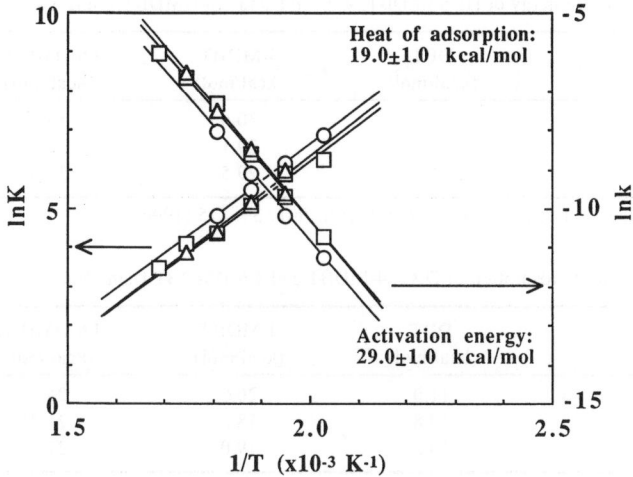

Fig. 2.18 Arrhenius and Van't Hoff plots in HDS reaction of 4-MDBT.
Catalyst: $\bigcirc$: Cat A; $\square$: Cat B; $\triangle$: Cat C.
(From T. Kabe, A. Ishihara *et al.*, *J. Jpn. Petrol. Inst.*, **39**, 414 (1996))

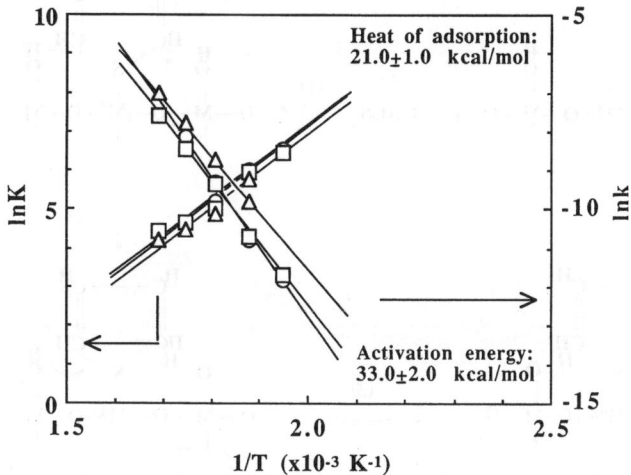

Fig. 2.19 Arrhenius and Van't Hoff plots in HDS reaction of 4,6-DMDBT.
Catalyst:$\bigcirc$: Cat A; $\square$: Cat B; $\triangle$: Cat C.
(From T. Kabe, A. Ishihara *et al.*, *J. Jpn. Petrol. Inst.*, **39**, 414 (1996))

Arrhenius and Van't Hoff plots in the cases of Cat. A, Cat. B and Cat. C, respectively. Activation energies of HDS of DBT, 4-MDBT and 4,6-DMDBT were calculated from slopes of Arrhenius plots, and were *ca.* 22±2, 29±1 and 33±2 kcal/mol, respectively, for all catalysts (Table 2.6). Heats of adsorption of DBT, 4-MDBT and 4,6-DMDBT were calculated from slopes of Van't Hoff plots, and were *ca.* 11±1, 19±1 and 21±1 kcal/mol, respectively, for all catalysts (Table 2.7). There was almost no difference in the activation energies and heats of adsorption among these three catalysts. This indicated that there was less difference in the mechanisms of HDS of DBTs among the three catalysts. As shown in Tables 2.6 and 2.7,

Table 2.6 Activation Energy of HDS of DBT, 4-MDBT and 4,6-DMDBT Using Various Co-Mo/Al$_2$O$_3$

Catalyst	DBT (kcal/mol)	4-MDBT (kcal/mol)	4,6-DMDBT (kcal/mol)
Cat A	22.7	30.0	33.6
Cat B	22.3	29.1	32.6
Cat C	21.0	29.5	31.0

(From T. Kabe, A. Ishihara *et al.*, *J. Jpn. Petrol. Inst.*, **39**, 415 (1996))

Table 2.7 Heat of Adsorption of DBT, 4-MDBT and 4,6-DMDBT Using Various Co-Mo/Al$_2$O$_3$

Catalyst	DBT (kcal/mol)	4-MDBT (kcal/mol)	4,6-DMDBT (kcal/mol)
Cat A	11.0	20.0	21.2
Cat B	10.8	18.8	21.2
Cat C	10.2	19.0	21.0

(From T. Kabe, A. Ishihara *et al.*, *J. Jpn. Petrol. Inst.*, **39**, 415 (1996))

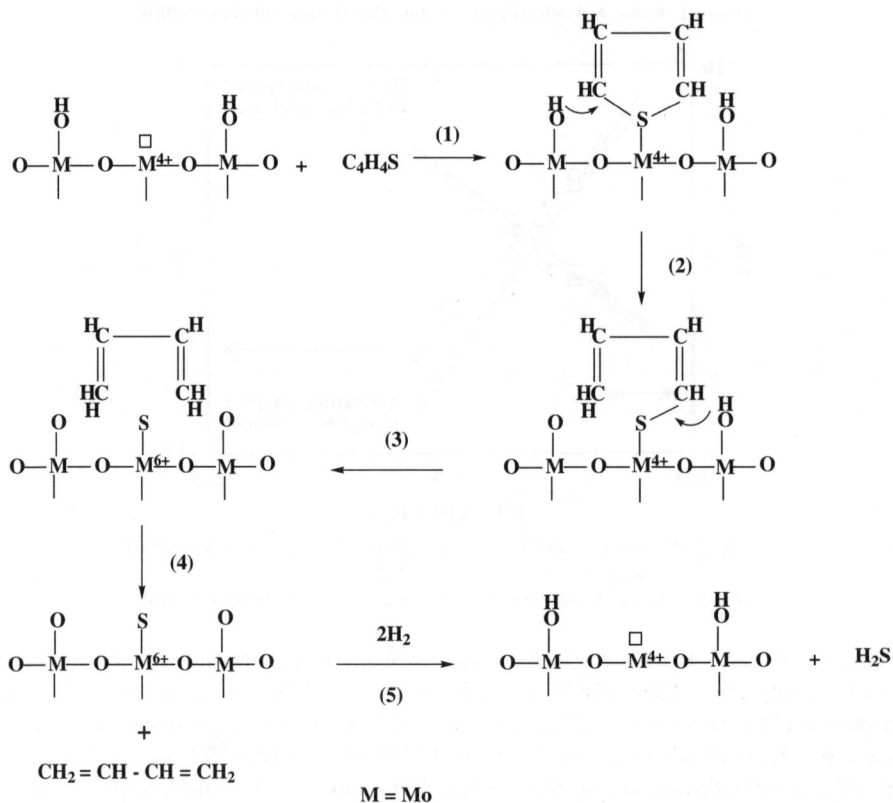

Fig. 2.20 HDS mechanism via end-on mode adsorption of thiophene.[147]
(From J. M. J. G. Lipsch and G. C. A. Schuit, *J. Catal.*, **15**, 187, (1969))

a

b

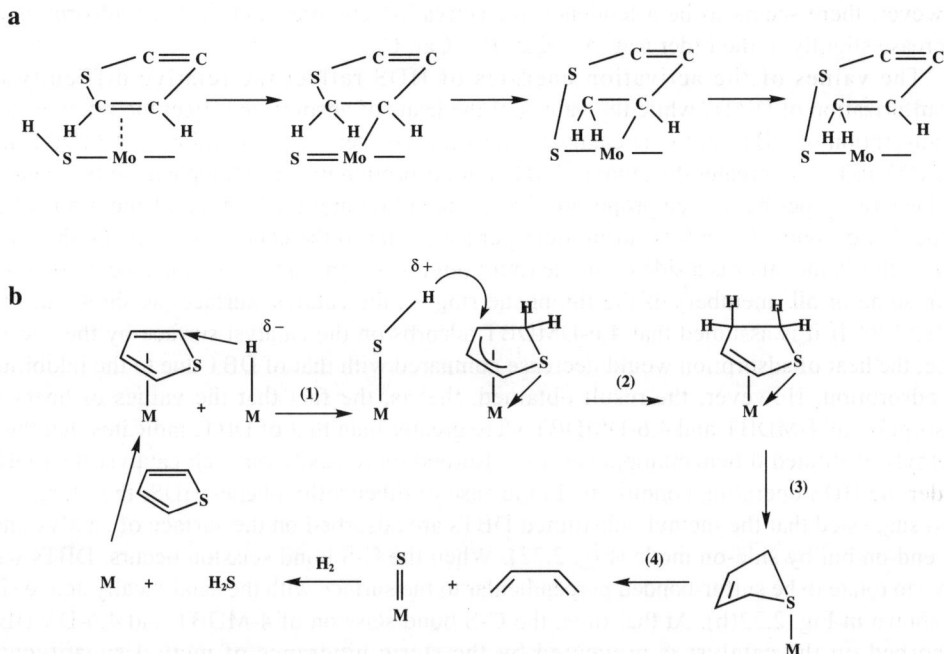

Fig. 2.21 HDS mechanism via side-on mode adsorption of thiophene.
(a) Ref. 148; (b) Ref. 149 and 150.
(a)(From H. Kwart, G. C. A. Shuit and B. C. Gates, *J. Catal.*, **61**, 133 (1980))
(b)(From N. N. Sauer, E. J. Market, G. L. Schrader and R. J. Angelici, *J. Catal.*, **117**, 296 (1989))

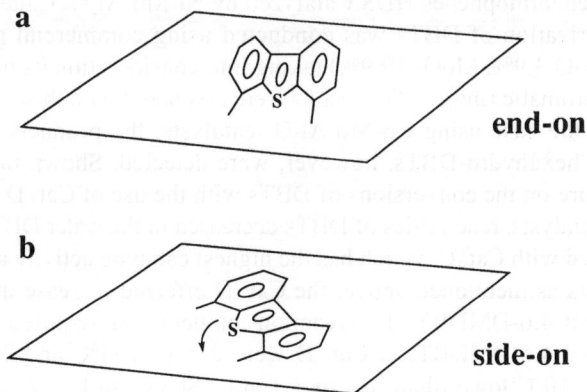

a

end-on

b

side-on

Fig. 2.22 Adsorption modes of 4,6-DMDBT.
(From T. Kabe, A. Ishihara *et al.*, *J. Jpn. Petrol. Inst.*, **39**, 415 (1996))

however, there seems to be a tendency for activation energies and heats of adsorption to decrease slightly in the order Cat. A > Cat. B > Cat. C.

The values of the activation energies of HDS reflect the relative difficulty of desulfurization of DBTs, while the values of the heats of adsorption reflect the relative ease of adsorption of DBTs onto the catalyst. The values of heats of adsorption of 4-MDBT and 4,6-DMDBT were greater than that of DBT. For adsorption mode of thiophene to the catalyst surface, two types have been proposed: One is an end-on mode where thiophene is adsorbed through a σ-bond with only a sulfur atom perpendicular to the catalyst surface, as shown in Fig. 2.20.[147]; the other is a side-on mode (multipoint adsorption) where thiophene is adsorbed with some or all members of the thiophene ring on the catalyst surface, as shown in Fig. 2.21.[148–150] If it is assumed that 4,6-DMDBT adsorbs on the catalyst surface by the end-on type, the heat of adsorption would decrease compared with that of DBT due to the inhibition of adsorption. However, the result obtained, that is, the fact that the values of heats of adsorption of 4-MDBT and 4,6-DMDBT were greater than that of DBT, indicates that these methyl-substituted dibenzothiophenes are adsorbed more easily on each catalyst than DBT under the HDS operating conditions. In the case of dibenzothiophenes HDS, therefore, it is also suggested that the methyl-substituted DBTs are adsorbed on the surface of catalyst, not by end-on but by side-on mode (Fig. 2.22). When the C-S bond scission occurs, DBTs will have to rotate to be sulfur-bonded perpendicular to the surface with the catalytically active site as shown in Fig. 2.22(b). At that time, the C-S bond scission of 4-MDBT and 4,6-DMDBT adsorbed on the catalyst is prevented by the steric hindrance of methyl substituents. Furthermore, if an aromatic ring is hydrogenated and a methyl group is located at an axial position in a cyclohexyl ring, the steric hindrance of a methyl group becomes trivial at the C-S bond scission, as shown in Fig. 2.16(b).

These results indicate that, if a catalyst shows both high selectivity for CHBs and high catalytic activity for HDS, deep desulfurization of 4,6-DMDBT could be achieved easily in the HDS of light gas oil.[32–34,37]

2. Kinetics of Dibenzothiophenes HDS Catalyzed by Ni-Mo/Al₂O₃ Catalysts

Hydrodesulfurization of DBTs was conducted using commercial grade Ni-Mo/Al$_2$O$_3$ catalyst (Cat. D: NiO 3.9%, MoO$_3$ 19.9%) taking into consideration its higher activity in the hydrogenation of aromatic rings.[68] The results were compared with those using Co-Mo/Al$_2$O$_3$ catalysts. As was the case using Co-Mo/Al$_2$O$_3$ catalysts, the products were also BPs and CHBs. Tetra- and hexahydro-DBTs, however, were detected. Shown in Fig. 2.13(d) is the effect of temperature on the conversions of DBTs with the use of Cat. D. Similar to the case of Co-Mo/Al$_2$O$_3$ catalysts, reactivities of DBTs decreased in the order DBT > 4-MDBT > 4,6-DMDBT. Compared with Cat. C, which has the highest catalytic activity among the three Co-Mo/Al$_2$O$_3$ catalysts as mentioned above, the Cat. D effected increase in all conversions of DBT, 4-MDBT and 4,6-DMDBT. The reaction temperatures required to convert 40% of DBT, 4-MDBT and 4,6-DMDBT on Cat. D were 235°C, 255°C and 290°C, respectively, each of which was 10°C lower than that over Cat.C. Shown in Fig. 2.14(d) is the effect of temperature on the conversions of DBTs to BPs. Below 290°C, the shapes of curves were significantly similar to those of Co-Mo/Al$_2$O$_3$ catalysts, and the reactivities decreased in the order DBT > 4-MDBT > 4,6-DMDBT. Above 290°C, the conversions of DBTs to BPs decreased quickly. It is assumed that a portion of the BPs formed were further hydrogenated to CHBs. On the other hand, as shown in Fig. 2.15(d), the conversions of DBTs to CHBs at

340 °C increased remarkably from 10% (Co-Mo/Al₂O₃) to 60% (Ni-Mo/Al₂O₃). This phenomenon can be interpreted by Fig. 2.10.

Regarding the Co-Mo/Al₂O₃ catalysts, the difference in reactivity among the DBTs was explained by the difference in conversions of DBTs to BPs. In the cited experimental conditions, the routes of DBTs to CHBs through BPs were not presented, and the CHBs were produced from the desulfurization of tetra- and hexahydro-DBTs, in the cases using Co-Mo/Al₂O₃ catalysts. When the Ni-Mo/Al₂O₃ catalyst was used, however, the conversions to CHBs increased significantly. In this case, two mechanisms leading to CHBs are considered possible: one is the formation of CHBs through hydrogenation of BPs, while the other is the formation of CHBs through desulfurization of tetra and hexahydro-DBTs. In the cited experiment, the tetra- and hexahydro-DBTs were detected as shown in Fig. 2.13(d). The latter route was probable under the cited experimental conditions. To prove the former route, biphenyl (BP) was added to DBT solution at steady state of HDS reaction. Increase in CHB formation was observed and conversion of BP to CHB was *ca.* 23%. This indicates that both routes were present in HDS reaction over Ni-Mo/Al₂O₃ catalyst.

From the point of view of catalyst structure, in the case of Co-Mo/Al₂O₃ catalysts, sulfides of active metals were able to exist on the surface of the catalyst, even at higher temperatures. On the contrary, in the case of Ni-Mo/Al₂O₃ catalyst, Ni sulfide was unstable at higher temperatures. This suggests that under pressure of hydrogen, the Ni-metal was deposited, and that this active metal provided the site to hydrogenate the aromatic ring of DBTs. Further, as shown in Fig. 2.13(d), the amounts of tetra- and hexahydro-DBTs increased below 260°C, while it decreased above 260°C. The amounts of tetra- and hexahydro-derivatives of 4-MDBT and 4,6-DMDBT were similar, however, and were more than that of

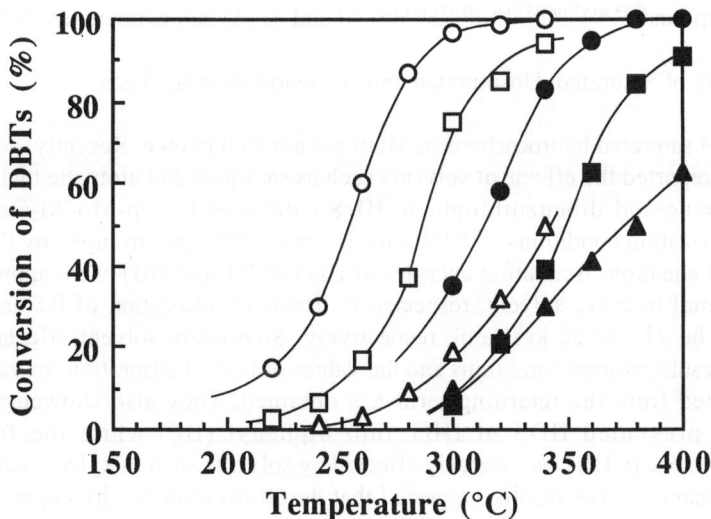

Fig. 2.23 Retarding effects of components in light gas oil on conversions of dibenzothiophenes.
Initial concentration of DBT, 4-MDBT or 4,6-DMDBT in decalin: 0.1 wt%; Initial concentrations of DBT, 4-MDBT and 4,6-DMDBT in light gas oil: 0.13, 0.17 and 0.08 wt%, respectively.
DBTs in decalin: ○: DBT; □: 4-MDBT; △: 4,6-DMDBT.
DBTs in light gas oil: ●: DBT; ■: 4-MDBT; ▲: 4,6-DMDBT.
(From A. Ishihara and T. Kabe, *Ind. Eng. Chem. Res.*, **32**, 754 (1993))

DBT. It is assumed that the hydrogenation of methyl-substituted dibenzothiophene is easier than that of DBT over Ni-Mo/Al$_2$O$_3$ catalyst.

2.1.4 Effects of Components on HDS Reaction

It is well known that oils derived from fossil fuels such as petroleum and coal usually consist of various components which include saturated and aromatic hydrocarbons, sulfur-containing organics, nitrogen-containing organics, products, hydrogen sulfide and ammonia, etc. and that hydrotreatment of these oils are retarded significantly by these components themselves. For example, in Fig. 2.23, HDS reactions of DBT, 4-MDBT and 4,6-DMDBT in decalin described in Section 2.1.3B1 have been compared with those in middle distillate described in Section 2.1.1B. In middle distillate, over 50°C higher temperature is required in order to obtain the same conversions as those of DBTs in decalin. In HDS of 4-MDBT in decalin, about 50°C higher temperature is also required to obtain the same conversions as those of DBT in decalin in the temperature range where the differential reactor can be applied. At 240°C, the rates of HDS of DBT and 4-MDBT in decalin using Cat B in Table 2.5 were 7.1 × 10^{-4} and 6.8 × 10^{-5} mol/g-cat/h, respectively. Thus, the rate of HDS of thiophenes in middle distillate decreases to less than 1/10 of that in decalin. This clearly shows the extent of the retarding effects of components in the hydroprocessing of oil. In this section, quantitative analysis of such retarding effects of components in hydroprocessing of oil are described. Effects of products, hydrogen sulfide, ammonia, etc. are also considered. There have been a number of reports concerning this subject,[50,53,77] and the components having effects can be classified as follows: 1) saturated, unsaturated and aromatic hydrocarbons,[50,53–55,57–59,93, 102,104,106,138,151–158] 2) sulfur compounds,[50,57–62,88,89,93,94,102,104–107,117,118,128,129,138,142,152,153,159–167] 3) nitrogen compounds[53,90,94,106,133,138,151,152,155,168–183] and 4) oxygen compounds.[152,184–187]

A. Effects of Saturated, Unsaturated and Aromatic Hydrocarbons

Effects of saturated hydrocarbons on HDS are not well known. Recently, however, Kabe and Ishihara reported the effects of solvents such as saturated and aromatic hydrocarbons on benzothiophene and dibenzothiophene HDS catalyzed by Co-Mo/Al$_2$O$_3$ under deep hydrodesulfurization conditions.[54,55] Data for BT and DBT were arranged by the Langmuir-Hinshelwood equation. Activation energies of HDS of BT and DBT were approximately 21 and 24 kcal/mol in every solvent, respectively. Heats of adsorption of BT and DBT were estimated to be 22 and 22 kcal/mol, respectively. Significant solvent effects were found under deep desulfurization conditions and the values of heat of adsorption for various solvent were estimated from the retarding term K_{sol} obtained. They also showed that solvents remarkably prevented HDS of DBT into biphenyl (BP) while the formation of cyclohexylbenzene (CHB) was scarcely affected by solvents such as xylene, decalin, tetralin and n-hexadecane.[156] The results suggested that desulfurization and hydrogenation proceed on different catalytic sites and that cyclohexylbenzene are formed by HDS of hexahydrodibenzothiophene as well as hydrogenation of BP. Nagai briefly reported the effects of the solvents tetralin and xylene on dibenzothiophene HDS catalyzed by Mo/Al$_2$O$_3$. It was indicated that the presence of a small amount of tetralin hindered the hydrogenation of dibenzothiophene, but a large amount of tetralin depressed C-S hydrogenolysis as well as the hydrogenation reaction.[155] Sohrabi and Zahedi reported that in HDS of thiophene in n-heptane catalyzed by cobalt and molybdenum oxides the diluent (*n*-heptane) acted as an inhibitor by occupying free active sites.[158]

The effects of solvent hydrocarbons on HDS are significant because a large amount of hydrocarbons are included in the system compared with that of a sulfur compound. When hydrocarbons are added as additives, however, the effect of those additives on HDS seems to be small. Although the effect of butene on thiophene HDS was investigated, there was very little or no effect, indicating that butene hydrogenation and thiophene HDS occur on different catalytic sites (catalysts: Co-Mo/Al$_2$O$_3$, Mo/Al$_2$O$_3$, etc.).[93,102,104,106] Competitive hydrodesulfurization of thiophene and hydrogenation of cyclohexene on a monolithic Co-Mo/Al$_2$O$_3$ were studied by Irandoust and Gahne and it was shown that cyclohexene and thiophene inhibited reaction rates of HDS and HYD, respectively.[154] The effects of aromatics are also weak. LaVopa and Satterfield reported mild inhibition effects of two hydrocarbons naphthalene and phenanthrene on thiophene conversion catalyzed by Ni-Mo/Al$_2$O$_3$ (vapor-phase flow reactor, 360°C, 69 atm).[151] It has been reported that naphthalene mildly retards both hydrogenation and HDS of DBT catalyzed by Ni-Mo/Al$_2$O$_3$[53] while phenanthrene does not retard BP formation but does retard CHB formation in DBT HDS catalyzed by Mo/Al$_2$O$_3$.[152] In DBT HDS catalyzed by Co-Mo/Al$_2$O$_3$, BP or naphthalene showed very little or no retarding effect,[50,57,59,153] suggesting that DBT HDS is only slightly or not at all inhibited by aromatic hydrocarbons.[50] Isoda et al. reported that the addition of naphthalene into decalin (10 wt%) markedly retarded HDS of 4,6-DMDBT catalyzed by Ni-Mo/Al$_2$O$_3$ although HDS of DBT was only slightly affected by naphthalene.[157] In this experiment, it was shown that naphthalene retarded hydrogenation route of 4,6-DMDBT, which is mainly desulfurized after hydrogenation.

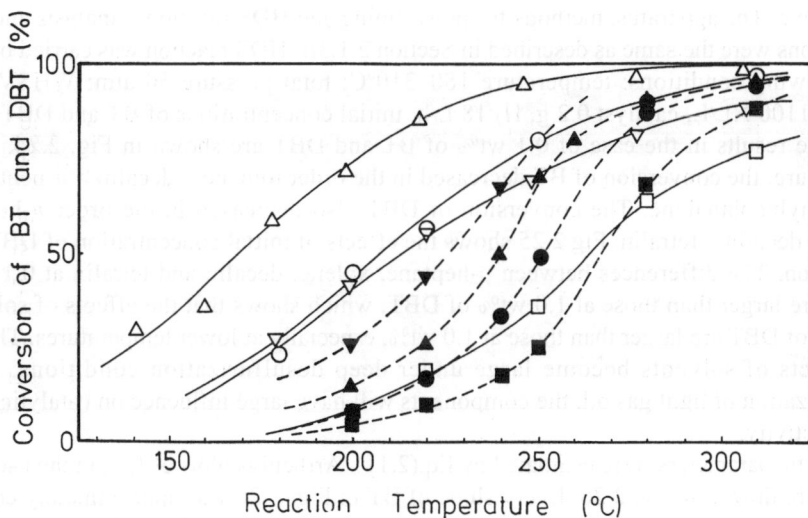

Fig. 2.24 Effects of solvents on conversion of benzothiophene and dibenzothiophene.
50 kg/cm^2; WHSV 70 h^{-1}; Catalyst 0.2 g; H$_2$ 18 l/h; Initial concentration of BT: 0.1 wt%; Solvent: △: toluene; ○: decalin; ▽: *n*-pentadecane; □: 1-methylnaphthalene; Initial concentration of DBT: 0.1 wt%; Solvent: ▼:*n*-heptane; ▲: xylene; ●: decalin; ■: tetralin.
(From A. Ishihara, T. Kabe et al., J. Catal., 140, 185 (1993))

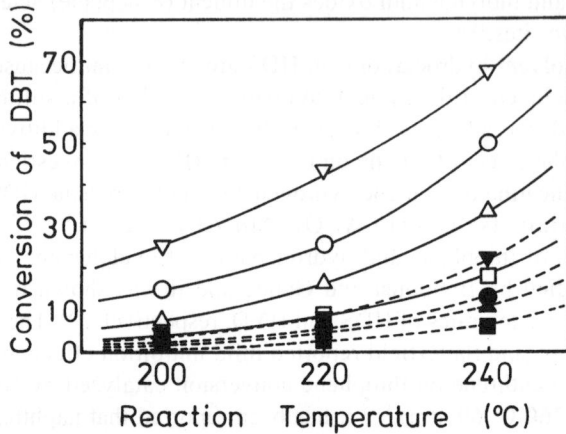

Fig. 2.25 Effect of initial concentration of dibenzothiophene on conversion of dibenzothiophene.
50 kg/cm²; WHSV 70 h⁻¹; Catalyst 0.2 g; H₂ 18 l/h; DBT 0.1 wt%: ∇: *n*-heptane; ○: xylene; △: decalin;
□: tetralin; DBT 1.0 wt%: ▼: *n*-heptane; ●: xylene; ▲: decalin; ■: tetralin.
(From A. Ishihara, T. Kabe *et al.*, *J. Catal.*, **140**, 186 (1993))

1. Effects of hydrocarbon solvents on model HDS reactions

As described above, the effects of saturated and aromatic hydrocarbons on HDS are significant when they are used for solvents. The example where the effects of solvent on benzothiophene and dibenzothiophene HDS catalyzed by Co-Mo/Al₂O₃ (CoO 3.8 wt%, MoO₃ 12.5 wt%) were determined are presented here in detail.[54,55] The solvents used were *n*-heptane, toluene, xylene, decalin, *n*-pentadecane, tetralin, 1-methylnaphthalene, and *n*-hexadecane. The apparatus, methods for presulfiding and HDS reactions, analysis and kinetic calculations were the same as described in Section 2.1.3B. HDS reaction was carried out under the following conditions: temperature 180–310°C; total pressure 50 atm; WHSV 70 h⁻¹; Gas/Oil 1100 NL/L, catalyst 0.2 g, H₂ 18 L/h, initial concentrations of BT and DBT 0.1–1.0 wt%. The results in the case of 0.1 wt% of BT and DBT are shown in Fig. 2.24. At each temperature, the conversion of BT decreased in the order toluene > decalin > *n*-pentadecane > 1-methylnaphthalene. The conversion of DBT also decreased in the order *n*-heptane > xylene > decalin > tetralin. Fig 2.25 shows the effects of initial concentration of DBT on the conversion. The differences between n-heptane, xylene, decalin and tetralin at 0.1 wt% of DBT were larger than those at 1.0 wt% of DBT, which shows that the effects of solvents at 0.1 wt% of DBT are larger than those at 1.0 wt%, especially at lower temperatures. Therefore, the effects of solvents become large under deep desulfurization conditions. In deep desulfurization of light gas oil, the components will have large influence on catalytic activity and selectivity.

Kinetic parameters were estimated by Eq.(2.11). Arrhenius plots of k_{HDS} in the use of each solvent are drawn in Fig. 2.26. Every slope of BT in Fig. 2.26 was approximately equal and the activation energy in the case of every solvent was calculated to be 21 kcal/mol. A similar result was also obtained in the case of DBT and the activation energy was calculated to be 24 kcal/mol. These results show that the activation energy is not affected by solvent. Heats of adsorption of BT and DBT were estimated from the Van't Hoff plots of K_{BT} and K_{DBT} and are summarized in Tables 2.8 and 2.9. When toluene in HDS of BT or *n*-heptane in HDS of DBT,

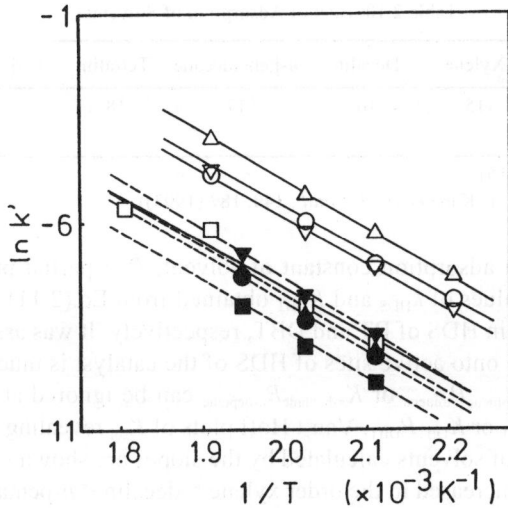

Fig. 2.26 Arrhenius plots of k'.
Solvents in BT HDS: △: toluene; ◯: decalin; ▽: n-pentadecane; ☐: 1-methylnaphthalene; Solvents in
DBT HDS: ▼: n-heptane; ▲: xylene; ●: decalin; ■: tetralin.
(From A. Ishihara, T. Kabe et al., J. Catal., 140, 186 (1993))

Table 2.8 Effects of Solvent on Heat of Adsorption of BT [a]

Solvent	Toluene	Decalin	n-Pentadecane	1-Methylnaphthalene
Q_{BT} (kcal/mol)	22	16	16	10

[a] Based on Eq.(2.11).
(From A. Ishihara, T. Kabe et al., J. Catal., 140, 186 (1993))

Table 2.9 Effects of Solvent on Heat of Adsorption of DBT [a]

Solvent	n-Heptane	Xylene	Decalin	Tetralin
Q_{DBT} (kcal/mol)	22	16	12	10

[a] Based on Eq.(2.11).
(From A. Ishihara, T. Kabe et al., J. Catal., 140, 187 (1993))

which can be regarded to have the least retarding effect on HDS among the four solvents, was
used, heats of adsorption of BT and DBT were 22 and 22 kcal/mol, respectively. In contrast
to the activation energy, the heat of adsorption seems to be affected by solvents and showed
a small value when a solvent which is adsorbed on the catalyst more strongly to retard the
HDS rate was used. It is suggested that the competitive adsorption between benzothiophene
and a solvent on the catalyst surface retards the HDS reaction.

For a better understanding these results, the retarding effects of solvents were estimated
by Eq.(2.15) where the retarding term, $K_{sol}P_{sol}$, was introduced into Eq.(2.11):

$$r_{HDS} = k_{HDS}K_{DBT}P_{DBT}/(1 + K_{DBT}P_{DBT} + K_{sol}P_{sol}),$$ (2.15)

Table 2.10 Heat of Adsorption of Solvents [a]

Solvent	Xylene	Decalin	n-Pentadecane	Tetralin	1-Methylnaphthalene
Q_{sol} (kcal/mol)	15	16	17	18	24

[a] Based on Eq.(2.15).
(From A. Ishihara, T. Kabe et al., J. Catal., **140**, 187 (1993))

where K_{sol}: equilibrium adsorption constant of solvent; P_{sol}: partial pressure of solvent. In determining K_{sol}, the values of k_{HDS} and K_{DBT} obtained from Eq.(2.11) in the case of toluene and n-heptane are used in HDS of BT and DBT, respectively. It was assumed that adsorption of toluene or n-heptane onto active sites of HDS of the catalyst is much weaker than that of BT or DBT, that is, $K_{toluene}P_{toluene}$ or $K_{n-heptane}P_{n-heptane}$ can be ignored at every temperature in comparison with $K_{BT}P_{BT}$ or $K_{DBT}P_{DBT}$. Van't Hoff plots of K_{sol} revealing the linear relationship and heat of adsorption of solvents calculated by the slopes are shown in Table 2.10. Heats of adsorption of solvents increased in the order xylene < decalin < n-pentadecane < tetralin < 1-methylnaphthalene, indicating that solvents such as 1-methylnaphthalene, which has high aromaticity, are adsorbed competitively with BT or DBT on Co-Mo/Al$_2$O$_3$ by a higher than 1:1 ratio.

In this study, activation energy of HDS was about 24 kcal/mol in each solvent. Some differences in activation energy appear between several reports, as mentioned in Section 2.1.2B3. Because solvents do not affect the activation energy, these differences are probably due to the reaction temperature. When the range of temperature is relatively high, the activation energy of HDS tends to be high. Since Kabe and Ishihara's[54,55] and Vrinat's[59]

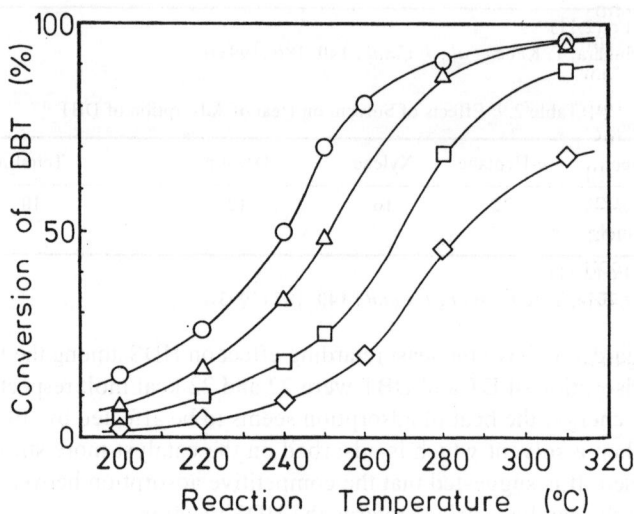

Fig. 2.27 Effects of solvents on conversion of dibenzothiophene.
50 kg/cm^2; WHSV 70 h^{-1}; Catalyst 0.2 g; H$_2$ 18 l/h; Initial concentration of DBT: 0.1 wt%; Solvent: ◯: xylene; △: decalin; ☐: tetralin; ◇: n-hexadecane.
(From A. Ishihara and T. Kabe, Ind. Eng. Chem. Res., **32**, 754 (1993))

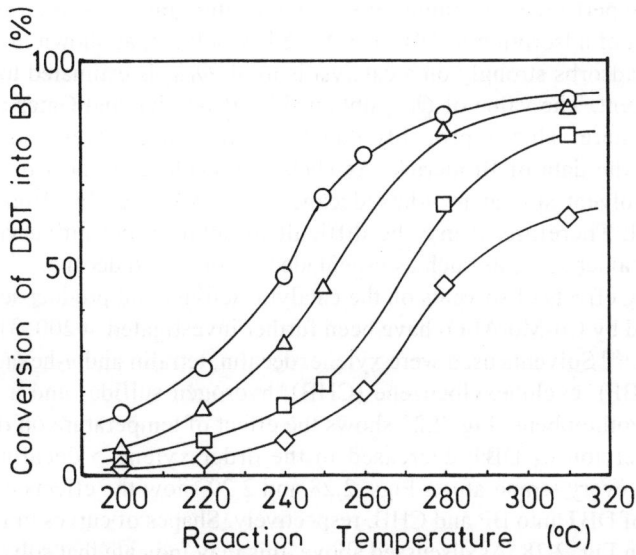

Fig. 2.28 Effects of solvents on conversion of dibenzothiophene into biphenyl.
50 kg/cm²; WHSV 70 h⁻¹; Catalyst 0.2 g; H₂ 18 l/h; Initial concentration of DBT: 0.1 wt%; Solvent: ○: xylene; △: decalin; □: tetralin; ◇: n-hexadecane.
(From A. Ishihara and T. Kabe, *Ind. Eng. Chem. Res.*, **32**, 754 (1993))

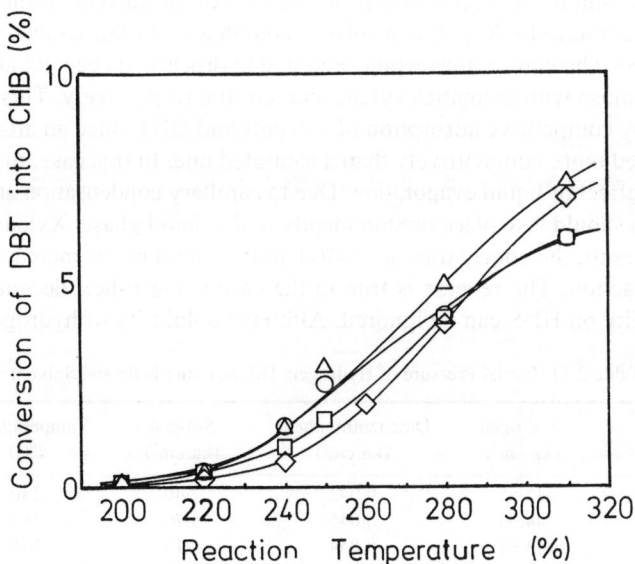

Fig. 2.29 Effects of solvents on conversion of dibenzothiophene into cyclohexylbenzene.
50 kg/cm²; WHSV 70 h⁻¹; Catalyst 0.2 g; H₂ 18 l/h; Initial concentration of DBT: 0.1 wt%; Solvent: ○: xylene; △: decalin; □: tetralin; ◇: n-hexadecane.
(From A. Ishihara and T. Kabe, *Ind. Eng. Chem. Res.*, **32**, 754 (1993))

experiments were performed in similar lowest temperature ranges, similar results seemed to be obtained. Heat of adsorption of DBT is affected by solvent, as shown in Table 2.9. When a solvent which adsorbs strongly on a catalyst is used, Q_{DBT} is estimated to be smaller than that without solvent. The value of Q_{DBT} obtained by Broderick and Gates was smaller than those presented here. This is probably due to the fact that solvents are not taken into consideration in the data of Broderick. As shown in Table 2.10, heat of adsorption of n-pentadecane, a solvent similar to n-hexadecane which was used by Broderick and Gates, was 17 kcal/mol. Therefore, it may be difficult to obtain a definitive value for Q_{DBT} in reactions with heavier solvents such as n-pentadecane or n-hexadecane.

The retarding effects of solvents on the catalytic activity and product selectivity in HDS of DBT catalyzed by Co-Mo/Al$_2$O$_3$ have been further investigated at 200–310°C, 50 atm and 0.1 wt% of DBT.[156] Solvents used were xylene, decalin, tetralin and n-hexadecane. Products were biphenyl (BP), cyclohexylbenzene (CHB), hydrogen sulfide and a trace amount of hexahydrodibenzothiophene. Fig. 2.27 shows the effect of temperature on the conversion of DBT. The conversion of DBT decreased in the order xylene > decalin > tetralin > n-hexadecane over every temperature. Figs. 2.28 and 2.29 show the effects of temperature on the conversions of DBT into BP and CHB, respectively. Shapes of curves in Fig. 2.27 are very similar to those in Fig. 2.28. As discussed above, this may indicate that solvents are adsorbed competitively at the active sites for desulfurization to prevent the adsorption of DBT and the formation of BP.

Another possibility is the effects of solvent evaporation on HDS of DBT reported by Kocis and Ho.[188] Table 2.11 shows the partial pressures of DBT, H$_2$ and a solvent. In the table, it was assumed that all components were gaseous. In order to make a solvent gaseous, vapor pressure of the solvent must exceed the partial pressure of the solvent. Table 2.11 also shows high enough temperature to evaporate a solvent completely. In this calculation, the Antoine equation was used. The conversions using saturated hydrocarbons decalin and n-hexadecane were lower than those with aromatics xylene and tetralin, respectively. This may be difficult to explain only by competitive adsorption of solvents and DBT since an aromatic compound would be adsorbed more competitively than a saturated one. In this case, the result may also be related to the effect of liquid evaporation. Due to capillary condensation and the conditions used, the reaction should take place predominantly in the liquid phase. Xylene is more volatile than DBT; as a result, the concentration of DBT in the liquid phase increases, which in turn speeds up the reaction. The reverse is true in the case using n-hexadecane. The effects of hydrogen solubility on HDS can be ignored. Although solubility of hydrogen into saturated

Table 2.11 Partial Pressure of Hydrogen, Dibenzothiophene and Solvent [a]

Solvent	Hydrogen (kg/cm^2)	Dibenzothiophene (kg/cm^2)	Solvent (kg/cm^2)	Temperature [b] (°C)
Xylene	42.55	0.043	7.40	240
Decalin	44.05	0.045	5.70	289
Tetralin	43.82	0.044	6.13	304 [c]
n-Hexadecane	46.18	0.047	3.78	358

[a] Total pressure 50 kg/cm^2; WHSH 70 h^{-1}; Catalyst 0.2 g; H$_2$ 18 l/h; Initial concentration of DBT: 0.1 wt%. [b] Temperature required to evaporate solvent completely. The Antoine equation was employed. [c] Calculated from data for vapor pressure of tetralin in the literature (Ref. 190).
(From A. Ishihara and T. Kabe, *Ind. Eng. Chem. Res.*, **32**, 754 (1993))

hydrocarbon *n*-hexadecane (ratio of dissolved hydrogen in *n*-hexadecane: 0.1007 at 269°C and 50 atm)[189] is higher than solubility into aromatic tetralin (0.0373 under the same conditions),[190] the conversion using *n*-hexadecane was lower than that using tetralin. In those reports, the solubility of hydrogen changed significantly depending on the pressure. When the partial pressure of hydrogen was changed to a range of 22 to 66 kg/cm², however, the conversion of DBT was hardly affected.

On the other hand, solvent effect on the formation of CHB in Fig. 2.29 was scarcely observed in comparison with the same effect on the formation of BP. This means that the formation of CHB is not affected by solvents. The fact that different solvents influence the desulfurization rate to BP but does not affect the formation of CHB supports the thesis in the literature[57,120] stating that desulfurization and hydrogenation proceed on different catalytic sites. The reaction scheme of HDS of DBT is shown in Fig. 2.10.[120] Although the BP to CHB pathway cannot be completely ruled out, the BP to CHB pathway would be kinetically insignificant under our experimental conditions. If CHB is mainly formed from hydrogenation of BP and hydrogenation of BP is not affected by solvents, the difference in the HDS rate of DBT into BP will generate a significant difference in the rate of CHB formation.

Although the effect of solubility of DBT into solvents is not discussed here, it may affect the catalytic activity of HDS. Retarding effects of components in light gas oil on the reactivity of DBTs are the sum of various factors such as competitive adsorption of components, liquid evaporation, solubility of DBTs, etc.

2. Effects of aromatic hydrocarbons on HDS of middle distillate

Effects of aromatics on HDS of DBT, 4-MDBT and 4,6-DMDBT included in middle distillate have also been investigated under practical conditions: fixed bed flow reactor; Co-

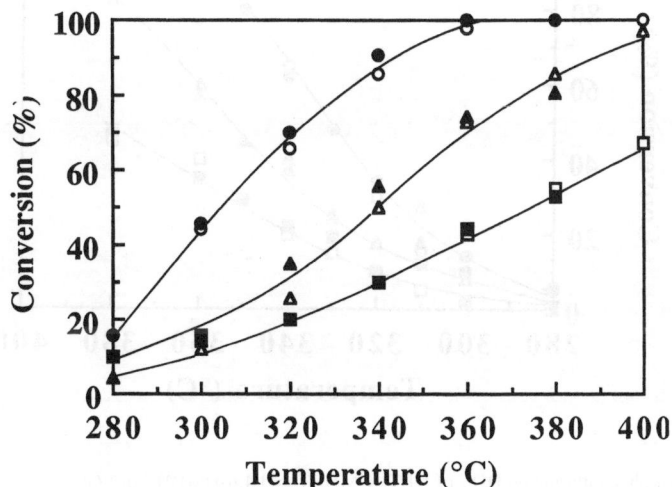

Fig. 2.30 Effect of methylnaphthalene on HDS of DBT, 4-MDBT and 4,6-DMDBT in LGO.
Gas/Oil = 125NL/L, Oil: LO-1.
LO-1: ○: DBT, △: 4-MDBT, □: 4,6-DMDBT.
5.0 wt% 1-methylnaphthalene + LO-1: ●: DBT, ▲: 4-MDBT, ■: 4,6-DMDBT.
(From T. Kabe, A. Ishihara *et al.*, *Ind. Eng. Chem. Res.*, **36**, 5150 (1997))

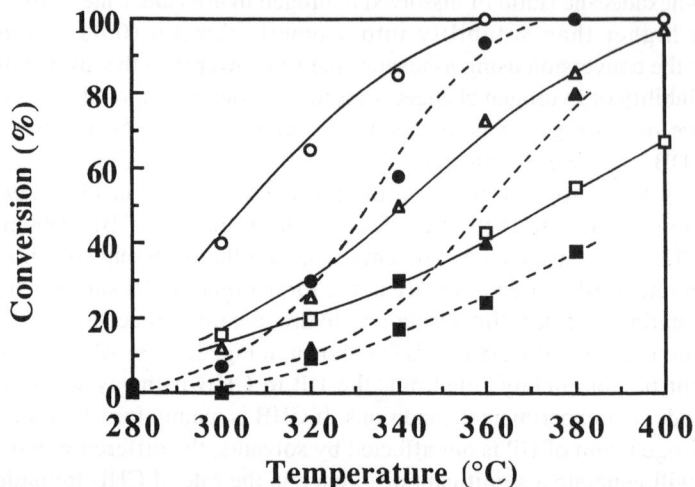

Fig. 2.31 Effect of phenanthrene on HDS of DBT, 4-MDBT and 4,6-DMDBT in LGO.
    Gas/Oil=125NL/L, Oil: LO-1.
    LO-1: ○: DBT, △: 4-MDBT, □: 4,6-DMDBT (———).
    5.0wt% phenanthrene + LO-1:●: DBT, ▲: 4-MDBT, ■: 4,6-DMDBT (----).
    (From T. Kabe, A. Ishihara *et al.*, *Ind. Eng. Chem. Res.*, **36**, 5150 (1997))

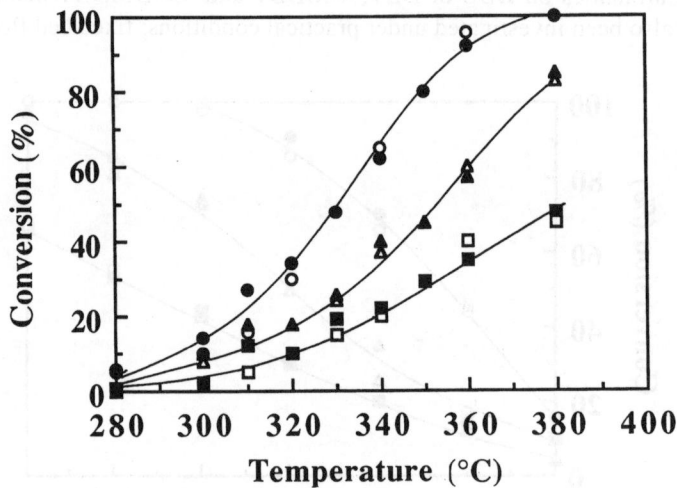

Fig. 2.32 Effect of phenanthrene on HDS of DBT, 4-MDBT and 4,6-DMDBT in LGO.
    Gas/Oil=125NL/L, Oil: LO-2.
    LO-2: ○: DBT, △: 4-MDBT, □: 4,6-DMDBT.
    5.0 wt% phenanthrene + LO-2: ●: DBT, ▲: 4-MDBT, ■: 4,6-DMDBT.
    (From T. Kabe, A. Ishihara *et al.*, *Ind. Eng. Chem. Res.*, **36**, 5150 (1997))

Mo/Al$_2$O$_3$: CoO: 4.5 wt%; MoO$_3$ 17.0 wt%; 30 atm; 265–380°C; H$_2$ 4 L/h; LHSV 4 h^{-1}, Gas/Oil ratio 125 NL/L.[138] Two kinds of light gas oils (LO-1, sulfur: 0.72 wt%; cut point 228–355°C (90 vol%); LO-2, sulfur: 1.5 wt%; cut point 165–371°C (90 vol%)) which were different in sulfur content and cut point were used. 5 wt% of 1-methylnaphthalene (MN) or phenanthrene (PN), one of components of LGO, was added to LO-1 or LO-2 as an additive. The addition of MN to LO-1 did not affect the conversions of DBT, 4-MDBT and 4,6-DMDBT, as shown in Fig. 2.30. In contrast to this, the addition of PN to LO-1 decreased the conversions of DBT, 4-MDBT and 4,6-DMDBT by 20–50%, as shown in Fig. 2.31. The result indicates that the aromatic compound with larger ring numbers has the stronger retarding effect on HDS activity. The extent of the retarding effect of PN on HDS of LO-1 decreased in the order DBT > 4-MDBT > 4,6-DMDBT. As shown in Table 2.7, heats of adsorption decreased in the order 4,6-DMDBT > 4-MDBT > DBT, indicating that PN is adsorbed competitively with each sulfur compound on the catalyst surface and that the sulfur compound with the lower heat of adsorption, that is lower adsorption ability, is retarded to a greater extent. As shown in Fig. 2.32, however, the addition of PN to LO-2 did not affect the conversions of DBT, 4-MDBT and 4,6-DMDBT probably because LO-2 has a higher sulfur content and cut point and includes heavier components than LO-1.

B. Effects of Sulfur Compounds

Among sulfur compounds included in the reaction system, hydrogen sulfide (H$_2$S) should be considered as the most important inhibitant because it is always present in HDS reactions. After Owens and Amberg[93] initially reported that H$_2$S significantly retards thiophene conversion, many researchers observed this effect in HDS of thiophene,[88,94,102,104–107,162,164,165,167] BT,[89,117,118,160] DBT,[57–61,128,129,138,140–142,152,153,161,166] substituted DBT[128,138,140–142,166] and distillates.[62,159,163] According to the rate expression of HDS reactions, H$_2$S inhibits HDS of thiophenic compounds through competitive adsorption with those thiophenic compounds at one kind of catalytic site. Further, the retarding effect of H$_2$S decreases with an increase in temperature because the adsorption parameter of H$_2$S generally decreases with an increase in temperature.[57–59,102,107] It has been shown that in HDS of thiophene, BT and DBT H$_2$S inhibits hydrogenolysis while it does not inhibit hydrogenation.[57,88,89] Yamada et al. reported the inhibition effect of H$_2$S on BT HDS catalyzed by Co-Mo/Al$_2$O$_3$, Co/Al$_2$O$_3$, Mo/Al$_2$O$_3$, and MoS$_2$.[117,118] It was observed that inhibition of H$_2$S was drastically suppressed over Co-Mo/Al$_2$O$_3$ compared with other catalysts examined and that the addition of H$_2$S inhibited hydrogenolysis but promoted hydrogenation. Nagai and Kabe[152] found that in DBT HDS catalyzed by Mo/Al$_2$O$_3$, the yields of BP and CHB decreased with increasing amounts of H$_2$S. Lamure-Meille et al.[140] studied the effect of H$_2$S on two different reaction pathways (direct desulfurization and hydrogenolysis through hydrogenation) in HDS of DBT and 4-MDBT catalyzed by Ni-Mo/Al$_2$O$_3$ and Co-Mo/Al$_2$O$_3$. It was found that the relative importance between the two routes is greatly modified by the presence of H$_2$S: the direct desulfurization route appears to be more inhibited by this compound. Isoda et al.[166] showed that in 4,6-DMDBT HDS catalyzed by Ni-Mo/Al$_2$O$_3$, by-product H$_2$S lowered hydrogenation activity and inhibited desulfurization. Zhang et al.[128] reported that in HDS of DBT and 4,6-DMDBT, the effects of H$_2$S on the catalytic activity and selectivity were investigated under deep desulfurization conditions (sulfur concentration < 0.05wt%) using a commercial Co-Mo/Al$_2$O$_3$ catalyst. The conversions of DBTs decreased with increasing partial pressure of H$_2$S. The

Table 2.12 Enthalpies of S-Containing Molecule Adsorption on SBMS of Different Compositions in Terms of the Method of Interacting Bonds [164]

SBMS	ΔH (kcal/mol) of molecule adsorbed		
composition	H_2S	Thiophene	THT
Co/MoS$_2$	11.5	21.2	35.1
Ni/MoS$_2$	12.2	22.1	36.1
Co/WS$_2$	11.9	21.7	35.8
Ni/WS$_2$	12.6	22.6	36.8

(From N. N. Bulgakov, A. N. Startsev et al., Mendeleev Commun., 98 (1991))

formations of both biphenyls (BPs) and cyclohexylbenzenes (CHBs) were inhibited by H_2S, but the former was inhibited more significantly. HDS reactions of DBTs were described using the Langmuir-Hinshelwood rate equation. Activation energies of the formation of BP and 3,3'-DMBP were 24±2 and 38±2 kcal/mol, respectively. Heats of adsorption of DBT, 4,6-DMDBT and H_2S for the formation of BPs were 10±2, 17±2 and 19±2 kcal/mol, respectively, and H_2S was adsorbed more strongly on the catalyst than DBT and 4,6-DMDBT, and inhibited the HDS of DBT and 4,6-DMDBT. Bulgakov and Startsev estimated the enthalpy of adsorption of thiophene, tetrahydrothiophene and hydrogen sulfide on the Ni(Co) atoms of the sulfide bimetallic species (SBMS) of HDS catalysts in terms of the Interacting Bonds Methods (IBM).[164] The heats of adsorption calculated are shown in Table 2.12. The smaller values for H_2S show that adsorbed H_2S may be supplanted by thiophene.

On the other hand, Broderick et al. reported that in dibenzothiophene HDS a structural change of the catalyst took place at low concentrations of hydrogen sulfide.[161] Further, it has also been reported that in HDS a certain partial pressure of hydrogen sulfide is often required to maintain the activity of sulfur catalysts.[50,191,192] Recently, Leglise et al. reported that at high temperature (400°C), the addition of hydrogen sulfide initially slightly increased thiophene conversion which then decreased. At lower temperatures, however, only a retarding effect was observed.[162] Pille et al. reported that in thiophene HDS catalyzed by Co-Mo/Al$_2$O$_3$, a kinetic model with interconversion of the active sites was superior to models based on a fixed number of active sites.[165] The proposed mechanism for the interconversion of active sites predicted that the hydrogenation and hydrogenolysis activities of a HDS catalyst could be substantially varied by H_2S. Gestel et al. studied thiophene HDS and cyclohexene hydrogenation catalyzed by Co-Mo/Al$_2$O$_3$ or Ni-Mo/Al$_2$O$_3$ simultaneously in a high pressure flow reactor.[167] Addition of small amounts of H_2S to the feed inhibited the activity of both catalysts. At high H_2S levels, however, the catalysts remained active, suggesting that a new type of site operates at high H_2S levels. The Ni-Mo catalyst was found to be far more active than the Co-Mo catalyst at low H_2S pressure, while the two catalysts became equivalent at high H_2S levels.

1. Effects of Hydrogen Sulfide in Model HDS Reaction

The above reports suggest that the effect of H_2S at lower concentrations of H_2S and higher temperatures or under deep desulfurization conditions where sulfur concentration is very low, may differ from that under normal conditions. Further, although HDS of 4,6-DMDBT becomes important under deep desulfurization conditions, the effect of H_2S on HDS of 4,6-DMDBT is not well known. In this section, the effects of H_2S on HDS of DBT and 4,6-DMDBT catalyzed by Co-Mo/Al$_2$O$_3$ under deep desulfurization conditions (S <

0.05%) are presented.[128] The differences in formation and selectivity of CHBs and BPs from DBT and 4,6-DMDBT are estimated in a wide range of H_2S concentrations.

The apparatus, methods for presulfiding and HDS reactions, analysis and kinetic calculations were the same as described in Section 2.1.3B. HDS reaction was carried out under the following conditions: Co-Mo/Al_2O_3 (CoO 4.0 wt%; MoO_3 12.0 wt%) 0.2 g, temperature 200–300°C, total pressure 50 atm, WHSV 70 h^{-1}, Gas/Oil 1100 NL/L, catalyst 0.2 g, H_2 18 L/h, initial concentration of DBTs 0.1–0.4 wt%, concentration of H_2S in H_2 0–2.0 vol% (partial pressure of H_2S 0–0.88 atm). The experiments for HDS of DBTs without the addition of H_2S have been described in Section 2.1.3B. In experiments on HDS of DBTs with the addition of H_2S, the retarding effects of hydrogen sulfide were estimated by Eq.(2.16) where the retarding term, $K_{H_2S}P_{H_2S}$, was introduced into Eq.(2.11):

$$r_{HDS} = k_{HDS}K_{DBT}P_{DBT}/(1 + K_{DBT}P_{DBT} + K_{H_2S}P_{H_2S}) \qquad (2.16)$$

where K_{H_2S}: equilibrium adsorption constant of H_2S; P_{H_2S}: partial pressure of H_2S. Eq.(2.17) can be obtained from taking the reciprocal of both sides of Eq.(2.16):

$$1/r_{HDS} = K_{H_2S}P_{H_2S}/k_{HDS}K_{DBT}P_{DBT} + (1 + 1/K_{DBT}P_{DBT})/k_{HDS}. \qquad (2.17)$$

The last term in Eq.(2.17) becomes a constant when the initial concentration of DBT or 4,6-DMDBT is constant by 0.1 wt %. The addition of H_2S was changed in the range of 0–2.0 vol% (0–0.88 atm). Figs. 2.33(a) and 2.33(b) show the plots of $1/r_{HDS}$ vs. P_{H_2S} for DBT and 4,6-

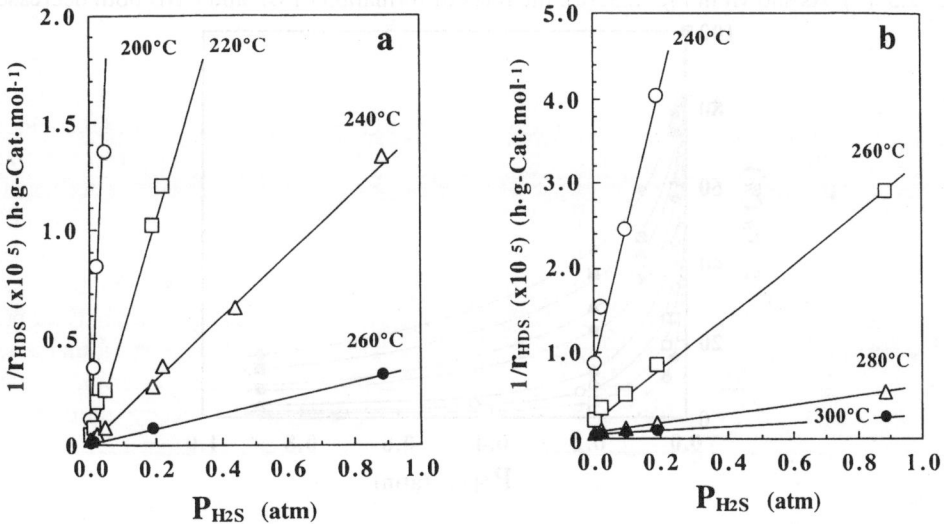

Fig. 2.33   Relations $1/r_{HDS}$ vs. P_{H_2S}.
(a) HDS of DBT,
(b) HDS of 4,6-DMDBT.
solvent: decalin; total pressure: 50 atm Catalyst: Co(4.0)–Mo(12.0).
(From T. Kabe, A. Ishihara et al., J. Jpn. Petrol. Inst., **40**, 187 (1997))

DMDBT, respectively. As shown in Figs. 2.33(a) and 2.33(b), the linear relationships were obtained both in $1/r_{HDS}$ vs. P_{H2S} for DBT and 4,6-DMDBT. K_{H2S} could be estimated by slopes while k_{HDS} and K_{DBT} calculated from Eq.(2.12) on the experiments without the addition of H_2S were used. To estimate the formation of BPs and CHBs separately, the same method was used for BPs or CHBs. Although they are not shown here, the linear relationships were also obtained both in $1/r_{BPs}$ or $1/r_{CHBs}$ vs. $1/P_{DBTs}$ and $1/r_{BPs}$ or $1/r_{CHBs}$ vs. P_{H2S} where r_{BPs} and r_{CHBs} were the formation rates of BPs and CHBs, respectively. Therefore, equations similar to Eqs.(2.17) and (2.12) were also used to determine k_{BPs} and k_{CHBs}, the formation rate constants of BPs and CHBs.

In Section 2.1.3B it has been shown that in HDS reactions of DBT and 4,6-DMDBT catalyzed by a Co-Mo/Al$_2$O$_3$ catalyst, the products were primarily BP and CHB in the case of DBT while in the case of 4,6-DMDBT, the products were 3,3'-DMBP and 3,3'-DMCHB. Fig. 2.34 shows the effects of partial pressure of H_2S on the HDS rates of DBT and 4,6-DMDBT. Both the HDS rates of DBT and of 4,6-DMDBT decreased with increasing partial pressure of H_2S, while that of DBT decreased more significantly than that of 4,6-DMDBT. As shown in Fig. 2.34, HDS rate of DBT under an H_2S partial pressure of 0.2 atm, decreased to *ca.* 13% and 8% of those without addition of H_2S at 240°C and 260°C, respectively. In contrast to this, HDS rate of 4,6-DMDBT under an H_2S partial pressure of 0.2 atm decreased to *ca.* 29% and 22% of those without addition of H_2S at 240°C and 260°C, respectively. This result indicates that H_2S inhibited more strongly HDS of DBT than 4,6-DMDBT.

For a better understanding of the effect of H_2S, the formation of BP and CHB in HDS of DBT at 240°C was investigated and the results obtained are shown in Fig. 2.35(a), and for 4,6-DMDBT, the effects of H_2S on the formation of 3,3'-DMBP and 3,3'-DMCHB is shown in Fig. 2.35(b). As shown in Fig. 2.35(a), the rates of formation of BP and CHB both decreased

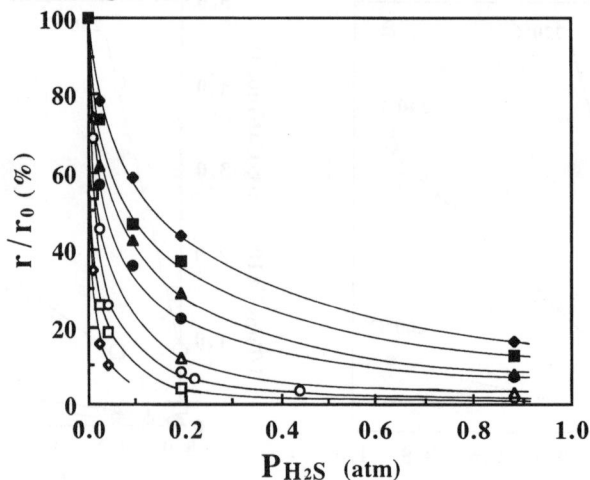

Fig. 2.34 Effects of H$_2$S partial pressures on HDS activities of DBTs.
r_0 = HDS rate of DBTs without the addition of H$_2$S.
r = HDS rate of DBTs with the addition of H$_2$S.
Open symbols for DBT: ◇: 200°C, ☐: 220°C, ○: 240°C, △: 260°C.
Closed symbols for 4,6-DMDBT: ●: 240°C, ▲: 260°C, ■: 280°C, ◆: 300°C.Total Pressure: 50 atm.
Catalyst: Co(4.0)–Mo(12.0). Solvent: Decalin.
(From T. Kabe, A. Ishihara *et al.*, *J. Jpn. Petrol. Inst.*, **40**, 187 (1997))

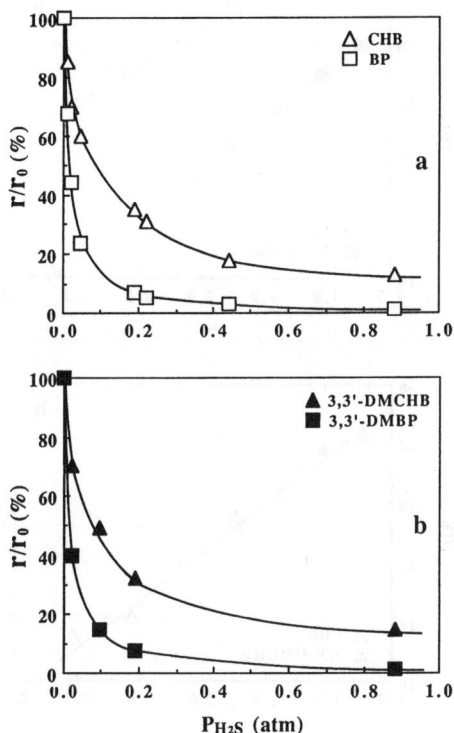

Fig. 2.35 Effects of H_2S partial pressures on the formation rates of BPs and CHBs.
(a) Formation rates of CHB and BP,
(b) Formation rates of 3,3'-DMCHB and 3,3'-DMBP.
r_0 = Formation rates of BPs or CHBs without the addition of H_2S.
r = Formation rates of BPs or CHBs with the addition of H_2S.
Total Pressure: 50 atm. Reaction Temperature: 240°C.
Catalyst: Co(4.0)–Mo(12.0). Solvent: Decalin.
(From T. Kabe., A. Ishihara et al., J. Jpn. Petrol. Inst., **40**, 188(1997))

with increasing partial pressure of H_2S. The rate of BP formation with addition of H_2S under H_2S partial pressure of 0.19 atm decreased to 7.5% of that without addition of H_2S at 240°C. On the other hand, the rate of CHB formation under the same partial pressure of H_2S decreased to 35% of that without addition of H_2S at 240°C. These results showed clearly that the retarding effect of H_2S on the formation of BP was much larger than that on the formation of CHB. Similar results were obtained in the case of 4,6-DMDBT, as shown in Fig. 2.35(b). The rates of formation of 3,3'-DMBP and 3,3'-DMCHB decreased with increasing partial pressure of H_2S. At 240°C, the rate of formation of 3,3'-DMBP with addition of H_2S under H_2S partial pressure of 0.19 atm decreased to 7.2% of that without addition of H_2S, while that of 3,3'-DMCHB decreased to 32.5%.

Comparing Fig. 2.35(a) and Fig. 2.35(b), the effect of H_2S on the rate of formation of either CHB or 3,3'-DMCHB was nearly the same for all H_2S partial pressures involved. In the case of BP and 3,3'-DMBP, the effect of H_2S on the rate of formation of either was also

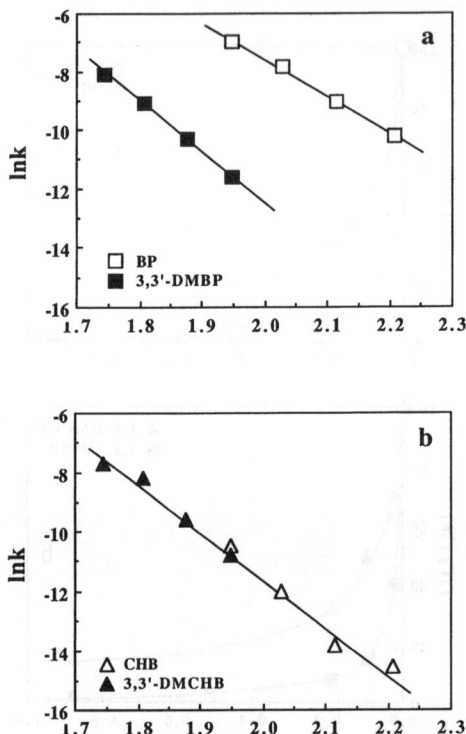

Fig. 2.36 Arrhenius plots of rate constants (k).
(a) $\square$: E_{BP} = 24 ± 2 kcal/mol,
$\blacksquare$: $E_{3,3'\text{-DMBP}}$ = 38 ± 2 kcal/mol,
(b) $\triangle$: E_{CHB} = 32 + 2 kcal/mol,
$\blacktriangle$: $E_{3,3'\text{-DMCHB}}$ = 32 ± 2 kcal/mol.
Catalyst: Co(4.0)–Mo(12.0). Solvent: Decalin.
(From T. Kabe., A. Ishihara *et al.*, *J. Jpn. Petrol.*, **40**, 189(1997))

nearly the same for all H_2S partial pressure. It can be assumed that HDS of 4,6-DMDBT also proceeds according to a pathway similar to that in Fig. 2.10, which shows the reaction network for DBT HDS. In HDS of DBT, the selectivity for BP was *ca.* 95% at 240°C, while, in the case of 4,6-DMDBT, the selectivity for 3,3′-DMBP was *ca.* 45%. Therefore, under the same partial pressure of H_2S, although the rates of formation of both BP and 3,3′-DMBP were inhibited by H_2S, the amount of decrease in the BP produced was more pronounced than that of 3,3′-DMBP. As a result, the HDS rate of DBT was inhibited by H_2S more strongly than that of 4,6-DMDBT.

In order to elucidate the differences in reactivities of DBTs into BPs and CHBs, the activation energies of the formation of BPs and CHBs, and the heats of adsorption of DBTs and H_2S on the catalyst were estimated using the Langmuir-Hinshelwood rate equation. The rate constants of HDS (k_{HDS}) and adsorption equilibrium constants of DBTs (K_{DBTs}) for BPs and CHBs formations were calculated using Eq.(2.12), and the adsorption equilibrium constant of hydrogen sulfide (K_{H2S}) was calculated using Eq.(2.17). With regard to the formation of BPs, Figs. 2.36(a) and 2.37(a) show Arrhenius and Van't Hoff plots. The activation energies

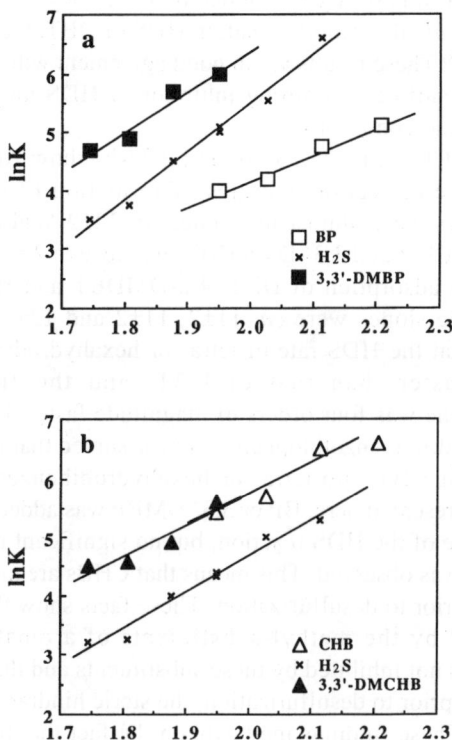

Fig. 2.37 Van't Hoff plots of adsorption equilibrium constants (K).
(a) $\square$: Q_{BP} = 10 ± 2 kcal/mol,
$\blacksquare$: $Q_{3,3'-DMBP}$ = 17 ± 2 kcal/mol,
X: Q_{H_2S} = 19 ± 2 kcal/mol,
(b) $\triangle$: Q_{CHB} = 11 ± 2 kcal/mol,
$\blacktriangle$: $Q_{3,3'-DMCHB}$ = 11 ± 2 kcal/mol.
X: Q_{H_2S} = 12 ± 2 kcal/mol.
Catalyst: Co(4.0)–Mo(12.0). Solvent: Decalin.
(From T. Kabe, A. Ishihara *et al.*, *J. Jpn. Petrol. Inst.*, **40**, 189 (1997))

of formation of BP and 3,3'-DMBP calculated from the slopes in Fig. 2.36(a) were *ca.* 24±2 and 38±2 kcal/mol, respectively. At 240°C, the rate constants of the formation of BP and 3,3'-DMBP were 9.1×10^{-4} and 8.3×10^{-6} (mol/h · g-Cat), respectively. The rate constant of the formation of BP was *ca.* 110 times larger than that of 3,3'-DMBP. In Fig. 2.37(a), the heats of adsorption of DBT, 4,6-DMDBT and H_2S for the formation of BPs calculated from slopes were 10±2, 17±1 and 19±2 kcal/mol, respectively.

The values of activation energies of HDS reflect the relative difficulty of desulfurization of DBTs, while the values of heats of adsorption reflect the relative ease of adsorption of DBTs or H_2S onto the catalyst. The heat of adsorption of 4,6-DMDBT was greater than that of DBT, and the heat of adsorption of H_2S was the greatest among them. This means that 4,6-DMDBT is adsorbed more easily on the catalyst than DBT, under HDS operating conditions, and that H_2S is adsorbed on the catalyst more strongly than DBT or 4,6-DMDBT and inhibits HDS. Broderick *et al.* investigated the effect of H_2S on HDS of DBT and showed that the conversion of DBT into BP was strongly inhibited by H_2S. They suggested that H_2S inhibited the HDS site.[57] Singhal *et al.* also reported that, in HDS of DBT, the conversion of DBT into BP was inhibited by H_2S.[58] These results are in good agreement with the work presented here, indicating that hydrogen sulfide is a strong inhibitor of HDS under deep desulfurization conditions as well as normal conditions.

For the formation of CHBs, Figs. 2.36(b) and 2.37(b) show Arrhenius and Van't Hoff plots. In Fig. 2.36(b), the activation energies of formation of CHB and 3,3'-DMCHB calculated from the slopes were almost the same, *ca.* 32 ± 2 kcal/mol. At 240°C, the rate constants of formation of CHB and 3,3'-DMCHB were close to 2.5×10^{-5} (mol/h · g-Cat). In Fig. 2.37(b), the heats of adsorption of DBT, 4,6-DMDBT and H_2S for the formation of CHBs calculated from the slopes were *ca.* 11 ± 2, 11 ± 2 and 12 ± 2 kcal/mol, respectively. Houalla *et al.* reported that the HDS rate of tetra- or hexahydrodibenzothiophene was one order of magnitude faster than that of DBT, and the HDS rate of tetra- or hexahydrodibenzothiophene was four orders of magnitude faster than the hydrogenation of DBT into tetra- or hexahydrodibenzothiophene. They assumed that the rate-determining step was the hydrogenation of DBT into tetra- or hexahydrodibenzothiophene.[120] Under the experimental conditions presented here, BP or 3,3'-DMBP was added to DBT or 4,6-DMDBT solution at the steady state of the HDS reaction, but no significant increase in the formation of CHB or 3,3'-DMCHB was observed. This means that CHBs are mainly formed through the hydrogenation of DBTs prior to desulfurization. These facts show that the formation of BPs is remarkably inhibited by the methyl-substituents of aromatic rings, although the hydrogenation of DBTs is not inhibited by these substituents and that when an aromatic ring in DBTs is hydrogenated prior to desulfurization, the steric hindrance of the methyl group is weakened significantly. These findings may explain the fact that the activation energies of formation of CHB and 3,3'-DMCHB were almost the same, *ca.* 32 ± 2 kcal/mol, and that the heats of adsorption of DBT and 4,6-DMDBT for the CHBs formation were almost the same *ca.* 11 ± 2 kcal/mol.

Broderick *et al.* reported that, on the effect of H_2S on HDS of DBT, the conversion of DBT into CHB was not inhibited by H_2S and that H_2S did not inhibit the active sites for hydrogenation.[57] Singhal *et al.* reported that, on the effect of H_2S on HDS of DBT, the conversions of DBT into BP and CHB were similarly inhibited by H_2S.[58] This is in disagreement with the work presented here where by comparing Fig. 2.35(b) and Fig. 2.35(a), both the formation rates of BPs and CHBs decreased with increasing partial pressure of H_2S, but the formation rates of BPs decreased more significantly than those of CHBs. This suggests that the extent of the retarding effect of H_2S on HDS differs between the HDS of DBTs and the hydrogenation of DBTs.

2. Effects of Hydrogen Sulfide on HDS of Middle Distillate

Before describing the effects of H_2S on HDS of middle distillate, initial kinetic experiments are presented here.[138] Pseudo-first-order plots of eleven dibenzothiophenes included in LO-2 (see Section 2.1.4A2) are shown in Fig. 2.38. HDS of LO-2 was performed under the following practical conditions: Fixed bed flow reactor; Co-Mo/Al_2O_3: CoO: 4.5 wt%; MoO_3 17.0 wt%; 30 atm; 310–370°C; Gas/Oil ratio 1000 NL/L; LHSV 2–10 h^{-1}. An approximate linear relationship was observed for each compound and HDS of each compound was treated as a pseudo-first-order reaction. As shown in Section 2.1.1B, dibenzothiophenes in middle distillate can be classified with respect to the HDS reactivities as follows: 1)

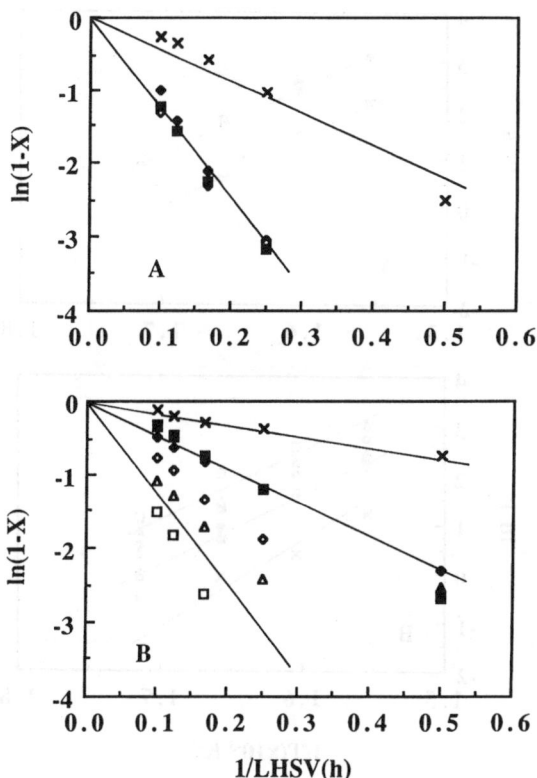

Fig. 2.38 Pseudo-first-order plots in HDS of DBTs.
Reaction temperature: 350°C, Gas/Oil: 1000NL/L, Oil: LO-2.
(a) DBT and C1-DBTs; ◆: 33(DBT), X: 41(4-MDBT), ◇: 42, ■ : 43.
(b) C$_2$-DBTs; ◆: 47, X: 48(4,6-DMDBT), ◇: 49, ■: 50, □: 51, ▲: 52, △: 54.
(From T. Kabe, A. Ishihara et al., J. Jpn. Petrol. Inst., **40**, 189 (1997))

nonsubstituted DBT and alkyl-substituted DBTs with no substituents at the 4- and 6-positions, 2) alkyl-substituted DBTs with a substituent at the 4-position and 3) alkyl-substituted DBTs with substituents at both 4- and 6-positions. The rate constants of DBT, 4-MDBT and 4,6-DMDBT, three representative compounds for this classification, were 12, 4.5 and 1.5 h^{-1}, respectively, at 350°C. The ratio of HDS rates of these compounds in model reactions was 100:10:1 at 240°C as shown in Table 2.5 (Cat B). The ratio predicted by Arrhenius plots of the rate constants was approximately 12:4:1 at 350°C. In HDS of LO-2, the ratio was 8:3:1 at 350°C, indicating that the differences between HDS rates of DBT, 4-MDBT and 4,6-DMDBT became small. LO-2 includes 1.5 wt% of sulfur and a main inhibitant can be assumed to be the H$_2$S formed. As described above (Section 2.1.4B1), H$_2$S retards the formation of BPs rather than the formation of CHBs. The selectivity for BP in HDS of DBT is higher than that for BPs in HDS of 4-MDBT or 4,6-DMDBT. Therefore, in HDS of middle distillate the conversion of DBT is more easily retarded by H$_2$S than that of 4-MDBT or 4,6-DMDBT and the differences between HDS rates of these compounds became small. Fig. 2.39 shows Arrhenius plots of pseudo-first-order rate constants of dibenzothiophenes in HDS

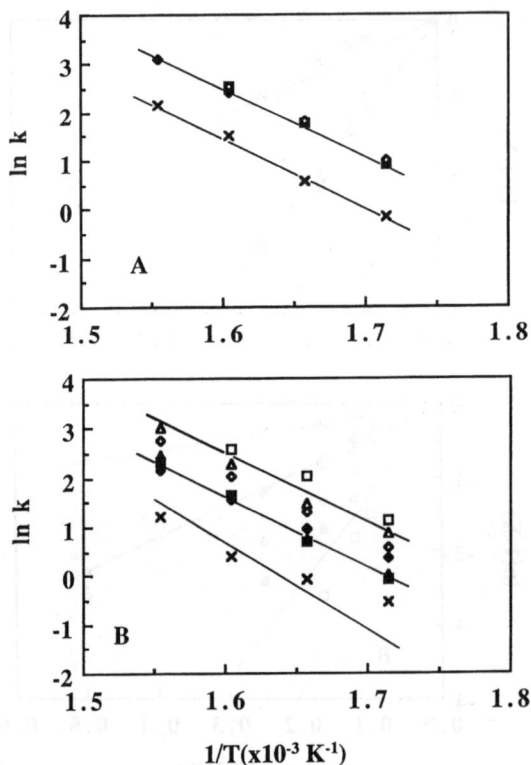

Fig. 2.39   Arrhenius plots of HDS rate constants of DBTs.
Reaction temperature: 350°C, Gas/Oil: 1000NL/L, Oil: LO-2, Symbols are the same as in Fig. 2.38.
(From T. Kabe, A. Ishihara *et al.*, *Ind. Eng. Chem. Res.*, **36**, 5148 (1997))

of LO-2. Activation energies for DBT, 4-MDBT and 4,6-DMDBT calculated from the slopes of straight lines were 26, 29 and 31 kcal/mol, respectively. In good agreement with the results from the model reactions, the values increased in the order DBT, 4-MDBT and 4,6-DMDBT while the differences between the values became small, indicating that the differences between the mechanisms of these compounds also became small in HDS of middle distillate compared with the model reactions. In the model reactions, the conversions of DBT, 4-MDBT and 4,6-DMDBT to CHBs were very similar. It was suggested that the selectivities for CHBs in HDS of DBT, 4-MDBT and 4,6-DMDBT became higher in HDS of LO-2 than in model reactions probably because H_2S retards the formation of BPs more markedly. At lower temperatures, activation energy of 4,6-DMDBT was lower than at higher temperatures, indicating that the mechanism at lower temperatures may be different from that at higher temperatures. This may be related to the fact that the selectivity for CHBs is higher at lower temperatures and the activation energy of CHBs formation is lower compared with the formation of BPs in the case of 4,6-DMDBT HDS.

In order to investigate the effects of components on HDS of middle distillate, experiments at Gas/Oil ratio=125 NL/L, where partial pressures of components become higher compared with those at Gas/Oil ratio=1000 NL/L, were performed. The reactions of each sulfur

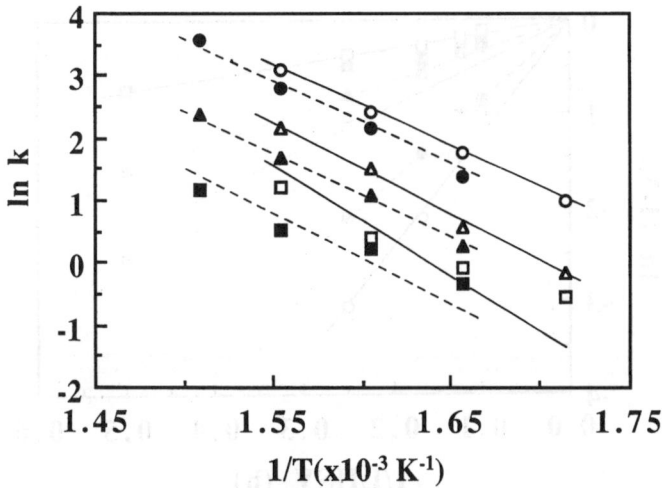

Fig. 2.40 Arrhenius plots of HDS rate constants of DBT, 4-MDBT and 4,6-DMDBT.
Oil: LO-2, Gas/Oil: 1000NL/L: ○: DBT; △: 4-MDBT; □:4,6-DMDBT(———);
Gas/Oil: 125NL/L: ●: DBT; ▲: 4-MDBT; ■:4,6-DMDBT(----).
(From T. Kabe, A. Ishihara *et al.*, *Ind. Eng. Chem. Res.*, **36**, 5149 (1997))

Table 2.13 Activation Energies of DBTs [a]

Dibenzothiophenes	Activation energies (kcal/mol)	
	Gas/Oil=1000	Gas/Oil=125
DBT	26	27
4-MDBT	29	28
4,6-DMDBT	31	27

[a] React. Temp. 280–400°C, Total pressure 30 kg/cm², LHSV2-10h⁻¹.

compound were treated as pseudo-first-order reactions. At 350°C, the rate constants for DBT, 4-MDBT and 4,6-DMDBT at Gas/Oil ratio=125 were 8.5, 3.0 and 1.3 h⁻¹, respectively, smaller than those at Gas/Oil ratio=1000. The ratio of the HDS rate of each compound at Gas/Oil ratio=125 to that at Gas/Oil ratio=1000 was in the range 0.7–0.9 at 350°C. The extent of decrease in HDS rate of 4,6-DMDBT became larger at 370°C than at lower temperatures because the activation energy of 4,6-DMDBT HDS in contrast to those of DBT and 4-MDBT decreased at Gas/Oil ratio=125 compared with that at Gas/Oil ratio=1000. The ratio of the rate constants for DBT, 4-MDBT and 4,6-DMDBT at Gas/Oil ratio=125 was 6.5:2.3:1, smaller than that (8:3:1) at Gas/Oil ratio=1000. As mentioned above, components retard BPs formation rather than CHBs formation. Therefore, the higher partial pressure of components at Gas/Oil ratio=125 decreased BPs formation to a greater extent compared with the cases at Gas/Oil ratio=1000. As a result, the selectivity for CHBs increased at Gas/Oil ratio=125. Further, the retarding effects of substituents were weakened in the formation of CHBs, as shown in Section 2.1.3B. This may explain the lower ratio of rate constants at Gas/Oil ratio=125. Arrhenius plots of the rate constants for DBT, 4-MDBT and 4,6-DMDBT

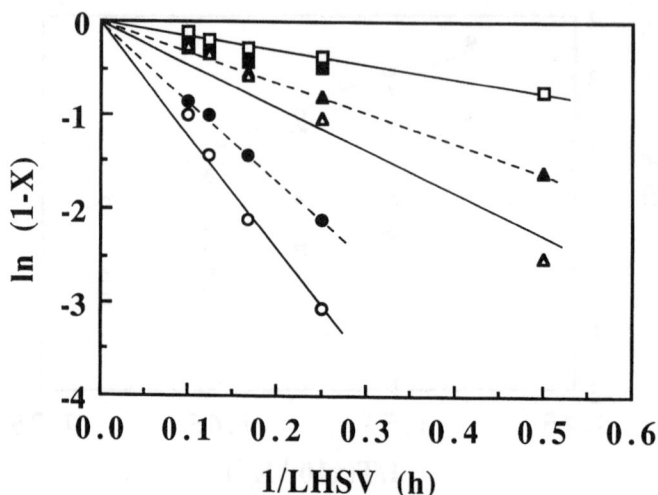

Fig. 2.41 Pseudo-first-order plots in HDS of DBT, 4-MDBT and 4,6-DMDBT.
Reaction temperature: 350°C, Gas/Oil: 1000NL/L, Oil: LO-2.
LO-2: ○: DBT; △: 4-MDBT; □:4,6-DMDBT (——); 3 vol% H$_2$S/H$_2$:
●: DBT; ▲: 4-MDBT; ■:4,6-DMDBT (----).
(From T. Kabe, A. Ishihara et al., Ind. Eng. Chem. Res., **36**, 5149 (1997))

Table 2.14 Pseudo-First-Order Rate Constants in HDS of Middle Distillate

Gas/Oil ratio	Rate constants k (h^{-1}) at 350°C			Rate constants k (h^{-1}) at 370°C	
	1000	125	1000 3vol% H$_2$S/H$_2$	1000	125
DBT	12	8.5 (0.71)	8.4 (0.70)	21.5	16.8 (0.78)
4-MDBT	4.5	3.0 (0.67)	3.2 (0.71)	8.6	5.4 (0.63)
4,6-DMDBT	1.5	1.3 (0.87)	1.5 (1.00)	3.3	1.7 (0.52)

The ratios of rate constants to rate constants at Gas/Oil=1000 without H$_2$S are given in parentheses.

at Gas/Oil ratio=125 NL/L are shown in Fig. 2.40 with the cases at Gas/Oil ratio=1000. As shown in Table 2.13, activation energies for DBT, 4-MDBT and 4,6-DMDBT calculated from the slopes of straight lines were 27, 28 and 27 kcal/mol, respectively. These values are very close to each other, indicating that the mechanisms of HDS are very similar to each other and that the route of CHBs formation for which no retarding effects were observed in HDS of DBT, 4-MDBT and 4,6-DMDBT may have a larger advantage at Gas/Oil ratio=125 than at Gas/Oil ratio=1000.

In experiments varying Gas/Oil ratio, it has been proved that the effect of the partial pressure of components on HDS rate is remarkable. Among components, H$_2$S seems to be significant in the case described above, and further experiments were performed. 3 vol% of H$_2$S was added to hydrogen gas and similarly the reactions of each sulfur compound could also be treated as pseudo-first-order reactions. Pseudo-first-order plots are shown in Fig. 2.41. The pseudo-first-order rate constants are listed in Table 2.14. At 350°C, the rate constants for DBT, 4-MDBT and 4,6-DMDBT at Gas/Oil ratio=1000 with 3 vol% of H$_2$S were 8.4, 3.2 and 1.5

h^{-1}, respectively, smaller than those at Gas/Oil ratio=1000 without H$_2$S. The ratio of the HDS rate for DBT and 4-MDBT at Gas/Oil ratio=1000 with 3 vol% of H$_2$S to that at Gas/Oil ratio=1000 without H$_2$S was about 0.7 and the addition of H$_2$S retarded the HDS of these compounds. In contrast, the ratio for 4,6-DMDBT was 1.0 and the HDS was not retarded by H$_2$S. As shown in Table 2.7, heats of adsorption of DBT, 4-MDBT and 4,6-DMDBT were about 11, 19 and 21 kcal/mol, respectively, indicating that DBT and 4-MDBT adsorb on the catalyst more weakly than 4,6-DMDBT. Further, as shown in Fig. 2.37, adsorption equilibrium constants for 4,6-DMDBT were higher than those for H$_2$S while adsorption equilibrium constants and heat of adsorption for DBT in BP formation were lower than those for H$_2$S. These results suggest that HDS of DBT and 4-MDBT was retarded by the competitive adsorption of H$_2$S while HDS of 4,6-DMDBT was not retarded. Further, hydrogen sulfide retards BPs formation rather than CHBs formation. In HDS of DBT or 4-MDBT, BPs formation was the major route. Therefore, HDS of DBT or 4-MDBT was retarded more significantly compared with that of 4,6-DMDBT. The ratio of the rate constants for DBT, 4-MDBT and 4,6-DMDBT at Gas/Oil ratio=1000 with H$_2$S was 5.6:2.1:1, smaller than that (8:3:1) at Gas/Oil ratio=1000 without H$_2$S and similar to that (6.5:2.3:1) at Gas/Oil ratio=125 without H$_2$S. As noted above, CHBs formation has an advantage in the presence of H$_2$S compared with BPs formation. Further, the retarding effects of substituents were weakened in the formation of CHBs, as shown in Section 2.1.3B. The above may also explain the lower ratio of rate constants of the three compounds at Gas/Oil ratio=1000 with H$_2$S.

The retarding effects of components at Gas/Oil ratio=125 are different from those at Gas/Oil ratio=1000 with 3 vol% of H$_2$S. That is, 4,6-DMDBT HDS was retarded under the former condition but not under the latter. In both cases, partial pressure of H$_2$S was similar. The extent of decrease in HDS rate of 4,6-DMDBT became larger at 370°C than at lower temperatures. In the experiment at Gas/Oil ratio=125, H$_2$S is directly generated on the catalyst surface. In contrast, in the experiment at Gas/Oil ratio=1000 with 3 vol% of H$_2$S, H$_2$S was added to the gas phase. Even if the partial pressure of hydrogen sulfide is similar in the system between the two conditions, the concentration of hydrogen sulfide on the local surface of the catalyst is higher in the former case. Such high concentrations of hydrogen sulfide may affect the structure of the catalyst related especially to HDS of 4,6-DMDBT.

C. Effects of Nitrogen Compounds

It is well known that nitrogen compounds inhibit the HDS reaction significantly and a number of attempts have been reported.[53,90,94,106,133,138,151,152,155,159,168–183] Reports on recent quantitative inhibition effects of nitrogen compounds on HDS reaction have been focused mainly on the differences in catalytic sites,[53,90,94,106,133,138,155,168–171,177–181] the effects of substituents in nitrogen compounds [151,152,172,173,175,176,182] and basicity of nitrogen compounds[151,152,172–175,182] which are related to one another.

Although nitrogen compounds generally retard the HDS of thiophenic compounds, the activity cannot be completely eliminated by the excessive addition of nitrogen compounds. The result shows that there are at least two kinds of catalytic sites: one is poisoned by a nitrogen compound and the other is not. In early works, the retarding effects of pyridine on HDS of thiophene and benzothiophene were reported.[90,106,168–171] The retarding effect of acridine, one of the components of LGO, on HDS of LGO catalyzed by Co-Mo/Al$_2$O$_3$ is presented here.[138]

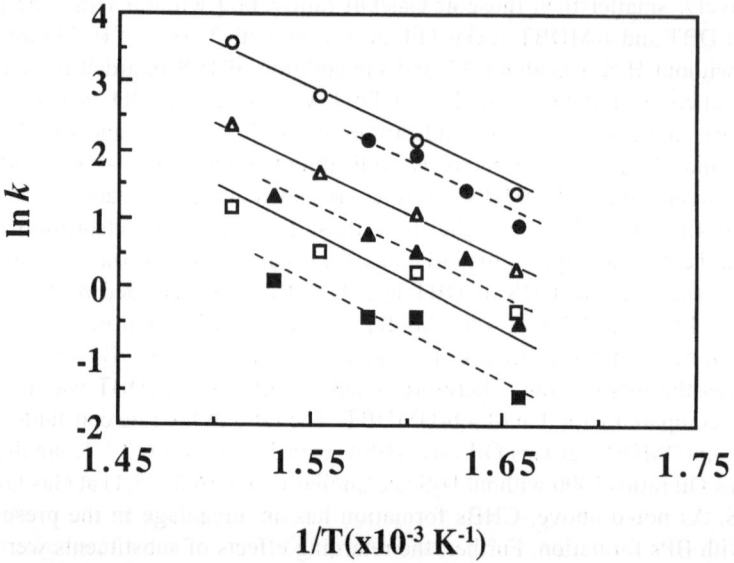

Fig. 2.42 Arrhenius plots of HDS rate constants of DBT, 4-MDBT and 4,6-DMDBT.
Gas/Oil: 125NL/L, Oil: LO-2.
LO-2: ○: DBT; △: 4-MDBT; □:4,6-DMDBT (——); 1.0 wt% acridine + LO-2:
●: DBT; ▲: 4-MDBT; ■:4,6-DMDBT (----).
(From T. Kabe, A. Ishihara et al., Ind. Eng. Chem. Res., 36, 5149 (1997))

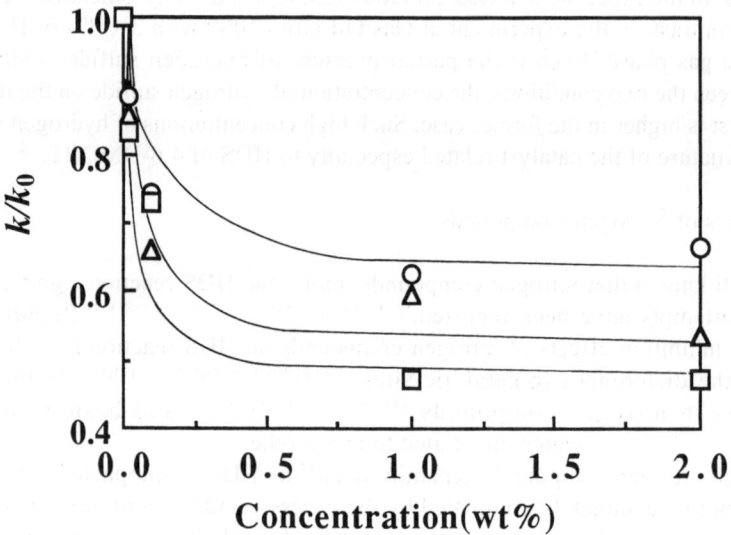

Fig. 2.43 Effect of concentration of acridine on HDS rate constants.
Reaction temperature: 345°C, k_0: LO-2; k: acridine + LO-2.
○: DBT; △: 4-MDBT; □: 4,6-DMDBT.
(From T. Kabe, A. Ishihara et al., Ind. Eng. Chem. Res., 36, 5151 (1997))

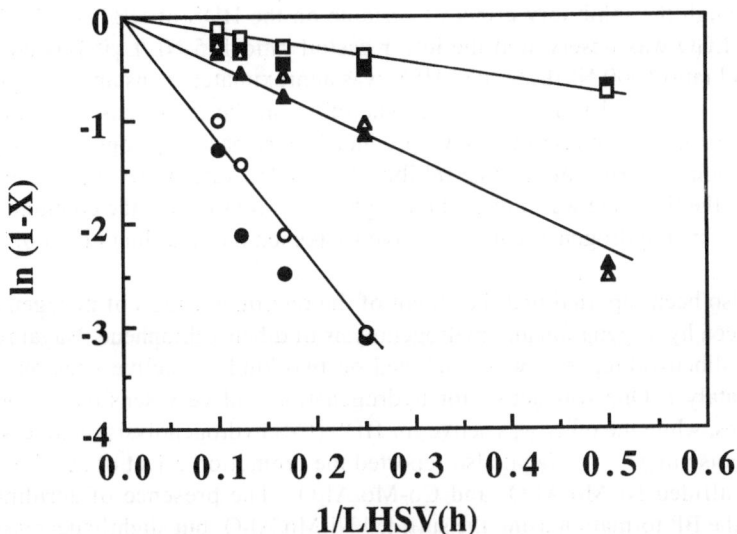

Fig. 2.44 Pseude-first-order plot in HDS of DBT, 4-MDBT and 4,6-DMDBT.
Reaction temperature: 350°C, Gas/Oil: 1000NL/L, Oil: LO-2.
LO-2: ○: DBT, △: 4-MDBT, □:4,6-DMDBT1.0 wt% Acridine + Lo-2;
●: DBT, ▲: 4-MDBT, ■: 4,6-DMDBT.
(From T. Kabe, A. Ishihara *et al.*, *Ind. Eng. Chem. Res.*, **36**, 5151 (1997))

Acridine was added to light gas oil, LO-2 (see Section 2.1.4A2). The reaction conditions were the same as that described in Section 2.1.4A2. The reaction of each sulfur compound was treated as a pseudo-first-order reaction. Fig. 2.42 shows the Arrhenius plots of pseudo-first-order rate constants of DBT, 4-MDBT, and 4,6-DMDBT. The activation energies of DBT, 4-MDBT, and 4,6-DMDBT were very similar to those in the HDS of LO-2. It appeared that the reaction mechanisms of DBTs did not significantly change with the addition of acridine. Although the addition of 5.0 wt% of phenanthrene to LO-2 did not affect the HDS activity, as shown in Fig. 2.32, the addition of 1.0 wt% acridine decreased remarkably the rate constants of these DBTs. Specifically the HDS of 4-MDBT and 4,6-DMDBT were more retarded than that of DBT. Fig. 2.43 shows the effect of the concentration of acridine on the ratio of the rate constants of DBT, 4-MDBT, and 4,6-DMDBT in the presence of acridine to those in LO-2. Only 0.1 wt% of acridine retarded HDS activity and further additions of acridine did not substantially decrease the HDS rate. This result may reveal the presence of at least two different kinds of sites as mentioned above. The extent of the retardation increased in the order DBT < 4-MDBT < 4,6-DMDBT, and the ratios of the rate constants of DBTs with 1.0 wt% of acridine to the rate constants without acridine were 0.65, 0.55, and 0.50 for DBT, 4-MDBT, and 4,6-DMDBT, respectively. It is obvious that the retarding effect of acridine is different from that of phenanthrene, where the extent of the retardation decreased in the order DBT > 4-MDBT > 4,6-DMDBT.

The effects of acridine on the HDS of DBT, 4-MDBT, and 4,6-DMDBT were also investigated under the following conditions: 350°C, LHSV 2–10 h^{-1}, and Gas/Oil ratio 1000

NL/L. Fig. 2.44 shows the pseudo-first-order plots of DBT, 4-MDBT, and 4,6-DMDBT at 350°C. Although the inhibitory effect of acridine on the HDS of DBT, 4-MDBT, and 4,6-DMDBT in LGO was observed at the lower gas/oil ratio 125 NL/L, it was not seen at the higher gas/oil ratio 1000 NL/L. Since LHSV was approximately constant, the hydrogen flow rate at the lower gas/oil ratio was lower. Therefore, at the lower gas/oil ratio, the longer contact time of the gas phase may have produced more basic hydrogenated derivatives of acridine, which in turn would have inhibited the HDS rate more significantly.[151,152,172] Furthermore, the flow rate was higher at the higher gas/oil ratio, and the continuous sweeping of the surface by a hydrogen stream may have weakened the retarding effect of the nitrogen compounds.

It has also been reported that the extent of the retarding effects of nitrogen compounds differs between hydrogenation and hydrogenolysis of dibenzothiophene. Nagai reported that the HDS of dibenzothiophene was catalyzed on two kinds of active sites on the sulfided Mo/Al$_2$O$_3$ catalyst: One was active for hydrogenation and very sensitive to poisoning by nitrogen bases, while the other was active for HDS (C-S hydrogenolysis), and less susceptible to acridine poisoning.[155,172] Nagai also reported the promotion effect of acridine on HDS of DBT over sulfided Ni-Mo/Al$_2$O$_3$ and Co-Mo/Al$_2$O$_3$. The presence of acridine markedly accelerated the BP formation using the sulfided Ni-Mo/Al$_2$O$_3$ but slightly decreased the CHB formation, indicating that hydrogenation and desulfurization took place at different sites on the catalyst. A similar effect using Co-Mo/Al$_2$O$_3$ was quite weak compared with Ni-Mo/Al$_2$O$_3$.[179-181] To consider the retarding effect of acridine on HDS of DBT, 4-MDBT and 4,6-DMDBT mentioned in the preceding paragraph, there is an alternative proposition. The products of HDS of these DBTs were corresponding BPs and CHBs. The selectivity for CHBs was higher in the order DBT < 4-MDBT < 4,6-DMDBT as mentioned before.[68] If acridine retards hydrogenation sites selectively, the extent of its retardation effect increases in the order DBT < 4-MDBT < 4,6-DMDBT, which is consistent with the observation.

Quinoline inhibited hydrogenation reactions more than hydrogenolysis in HDS of DBT catalyzed by Ni-Mo/Al$_2$O$_3$. As an approximate two fold difference in the decrease of rate constants was observed, it was regarded as evidence of separate sites for hydrogenation and hydrogenolysis.[53] In another approach using these data, a model based on Langmuir-Hinshelwood rate expressions for each reaction of the dibenzothiophene network was developed. In this simultaneous modeling of HDS and HDN, it was found that a two-site model gave a better fit rather than a single-site model for the hydrogenolysis reactions in the DBT network.[53] The result is similar to those proposed in the effect of pyridine or acridine on HDS of thiophene, BT or LGO mentioned above. One is a more active hydrogenolysis site that is more susceptible to poisoning and the other is a less active site that is more resistant to poisoning. The values of the adsorption parameters for the latter sites were found to be more than an order of magnitude smaller than those for the former sites.[53] Kaernbach et al.[177] studied the influence of nitrogen compounds in a crude oil on HDS of a vacuum distillate and DBT catalyzed by Co-Mo/Al$_2$O$_3$. The nitrogen compounds in the vacuum distillate were separated to acidic, basic and neutral concentrates. These nitrogen compounds added to the raw material decreased the degree of desulfurization which decreased in the order acidic > neutral > basic. In the experimental conditions used the deactivation of catalyst by nitrogen compounds did not have a permanent effect, and disappeared after introduction of the raw material in the absence of nitrogen compounds. Although the nitrogen compounds added decreased the catalyst's HDS properties, they did not influence its hydrogenation properties.

Vladov et al.[178] studied the effect of piperidine on a commercial Co-Mo/Al$_2$O$_3$ catalyst upon its sulfidation and during thiophene HDS under atmospheric pressure. Upon sulfidation of the catalyst, in the presence of piperidine, a reversible deactivation of HDS active sites and/or an irreversible deactivation of the hydrogenation and isomerization active sites were observed.

The investigation of the effects of substituents in nitrogen compounds on HDS have given insights into not only the structure and nature of catalytic sites but also mechanisms of inhibition. A number of attempts have been made and the representative examples using substituted pyridines reported. LaVopa and Satterfield[151] reported the effects of hindered pyridines on thiophene HDS catalyzed by Ni-Mo/Al$_2$O$_3$. To determine the inhibition effect of pyridines, the adsorption parameter of an inhibitor, K_I, was estimated by using the following equation:

$$k'=k/(1 + K_I P_I) \tag{2.18}$$

where k' represents the pseudo-first-order rate constant for thiophene disappearance in the presence of an inhibitant, k the pseudo-first-order rate constant for thiophene disappearance in the absence of an inhibitant, and P_I partial pressure of an inhibitant. At 360°C, the values of K_I for 4-methylpyridine, pyridine and 2,6-dimethylpyridine were 680, 430 and 110 atm^{-1}, respectively. The results indicate that 2,6-dimethylpyridine is the most difficult to be adsorbed on the catalyst surface among the three and that these pyridines may be adsorbed on the catalyst not parallel but perpendicular to the surface plane through nitrogen atom, that is, not by an side-on mode but by an end-on mode. The adsorption mode of a hindered nitrogen compound on HDS catalyst is different from that of a hindered sulfur compound such as 4-MDBT or 4,6-DMDBT, which was described in Section 2.1.3B and illustrated in Fig. 2.22.

Miciukiewicz et al. reported that the poisoning effect of 2,6-dimethylpyridine on thiophene HDS catalyzed by Co-Mo/Al$_2$O$_3$ was weaker than that of pyridine, piperidine and 3,5-dimethylpyridine.[173] A similar result was reported for 2-methylthiophene HDS.[174] Gutberlet and Bertolacini investigated the effect of various methyl- and ethyl-substituted pyridines on naphtha HDS catalyzed by Co-Mo/MgO.[175] Adsorption parameters were estimated using a model based on the Langmuir-Hinshelwood rate expression. Hindered pyridines showed relatively low values of adsorption parameter and the value for 2,6-dimethylpyridine was zero, indicating that the poisoning effect of this compound was the weakest and did not affect the HDS reaction. All the results were consistent with the above-mentioned suggestion that the nitrogen compounds retard HDS by adsorption through the nitrogen atom. In contrast to this, Mathur et al. reported that heterocyclic nitrogen compounds are adsorbed parallel to the surface through their π-electrons.[193,194] The adsorption of nitrogen compounds through π-electrons may be important for hydrogenation of their aromatic ring and such adsorption may be present on the surface. Miciukiewicz et al. also reported that the hindered nitrogen compounds inhibited 1-hexene hydrogenation more than thiophene HDS.[173] In this case, there is a possibility that the hindered nitrogen compounds are adsorbed through π-electrons and retard hydrogenation although they are not hydrogenated themselves at the temperature employed. In the inhibition experiments described above, hydrogenation of the nitrogen compounds did not occur to a significant extent. These results indicate that catalytic hydrogenation of the inhibitors may occur in a different temperature range from those used for the inhibition experiments of HDS and that the adsorption of inhibitors for HDS retardation is different from that for hydrogenation of the inhibitors themselves. This proposition is supported by the report of Kwart *et al.* that 2,6-dimethylpyridine is hydrogenated at higher temperatures and pressures than those used in the HDS inhibition experiments and that the 2,6-dimethylpiperidine formed is a strong inhibitor of HDS because the steric hinderance of methyl groups at the 2- and 6-positions are weakened in the adsorption of this compound through the nitrogen atom.[176)]

The basicity of nitrogen compounds has also been correlated with the strength of the inhibition of HDS, and the nature of catalytic sites as well as mechanisms of inhibition have been considered. LaVopa and Satterfield[151)] also reported the effects of unhindered nitrogen compounds on thiophene HDS catalyzed by Ni-Mo/Al$_2$O$_3$. Eq.(2.18) was also used to determine the inhibition effect of those compounds. Adsorption parameters K_I (atm^{-1}) estimated for unhindered or nonsubstituted nitrogen compounds, ammonia, aniline, pyridine, and quinoline at 360 °C were 65, 95, 430, and 990, respectively, indicating that the extent of poisoning effect roughly increased with an increase in the basicity of the nitrogen compound. Nagai *et al.*[172)] reported that nitrogen compounds did not retard hydrogenolysis but did retard hydrogenation of DBT to hexahydrodibenzothiophene (HHDBT) in HDS of DBT catalyzed by Ni-Mo/Al$_2$O$_3$. To determine the inhibition effect of nitrogen compounds on HHDBT formation, the rate equation (2.19) based on the Langmuir-Hinshelwood rate expression was used:

$$r_{6H} = kK_DP_DK_HP_H/(1 + K_NP_N)$$ (2.19)

where r_{6H} is the rate for HHDBT formation; P_i is the partial pressure; K_i and k are constants; D, H and N denote DBT, hydrogen and nitrogen compound, respectively. It was assumed that the nitrogen compound adsorbed more strongly on the hydrogenation sites than either DBT or the hydrogenated compounds. Adsorption parameters K_N (atm^{-1}) estimated for aniline, pyridine, quinoline and acridine at 260°C were 0.19, 1.07, 1.31, and 2.00, respectively. The poisoning effect of the nitrogen compounds was not correlated with their solution basicities but with their gas-phase proton affinities. Adsorption parameters for hindered nitrogen compounds were not correlated with the proton affinities. Since the gas-phase proton affinities represent enthalpy changes for proton-transfer reactions and is a direct measure of the Brønsted base strength of the molecule, it is suggested that the sites for hydrogenation poisoned by nitrogen compounds may be Brønsted acid sites.

The relatively strong inhibition effects of the nonbasic carbazole have been reported although the basicity of nitrogen compounds was correlated with the strength of the inhibition of HDS. In thiophene HDS catalyzed by Ni-Mo/Al$_2$O$_3$, the value of the adsorption parameter for carbazole (K_N: 510 atm^{-1}) was comparable to that of the much more basic piperidine (K_N: 590 atm^{-1}).[151)] Nagai and Kabe investigated the effect of nitrogen compounds on HDS of DBT catalyzed by a presulfided Mo/Al$_2$O$_3$ in a high-pressure flow microreactor.[152)] The nitrogen compounds inhibited the hydrogenation of DBT because they adsorbed more strongly than did DBT at lower temperatures. At higher temperatures, the nitrogen compounds also hindered the desulfurization together with the hydrogenation of DBT. In this study, the inhibition effect of nitrogen compounds on hydrogenation of DBT to tetrahydrodibenzothiophene (THDBT) was determined using an equation similar to Eq.(2.19). The adsorption parameters for acridine, carbazole, phenothiazine, and dicyclohexylamine were 690, 650, 62, and 41 atm^{-1}, respectively. The high value for carbazole was attributed to the basic product from hydrogenation of carbazole although the hydrogenated product was not identified. Similar results observed in the investigation of the effect of acridine on HDS of DBTs in middle distillate were mentioned above in this section.[138]

Nagai and Kabe also determined the adsorption parameter of phenanthrene.[152] The value of K was 10 atm^{-1} lower than that of acridine. As described above (Section 2.1.4A, Figs. 2.31–2.32), the addition of phenanthrene did not retard HDS of heavier LGO (LO-2) but did retard HDS of lighter LGO (LO-1), while, as shown in Figs. 2.42 and 2.43 of this section, the addition of acridine retarded not only HDS of lighter LGO but also HDS of heavier LGO. These results can be explained by the difference in the values of the adsorption parameters of phenanthrene and acridine. On the other hand, the weak inhibition effect of dicyclohexylamine in this work may be explained by the steric hindrance of two cyclohexyl groups.

D. Effects of Oxygen Compounds

Nagai and Kabe investigated the effect of oxygen compounds on dibenzothiophene HDS catalyzed by a presulfided Mo/Al$_2$O$_3$ in a high-pressure flow microreactor.[152] Oxygen compounds examined were dibenzofuran, xanthene and phenol. The oxygen compounds inhibited both the formation of BP and CHB. In this study, the inhibition effects of oxygen compounds on the formation of BP and CHB were determined using the rate equations (2.20) and (2.21), respectively, based on the Langmuir-Hinshelwood rate expression:

$$r_{BP} = kK_D P_D K_H P_H/(1 + K_O P_O)^2 \qquad (2.20)$$

$$r_{CHB} = kK_{6H} P_{6H} K_H P_H/(1 + K_O P_O)^2 \qquad (2.21)$$

where r_{BP} and r_{CHB} are the rates for BP and CHB formation; P_i is the partial pressure; K_i and k are constants; D, 6H, H and O denote DBT, HHDBT, hydrogen and oxygen compound, respectively. It was assumed that the oxygen compound was adsorbed more strongly at the desulfurization sites than either DBT or the hydrogenated compounds. For the BP formation, adsorption parameters K_O (atm^{-1}) estimated for dibenzofuran, xanthene and phenol at 300°C were 11.7, 11.2, and 9.2, respectively. For the CHB formation, adsorption parameters K_O (atm^{-1}) estimated for dibenzofuran, xanthene and phenol at 300°C were 6.9, 4.7, and 4.7, respectively. K_O for Eq.(2.20) was approximately one-half K_O for Eq.(2.21), but the degrees for two equations were similar. It was suggested that the retarding effects by one-ring and multiring oxygen compounds were similar because the values of K_O in the formation of BP or CHB were very near each other. At lower concentrations of oxygen compounds added, the ratio of BP to CHB was constant, indicating that the oxygen compounds are adsorbed competitively at the desulfurization site then retarded the desulfurization of DBT and the hydrogenated compounds. When higher concentration of the oxygen compounds were added, however, the BP formation was retarded to a larger extent than the CHB formation, indicating that the surface structure of the catalyst changes by the oxidation with the oxygen compounds or water formed in the hydrogenolysis of the oxygen compound added.

The effects of oxygen compounds on HDS using Co-Mo/Al$_2$O$_3$ in a trickle bed reactor have been reported.[184,185] Odebunmi and Ollis investigated the effect of m-cresol on benzothiophene HDS.[184] The inhibition of the HDS reaction was represented by a rate equation similar to Eq.(2.15). The values of adsorption parameters estimated for BT and cresol were 16.5 and 8.4 L/mol, respectively, at 375°C. Ollis and Lee investigated the effect of benzofuran on HDS of DBT.[185] The inhibition of the HDS reaction was also represented by a rate equation similar to Eq.(2.15). The values of the adsorption parameters estimated for DBT and benzofuran were 11.9 and 61.7 L/mol, respectively, at 325°C. The effect of dibenzofuran and water on HDS of DBT catalyzed by Ni-Mo/Al$_2$O$_3$ in a batch reactor has also been reported by Krishnamurthy and Shah.[186] These oxygen compounds revealed weak inhibition of HDS.

The effects of other oxygen compounds appear in the literature. Lee et al.[187] reported thiophene HDS catalyzed by Co-Mo/Al$_2$O$_3$ using hydrogen which was generated by a water gas shift reaction (WGSR) in the same catalyst bed. The activity using this method was inferior to that using pure hydrogen. The lower efficiency was attributed to water, carbon monoxide and carbon dioxide generated in WGSR. The retarding effects of these oxygen-containing compounds were estimated using a Langmuir-Hinshelwood rate equation including their retarding terms.

2.1.5 Model Studies of Hydrodesulfurization Mechanisms

A. Mechanisms of Hydrodesulfurization Reactions

A number of mechanisms have been proposed for hydrodesulfurization reactions. In the mechanisms, different types of bonding and C-S bond scission of sulfur compounds on the active sites, the reactions of various intermediates with H, SH or OH species, etc. appear. Some of these are presented here. It is reasonable to assume that sulfur anion vacancies are the active sites for HDS because the coordinatively unsaturated sites are needed for the adsorption of thiophenes to occur. As shown in Figs. 2.20 and 2.21, adsorption of thiophene is divided roughly into two modes: It takes place via one-point (end-on) mode with sulfur atom[147,195-202] or via multipoint (side-on) mode.[148-150,202-210] The latter may involve adsorption with all or some members of the thiophene ring. In Fig. 2.20, desulfurization begins with adsorption of the sulfur atom of thiophene on the sulfur vacancy. Hydrogen is supplied to the adsorbed thiophene from the OH species to give an adsorbed butadienethiolate. When a sulfided catalyst is used, OH can be replaced by SH. Mo^{4+} is oxidized to Mo^{6+} for the formation of butadiene. After that, the surface having sulfur from thiophene is reduced by a reductive addition of hydrogen and Mo^{6+} is reduced to Mo^{4+} for the formation of hydrogen sulfide.[147,195] In a catalyst promoted by cobalt or nickel, the redox process of the promoter is assumed.[49] Another mechanism includes one-point adsorption of thiophene and a subsequent dehydrodesulfurization reaction which yields adsorbed H$_2$S and a diacetylenic residuum through direct β-hydrogen transfer from carbon to sulfur.[196,211] Further hydrogenation of the latter product gives butenes and butane. In contrast, Kwart et al. reported another kind of mechanism which starts by multipoint adsorption, that is, the adsorption of carbon-carbon double bonds of thiophene on a sulfur anion vacancy.[148] As shown in Fig. 2.21(a), after transfer of hydrogen from an adjacent SH group to an α carbon, sulfur in thiophene interacts with a neighboring sulfur ion at the catalyst surface. Angelici and coworkers have reported the partial hydrogenation of π-bonded thiophene to dihydrothiophene intermediate. As shown in Fig. 2.21(b), adsorbed thiophene is attacked by a metal hydride and then by an SH group to give S-bonded dihydrothiopehene which undergoes rapid conversion to butadinene.[149,150] The total hydrogenation of thiophene to yield adsorbed tetrahydrothiophene which undergoes stepwise scission of the two C-S bonds with metal hydride or surface SH species has also been reported.[212,213] Theoretical studies for surfaces,[197,214-218] as well as other experimental studies[94,95,170,219-224] have complemented the above experimental studies.

For HDS of BT, DBT or their substituted derivatives, similar mechanisms, where the adsorption on anion vacancies, hydrogenolysis and hydrogenation with metal hydride or SH species, desorption of products and hydrogen sulfide, and the regeneration of anion vacancies proceed, may be proposed. As shown in Section 2.1.3, the mechanism for HDS of DBT and its substituted derivatives was different from those mentioned above. DBTs are adsorbed on the surface through aromatic rings. It was proposed that DBTs rotated around sulfur atom from parallel to perpendicular configuration when the C-S bond scission occurred (Fig. 2.22). It was concluded that HDS of 4,6-DMDBT and 4-MDBT was retarded in comparison with that of DBT because the rotation of 4,6-DMDBT and 4-MDBT during the C-S bond scission was inhibited due to the steric hindrance of the methyl group.[68,122] The studies for HDS mechanisms of these substituted compounds have just begun and further work is needed to elucidate the mechanisms in detail.

B. Approach by Organometallic Chemistry

One way to understand the mechanism of thiophenes HDS is to investigate the structures and the reactivities of thiophenes on the organometallic complexes. The reactions of thiophenes occurring on the complexes at lower temperatures near room temperatures may not directly be related to those on heterogeneous catalysts at higher temperatures near 300°C. However, organometallic chemistry provides not only visual pictures of the coordination and reaction of thiophenes and thiophene derivatives but also the possibility that reactions similar to those pictured may actually occur on catalyst surface. Organometallic chemistry has been utilized in examining the mechanisms of other reactions in this manner.[225,226]

The thiophene ring is conjugated with π-electrons and its C-S bonds have a double bond character. Therefore, it would appear difficult to cleave the C-S bond compared with thiols or sulfides. In an early work on thiophene reactivity on an organometallic complex, however, thiophene easily reacted with $Fe_3(CO)_{12}$ to lose sulfur and give a ferrole, $[Fe(CO)_3]_2(\mu-\eta^2:\eta^4-C_4H_4)$, where sulfur was replaced by an iron atom and a remaining unsaturated hydrocarbon part combined with another iron atom through coordinate bonds with π electrons, as shown in Eq.(2.22).[227-229] Recent heightening of interest in environmental problems, especially deep HDS of middle distillates, has given rise to an increase in the number of reports related to HDS in organometallic chemistry where the C-S bond scission of thiophenes is often included. Angelici,[230-232] Rauchfuss,[233] Wiegand and Friend,[234] and Sanchez-Delgado,[235] have reviewed model studies of HDS reactions using organometallic compounds. The C-S bond scission reactions in organometallic compounds are presented here only briefly.

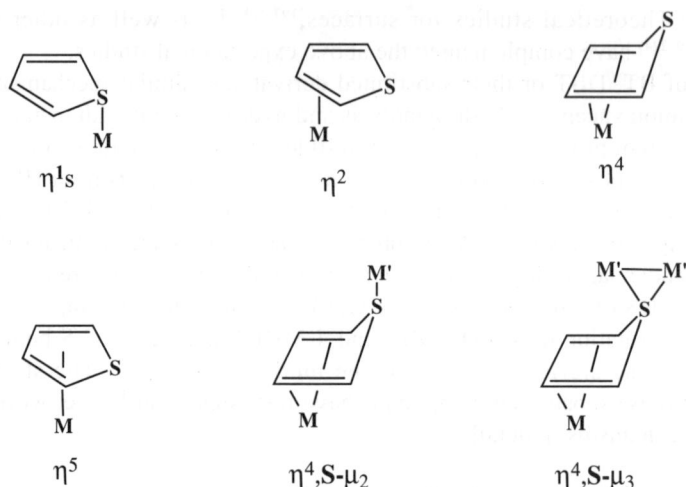

Fig. 2.45 Coordination modes of thiophene.

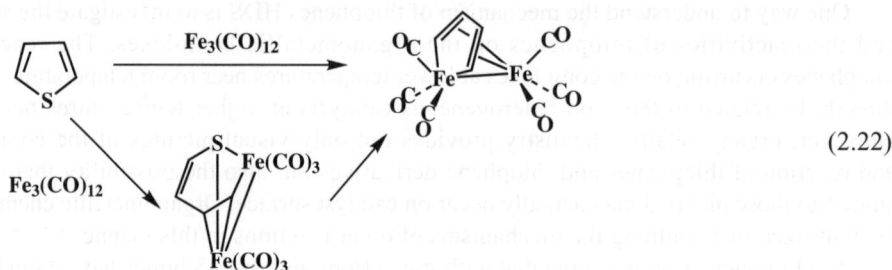

(2.22)

(From H. D. Kaesz et al., J. Am. Soc., **82**, 4749 (1960))

1. Reactions of Coordinated Thiophenes on Organometallic Complexes

Thiophenes may coordinate to metal centers in organometallic complexes before C-S bond scission occurs. Several coordination modes of thiophenes on complexes which could be models of adsorption modes of thiophenes on heterogeneous surface have been reported, and Fig. 2.45 shows some known coordination modes of thiophene in organometallic complexes (η^1-S,[236-238] η^2-C = C,[239] η^{4},[240-243] η^5,[240,244-253] η^4,S-μ_2,[242,254-256] η^4,S-μ_3,[256,257]). In the cases of BT and DBT, η^6 through the benzene ring is the major coordination mode. A few examples of η^1S-BT and η^1S-DBT are also known.[230-233] The η^5-coordination of BT, which is important as a model intermediate of heterogeneous HDS, has not yet been experimentally observed. However, some indirect experimental evidence as well as theoretical studies which suggest the importance of η^5-coordination through thiophene ring in BT have been published.[235]

Although C-S bond scission of coordinated thiophenes does not occur by itself, those coordinated thiophenes reacted with various nucleophiles to give products in which the C-S bond of thiophenes was cleaved. Angelici and coworkers[248,250] reported that nucleophiles such as OMe$^-$, SEt$^-$, CH(COOMe)$_2^-$ or H$^-$ attacked on CpRu(η^5-Th)$^+$ (Th = thiophene or a methyl-substituted thiophene) to lead to the C-S bond scission of coordinated thiophenes according to Eq.(2.23). It has been suggested that the reactions proceed as follows: The

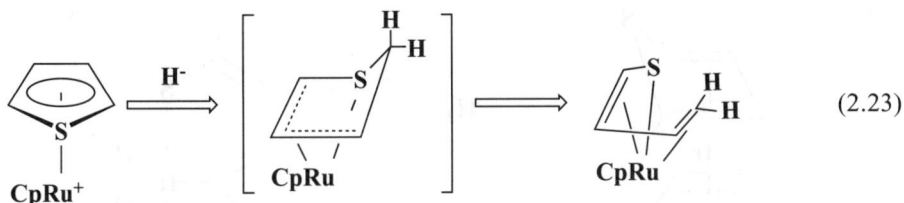

$$(2.23)$$

(From G. H. Spies and R. J. Angelici, *Organometallics*, **6**, 1899–1900 (1987))

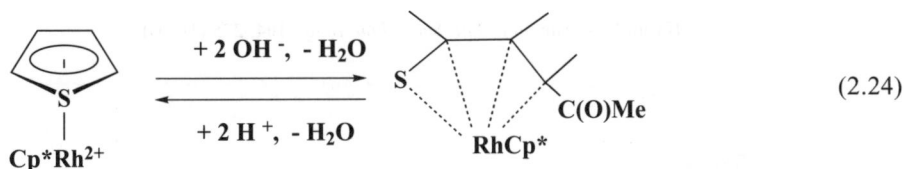

$$(2.24)$$

(From A. E. Skaugst, T. B. Rauchfuss and S. R. Wilson, *Organometallics*, **9**, 2876 (1990))

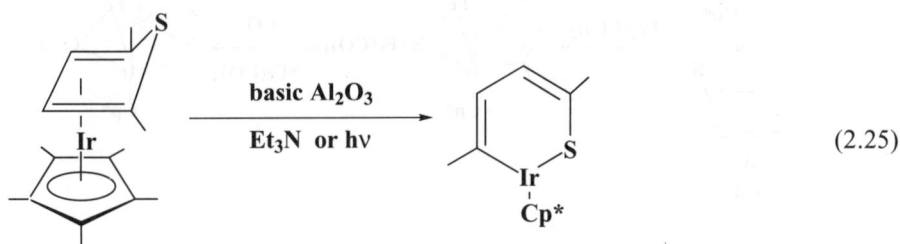

$$(2.25)$$

(From J. Chen *et al.*, *J. Am. Chem. Soc.*, **112**, 202 (1990))

addition of the nucleophile, hydride in this case, to the exo-C_2 position of thiophene yielded an allylsulfide intermediate which further reacted to give a butadienethiolate complex. In the reaction of [Cp*Rh(η^5-TMT)]$^{2+}$ (TMT = tetramethylthiophene) with aqueous KOH, Rauchfuss and coworkers reported the hydrolytic cleavage of the C-S bond. Thermolysis of the product shown in Eq.(2.24) produced desulfurized tetramethylfuran and [Cp*RhS].[258] The C-S bond scission of coordinated BT or DBT with nucleophiles is not well known.

Angelici and coworkers reported another type of C-S bond scission of coordinated thiophenes on iridium complexes.[255] As shown in Eq.(2.25), the reaction of Cp*Ir(η^4-2,5-Me$_2$T) catalyzed by suspended basic Al$_2$O$_3$ (but not neutral Al$_2$O$_3$) gave Cp*Ir(η^2-C,S-2,5-Me$_2$T) where the Ir was inserted into C-S bond of the thiophene. This reaction is catalyzed more slowly by Et$_3$N and by ultraviolet light.[259] X-ray and NMR studies support a structure for Cp*Ir(η^2-C,S-2,5-Me$_2$T) in which the π-system of the six-membered ring is delocalized and may best be described as an iridathiabenzene. The mechanism of the reaction is not known but such a ring opening could be catalyzed by basic sites on a catalyst surface. As shown in Eq.(2.26), the η^4 complex also reacted with H$_2$ to give the dihydride complex.[260] However, attempts to promote the migration of the hydride ligands to the thiophene yielded only unidentified products. Further, Cp*Ir(η^4-2,5-Me$_2$T) or Cp*Ir(η^2-C,S-2,5-Me$_2$T) reacted with Fe$_3$(CO)$_{12}$ to give a sulfide cluster in which a sulfur atom was released from the thiophene

a   s

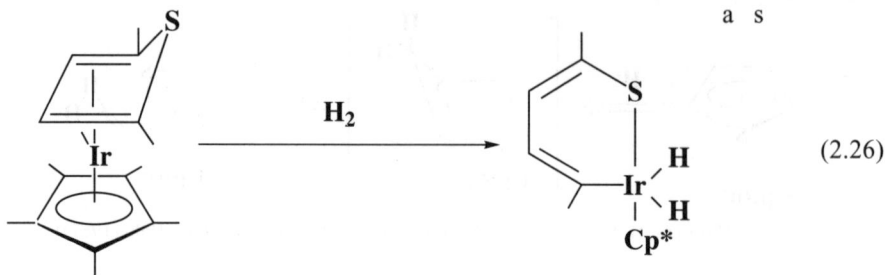

$$(2.26)$$

(From R. J. Angelici, *Bull. Soc. Chim. Belg.*, **104**, 275 (1995))

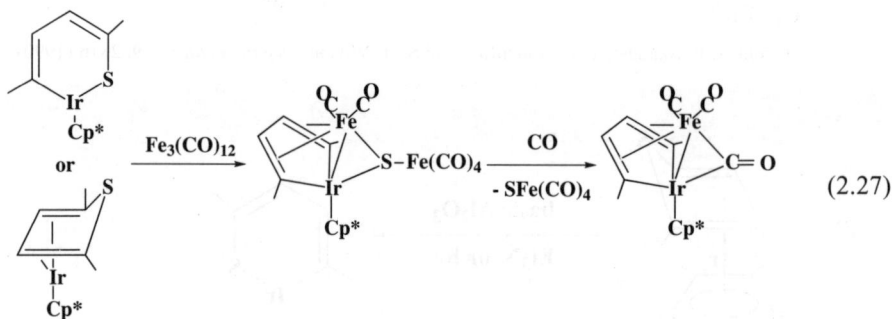

$$(2.27)$$

(From J. Chen *et al.*, *J. Am. Chem. Soc.*, **113**, 2544 (1991))

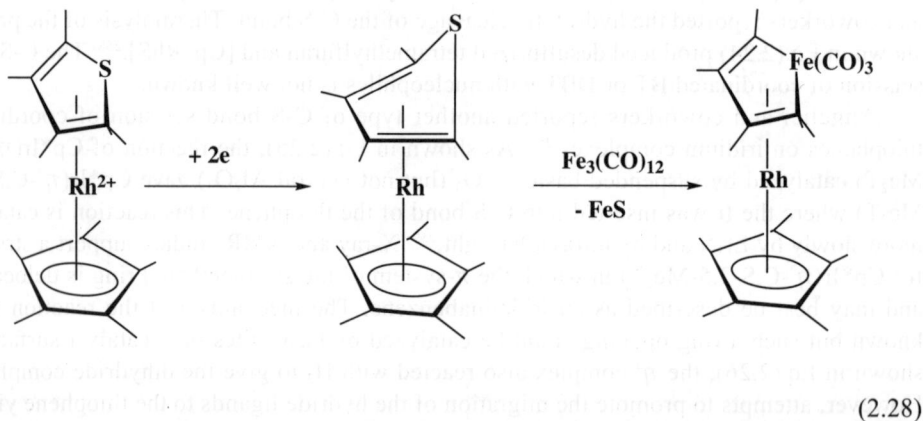

$$(2.28)$$

(From A. E. Ogilvy *et al*, *Organometallics*, **8**, 2740 (1989))

(2.29)

L=CO, P(OMe)₃, P(OEt)₃

(From C. A. Dullaghan *et al.*, *Angew. Chem. Int. Ed. Engl.*, **35**, 212 (1996))

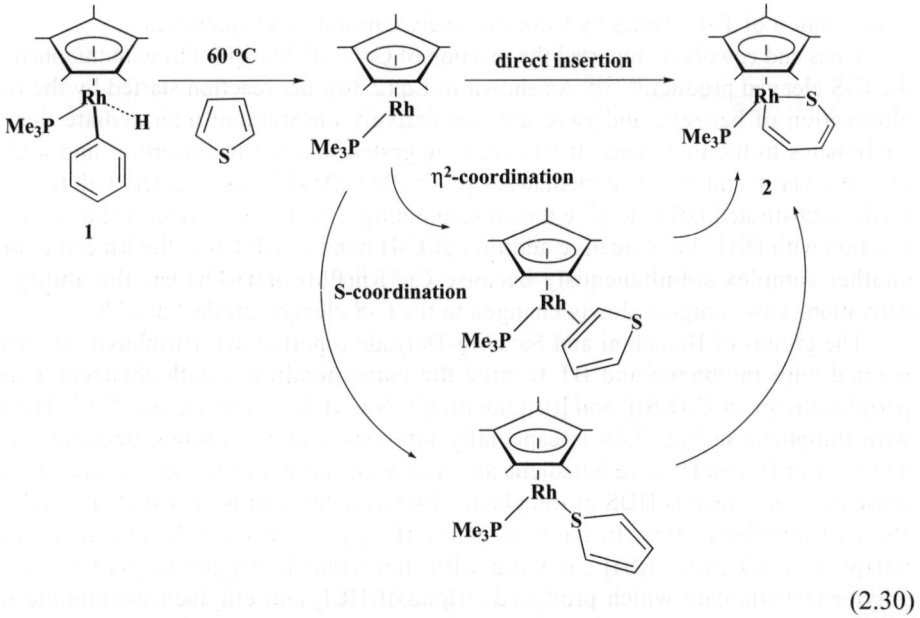

(2.30)

(From W. D. Jones and L. Dong, *J. Am. Chem. Soc.*, **113**, 562 (1991))

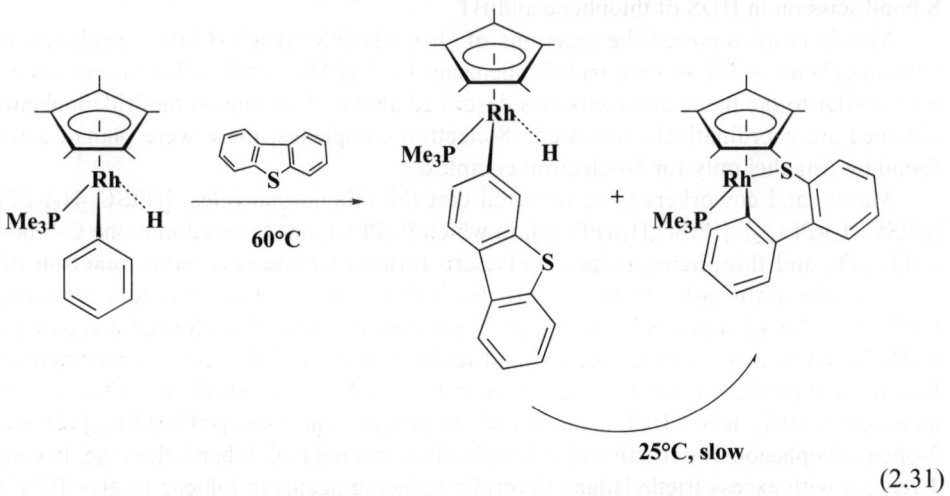

(2.31)

(From W. D. Jones and L. Dong, *J. Am. Chem. Soc.*, **113**, 561 (1991))

shown in Eq.(2.27).[256] This cluster reacted with carbon monoxide to give a complex free of sulfur. Rauchfuss and coworkers reported that $Cp*Rh(\eta^4-Me_4T)$ formed by the two-electron reduction of $[Cp*Rh(\eta^5-Me_4T)]^{2+}$ reacted with $Fe_3(CO)_{12}$ to give a bimetallic complex free of sulfur as shown in Eq.(2.28).[240] These reactions are the basis for a proposed thiophene HDS mechanism.

Recently, Dullaghan et al. reported that a highly nonplanar metallacyclic complex with a nucleophilic S atom was formed by reduction of the BT complex $[Mn(CO)_3(\eta^6-BT)]BF_4$ in the presence of additional ligands (L=CO, $P(OMe)_3$, $P(OEt)_3$) (Eq.(2.29)).[261]

2. Activation of Thiophenes by Coordinatively Unsaturated Complexes

Jones and coworkers reported the reaction of $Cp*Rh(PMe_3)(H)(Ph)$ with thiophene to give the C-S cleaved product.[262-264] As shown in Eq.(2.30), the reaction started by the reductive elimination of benzene and gave a coordinatively unsaturated intermediate that rapidly coordinates to the thiophene. It has been suggested that the Rh insertion into a C-S bond proceeds via η^2 and η^1-S intermediates. $Cp*Rh(PMe_3)(H)(Ph)$ also reacted with BT, DBT and methyl-substituted DBTs to give the corresponding C-S cleaved products (Eq.(2.31)). In the reaction with DBT, the oxidative addition of C-H bond of DBT into the Rh center produced another complex simultaneously because $Cp*Rh(PMe_3)(H)(Ph)$ has the ability of C-H activation. This complex slowly changed to the C-S cleaved product at 25°C.

The groups of Bianchini and Sanchez-Delgado reported that $[(triphos)Ir(\eta^4-benzene)]^+$ reacted with thiophene and BT to give the corresponding iridathiabenzene complexes $[(triphos)Ir(\eta^2C,S-C_4H_4S)]^+$ and $[(triphos)Ir(\eta^2C,S-C_8H_6S)]^+$, respectively.[265-267] The reaction with thiophene in Fig. 2.46 is especially interesting and important because successive additions of H^- and H^+ gave butadiene and hydrogen sulfide, which are the same products as those in heterogeneous HDS of thiophene. Therefore this can be an excellent model for the study of thiophene HDS. In the case of BT (Fig. 2.47), the C-S bond cleaved complex, $[(triphos)Ir(\eta^2C,S-C_8H_6S)]^+$, reacted with molecular hydrogen to give a coordinated ethylbenzenethiolate which produced $(triphos)IrHCl_2$ and ethylbenzenethiolate with the addition of HCl. These results suggest that hydrogenation does not need to occur prior to C-S bond scission in HDS of thiophene and BT.

Merola et al. reported the reactions of $[Ir(COD)(PMe_3)_3]Cl$ (COD = cyclooctadiene) with thiophene or BT to give iridathiabenzene $Ir(Cl)(PMe_3)_3(\eta^2C,S-Ts)$ (Ts=thiophene or BT) similar to the metallathiabenzenes described above.[268] Although the iridathiabenzenes obtained are coordinatively saturated 18-electron complexes, these were planar, a feature found by Angelici only for 16-electron complexes.

Maitlis and coworkers have reported that the thiaplatinacycles, $[(PtSC_{12}H_8)(PEt_3)_2]$, $[(PtSC_8H_6)(PEt_3)_2]$, $[(PtSC_4H_4)(PEt_3)_2]$, in which $Pt(PEt_3)_2$ has inserted into one C-S bond of DBT, BT, and thiophene, respectively, are formed by the reversible reaction of tris (triethylphosphine)platinum(0) with the thiophene.[269] The structure of complex $[(PtSC_8H_6)(PEt_3)_2]$ was confirmed by an X-ray determination that showed a square planar Pt(II) bound to two cis PEt3 ligands, and to the S and the CH=, in a six-membered ring. Reactions of $[(PtSC_{12}H_8)(PEt_3)_2]$ are shown in Fig. 2.48. $[(PtSC_{12}H_8)(PEt_3)_2]$ (1) reacted with an excess of HCl gas in $CDCl_3$ quantitatively to give cis- and trans-$[PtCl_2(PEt_3)_2]$ (2) and free 2-phenylthiophenol. The reaction of acid led to Pt-C but not to C-S bond cleavage. In contrast, 1 reacted with excess triethylsilane (hydridic reducing agent) in toluene to give BP (56%), DBT (32%), trans-$[Pt(SH)(H)(PEt_3)_2]$ (3, 50%), and cis-$[PtH(SiEt_3)(PEt_3)_2]$ (4, 31%). It has

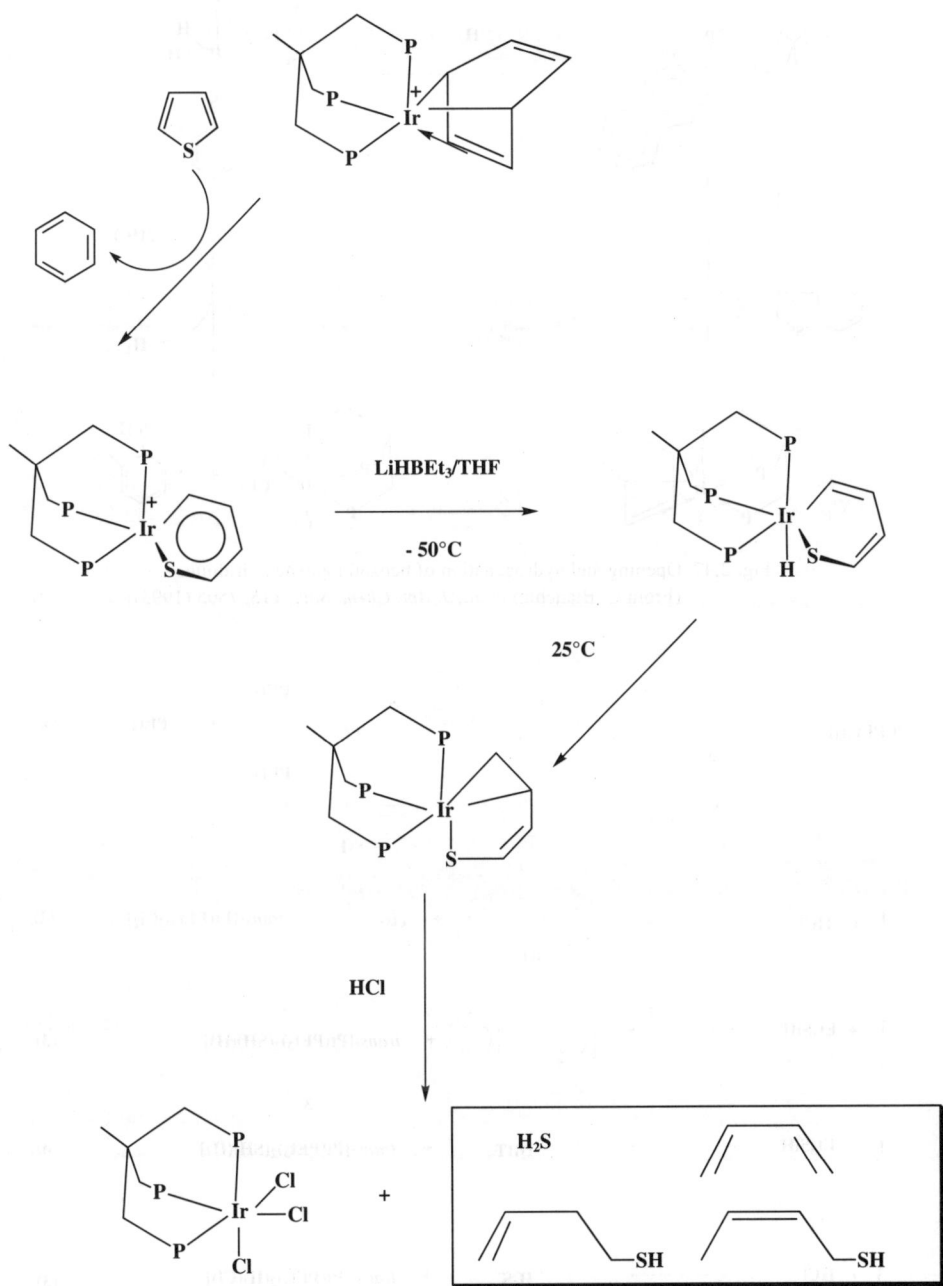

Fig. 2.46 Opening, hydrogenation and desulfurization of thiophene at iridium.
(From C. Bianchini *et al.*, *J. Am. Chem. Soc.*, **115**, 2735 (1993))

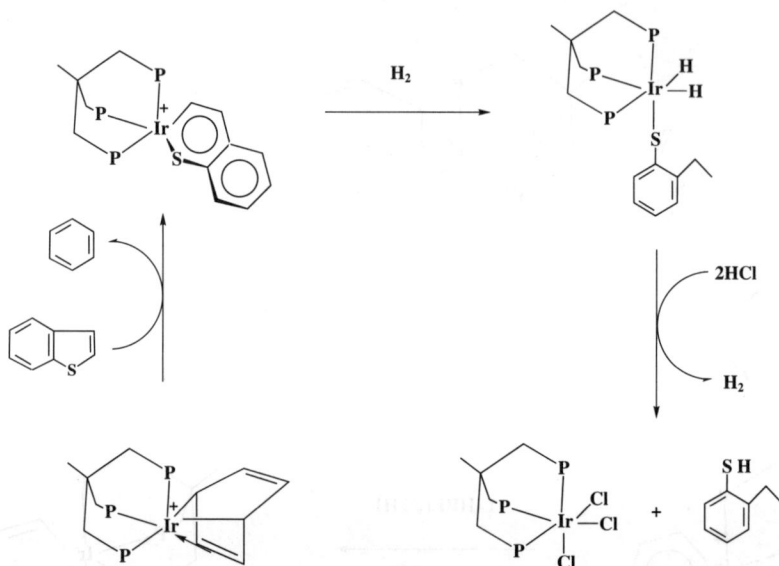

Fig. 2.47 Opening and hydrogenation of benzothiophene at iridium.
(From C. Bianchini *et al.*, *J. Am. Chem. Soc.*, **115**, 7505 (1993))

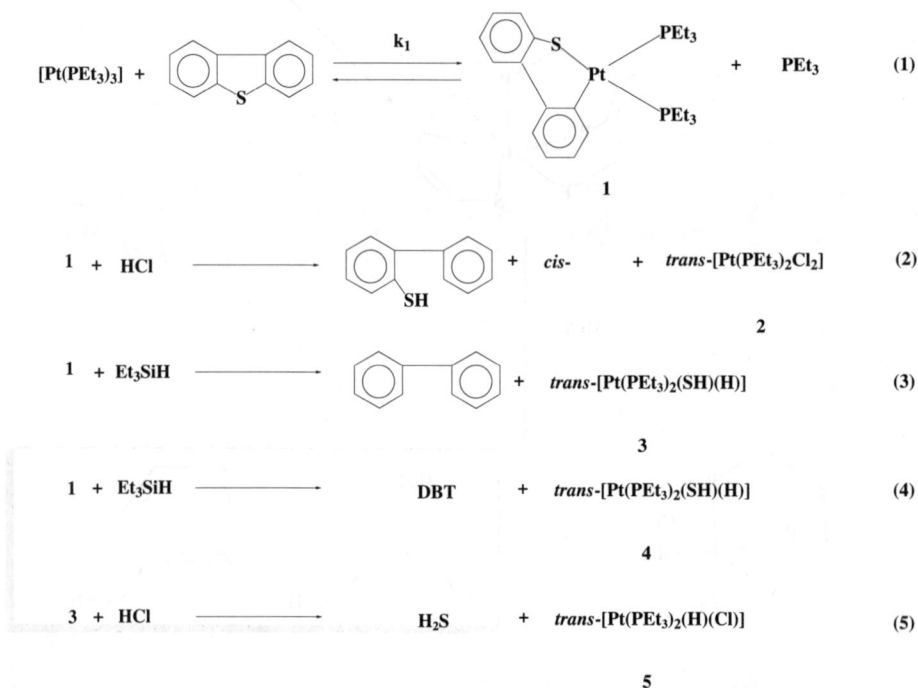

Fig. 2.48 Opening, hydrogenation and desulfurization of dibenzothiophene at platinum.
(From J. J. Garcia and P. M. Maitlis, *J. Am. Chem. Soc.*, **115**, 12200 (1993))

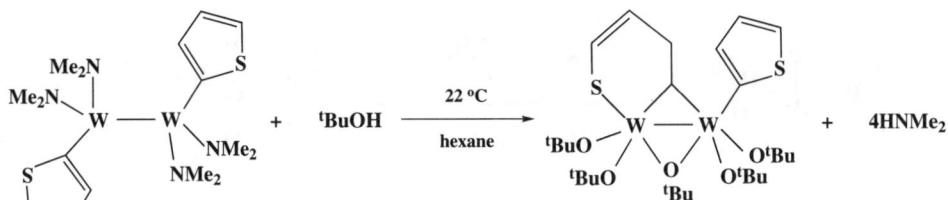

$$(2.32)$$

(From M. H. Chisholm et al., J. Am. Chem. Soc., **119**, 1636–1638 (1997))

been proposed that BP and **3** arise from HDS, while DBT and **4** arise via reversal of reaction 1. Addition of a saturated solution of HCl gas in CDCl$_3$ released H$_2$S (47%) from **3** and converted the complexes into a mixture of [Pt(Cl)(H)(PEt$_3$)$_2$] (**5**, 10%) and **2** (90%). The series of reactions with DBT in Fig. 2.48 are very similar to those in Fig. 2.46 because the successive additions of H$^-$ and H$^+$ gave BP and H$_2$S, which are the same products as those in heterogeneous HDS of DBT. Therefore this can also be an excellent model for the study of DBT HDS.

Chisholm et al. prepared a series of compounds of formula 1,2-M$_2$(σ-Th)$_2$(NMe$_2$)$_4$, **1**, where M = Mo and/or W and Th = 2-thienyl[2-Th], 3-thienyl[3-Th], 5-methyl-2-thienyl[2,5-MeTh], and 2-benzothienyl[2-BTh].[270] Addition of ^{t}BuOH or CF$_3$Me$_2$COH to hydrocarbon solutions of **1**, where M = W, leads to ring-opened products, **2**, when the thienyl ligand is attached via 2-carbon position. W$_2$(OR)$_5$(μ-CCH$_2$CHCHS)(σ-2-Th), **2**, where one of the 2-thienyl rings has been opened, has been fully characterized and shown to be derived from a

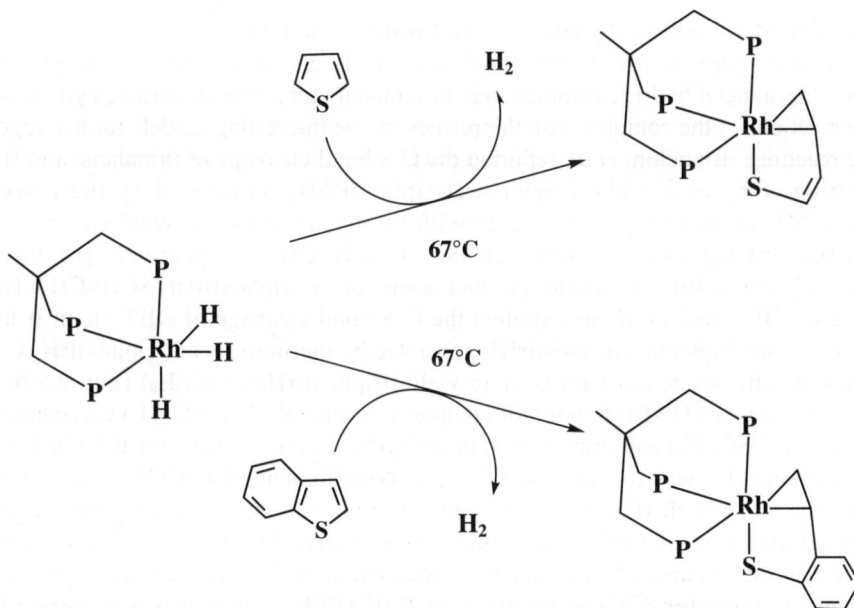

Fig. 2.49 Ring opening of thiophene and benzothiophene at rhodium.
(From C. Bianchini et al., J. Am. Chem. Soc., **117**, 4336 (1995))

C_2H_6

1   THF, 160°C, 4h   **2**

H_2S

(40%)

(60%)

SH

2 H_2(30atm)   170°C   **3**   +   **4**

12%   88%

ᵃ Conversion: 24h, 10mol of DBT/mol of 2

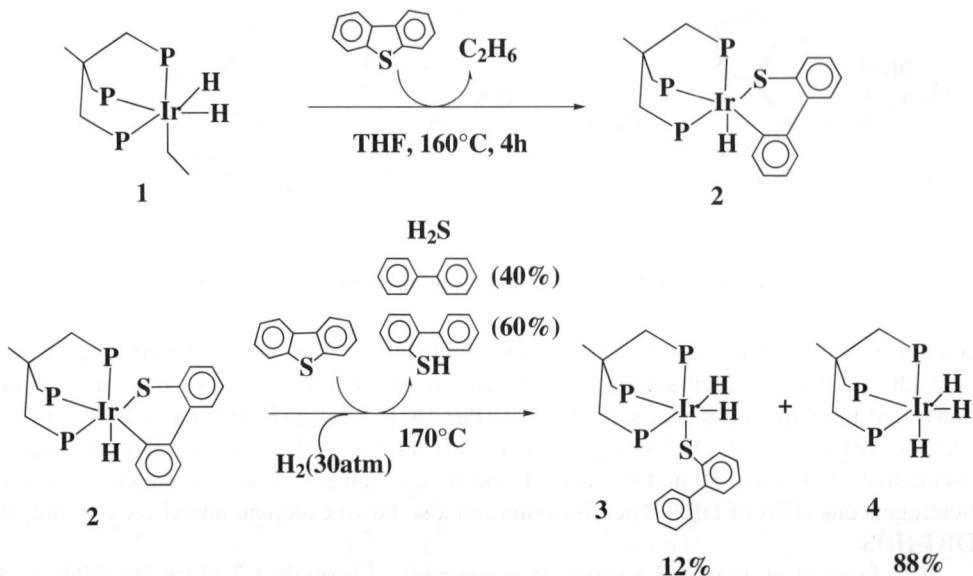

Fig. 2.50 Homogeneous hydrodesulfurization catalyzed by iridium complex.
(From C. Bianchini et al., *Organometallics*, **14**, 2347, 2350 (1995))

ring-opened μ-vinylidene intermediate $W_2(OtBu)_4(\mu\text{-CCHCHCHS})(\sigma\text{-2-Th})$ (Eq.(2.32)).

3. Activation of Thiophenes by Hydride and Cluster Complexes

Molecular hydrogen readily adsorbs dissociatively on surfaces to form surface hydride species. Thus a metal hydride complex may be a model compound of surface hydride species and the reactions of the complex with thiophenes can be interesting models for heterogeneous surface reactions. Bianchini et al. reported the C-S bond cleavage of thiophene and BT on a Rh hydride complex.[271] The fragment [(triphos)RhH], generated by thermolysis of (triphos)RhH₃ in refluxing THF, reacts with thiophene or BT to yield (triphos)Rh(η^3-SCH=CH-CH=CH₂) and (triphos)Rh{η^3-S(C₆H₄)CH=CH₂}, respectively [triphos=MeC(CH₂PPh₂)₃] (Fig. 2.49). The former product is similar to (triphos)Ir(η^3-SCH=CH-CH=CH₂) in Fig. 2.46. Bianchini et al. also studied the C-S bond cleavage of DBT on an Ir hydride complex.[272] The fragment [(triphos)IrH], generated by thermolysis of (triphos)IrH₂(C₂H₅) in THF at 170°C for 4h, reacts with DBT to yield (triphos)IrH(η^2-SC₁₂H₈) (Fig. 2.50). In the presence of excess of DBT (10 mol/mol-complex), (triphos)IrH(η^2-SC₁₂H₈) was converted to (triphos)IrH₂(η^1-SC₁₂H₉) and (triphos)IrH₃ under hydrogen pressure 30 atm and 170°C for 24h. In this reaction, DBT was converted to BP and 2-phenylthiophenol in yields of 40% and 60%, respectively, along with H₂S. This represents the first example for homogeneous catalytic hydrodesulfurization of DBT where the (triphos)IrH(η^2-SC₁₂H₈) complex served as an effective catalyst precursor. Rosini and Jones reported the C-S bond cleavage of thiophene on a Re hydride complex.[273] The reaction of ReH₃(PPh₃)₂, which was generated from ReH₇(PPh₃)₂ and 3,3-dimethyl-1-butene, with thiophene yielded a stable thioallyl complex through endo attack of one hydride to the C₂ position as shown in Eq.(2.33). The photolysis

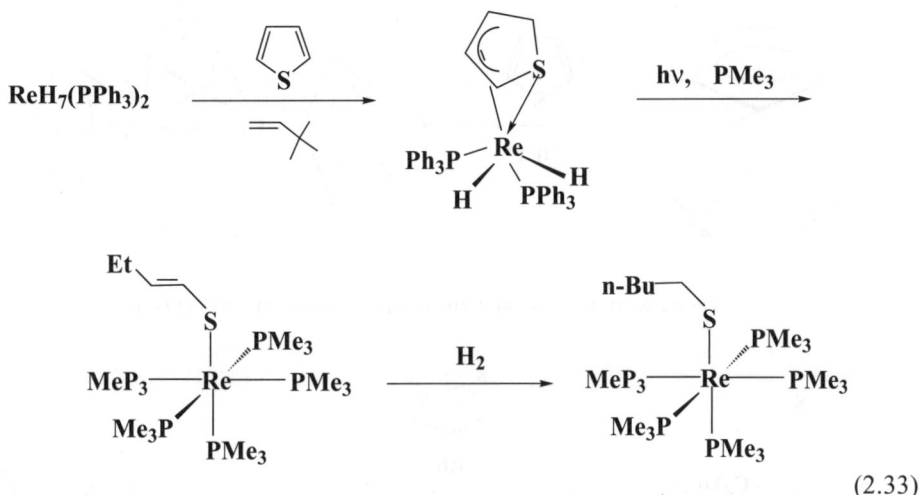

$$(2.33)$$

(From G. P. Rosini and W. D. Jones, *J. Am. Chem. Soc.*, **114**, 10768, 10771, 10772 (1992))

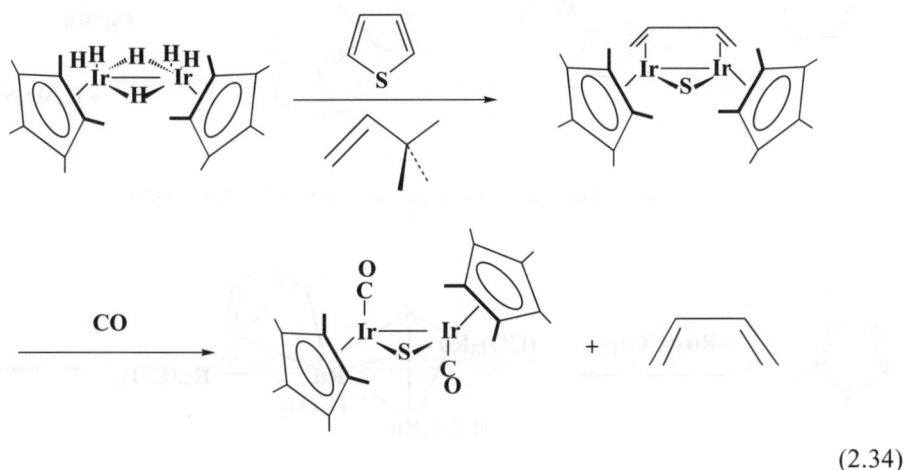

$$(2.34)$$

(From W. D. Jones and R. M. Chin, *J. Am. Chem. Soc.*, **116**, 201 (1994))

of the thioallyl complex in the presence of excess of PMe$_3$ resulted in the C-S bond scission producing 1-butenethiolate complex which underwent further hydrogenation yielding 1-butanethiolate complex. However, neither the elimination of 1-butanethiol from the Re center nor desulfurization of the species was possible and only one C-S bond of thiophene was activated in the mononuclear complex. Thus Jones and coworker used dinuclear iridium hydride complex.[274] As shown in Eq.(2.34), [Cp*Ir(H)$_2$(μ-H)]$_2$ reacted with thiophene in the presence of excess of 3,3-dimethyl-1-butene at 60°C to give Cp*$_2$Ir$_2$(μ-C$_4$H$_6$)(μ-S) in moderate yield in which butadiene and sulfide were separated with each of them bridging two iridium atoms. According to the isolated complex, only added olefin removed hydrides and hydrogenation of butadiene derived from thiophene does not seem to occur. Further, formation of tetrahydrothiophene was not observed. This butadiene-sulfide Ir complex reacted with

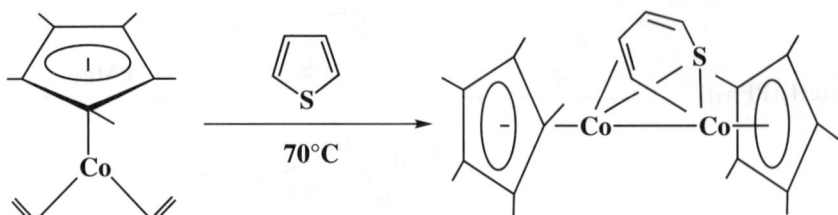

(2.35)

(From W. D. Jones, R. M. Chin, *Organometallics*, **11**, 2698 (1992))

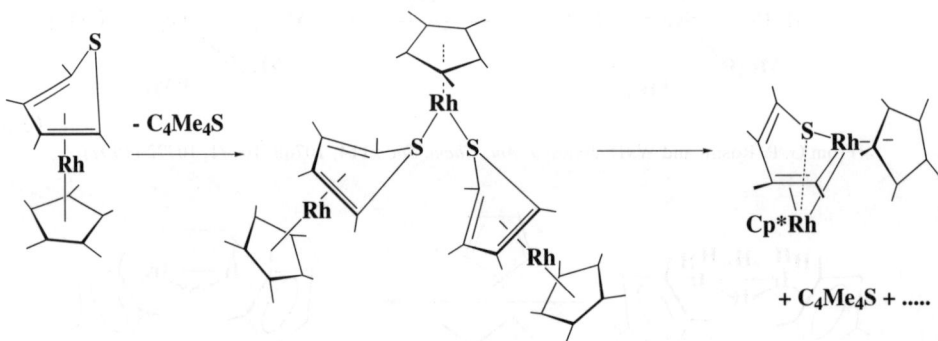

(2.36)

(From S. Luo *et al*, *J. Am. Chem. Soc.*, **114**, 1733–1734 (1992))

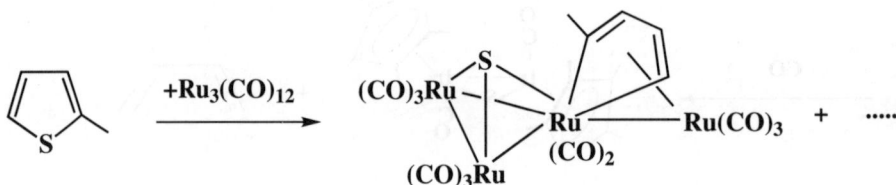

(2.37)

(From A. Arce *et al.*, *J. Chem., Dalton Trans.*, 2423 (1992))

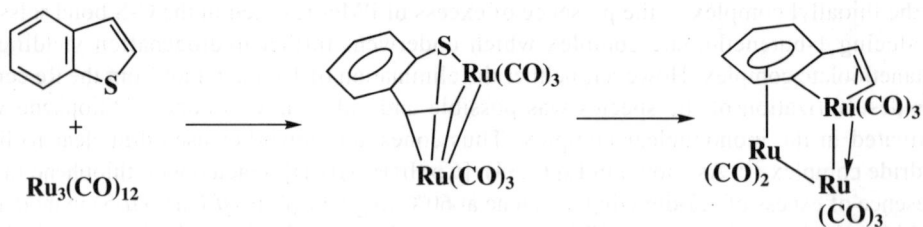

(2.38)

(From R. A. Sánchez-Delgado, *J. Mol. Catal.*, **86**, 303 (1994))

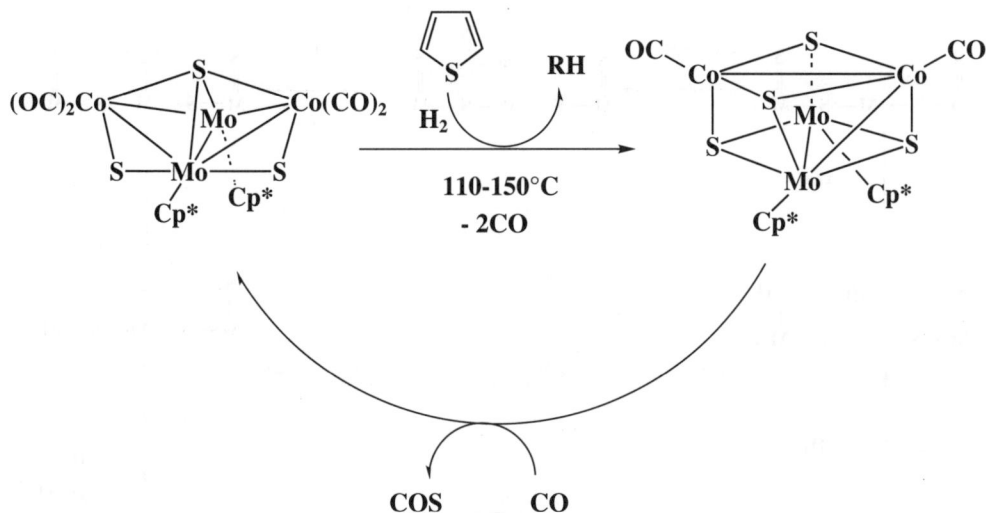

Fig. 2.51 Desulfurization of thiophene by a Co-Mo cluster.
(From U. Riaz et al., J. Am. Chem. Soc., **113**, 1416–1417 (1991))

carbon monoxide to give sulfide-carbonyl complex and butadiene at the ratio 1:1.

Reactions of thiophenes with metal clusters represent a homogeneous analogue for heterogeneous HDS catalysis. In cluster complexes, thiophenes may react simultaneously with more than one metal center; this is similar to several possible surface reactions. The studies of the reactivities of the sulfur-containing compounds on metal clusters shed light on our consideration of HDS mechanisms on heterogeneous surface. Some examples in which C-S bond scission of thiophenes occurs on more than one metal center are presented here. As shown in Eq.(2.35), CpCo(C$_2$H$_4$)$_2$ reacted with thiophene thermally to give the unusual dimer Cp$_2$Co$_2$(μ-T), in which one C-S bond was cleaved on one metal center, one olefinic part including α-carbon coordinated the other metal center and the sulfur bridged the two metal centers.[275] Rauchfuss and coworkers[276] reported that the thermal decomposition of Cp*Rh(η^4-C$_4$Me$_4$S) gave [Cp*Rh]$_3$(η^4,η^1-C$_4$Me$_4$S)$_2$ which decomposed to give Cp*$_2$Rh$_2$(μ-C$_4$Me$_4$S) with a structure similar to that of Cp$_2$Co$_2$(μ-T) in Eq.(2.35) (Eq.(2.36)). Rauchfuss and coworkers[277]

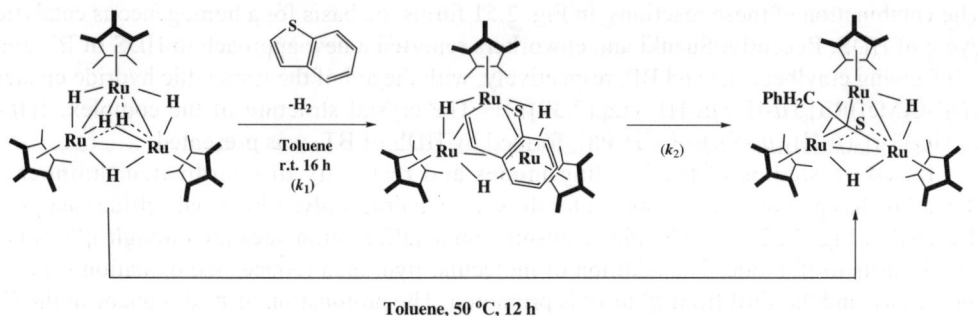

(2.39)

(From K. Matsubara et al., J. Am. Chem. Soc., **120**, 1108 (1998))

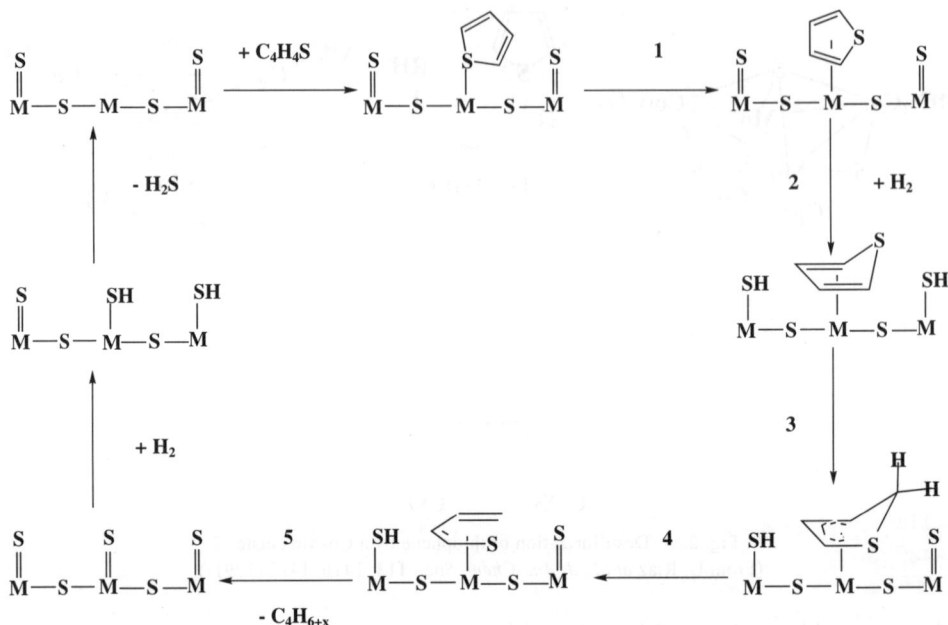

Fig. 2.52 Proposed mechanism for thiophene HDS by metal sulfide catalysts.
(From H. Topsøe, B. S. Clausen and F. E. Massoth, *Hydrotreating Catalysis*, p.125 (1996))

also studied the reactions of thiophene and BT with $Fe_3(CO)_{12}$ in detail, one of which was previously reported by Stone and coworkers as shown in Eq.(2.22).[227-229] Although the benzothiaferrole derivative was more readily produced, it did not convert to a benzoferrole. In contrast, a ferrole was readily obtained in the case of thiophene. Arce *et al.* reported that the reactions of thiophenes and BT with $Ru_3(CO)_{12}$ also gave desulfurized products as shown in Eqs.(2.37) and (2.38).[235,278] Curtis and coworkers reported desulfurization of thiophene on a soluble hetero bimetallic cluster $Cp^*_2Mo_2Co_2S_3(CO)_4$ (Fig. 2.51).[279,280] The complex reacted with thiophene under 15 atm of H_2 at 150°C to produce alkanes and alkenes with the carbon numbers of C_2-C_4 and $Cp^*_2Mo_2Co_2S_4(CO)_2$, which included sulfur from thiophene. This new cluster can be converted to the original cluster through the reaction with carbon monoxide. The combination of these reactions in Fig. 2.51 forms the basis for a homogeneous catalytic cycle of HDS. Recently, Suzuki and coworkers reported a new approach to HDS of BT and DBT giving ethylbenzene and BP, respectively, with the aid of the trimetallic hydride cluster $\{(\eta^5\text{-}C_5Me_5)Ru\}_3(\mu\text{-}H)_3(\mu_3\text{-}H)_2$ (Eq.(2.39)).[281] The crystal structure of the complex, $\{(\eta^5\text{-}C_5Me_5)Ru\}_3(\mu\text{-}H)_2(\mu_3\text{-}S)(\mu_3\text{-}CCH_2Ph)$, formed by HDS of BT, was presented.

Based on studies of the bonding modes and reactivity of coordinated thiophene, Rauchfuss has proposed a mechanism for thiophene hydrogenolysis by metal sulfide catalysts, depicted in Fig. 2.52.[77,233] Thiophene adsorbs on a sulfur anion vacancy through $\eta^1 S$ mode which shifts to η^5 mode. The addition of molecular hydrogen reduces the oxidation state of the surface and the shift from η^5 to η^4 is promoted. The protonation of η^4-thiophene at the C_2 position by a surface SH group cleaved the C-S bond to give butadienethiolate adsorbed on surface. Further, C-S cleavage with SH group forms C_4 products and the surface is oxidized to the original oxidation state. Gaseous hydrogen reacted with surface sulfide to give adsorbed hydrogen sulfide which desorbs from the surface to form sulfur anion vacancy again.

C. Approach by Metal Surface Chemistry

Surface science studies of the structures and the reactivities of thiophenes on metal single-crystal surfaces also provide one of the methods for understanding the mechanism of thiophenes HDS.[195,234] Salmeron and Somorjai and coworkers studied the temperature-programmed desorption (TPD) of thiophene and other molecules from clean, sulfided, and carbided Mo(100) surfaces.[282–286] The heat treatment of thiophene adsorbed on clean Mo(100) decomposed to carbon and sulfur deposits on the surface while preadsorbed sulfur blocked the decomposition of thiophene.

The related reactions of thiophene on various metal surfaces using extended X-ray adsorption fine structure (EXAFS), high-resolution electron-energy-loss spectroscopy (HREELS), TPD, etc. have been reported.[287–306] The nonselective decomposition of thiophene to atomic sulfur, carbon and gaseous dihydrogen also occurred on Re(0001),[287] Mo(100),[288] Mo(110),[289] and Ru(0001).[290] The formation of butadiene on Pd(111),[291] Pt(111),[292,293] and Rh(111),[294] and various products on Ni(111)[295] and W(211)[296] was observed during the reaction of thiophene. Thiophene was not activated on Cu(100).[297]

The coverage of thiophene and coadsorbed hydrocarbons and sulfur affects the orientation of thiophene with respect to metal surfaces. On most surfaces, thiophene usually adsorbs parallel to the metal surface at low coverage, while at high coverage it adsorbs perpendicular to the surface, or in a somewhat tilted configuration. These coverage dependences are probably due to intermolecular interactions.

Although the hydrogenation of thiophene has not been observed under ultrahigh vacuum conditions, butadiene was produced on Mo(110)[298] during the temperature-programmed reaction of 2,5-dihydrothiophene which may be formed by partial hydrogenation of thiophene under hydrogen pressure. In contrast, butadiene was formed in the reactions of 2,5-dihydrothiophene and thiophene on Rh(111) even under ultrahigh vacuum conditions, indicating that 2,5-dihydrothiophene may be an intermediate in the desulfurization of thiophene and that Rh(111) has a higher hydrogenation activity for ring than Mo(110).[294] Roberts and Friend reported that in TPD of tetrahydrothiophene and butanethiol on Mo(110), the same butylthiolate intermediate which gave either butane by hydrogenation at the C_α position or butene by β-H elimination was found for both adsorbates.[302]

It has been reported that the metal surface acts as not only an electron donor[303] but also as an electrophile in the reaction of sulfur compounds.[295,296] Benziger and Preston investigated the TPD of methanethiol and methanol on clean, oxidized, carbided, and sulfided W(211) surfaces.[303] A clean metal surface completely decomposed the adsorbate to form a metal oxide, carbide, or sulfide while partial oxidation of the metal surface with oxygen, carbon, or sulfur decreased the surface reduction potential and stabilized CH_3-O and CH_3-S species. The results show that a clean metal surface can act as a strong reducing agent. In contrast, it was indicated by same authors that the highest occupied molecular orbital of thiophene had the highest electron density on the α-carbon atom and that sulfur anion vacancies would promote electrophilic reactions because of their electron deficiency.[296]

Somorjai and coworkers studied thiophene desulfurization on Mo(100) single-crystal surfaces using surface science techniques for the characterization of substrate and adsorbates.[304–306] The products were butadiene, butene and butane. H₂S inhibited HDS while butene did not affect the catalytic activity. These results from reactions on a single-crystal surface were very similar to those on supported MoS_2, indicating that single-crystal surface studies can be regarded as a good model reaction of actual heterogeneous HDS.

2.2 Hydrodenitrogenation

2.2.1 Kinetics of Catalytic Hydrodenitrogenation Reaction

The increasing demand for clean oils such as gasoline, kerosene and light gas oil is giving rise to an increase in the amounts of much heavier feedstocks such as vacuum gas oil (VGO) and residue, and light cycle oil (LCO) produced from VGO and residue. This means that not only hydrodesulfurization (HDS) but also hydrodenitrogenation (HDN) of those heavier feedstocks is required to obtain the lighter clean oils which can meet strict environmental regulations. Especially in the hydroprocessing of heavier feedstocks, HDN becomes more important because the concentration of the nitrogen-containing compounds in the heavier feedstocks is much higher than that in straight run distillates, and the acidic catalysts used for catalytic cracking of the heavier feedstocks are poisoned by nitrogen compounds and polyaromatics. Many reviews on this subject have been published for different aspects of HDN.[53,77,307–314]

Nitrogen compounds included in the feedstocks are divided into two types: heterocycles and nonheterocycles. In general, nonheterocyclic nitrogen compounds such as aliphatic amines and nitriles are present in smaller content and are denitrogenated more easily than heterocyclic compounds. Among nonheterocyclic compounds, aniline derivatives are important in HDN because they always appear in the HDN network of heterocycles and are more difficult to denitrogenate than aliphatic ones. Heterocyclic nitrogen compounds are most important in HDN because they are included in feedstocks in larger amounts and are most difficult to remove. Heterocyclic nitrogen compounds can be divided into basic compounds and nonbasic compounds. Basic compounds include six-membered heterocycles such as pyridine, quinoline, acridine, etc. The lone-pair electrons on the nitrogen atom of these compounds are not tied up in the π-cloud of hetrocyclic ring and are available for sharing with acids. Nonbasic compounds include five-membered heterocycles such as pyrrole, indole, carbazole, etc. The lone-pair electrons on the nitrogen atom of these compounds, in contrast, are delocalized around the aromatic ring and are not available for donation to acids.

Conventional HDS catalysts, $Ni-Mo/Al_2O_3$ and $Co-Mo/Al_2O_3$, are also used for HDN, and $Ni-Mo/Al_2O_3$ is usually more active for HDN as hydrogenation of a ring containing nitrogen is needed prior to C-N bond scission. Under industrial conditions, hydrogen sulfide is always present because HDS always occurs under HDN conditions. Therefore, HDN on sulfided catalysts was performed in the absence and presence of hydrogen sulfide to elucidate the kinetics and reaction networks. When hydrogen sulfide is absent, the reaction profiles may be changed because sulfided catalysts are reduced by hydrogen especially under high pressure.

A. Thermodynamics and Reactivities

Hydrogenation of a N-containing ring occurs prior to C-N bond scission over conventional catalysts. This means that in order to reduce the strong bond energy of C-N bond in N heterocycles (C = N, 615 kJ/mol, C-N, 305 kJ/mol),[315] ring hydrogenation is needed in HDN of N heterocycles. When the C-N bond in a heterocycle has higher bond energy, the activation energy of C-N bond-breaking may be higher. The HDN rate can be affected by the equilibrium of N-ring hydrogenation because N-ring hydrogenation occurs before nitrogen removal. Lower equilibrium constant for hydrogenation, or lower concentrations of nitrogen compounds and hydrogen pressure reduce the equilibrium concentrations of hydrogenated N-ring compounds and thus decrease the HDN rate.

Cocchetto and Satterfield reported thermodynamic equilibrium calculations for a number of HDN reactions.[316,317] In general, at higher temperatures the equilibrium between ring hydrogenation and dehydrogenation shifts to dehydrogenation leading to a drop in the equilibrium concentration of a hydrogenated compound and thus a decrease in nitrogen removal rate. Satterfield and Cocchetto[318] observed that under such a condition, the overall rate of pyridine HDN showed a maximum with an increase in temperature. In the reactions occurring in quinoline and indole reaction networks, the equilibrium constants are less than unity for ring hydrogenation and decrease with increasing temperature, indicating that ring hydrogenation and hydrogenolysis or overall HDN are exothermic. However, the hydrogenolysis and overall HDN equilibria are favorable even at higher temperatures.[316,317] Therefore, HDN is virtually irreversible under the reaction conditions used in current commercial hydrotreating reactors and is generally controlled by kinetics rather than by thermodynamics.

Hydrodesulfurization does not always require ring hydrogenation. As mentioned above, however, N heterocycles are hydrogenated then undergo C-N bond scission and overall HDN. In general, since HDN is more difficult to occur than HDS and heavier feedstocks are treated in HDN, the HDN reactions are operated under more severe conditions, i.e. higher temperature and hydrogen pressure, than HDS. Hydrogenation reactions are more sensitive to hydrogen pressure than hydrogenolysis.[319] These facts result in the formation of various products through ring hydrogenation as well as those through hydrogenolysis. Therefore, the conversions of N heterocycles must be distinguished between those for ring hydrogenation, hydrogenolysis of C-N bond and overall HDN forming hydrocarbons and ammonia. The reactivities for N-ring hydrogenation have been reported by several authors. Schulz et al. reported that the reactivities for N-ring hydrogenation decreased in the order quinoline > pyridine > isoquinoline > indole > pyrrole, and that the reactivities for ring hydrogenation decreases in the order N-ring > aniline-like compounds > comparable aromatic.[309] Cocchetto and Satterfield reported that although in the thermodynamic analysis of the N heterocycles benzenoid ring hydrogenation is slightly more favored than N-ring hydrogenation,[317] the latter is always faster than the former due to kinetic factors.[317,320] The hydrogenation rates of pyrrole, indole and carbazole have been compared using Ni-Mo/Al$_2$O$_3$ and n-hexadecane solvent in a batch reactor at 350°C and 68 atm.[320] The first-order rate constants decreased with an increase in the number of the ring in the order pyrrole > indole > carbazole, but an explanation was not given. The effects of methyl substituents and position of benzenoid ring on hydrogenation of quinoline and benzoquinoline derivatives have also been reported and the steric effect as well as electronic effect has been discussed.[321,322] However, the effects of reaction conditions, kinds of catalysts and the relationship between hydrogenation and hydrogenolysis need to be investigated before a conclusion is reached on this subject.

In general, the steric effect is not observed in HDN reaction so much as HDS probably because C-N bond scission and successive N removal occur only after hydrogenation of N-

Table 2.15 Reactivities of Several Six-membered Heterocyclic Nitrogen Compounds [322]

Reactant	Structure	Pseudo-first-order rate constants (min^{-1})
Pyridine		20.98
Quinoline		6.62
Acridine		2.20
Benz[a]acridine		0.97
Benz[c]acridine		1.54
Dibenz[c,h]acridine		4.21

Batch reactor, React. Temp. 367°C, 136 atm, paraffinic white oil solvent, concentration of reactant: 0.5 wt%, sulfided Ni-Mo/Al$_2$O$_3$.
(From Z. Sarbak, *Acta Chim. Hung.*, **127**, 374 (1990))

ring, and the steric effect is extremely weakened. It is also suggested that N compounds may be adsorbed on the catalyst surface not with the end-on mode through N atom but with the side-on mode through π electrons of N-ring. As reactions of basic nitrogen compounds with Lewis acids are affected by steric hindrance,[310] strong steric hindrance may be observed with the end-on adsorption mode. In contrast, hydrogenation precedes hydrogenolysis in HDN and the steric effect is not important with the side-on adsorption mode. Therefore, the side-on mode is supported in HDN reaction. This is consistent with the fact that when 4-methyldibenzothiophene and 4,6-dimethyldibenzothiophene were hydrogenated prior to hydrogenolysis, the steric effect disappeared and cyclohexylbenzene derivatives were formed in the same yield as in the case of nonsubstituted dibenzothiophene (see Section 2.1.3).

HDN of six-membered nitrogen heterocyclic compounds has been compared using sulfided Ni-Mo/Al$_2$O$_3$ and paraffinic white oil solvent in a batch reactor at 367°C and 136 atm.[193,194,323] As shown in Table 2.15, the pseudo-first-order rate constants for HDN decreased in the order pyridine (20.98 min^{-1}) > quinoline (6.62) > dibenz[c,h]acridine (4.21) > acridine (2.20) > benz[c]acridine (1.54) > benz[a]acridine (0.97). The difference between each was very small especially among 3- and 4-ring compounds. It was inferred that the rates of hydrogenation were higher than those of C-N bond scission. The results suggest that the rates of the C-N bond scission were similar to each other because the steric effect was nearly equivalent after N-ring hydrogenation. The effect of methyl substituents on HDN rates of quinolines has been reported using Ni-Mo/Al$_2$O$_3$ and n-hexadecane solvent in a batch reactor at 350°C and 34 atm.[324] Although the pseudo-first-order rate constant of nonsubstituted quinoline was slightly higher than that of dimethyl-substituted quinolines, the ratio of rate constants was less than 2, indicating that the steric effect was equivalent. There are researches concerning the substituents effect on pyridine HDN.[325,326] However, it is difficult to compare the data because the reaction conditions and catalysts may be different or unclear.

B. Reaction Networks and Kinetics

Reaction networks and kinetics for HDN of pyridine, quinoline, acridine, indole, several anilines and their derivatives are described here in order. Major common pathways of N heterocycles are as follows: 1) hydrogenation of N-ring, 2) C-N bond scission to an amine, and 3) hydrogenolysis of amine to hydrocarbons and ammonia.

1. Pyridine

The reaction network of pyridine in Fig. 2.53 has been reported using Ni-Mo/Al$_2$O$_3$ in a vapor-phase reactor.[327–331] Pyridine is hydrogenated to piperidine of which one C-N bond is cleaved to form n-pentylamine. n-Pentane and ammonia are formed by hydrogenolysis of the amine. The HDN rate of n-pentylamine is one order of magnitude faster than that of pyridine. N-alkylation of piperidine occurs to a minor extent.[328,329] The reaction of piperidine does not form pyridine under the conditions used (for example, vapor-phase flow reactor, 310°C, 12–99 atm, Ni-Mo/Al$_2$O$_3$ [328]), indicating that hydrogenation of pyridine is irreversible. Hanlon performed experiments with pyridine, piperidine and n-pentylamine and determined the adsorption parameters of these compounds and ammonia using a one-site Langmuir-Hinshelwood rate equation.[328] The values of pyridine, piperidine, n-pentylamine and ammonia

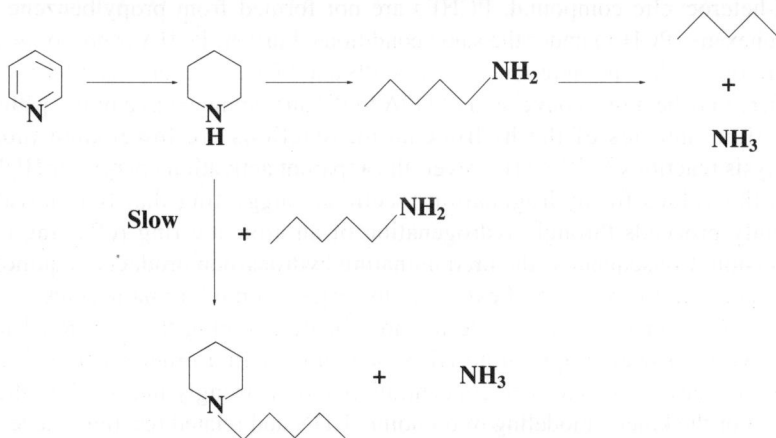

Fig. 2.53 Reaction pathways for pyridine HDN.
(From R. T. Hanlon, *Energy & Fuels*, **1**, 424 (1987))

were 3×10^4, 6×10^4, 10^5 and 5000 atm^{-1}, respectively. The value of the adsorption parameter for n-pentylamine was the largest among the four. That of ammonia was somewhat lower than those for the others. These results suggest that the inhibition effects by produced amine derivatives as well as the self-inhibition by pyridine are quite large. In contrast, McIlvried estimated the retarding effect of ammonia using different rate equations in hydrogenolysis of piperidine and hydrogenation of pyridine.[327] HDN rate of piperidine decreased with increasing concentration of piperidine due to self-inhibition. The adsorption parameters for piperidine and ammonia were the same, 23 atm^{-1} at 316°C. It was assumed that hydrogenation of pyridine was retarded by ammonia and the estimated adsorption parameter was 150 atm^{-1} at 316°C. The adsorption parameters of ammonia for these two reactions are different from each other, indicating that hydrogenation and hydrogenolysis proceed on separate sites. In these calculations, the adsorption of n-pentylamine was ignored.

2. Quinoline

The behavior of quinoline to HDN is representative of that of many nitrogen heterocycles contained in the products of petroleum. A representative reaction network for quinoline HDN has been proposed in Fig. 2.54 by Satterfield and Yang.[317,332] Quinoline HDN begins with the initial hydrogenation of one or both of N-ring and benzenoid ring. Hydrogenation of quinoline to 1,2,3,4-tetrahydroquinoline (14TQ) is much more rapid than that to 5,6,7,8-tetrahydroquinoline (58TQ). Hydrogenation of quinoline to 14TQ is also much faster than hydrogenolysis of 14TQ to o-propylaniline (OPA) and hydrogenation of 14TQ to decahydroquinoline (DHQ). Hydrogenolysis of 14TQ to OPA is slower than its hydrogenation to DHQ. As a result, the saturation of the nitrogen heteroring will approach equilibrium and further nitrogen removal proceeds preferentially through DHQ. Nitrogen removal of OPA generally requires hydrogenation of the aromatic ring to eliminate the resonance interaction between the aromatic ring and the nitrogen-lone-pair electrons.

OPA is hydrogenated to give a reactive propylcyclohexylamine intermediate[320,333,334] which yields propylcyclohexenes (PCHEs) and ammonia. The conversion of propylcyclohexylamine is very fast while hydrogenation of OPA is slower than that of the original N-heterocyclic compound. PCHEs are not formed from propylbenzene (PB) and propylcyclohexane (PCHA) under the same conditions. Further, PCHA is not formed from PB either. Therefore, hydrocarbon products such as PB and PCHA, etc. are mainly produced from PCHEs. There may be direct conversion of OPA to PB and ammonia as a minor pathway.[335,336] The activation energies of the hydrogenation reactions are lower than those of the hydrogenolysis reactions.[319,332,337] However, the apparent activation energy for HDN of OPA is close to the values for hydrogenation reactions, suggesting that N removal of OPA predominantly proceeds through hydrogenation of an aromatic ring reflecting the rate of OPA conversion. Consequently, the predominating hydrocarbon products in quinoline HDN is PCHA rather than PB because of extensive hydrogenation of aromatic rings.[92,319,320,338,339]

Pseudo-first-order kinetics are one method for determining the individual rate in the reaction network. However, quinoline HDN does not fit first-order kinetics because it is affected by the conversion and feed concentration and is strongly inhibited by the nitrogen compounds. For the kinetic modeling of quinoline HDN and related reactions, several authors correlated their data with Langmuir-Hinshelwood (L-H) rate equations. Satterfield and Yang[332] performed HDN reactions of quinoline and various intermediate reaction products over a commercial Ni-Mo/Al$_2$O$_3$ (350°C, 375°C and 390°C and 6.9 MPa). The overall HDN

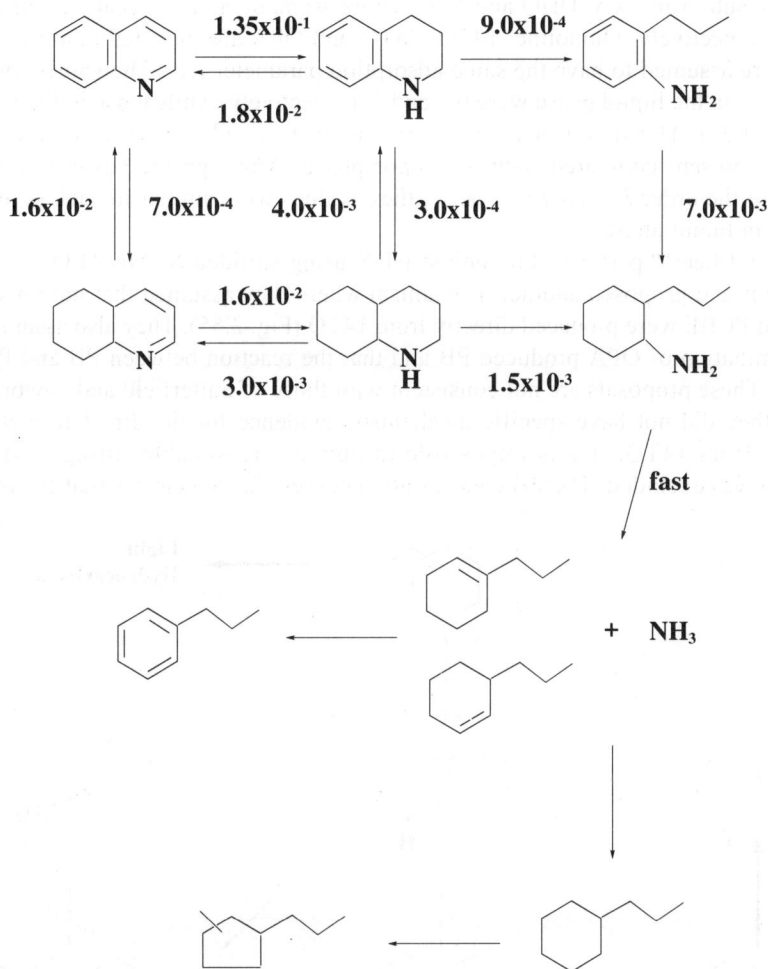

Fig. 2.54 Reaction pathways for quinoline HDN.
The numbers next to the arrows are rate constants for the liquid phase reaction at 375 °C with 5 wt%
quinoline and 0.59 wt% CS_2 in the feed in units of mol/(g of catalyst · h).
(From C. N. Satterfield and S. H. Yang, *Ind. Eng. Chem., Process Des. Dev.*, **23**, 13 (1984))

reaction is essentially zero order under the conditions probably because the surface coverage
of N compounds is very high. The rate constants in the reactions included in the network were
estimated. The values in the liquid phase reaction which are shown in Fig. 2.54 were very
similar to those in the vapor phase reaction. The two hydrogenolysis rate constants were
smaller than those for the hydrogenation by factors of 2.7 to 150. Although the formation of
14TQ is faster than that of any other N compound, its reactivities for hydrogenation and C-
N bond scission are relatively slow compared with other compounds included in the network.
This shows that the control of 14TQ HDN is important to control quinoline HDN. The same
authors also estimated the ratios between the adsorption equilibrium constants, K_{AA}, K_{DHQ} and

K_{NH_3}, where subscripts AA, DHQ and NH_3 denote aromatic amine, decahydroquinoline and ammonia, respectively. Quinoline, 14TQ, 58TQ and OPA are included in aromatic amines (AA) and are assumed to have the same adsorption parameter K_{AA}. The values for K_{NH_3}/K_{AA} and K_{DHQ}/K_{AA} in the liquid phase were 0.7 and 2, respectively, while those in the vapor phase were 0.25 and 6. The liquid phase tends to equalize the adsorptivities of the various N compounds present compared with the vapor phase. Although the adsorption parameter decreased in the order $K_{DHQ} > K_{AA} > K_{NH_3}$, these values were similar to each other within a factor of 3 in liquid phase.

Gioia and Lee[340] performed quinoline HDN using sulfided Ni-Mo/Al$_2$O$_3$ at 350°C in a batch reactor and proposed another mechanism where they assumed that not only OPA but also PB and PCHE were produced directly from 14TQ (Fig. 2.55). They also assumed that the direct deamination of OPA produced PB and that the reaction between PB and PCHA was reversible. These proposals are not consistent with those of Satterfield and coworkers.[332,339] Although they did not have specific mechanistic evidence for the direct formation of PB and PCHE from 14TQ, it was impossible to obtain a reasonable fitting of data if this assumption was excluded. The PB concentration curves showed clearly that PB is produced

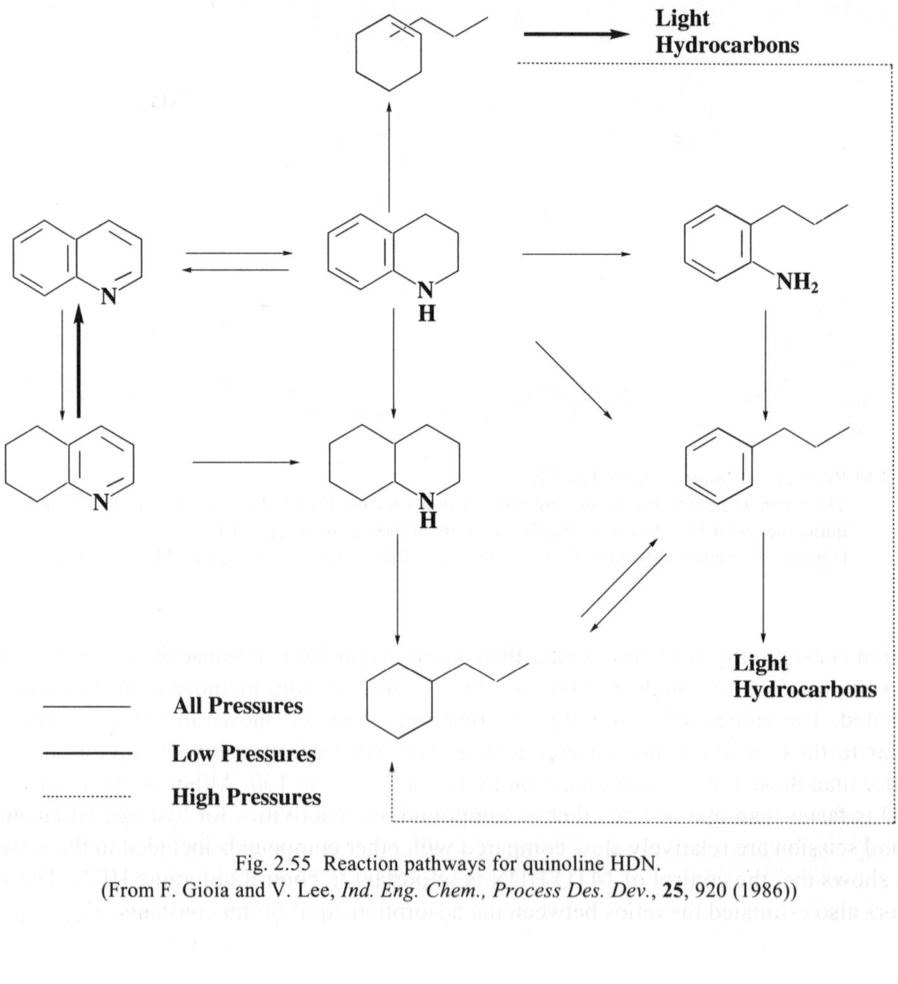

Fig. 2.55 Reaction pathways for quinoline HDN.
(From F. Gioia and V. Lee, *Ind. Eng. Chem., Process Des. Dev.*, **25**, 920 (1986))

at an early stage of the reaction. The direct formation of PB from OPA without a ring hydrogenation of OPA was also strongly supported by data fitting. Not all the reactions occurred at each pressure and the networks were different between low (< 31 atm) and high (> 78 atm) pressures. They also pointed out that under practical conditions, only the dehydrogenation route of 14TQ to quinoline was significant. Further, significant cracking of PCHA and PB to light hydrocarbons was observed at low pressures. Some investigators have also correlated their data successfully by neglecting the dehydrogenation reactions.[324,338]

Shih et al.[319] performed quinoline HDN using sulfided Ni-Mo/Al$_2$O$_3$ in a batch reactor and assumed two kinds of catalytic sites for hydrogenation and hydrogenolysis. The pseudo-first-order rate equations were correlated with L-H rate equations where all nitrogen-containing compounds were assumed to have the same adsorption parameter. The adsorption parameters estimated for hydrogenation and hydrogenolysis separately were different by a factor of 100, indicating that two kinds of sites are actually present.

Miller and Hineman [341] performed quinoline HDN using Co-Mo/Al$_2$O$_3$ and Co-Mo/USY-Al$_2$O$_3$ at 330°C and 375°C, and 65 atm in a trickle-bed reactor. To estimate the overall HDN reaction, they also used a L-H rate expression where quinoline, 14TQ and DHQ were assumed to have the same adsorption parameter, but adsorption of ammonia was neglected. Although the kind of catalyst and the reactor system in this experiment were different from those of Shih et al.[338] and Gioia and Lee,[340] the differences between the values for adsorption parameters were within one order of magnitude.

As shown above, several investigators tried to determine kinetics in complicated reaction networks by using considerably simplified L-H equations. However, it is still difficult to establish a firm quantitative picture of the network and to generalize the rate expressions. There are many stable reaction intermediates and products and these compounds often have very different reactivities. Further, the rate expressions have been estimated separately and the differences between those expressions have not been considered in one investigation. For example, although the experimental conditions were very similar except reactors, Satterfield and Yang[332] neglected the term of unity in the denominator of the L-H equation while Gioia and Lee [340] did not remove this term. One-site rate expressions[332,340,341] were usually used for kinetic analysis in quinoline HDN while Shih et al.[319] used a two-site L-H rate expression where the denominator is a square form. The difficulties for the determination of the kinetics results from these factors. Carefully selected kinetic experiments and calculations taking into consideration the effects of various factors such as concentrations of organic reactants, hydrogen, and hydrogen sulfide, kinds of catalysts, reactor systems and reaction conditions may be required for a better understanding of the reactions.

3. Acridine and Benzoquinolines

HDN reactions of 3-ring compounds with a six-membered nitrogen heterocycle have been reported for acridine,[342,343] 5,6-benzoquinoline,[335,344] and 7,8-benzoquinoline.[322,335,345] Many reactions and intermediates exist in the networks, which are much more complicated than those mentioned above. Pseudo-first-order kinetics was assumed to determine the reaction rates although the individual reaction rates and overall nitrogen removal are affected by the concentration of nitrogen reactant. Kabe and coworkers reported acridine HDN catalyzed by MoO$_3$/Al$_2$O$_3$ in a flow reactor at 280–360 °C and 10.1 MPa. They initially showed that HDN giving dicyclohexylmethane occurred only after aromatic rings in acridine were completely hydrogenated to give perhyroacridine.[342] They concluded that, under the conditions,

0.13   0.132

0.503   0.13

fast   0.127

0.0190   0.0440

0.127   Hydrocarbons + NH₃

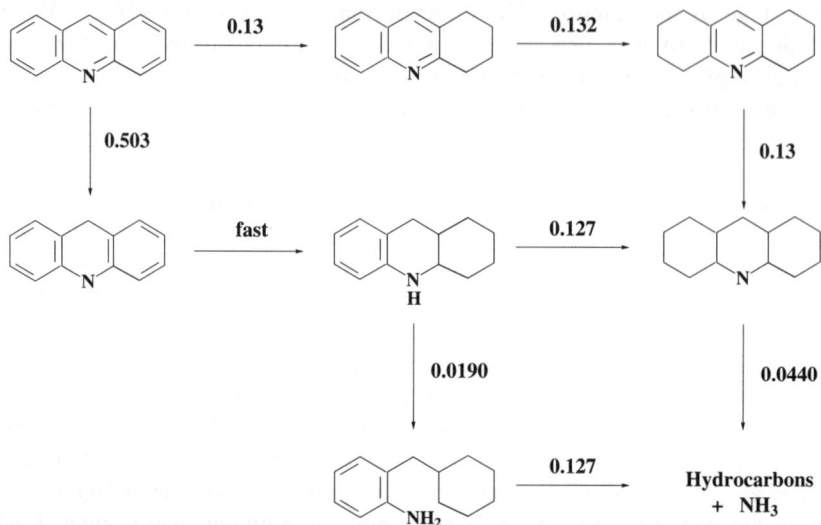

Fig. 2.56 Reaction pathways for acridine HDN.
The numbers next to the arrows are pseudo-first order rate constants in (g of carrier oil)/(g of catalyst · s).
at 367°C, 137 atm, and 0.5 wt% acridine in the feed.
(From M. J. Girgis and B. C. Gates, *Ind. Eng. Chem. Res.*, **30**, 2038 (1991))

hydrogenation equilibrium between acridine and perhydroacridine was achieved and that hydrogenolysis of perhydroacridine was the rate-determining step. Zawadski *et al.*[343] performed acridine HDN using Ni-Mo/Al₂O₃ at 317–365°C and 54–172 atm in a batch reactor. The reaction network proposed is shown in Fig. 2.56. Pseudo-first-order kinetics was assumed in each reaction and overall nitrogen removal also followed first-order kinetics. Both hydrogenation and hydrogenolysis were kinetically important, consistent with the results from quinoline HDN. Pyridine ring was hydrogenated more rapidly than benzenoid ring in acridine hydrogenation probably because of higher π-electron density on the heteroring. However, hydrogenolysis of the product asym-octahydroacridine was much slower than its hydrogenation and this pathway with low hydrogen consumption is not favored kinetically. Compared with hydrogenation of acridine forming asym-octahydroacridine, hydrogenation of N heteroring in 1,2,3,4-tetrahydroacridine or sym-octahydroacridine was insignificant or much slower. This effect was attributed to the presence of a fused cyclohexane ring.

Shabtai *et al.*[344] studied HDN of 5,6-benzoquinoline in a batch reactor under the following conditions: temperature 330°C, H₂ 2500 psig, sulfided Co-Mo/Al₂O₃ and Ni-Mo/Al₂O₃. The reaction network is shown in Fig. 2.57. Hydrogenation of 5,6-benzoquinoline gave only one product, 1,2,3,4-tetrahydro-5,6-benzoquinoline which underwent two competing reactions, that is, slow C-N hydrogenolysis giving 1-propylnaphthalene and fast ring hydrogenation giving 1,2,3,4,7,8,9,10-octahydro-5,6-benzoquinoline. The rate of the former was much slower than that of the latter, indicating that the C-N bond was strengthened by the effect of resonance stabilization while aniline derivatives were not detected in contrast to the case of quinoline. 1,2,3,4,7,8,9,10-Octahydro-5,6-benzoquinoline also underwent competing C-N hydrogenolysis giving propyltetralin and ring hydrogenation giving perhydro-

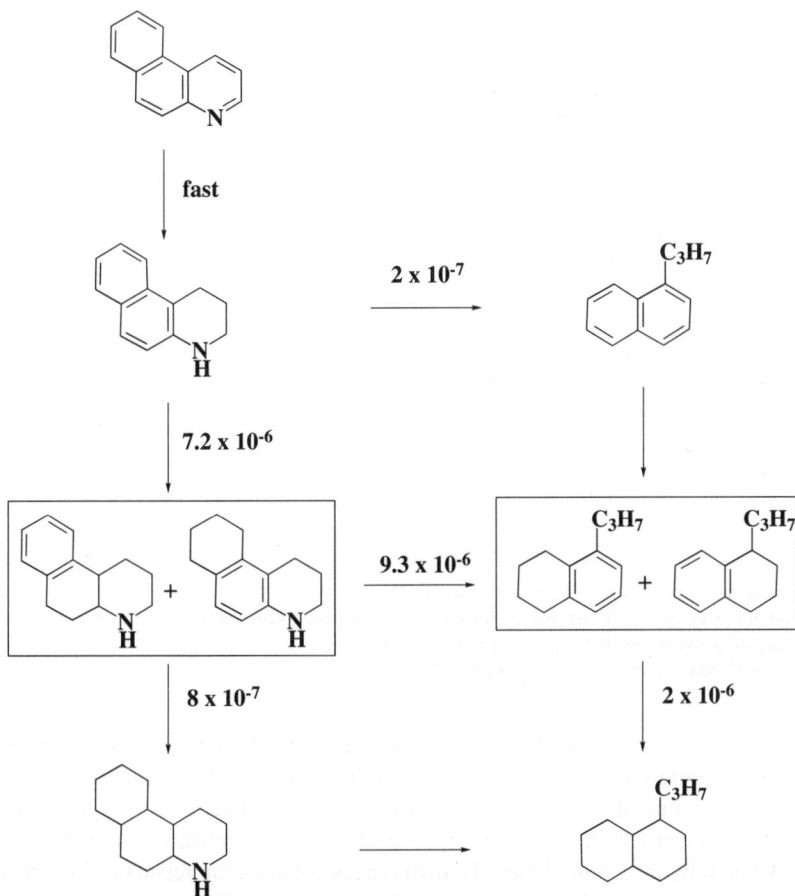

Fig. 2.57 Reaction pathways for 5,6-benzoquinoline HDN.
The numbers next to the arrows are pseudo-first-order rate constants in L/(g of catalyst) at 330°C, 79–171 atm, n-dodecane solvent, and 0.03 mol/L reactant concentration.
(From J. Shabtai et al., Ind. Eng. Chem. Res., **28**, 142 (1989))

5,6-benzoquinoline. In this case, the rate of the former was faster than that of the latter. These results indicate that C-N bond hydrogenolysis with nitrogen attached to aromatic carbon is much slower, whereas C-N bond hydrogenolysis with nitrogen attached to aliphatic carbon is faster as compared with competing ring hydrogenation. Activation energies determined for the hydrogenation of a conjugated pyridine ring, a conjugated benzene ring and a nonconjugated (single) benzene ring were 12.4, 12.7 and 39.4 kcal/mol, respectively. Activation energies determined for C-N bond hydrogenolysis with nitrogen attached to aliphatic carbon and C-N bond hydrogenolysis with nitrogen attached to aromatic carbon were 24.4 and 49.4 kcal/mol, respectively. These values were consistent with the results from quinoline HDN.[319,332,337]

Moreau et al.[335] studied HDN of 5,6-benzoquinoline in a batch reactor under the following conditions: temperature 340°C, H₂ 70 atm, sulfided Ni-Mo/Al₂O₃. The reaction network reported was different from that shown in Fig. 2.57 by Shabtai et al. Hydrogenation

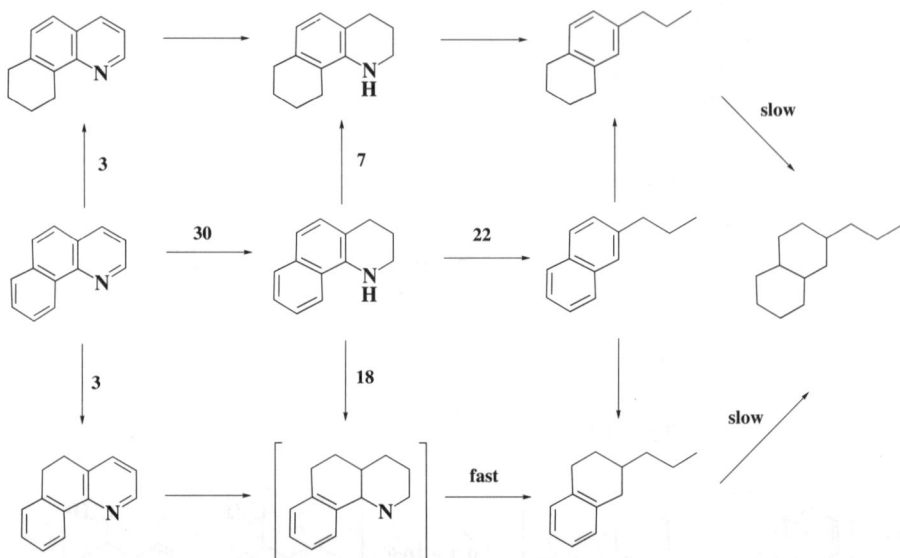

Fig. 2.58 Reaction pathways for 7,8-benzoquinoline HDN.
        The numbers next to the arrows are relative values of pseudo-first order rate constants at 340 °C, 70 atm,
        n-decane solvent, and 0.1 mol/L reactant concentration.
        (From C. Moreau et al., J. Catal., 112, 412 (1988))

of 5,6-benzoquinoline gave not only 1,2,3,4-tetrahydro-5,6-benzoquinoline but also 7, 8-dihydro-5,6-benzoquinoline. In contrast to the results of Shabtai et al., the first denitrogenated product was 1-propylnaphthalene by hydrogenolysis of 1,2,3,4-tetrahydro-5,6-benzoquinoline without aniline intermediate, followed by 1-propyl-1,2,3,4-tetrahydronaphthalene and 5-propyl-1,2,3,4-tetrahydronaphthalene. The differences between the results of Shabtai et al. and those of Moreau et al. may be due to the difference in reaction conditions, especially the different partial pressure of hydrogen.

Moreau et al.[335] also studied HDN of 7,8-benzoquinoline under the same conditions as those for 5,6-benzoquinoline. The reaction network is shown in Fig. 2.58. Three hydrogenated products were observed and the major route was the formation of 1,2,3,4-tetrahydro-7,8-benzoquinoline through hydrogenation of pyridine ring. The results were consistent with those of Shabtai et al.[322] and Malakani et al.[345] In this case, the first denitrogenated product was 2-propylnaphthalene by hydrogenolysis of 1,2,3,4-tetrahydro-7,8-benzoquinoline without aniline intermediate, followed by 2-propyl-1,2,3,4-tetrahydronaphthalene and 6-propyl-1,2,3,4-tetrahydronaphthalene. Moreau et al. assumed not aniline derivatives but naphthylpropylamine intermediates for the formation of propylnaphthalene products in HDN of 5,6-benzoquinoline and 7,8-benzoquinoline. It was concluded that the presence of an important and unusual degree of C_{sp^2}-N bond cleavage from 1,2,3,4-tetrahydrobenzoquinolines is related to the aromatic character of the ring adjacent to the C_{sp^2}-N bond to be cleaved and that a low aromatic character favors both hydrogenation of N-rings and cleavage of C_{sp^2}-N bonds.

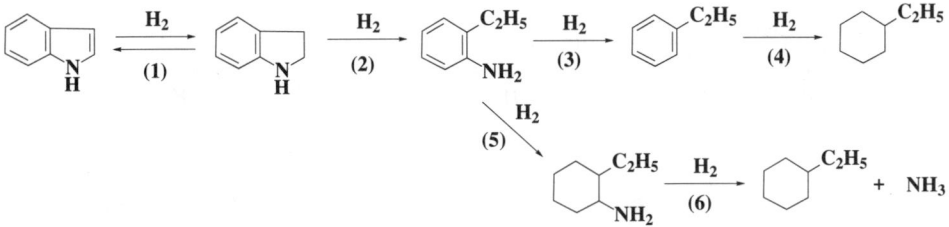

Fig. 2.59 Reaction pathways for indole HDN.
(From E. O. Odebunmi and D. F. Ollis, *J. Catal.*, **80**, 87 (1983))

4. Indole and Other Nonbasic Compounds

There are few examples of quantitative analysis of HDN of nonbasic five-membered nitrogen heterocycles. Only indole HDN has been reported by several researchers. [92,320,324,333,334,346–348] The typical reaction network proposed for indole HDN is shown in Fig. 2.59. In this pathway, the reaction proceeds via the intermediates, 2,3-dihydroindole and o-ethylaniline (OEA) and hydrocarbons such as ethylcyclohexane, ethylbenzene, etc. are formed exclusively through direct hydrogenolysis of OEA or ring hydrogenation of OEA and ammonia elimination. Stern[320] reported indole HDN using Ni-Mo/Al$_2$O$_3$ and Co-Mo/Al$_2$O$_3$ at 350°C and 68 atm in a batch reaction. Olivé et al.[334] studied indole HDN catalyzed by sulfided Ni-Mo/Al$_2$O$_3$, Co-Mo/Al$_2$O$_3$ and Ni-W/Al$_2$O$_3$ in a batch reactor at 250–350°C and 34–70 MPa. In these studies, direct hydrogenolysis of OEA to ethylbenzene occurred to a lesser extent. Further, because hydrogenation of ethylbenzene was slow under the conditions used, it was assumed that ethylcyclohexane was mainly formed through 2-ethylcyclohexylamin. Stern observed that 2-ethylcyclohexylamine was rapidly converted to ethylcyclohexene, ethylbenzene and ethylcylcohexane and that indoline was easily dehydrogenated to give indole under the conditions showing the reversibility of indole hydrogenation to indoline. Bhinde[324] reported indole HDN catalyzed by Ni-Mo/Al$_2$O$_3$ at 350°C and 34 atm in a batch reactor. Although a similar reaction pathway was proposed, the initial rate of formation of ethylcyclohexane could not be explained only by the disappearance rate of OEA. Therefore, it was assumed that some ethylcyclohexane was formed through a perhydroindole intermediate.

Odebunmi and Ollis[333] studied simultaneous hydrodenitrogenation of indole and hydrodeoxygenation of m-cresol using reduced and subsequently sulfided Co-Mo/Al$_2$O$_3$ in a trickle-bed reactor at 250–350°C and 69 atm. Under the conditions used, the indole-indoline hydrogenation equilibrium was essentially achieved, consistent with the case of the batch system, and hydrogenolysis of indoline was the rate-limiting step. At lower HDN temperatures, the intermediate o-ethylaniline is converted to ethylbenzene, then ethylcyclohexane while at or above 300°C hydrogenation to 2-ethylcyclohexylamine preceded HDN. In an early study using a Co-Mo/Al$_2$O$_3$ oxide catalyst (350–400°C and 21 MPa) by Aboul-Gheit,[347] hydrogenation of indole to indoline was slowest in indole HDN.

Very little information is available for pyrrole[320,349] and carbazole[310,320,350–352] HDN. Morávek et al.[349] studied pyrrole and pyridine HDN using sulfided Ni-W/Al$_2$O$_3$ and Ni-Mo/Al$_2$O$_3$ in a flow reactor at 300°C and 3 MPa. The kinetic scheme of pyrrole HDN at low initial partial pressure was simplified to a one-step reaction of pyrrole to NH$_3$ + C$_4$-hydrocarbons because of the higher relative reactivity of pyrrolidine and 1-butylamine. In

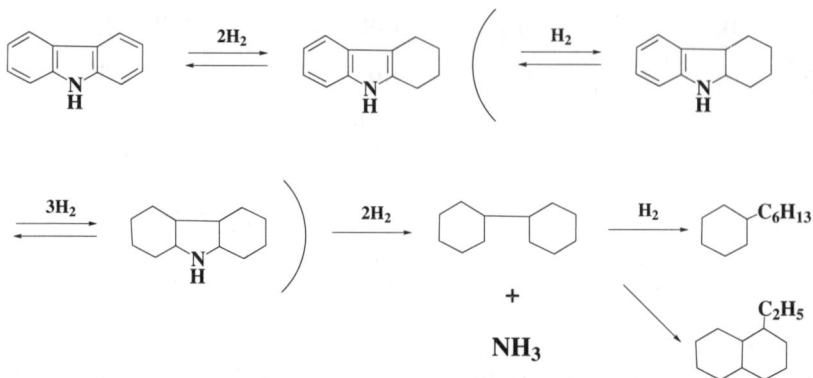

Fig. 2.60 Reaction pathways for carbazole HDN.
(From M. Nagai *et al.*, *Energy & Fuels*, **2**, 647 (1988))

contrast to this, pyridine HDN proceeded via a consecutive reaction scheme including pyridine, piperidine, *n*-pentylamine and NH_3 + C_5-hydrocarbons. In pyrrole HDN, the activity for the most active Ni-W/Al_2O_3 reached that for the standard Ni-Mo/Al_2O_3. In a competitive reaction with pyridine, pyrrole HDN was inhibited to a larger extent on the Ni-W/Al_2O_3 than on the standard Ni-Mo/Al_2O_3. The low reactivity of pyrrole was due to its poorer adsorptivity compared with the other nitrogen bases.

Nagai *et al.*[350,351] reported carbazole HDN using reduced and sulfided Mo/Al_2O_3 in a continuous flow microreactor at 280–360°C and 10.1 MPa. The reaction pathway proposed for carbazole HDN on the Mo/Al_2O_3 catalyst is shown in Fig. 2.60. The major product, bicyclohexyl, was formed from the C-N bond cleavage of perhydrocarbazole via the successive hydrogenation of carbazole although the hydrogenated compounds were hardly observed except for tetrahydrocarbazole. α-Ethylbicyclo[4.4.0]decane and hexylcyclohexane were also produced in the hydrogenolysis of bicyclohexyl. They found that the sulfidation of the reduced catalyst enhanced the hydrogenation of carbazole and the formation of denitrogenated compounds such as bicyclohexyl, α-ethylbicyclo[4.4.0]decane and hexylcyclohexane at all temperatures used. Sarbak[352] also reported carbazole HDN using sulfided Ni-Mo/Al_2O_3 in a batch reactor at 300–367°C and 34–136 atm. Several perhydrocarbazoles other than tetrahydrocarbazole, and BP and CHB as well as bicyclohexyl were observed in this case. The interaction between a nitrogen-containing compound and a pair of acid and basic sites was assumed for the mechanism of carbazole HDN.

Ho[310] has studied HDN of 3-ethylcarbazole catalyzed by sulfided Ni-Mo/Al_2O_3 at 290°C, 7.0 MPa and 1.4 LHSV. Nitrogen-containing products were partially and fully hydrogenated 3-ethylcarbazoles and the dominant product was tetrahydro-3-ethylcarbazole, suggesting that further hydrogenation of the latter is very slow. Neither amines nor anilines were formed, indicating that hydrogenolysis was relatively fast. The result provides evidence suggesting that hydrogenation may be the slowest step in HDN of higher molecular weight nitrogen compounds.

5. Aniline and Related Compounds

As mentioned above, aniline intermediates appear in almost all the reaction networks of HDN of N heterocycles such as quinoline, acridine and indole. Therefore, nitrogen removal

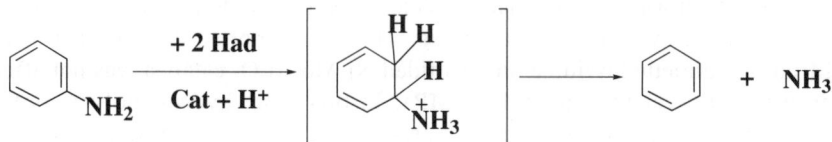

Fig. 2.61 Reaction pathways for aniline HDN.
(From M. J. Girgis, B. C. Gates, *Ind. Eng. Chem. Res.*, **30**, 2040 (1991))

from aniline intermediates has often been one of the most important reactions in a network. Although hydrogenation of aromatic ring is generally required before C-N bond scission,[320,332,334] the direct formation pathway of alkylbenzene which can reduce hydrogen consumption is still valid. If one can find a selective catalyst for C-N bond scission without ring hydrogenation, a large amount of hydrogen can be saved.

C-N bond scission with and without ring hydrogenation has been reported in anilines HDN. Mathur *et al.*[194] studied aniline, o-ethylaniline and diphenylaniline HDN catalyzed by Ni-Mo/Al$_2$O$_3$ in a batch reactor at 320°C and 350°C and 34 atm. In these reactions, major products of hydrogenolysis were aromatic compounds. The authors proposed the formation of dihydroaniline intermediate for the formation of major aromatic products in Fig. 2.61 instead of direct hydrogenolysis of C-N bond. The aromatic resonance interaction with a lone-pair nitrogen atom disappeared by the formation of this intermediate and the C-N bond was weakened. C-N bond scission rapidly occurred through β-elimination in which the recovery of aromatic resonance interaction by the formation of an aromatic was the driving force. Finiels *et al.*[353] studied secondary amines HDN catalyzed by Ni-W/Al$_2$O$_3$ in a batch reactor at 350°C and 59 atm. In diphenylamine HDN, a significant amount of cyclohexane was produced and the yield of aniline was higher than that of benzene. The results suggest that the major pathway for nitrogen removal from diphenylamine included hydrogenation of one aromatic ring before C-N bond scission.

2.2.2 Effects of Components on Reactions Included in HDN

A. The Effects of Hydrogen, Hydrogen Sulfide and Organic Sulfur Compounds on Reactions Included in HDN

There are very few studies dealing with the effect of hydrogen pressure on HDN. Gioia and Lee[340] investigated this effect of hydrogen on quinoline HDN. The hydrogen pressure was varied in the range 10.5 bar $\leq P_{H_2} \leq$ 151.6 bar. The results were interpreted according to the L-H rate expression where nitrogen compounds and hydrogen adsorbed separately on two different sites of a catalyst. It was found that the rate constants for hydrogenolysis were independent of hydrogen pressure while those for hydrogenation increased with increasing hydrogen pressure except for hydrogenation of quinoline to 58TQ. The result for hydrogenation of quinoline is consistent with that of pyridine.[328] In quinoline HDN, the rates of hydrogenation and hydrogenolysis steps were of similar order of magnitude,[354,355] and a single rate-determining step cannot be given. However, both the hydrogen consumption and the overall rate of HDN increased with pressure, indicating that the contribution of the pathway through DHQ increased in the network. If hydrogenation in a network is the rate-determining step, the overall HDN rate will reflect a hydrogenation rate rather than a hydrogenolysis rate. It was reported that in indole HDN the rate-determining step in HDN was hydrogenation of o-ethylaniline.[92] In contrast, Ho et al. reported that the rate of nitrogen removal from 2,4-dimethylpyridine on a sulfided Ni-Mo/Al$_2$O$_3$ catalyst was not affected by hydrogen pressure over the range 4-13 MPa. In this case, hydrogenolysis is the rate-determining step.[356] The effect of hydrogen pressure on acridine HDN was also reported by Zawadski et al.[343] HDN rate remarkably increased with increasing hydrogen pressure at lower pressures of hydrogen below 100 atm, while it approached zero order with respect to hydrogen pressure over 100 atm, indicating that adsorption of hydrogen on the catalyst reached saturation at this pressure range. Further, higher pressure is required for acridine HDN to obtain a similar reaction rate than for quinoline HDN.

The effects of hydrogen and hydrogen sulfide on hydrogenation and hydrogenolysis have been reported by Hanlon.[328] Pyridine hydrogenation increased with increasing hydrogen pressure but was not affected by hydrogen sulfide. In contrast, piperidine hydrogenolysis decreased with increasing hydrogen pressure in the range 1 to 10 atm under constant hydrogen sulfide partial pressure (0.08 atm) but increased by a factor of 2 with increasing hydrogen sulfide partial pressure by a factor of 6 under constant hydrogen pressure. Although an equilibrium including two catalytic sites, hydrogen and hydrogen sulfide was proposed, quantitative evidence was not given. The determination of the effects of hydrogen and hydrogen sulfide using the Langmuir-Hinshelwood (L-H) rate expression is needed, but another rate expression may be necessary for the estimation of promoting effect of hydrogen sulfide. Determination of the effects of reactants in pyridine HDN network may provide a guideline for the kinetic modeling of more complicated HDN networks because the pyridine HDN network is much simpler than those of polyaromatic N compounds. In pyridine HDN catalyzed by Ni-Mo/Al$_2$O$_3$ using a batch reactor, both the promoting and retarding effects of H$_2$S on HDN were also reported.[183] Increase in partial pressure of H$_2$S increased HDN activity. After reaching the maximum, however, the activity decreased with further increase of H$_2$S.

Satterfield and coworkers[332,337,354] estimated the effect of H$_2$S on quinoline HDN. H$_2$S has a slight inhibiting effect on intermediate hydrogenation and dehydrogenation but a marked accelerating effect on the overall HDN rate. These effects are observed for vapor-phase reactions. H$_2$S has little effect on the activation energies for hydrogenation and dehydrogenation reactions, but it significantly reduces them for hydrogenolysis reactions. Similar results were also reported for pyridine and quinoline HDN,[328,355,357] suggesting that different kinds of sites exist in quinoline HDN. In the reaction of 5,6-benzoquinoline,[344] increase in the concentration of H$_2$S decreased the hydrogenation rate and increased the dehydrogenation rate, consistent with the results from the reaction of quinoline.[332] In indole HDN, Massoth et al.[348] reported that the conversion to o-ethylaniline increased with increase in H$_2$S concentration at constant hydrogen pressure. It was found that the rate in the presence of H$_2$S was proportional to the square root of H$_2$S partial pressure. In HDN of aromatic amines, increase in H$_2$S decreased C-N bond scission.[358-360]

Satterfield et al.[168,169] studied the simultaneous catalysis of pyridine HDN and thiophene HDS by Ni-Mo/Al$_2$O$_3$ and Co-Mo/Al$_2$O$_3$ in a vapor-phase flow reactor under the conditions 1.1–7.0 MPa and 200–425°C. Pyridine inhibited the HDS reaction as described in the previous section. Thiophene had a dual effect on HDN. At low temperatures, the competitive adsorption of thiophene for hydrogenation sites on the catalyst retarded the hydrogenation of pyridine to piperidine, reducing the overall reaction rate. At higher temperatures, hydrogen sulfide produced from thiophene HDS increased the rate of piperidine hydrogenolysis and enhanced the overall rate of HDN. Ozkan et al.[361,362] also studied the effect of thiophene on pyridine HDN catalyzed by Ni-Mo/Al$_2$O$_3$, Mo/Al$_2$O$_3$ and Ni/Al$_2$O$_3$ in a flow reactor under the conditions 100 psig and 320–400°C. They examined the role of Ni as a promoter in the Ni-Mo catalyst system and suggested that Mo active species promoted hydrogenolysis of piperidine while Ni-Mo species enhanced the hydrogenation of pyridine. The presence of thiophene enhanced pyridine HDN, especially when Ni was combined with Mo. Bhinde[324] reported the effect of DBT on quinoline HDN catalyzed by Ni-Mo/Al$_2$O$_3$ in a batch reactor under the conditions 35 atm and 350°C. The pseudo-first-order rate constants determined for the reactions in the network were not affected by the presence of DBT. The addition of DBT increased only the rate of hydrogenolysis of decahydroquinoline by 25%. Later it was shown using same data that the inhibition parameter for quinoline was two orders of magnitude greater than that for DBT.[363]

Nagai et al.[364] studied the effect of several sulfur compounds on acridine HDN catalyzed by reduced and sulfided Mo/Al$_2$O$_3$ in a continuous flow microreactor under the conditions 280–380°C and 10.1 MPa. When the concentration of sulfur compounds increased, the yield of dicyclohexylmethane decreased and HDN of acridine was depressed. In kinetic calculation, Eq.(2.37) was used:

$$r_{DCM} = kK_{PA}P_{PA}K_HP_H/(1 + K_{PA}P_{PA} + K_HP_H + K_SP_S)^2 \tag{2.40}$$

where r_{DCM} is the rate for dicyclohexylmethane formation, P_i is the partial pressure, K_i and k are constants, PA, H and S denote perhydroacridine, hydrogen and sulfur compound, respectively. It was assumed that the rate-determining step was the hydrogenolysis of perhydroacridine to give dicyclohexylmethane and that perhydroacridine was the sole nitrogen compound adsorbed on the hydrogenolysis sites. Adsorption parameters K_S ($\times$ 10^{-2} kPa^{-1}) estimated for DBT, BT, thiophene, dimethyl sulfide, and ethanethiol at 360°C were 16.7, 6.8, 5.1, 3.2, and 1.0, respectively. Adsorption parameters K_{PA} (kPa^{-1}) which were also estimated for DBT, BT, thiophene, dimethyl sulfide, and ethanethiol at 360°C were different from each other and were 13.3, 7.1, 5.5, 9.0, and 8.2, respectively. K_{PA} was two orders of magnitude larger than K_S for each compound, consistent with that in the effect of DBT on quinoline HDN described above.[363] These results indicate that sulfur compounds are weak inhibitors of HDN. The values of K_S for sulfur heterocycles were larger than those for thiol and sulfide and that for three-ring DBT was the largest among those examined. The values of K_H for hydrogen were three orders of magnitude smaller than those of K_S.

Similar effects were reported for carbazole HDN catalyzed by Mo/Al$_2$O$_3$[351] and reduced Ni-Mo/Al$_2$O$_3$[365] catalysts. Further, the presence of hydrogen sulfide depressed the total C-N hydrogenolysis at a low P_{H_2S}/P_{H_2} ratio of 1.1–79 $\times$ 10^{-5} in acridine HDN catalyzed by sulfided Ni-Mo/Al$_2$O$_3$ while it hardly affected the C-N hydrogenolysis at a ratio above 0.003.[366] These results for acridine and carbazole HDN using reduced and sulfided Mo/Al$_2$O$_3$ and Ni-Mo/Al$_2$O$_3$ catalysts are different from those for pyridine and quinoline HDN using sulfided Ni-Mo/Al$_2$O$_3$ or Co-Mo/Al$_2$O$_3$ in which formed or added hydrogen sulfide usually promoted C-N bond hydrogenolysis and enhanced the overall rate of HDN. Possible reasons may be as follows: 1) A Ni-Mo/Al$_2$O$_3$ catalyst is reduced at extremely low concentration of H$_2$S even if it is initially used in sulfided form, 2) the presence of Ni or Co in the sulfided form generates the promotion effect of H$_2$S, 3) the adsorption of only small amounts of H$_2$S on Mo/Al$_2$O$_3$ and

Ni-Mo/Al$_2$O$_3$ catalysts remarkably inhibits the adsorption and hydrogenation of three ring heterocycles, acridine, carbazole and their hydrogenated derivatives, and 4) the steric hindrance with hydrogenated side rings adjacent to the N-ring inhibits the approach of the N-ring to the catalyst surface (see Section 2.2.1B3).

Although the promotion effects of nickel and cobalt in molybdenum-based catalysts on HDS are well known, similar promotion effects on HDN are not well established. HDN catalyzed by Mo, Co-Mo and Ni-Mo catalysts in the absence and presence of sulfur compounds has been compared to clarify the effect. Under industrial conditions, the effect was observed for a Ni-Mo catalyst,[367] while the effect was not observed in pyridine HDN under low hydrogen pressure.[368] On the other hand, Perot *et al.* studied HDN of 14TQ catalyzed by sulfided Ni-Mo catalysts under high hydrogen pressure and observed the effect in the presence of sulfur but not in the absence of sulfur.[369] Ledoux and Djellouli reported a similar effect in pyridine HDN catalyzed by alumina-supported Mo, Ni-Mo and Co-Mo under high hydrogen pressure. Although the three catalysts exhibited the same activity in the absence of sulfur, the Ni-Mo catalyst was 2.5 more active than Mo and Co-Mo catalysts. The difference in activity between the catalysts was attributed to the difference in acidity of their surfaces.[370] These results indicate that the presence of sulfur plays an important role in the manifestation of the high HDN activity of Ni-Mo catalysts.

B. The Effects of Aromatic Hydrocarbons on Reactions Included in HDN

There are few examples of studies to determine the effect of aromatic hydrocarbons on HDN. Ho *et al.* studied the effect of 2-methylnaphthalene (MN: 20 wt%) on 2,4-dimethylpyridine (DMP: 1 wt%) HDN catalyzed by Ni-Mo/Al$_2$O$_3$ in a flow reactor under the conditions, 280–360°C, 4.0–13.0 MPa and 1.4–5.5 LHSV.[356] Although the hydrogenation of MN was severely inhibited by nitrogen compounds, the hydrogenation of DMP was virtually not affected by MN. The C-N bond scission of 2,4-dimethylpiperidine (DMPP) was moderately reduced by MN. A Langmuir-Hinshelwood (L-H) rate equation was used to analyze the data. It was assumed that hydrogenation and hydrogenolysis occurred on the same site and that hydrogen adsorption occurred on a different site. Since the hydrogenation of DMP to DMPP was much faster than the C-N bond hydrogenolysis of DMPP and dehydrogenation of DMPP was kinetically insignificant, DMP and DMPP were lumped and treated as a single species in the kinetic analysis for nitrogen removal. The adsorption parameters K_N and K_M calculated for nitrogen compound and MN were 50.6 and 0.085 L/mol, respectively, at 310°C and 7.0 MPa. The result indicates that the catalyst surface was almost exclusively covered by nitrogen compounds.

Bhinde[324] and Lo[363] investigated the retarding effect of quinoline on naphthalene hydrogenation catalyzed by Ni-Mo/Al$_2$O$_3$ in a batch reactor at 350°C and 35 atm. For the kinetic analysis, a L-H rate expression was used and three adsorption parameters, K_{HC}, K_{QBN}, and K_A, introduced to retarding terms in denominator represented hydrocarbons (naphthalene and its products), organic nitrogen compounds (quinoline, its hydrogenated derivatives and o-propylaniline) and ammonia, respectively. The values calculated for K_{HC}, K_{QBN}, and K_A, were 2.7, 1000 and 400 L/mol, respectively. In the same experiment, it was found that when 5.9 wt% naphthalene was added to 0.2 wt% quinoline in *n*-hexadecane, the pseudo-first-order rate constants decreased by 10–15%. The results indicate that the inhibition of both the hydrogenation and the hydrogenolysis reactions of the quinoline network by aromatic hydrocarbons was minimal.

C. The Effects of Oxygen Compounds on Reactions Included in HDN

Oxygen compounds have moderate effects on reactions included in HDN. Satterfield and Yang[371] investigated the effects of organic oxygen compounds on the HDN of quinoline and o-ethylaniline catalyzed by Ni-Mo/Al$_2$O$_3$ in a trickle-bed reactor at 375°C and 69 atm. Organic oxygen compounds were m-ethylphenol, o-ethylphenol, benzofuran and dibenzyl ether. Quinoline HDN was promoted by the addition of the oxygen compounds and the extent of the promotion was similar for each oxygen compound. It was suggested that the water formed with rapid HDO caused the HDN promotion. This was confirmed by an experiment using water as an additive to the feed. This experiment revealed the same effect as the cases of organic oxygen compounds. The effect of the oxygen compounds on o-ethylaniline HDN was somewhat different. o-Ethylphenol and m-ethylphenol increased the HDN conversion while benzofuran decreased it. Water did not affect the HDN conversion. It was concluded that water did not promote the hydrogenation of aromatic ring which affects the HDN rate of o-ethylaniline.

Satterfield and coworkers[372,373] also investigated the effects of water and hydrogen sulfide on the HDN of quinoline catalyzed by Ni-Mo/Al$_2$O$_3$ in a trickle-bed reactor at 375°C and 69 atm. The HDN rate increased in the presence of either water or hydrogen sulfide while the HDN rate increased remarkably in the presence of both water and hydrogen sulfide. Water increased only the hydrogenolysis of decahydroquinoline. Hydrogen sulfide increased not only hydrogenolysis but also hydrogenation. When both water and hydrogen sulfide were present, the rate constants for all the reactions were larger than when either water or hydrogen sulfide was present alone. In the presence of water, hydrogen sulfide, and both water and hydrogen sulfide, hydrogenolysis rate of decahydroquinoline, which was most sensitive to additives, increased by 20%, 167% and 197%, respectively. This suggests that the effects of water and hydrogen sulfide on hydrogenolysis and hydrogenation in HDN is related to the increase in the surface acidity of the catalyst.

Mild inhibition effects of oxygen compounds on HDN have been reported by several authors. Nagai et al.[364] studied the effect of xanthene on acridine HDN catalyzed by reduced and sulfided Mo/Al$_2$O$_3$ in a continuous flow microreactor under the conditions, 280–380 °C and 10.1 MPa. Although xanthene as well as sulfur compounds retarded the acridine HDN, the adsorption parameter of xanthene $K_{xanthene}$ estimated using Eq.(2.37) was 0.18 ($\times$ 10^{-2} kPa^{-1}) more than three orders of magnitude less than that of perhydroacridine. A similar result was reported for carbazole HDN catalyzed by a reduced Ni-Mo/Al$_2$O$_3$ catalyst.[365] Krishnamurthy and Shah[186] reported the effect of dibenzofuran and o-cyclohexylphenol on 7,8-benzoquinoline HDN catalyzed by Ni-Mo/Al$_2$O$_3$ in a batch reactor under the conditions, 350°C and 102 atm. Dibenzofuran and o-cyclohexylphenol decreased the HDN rate by 5 and 15 wt%, respectively. Odebunmi and Ollis[333] reported the effect of m-cresol on indole HDN catalyzed by Co-Mo/Al$_2$O$_3$ in a trickle-bed reactor under the conditions, 250–350 °C and 69 atm. m-Cresol slightly retarded the HDN rate. These inhibition effects of oxygen compounds were not consistent with the promotion effects reported by Satterfield and coworkers probably because in these studies the promotional effect of water was minimized at lower conversions and competitive adsorption between nitrogen compounds and oxygen compounds occurred.

D. The Effects of Nitrogen Compounds on Reactions Included in HDN

Bhinde[324] studied the effect of quinoline on indole HDN and the effect of indole on quinoline HDN catalyzed by Ni-Mo/Al$_2$O$_3$ in a batch reactor at 350°C and 35 atm. The pseudo-first-order rate constants calculated for the reactions included in the indole HDN network decreased by one-third with the addition of quinoline. In contrast, the pseudo-first-order rate constants calculated for the reactions included in the quinoline HDN network decreased by less than 10% with the addition of indole. This result indicates that basic quinoline and its derivatives were more strongly adsorbed on the catalyst surface than nonbasic indole. However the pseudo-first-order rate constants calculated for the reactions of naphthalene decreased by one order of magnitude with the addition of quinoline. The extent of this inhibition effect on reactions in the naphthalene network was larger than that on reactions in the indole network, indicating that indole and its nitrogen-containing derivatives (for example, basic indoline) may be more strongly adsorbed than aromatic hydrocarbons.

Although alkylanilines are known to be relatively reactive when they are pure, it was reported that alkylanilines became significantly unreactive in the presence of polyaromatic nitrogen heterocycles.[312,313,339,374–380] Satterfield and Cocchetto[339] initially pointed out this effect. Perot et al.[376] investigated the inhibiting effect by comparing pure anilines HDN with anilines HDN in the presence of quinoline type compounds. It was found that OPA was almost completely unreactive in the presence of 14TQ. Further, to confirm this effect more precisely, they investigated HDN of OPA and 6-methylquinoline (6MQ) catalyzed by Ni-

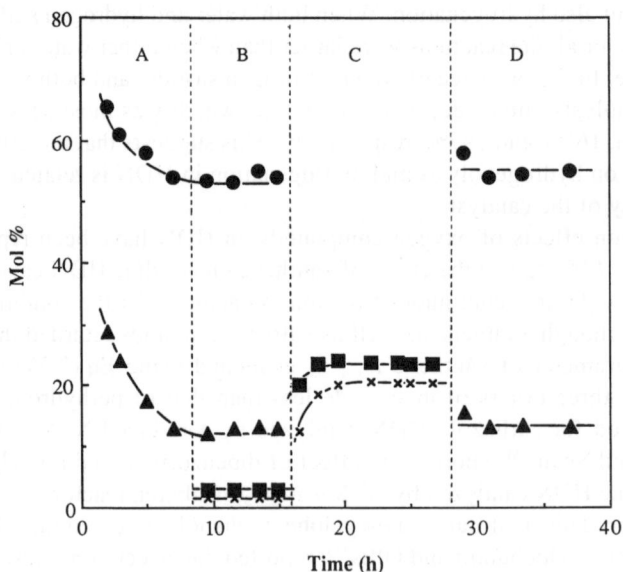

Fig. 2.62   Transformation of 6-methylquinoline and of orthopropylaniline over sulfided Ni-Mo/Al$_2$O$_3$ (623K, 7 MPa). Reactants: A, D: pure 6-methylquinoline; B: 6-methylquinoline + orthopropylaniline; C: pure orthopropylaniline.
● : 6-methylquinoline conversion; ▲: 1-methyl-3-propylcyclohexane; ■: orthopropylaniline conversion; X: propylcyclohexane.
(From G. Perot, *Catal Today*, **10**, 460 (1991))

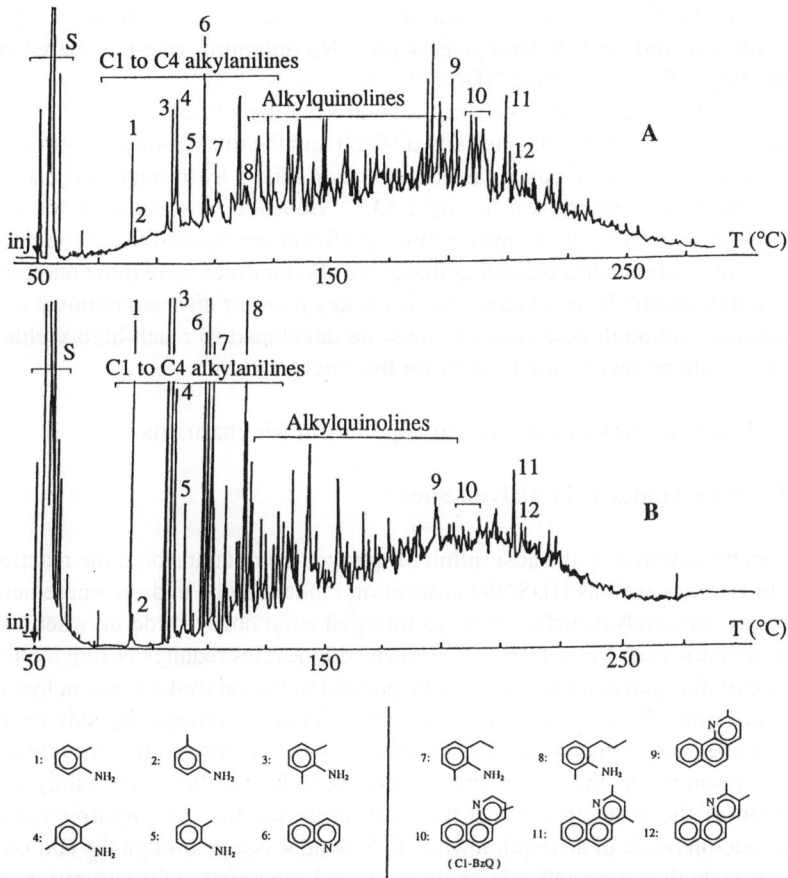

Fig. 2.63 Variation of the basic nitrogen fraction of the coker gas oil subjected to hydrotreatment on a Ni-Mo catalyst on an alumina support.
A: Basic nitrogen fraction of the feedstock.
B: Basic nitrogen fraction of the effluent.
C: Solvent peaks.
(From S. Kasztelan et al., Catal Today, **10**, 439 (1991))

Mo/Al$_2$O$_3$ at 350°C and 7 MPa. They carried out four consecutive experiments using reactants in the following order: A) pure 6MQ, B) an equimolar mixture of 6MQ and OPA, C) pure OPA and D) 6 MQ in order to check the activity of the catalyst. Fig. 2.62 shows the changes in the conversions of reactants and the yields of products with reaction time. When experiments B and C were compared, the presence of 6MQ inhibited the conversion of OPA in B while OPA showed remarkably higher conversion in the absence of 6 MQ in C. Similar effects were observed in HDN reactions of 2,6-diethylaniline and 14TQ.[376–379] These effects are most likely due to the competitive adsorption onto the catalyst surface between anilines and quinoline derivatives. La Vopa and Satterfield determined the Langmuir adsorption parameter of nitrogen compounds in the investigation of the effect of nitrogen compounds on thiophene HDS.[151] The adsorption parameter of 14TQ (0.46 kPa^{-1}) was much larger than that of aniline (0.094 kPa^{-1}). This is consistent with the fact that HDN of aniline derivatives is always inhibited by the presence of quinoline derivatives. It has also been reported that pyrrole, indole and indoline, heterocycles with a N-containing five-membered ring, have similar inhibiting effects on aniline HDN.[380]

A similar effect is also observed in the hydrotreating of heavy fuels.[313,374,375] In this hydroprocessing nitrogen removal is usually difficult and the total amount of nitrogen changes very little. The composition of coker gas oil before and after a hydrotreating process has been analyzed and the results are shown in Fig. 2.63.[313] Before hydrotreatment, the amounts of alkylanilines were relatively very small while significant amounts of these compounds were observed after hydrotreatment, indicating that these alkylanilines were quite unreactive in the presence of polyaromatic heterocycles. This is the key point of nitrogen removal in heavy oil hydroprocessing. Although new catalysts must be developed to reach high yields in HDN, these effects should be taken into account for the catalyst design.

2.2.3 Model Studies of Hydrodenitrogenation Mechanisms

A. Adsorption Modes and Catalytic Sites

HDN can be initiated by the adsorption of nitrogen compounds onto the reaction sites of a catalyst. In HDN as well as HDS, the adsorption mode may be end-on where heterocycles are adsorbed on the catalyst surface through nitrogen atom only or side-on where this occurs through the aromatic ring. Since HDN of nitrogen heterocycles requires N-ring hydrogenation, the ring plane of nitrogen heterocycles may be parallel to the catalyst surface in hydrogenation of the aromatic ring. When a nitrogen heterocycle adsorbs through the side-on mode, the compound can be hydrogenated directly. In contrast, when a nitrogen heterocycle is adsorbed through the end-on mode, the compound must rotate to be parallel to the catalyst surface in hydrogenation of the aromatic ring. It has been proposed that ring hydrogenation occurs through the side-on mode of adsorption while C-N bond scission through the end-on mode.[311]

In literature, both end-on and side-on modes have been reported for adsorption of nitrogen compounds. Some of these are related to the inhibition of HDS by the adsorption of nitrogen compounds on sulfided catalysts and have already been described in Section 2.1.4C. LaVopa and Satterfield[151] reported the effects of hindered pyridines on thiophene HDS catalyzed by sulfided Ni-Mo/Al$_2$O$_3$. The values of adsorption parameter decreased in the order 4-methylpyridine > pyridine > 2,6-dimethylpyridine. This indicates that 2,6-dimethylpyridine is the most difficult to be adsorbed on the catalyst surface among the three and that these pyridines may be adsorbed on the catalyst not by the side-on mode but by the end-on mode. Similar results were reported regarding the poisoning effect of substituted pyridines on HDS catalyzed by catalysts other than Ni-Mo/Al$_2$O$_3$.[173-175] On a sulfided Ni-W/Al$_2$O$_3$, 5,6-benzoquinoline was hydrogenated more rapidly than 7, 8-benzoquinoline. This difference in the reactivity was correlated to the steric effect on the adsorption of these compounds onto the catalyst surface through a nitrogen atom.[322] In contrast, it has been reported that heterocyclic nitrogen compounds are adsorbed parallel to the surface through their π-electrons.[193,194] Further, the hindered nitrogen compounds inhibited 1-hexene hydrogenation more than thiophene HDS.[173] This result indicates the possibility that the hindered nitrogen compounds are adsorbed through the side-on mode. The adsorption and reaction of nitrogen compounds on metals,[381-387] γ-Mo$_2$N[388] and a reduced Mo/Al$_2$O$_3$ [389,390] has also been reported and end-on

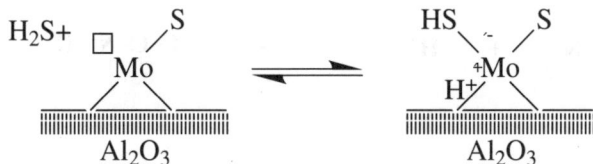

Fig. 2.64 Interconversion between sulfur vacancy and Brønsted acid site.
(From S. H. Yang and C. N. Satterfield, *Ind. Eng. Chem. Res.*, **23**, 25 (1984))

and/or side-on adsorption modes have been proposed.

The inhibition of HDS by the adsorption of nitrogen compounds was correlated with the adsorption of these compounds to Brønsted acid sites as discussed in Section 2.1.4C.[151,172] In pyridine HDN catalyzed by nonsulfided Mo oxide supported on Al_2O_3, SiO2 and SiO2-Al_2O_3, the functions of the coordinatively unsaturated sites of Mo and Brønsted acid sites have been discussed.[391] It was suggested that in HDN of quinoline the catalyst surface contained vacancies and Brønsted acids as catalytically active sites,[332] while Brønsted acid sites facilitated hydrogenolysis of the C-N bond in HDN.[392] Brønsted acid sites related to supports may be involved in HDN.[341,355] In HDN of indole, however, SiO2-Al_2O_3 was inactive and the activity of Co-Mo/SiO2-Al_2O_3 in C-N hydrogenolysis did not exceed that of Co-Mo/Al_2O_3.[346] The above suggests that Brønsted acid sites on the active phase such as metal sulfide are more important than those on the support itself. To explain the effect of H_2S on reactions in HDN, Satterfield and Yang[337,392] postulated the existence of two kinds of catalytic sites: Site I is a vacancy responsible for hydrogenation and hydrogenolysis while site II is a Brønsted acid site responsible for hydrogenolysis and ring isomerization. As shown in Fig. 2.64, a sulfur vacancy (site I) can be converted to a Brønsted acid site and an SH group (site II) in a reversible process that includes the adsorption and dissociation of an H_2S molecule. There are also a number of studies which address the mechanistic role of acidic functionalities in HDS and HDN using promoted and sulfided Co- or Ni-Mo catalysts.[393-398]

B. Mechanism of C-N Bond Scission

Several mechanisms of C-N bond scission [312,319,352,389,399-403] have been proposed and some of these have been explained based on classical Hofmann-type elimination (Fig. 2.65)

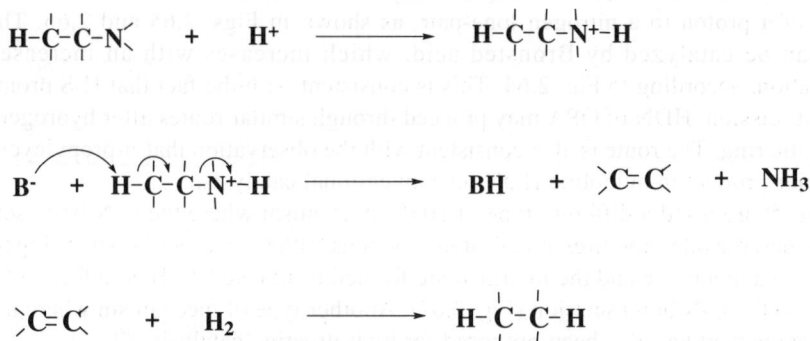

Fig. 2.65 E_2 Hofmann-type elimination for HDN reaction.
(From G. Perot, *Catal Today*, **10**, 464 (1991))

$$-\overset{|}{\underset{|}{C}}-\overset{|}{\underset{|}{C}}-N \diagup \quad + \quad H^+ \quad \longrightarrow \quad -\overset{|}{\underset{|}{C}}-\overset{|}{\underset{|}{C}}-N^+\!-H$$

$$SH^- \quad + \quad -\overset{|}{\underset{|}{C}}-\overset{|}{\underset{|}{C}}-N^+\!-H \quad \longrightarrow \quad -\overset{|}{\underset{|}{C}}-\overset{|}{\underset{|}{C}}-S-H \quad + \quad NH_3$$

$$-\overset{|}{\underset{|}{C}}-\overset{|}{\underset{|}{C}}-S-H \quad + \quad H_2 \quad \longrightarrow \quad -\overset{|}{\underset{|}{C}}-\overset{|}{\underset{|}{C}}-H \quad + \quad H_2S$$

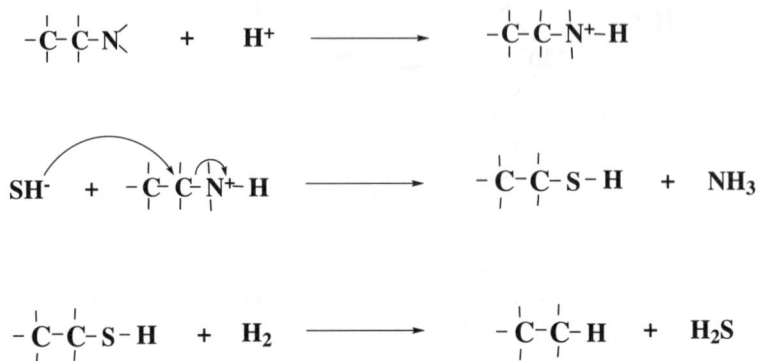

Fig. 2.66 S_N nucleophilic substitution mechanism for HDN reaction.
(From G. Perot, *Catal Today*, **10**, 464 (1991))

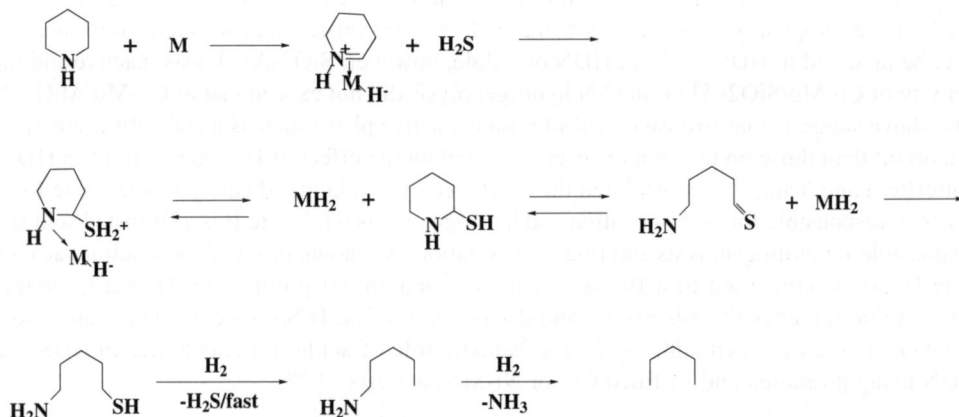

Fig. 2.67 Reaction pathways for piperidine HDN.
(From R. M. Laine, *Catal. Rev.-Sci. Eng.*, **25**, 467 (1983))

and nucleophilic substitution (Fig. 2.66). Nelson and Levy[399] suggested these mechanisms for C-N bond scission of 14TQ where a quaternary ammonium salt is initially formed with the addition of a proton to a nitrogen lone-pair, as shown in Figs. 2.65 and 2.66. These two routes can be catalyzed by Brønsted acid, which increases with an increase in H₂S concentration, according to Fig. 2.64. This is consistent with the fact that H₂S promotes the C-N bond scission. HDN of OPA may proceed through similar routes after hydrogenation of the aromatic ring. The route is also consistent with the observation that *n*-propylcyclohexane is the major product in quinoline HDN on conventional catalysts.

Laine[404] proposed a different type of HDN mechanism where the C-N bond scission in saturated heterocycles requires metal atoms or ions rather than acidic sites. Piperidine is activated by a metal site and the intermediate formed is attacked by H₂S to form a C-S bond and leads to the C-N bond scission (Fig. 2.67). Another type of mechanism where hydride is added to a reactant has also been proposed for hydrotreating catalysis.[405]

In order to save hydrogen in HDN, it is important to know the mechanism of HDN of aniline derivatives. However, there are few examples which address this mechanism. Aromatic

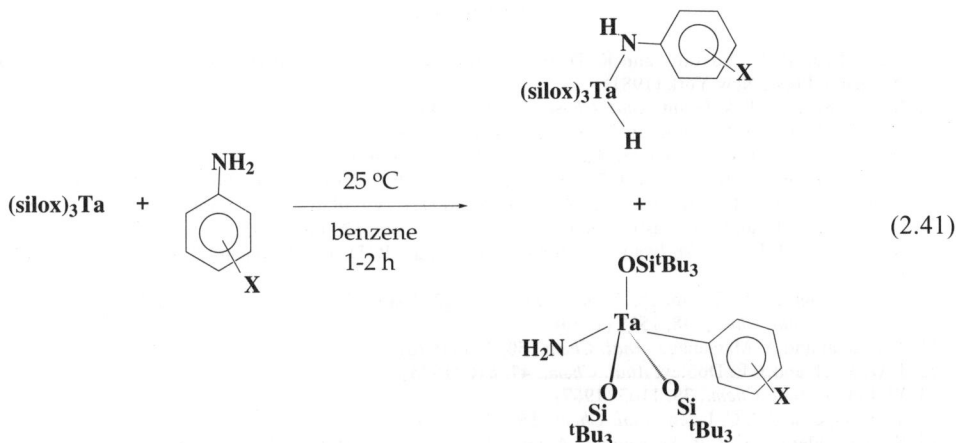

$$(silox)_3Ta \quad + \quad \text{(aniline-X)} \xrightarrow[\substack{\text{benzene} \\ \text{1-2 h}}]{25\ ^\circ C} \quad + \quad \text{(2.41)}$$

(From J. B. Bonanno et al., J. Am. Chem. Soc., **118**, 5133 (1996))

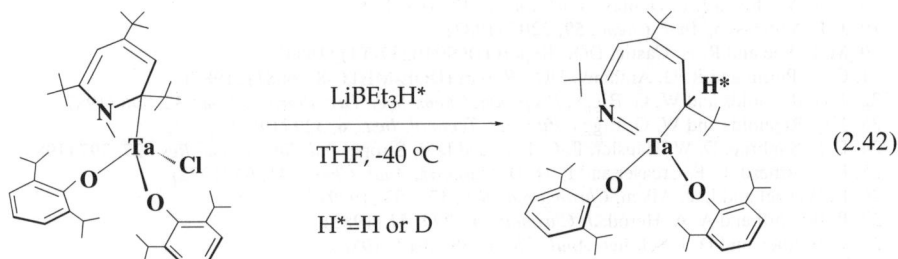

$$\xrightarrow[\substack{THF,\ -40\ ^\circ C \\ H^* = H\ or\ D}]{LiBEt_3H^*} \quad \text{(2.42)}$$

(From S. D. Gray et al., J. Am. Chem. Soc., **117**, 10679 (1995))

products were produced in HDN of quinoline,[340] indole,[334] 14TQ[402] and aniline derivatives.[194] As shown in Fig. 2.61 (Section 2.2.1B5), Mathur et al.[194] proposed the dihydroaniline intermediate for the formation of major aromatic products in HDN of aniline derivatives instead of direct hydrogenolysis of C-N bond, which may proceed via nucleophilic substitution. Recently, oxidative addition of arylamine ($H_2NC_6H_4X$, X = CF_3, F or Ph) C-N bond to $(silox)_3Ta$ (silox =$^tBu_3SiO$) has been reported (Eq.(2.41)).[406] This study suggests that low-energy arylamine C-N bond cleavage pathways are available to heterogeneous surfaces of suitable nucleophilicity, such as the relatively electron-rich catalysts involved in hydrotreating.

Another type of C-N bond scission reaction has also been reported in the organometallic chemistry. Gray et al. found that the reaction of the η^2(N,C)-pyridine complex [η^2(N,C)-2,4,6-$NC_5{}^tBu_3H_2$]$Ta(OAr)_2Cl$ (Ar = 2,6-$C_6H_3{}^iPr_2$) with $LiBEt_3H$ affords the C-N bond scission product Ta(= NC^tBu = CHC^tBu = $CHCH^tBu$)-$(OAr)_2$ (Eq.(2.42)).[407,408] The simplest mechanism of this reaction involves a direct, exo hydride attack on the bound carbon of the pyridine complex. Nucleophilic attack of the hydride at the metal to form an unstable hydride complex, followed by an endo hydride transfer from metal to the pyridine ligand also represents a viable pathway for C-N bond scission. It should be noted that, in this reaction, hydrogenation of pyridine ring to give piperidine, which is often proposed in the heterogeneous HDN process (Fig. 2.53), is not required before C-N bond scission occurs.

References

1. M. L. Lee, M. V. Novotny and K. D. Bartle, *Analytical Chemistry of Polycyclic Aromatic Compounds*; Academic Press: New York (1981).
2. R. L. Martin and J. A. Grant, *Anal. Chem.*, **37**, (6), 644 (1965).
3. R. L. Martin and J. A. Grant, *Anal. Chem.*, **37**, (6), 649 (1965).
4. H. V. Drushel and A. L. Sommer, *Anal. Chem.*, **39**, 1819 (1967).
5. R. G. Jewell, R. G. Rubelto and J. T. Swansiger, *Prepr., Div. Petrol. Chem., Am. Chem. Soc.*, **20**, 19 (1975).
6. D. W. Later, M. L. Lee, K. D. Bartle, R. C. Kong and D. L. Vassilaros, *Anal. Chem.*, 53, 1612 (1981).
7. C. Willey, M. Iwao, R. N. Castle and M. L. Lee, *Anal. Chem.*, **53**, 400 (1981).
8. R. C. Kong, M. L. Lee, M. Iwao, Y. Tominaga, R. Pratap, R. D. Thompson and R. N. Castle, *Fuel*, **63**, 702 (1984).
9. M. A. Poirier and G. T. Smiley, *J. Chromatog. Sci.*, **22**, 304 (1984).
10. W. L. Orr, *Anal. Chem.*, **38**, 1558 (1966).
11. T. Kaimai and A. Matsunaga, *Anal. Chem.*, **50**, 268 (1978).
12. J. W. Vogh and J. E. Dooley, *Anal. Chem.*, **47**, 816 (1975).
13. W. L. Orr, *Anal. Chem.*, **39**, 1163 (1967).
14. W. F. Joyce and P. C. Uden, *Anal. Chem.*, **55**, 540 (1983).
15. K. D. Gundermann, H. P. Ansteeg and A. Glitsch, *Proc. Int. Conf. Coal Sci.*, 1983, 631.
16. M. Nishioka, *Energy & Fuels*, **2**, 214 (1988).
17. M. Nishioka, R. M. Campbell, M. L. Lee and R. N. Castle, *Fuel*, **65**, 270 (1986).
18. J. W. Vogh and J. E. Dooley, *Anal. Chem.*, **47**, 816 (1975).
19. J. T. Andersson, *Anal. Chem.*, **59**, 2207 (1987).
20. M. L. Lee and R. N. Castle, DOE Report (DOE/10237-T1) (1984).
21. C. V. Philip and R. G. Anthony, DOE Report (DOE-METC-87-6081) (1987).
22. J. G. Reynolds and W. G. Biggs, *Prep. Am. Chem. Soc., Div. Petrol. Chem.*, **32**, 398 (1987).
23. J. G. Reynolds and W. G. Biggs, *Fuel Sci. Technol. Inst.*, **6**, 329 (1988).
24. E. L. Sughrue, D. W. Hausler, P. C. Liao and D. J. Strope, *Ind. Eng. Chem. Res.*, **27**, 397 (1988).
25. E. R. Adlard, L. F. Creaser and P. H. D. Mathews, *Anal. Chem.*, **44**, 64 (1972).
26. B. Wenzel and J. Aiken, *Chromatogr. Sci.*, **17**, 503 (1979).
27. P. Burchill and A. A. Herod, *J. Chromatogr.*, **242**, 51 (1982).
28. C. Bradley and D. J. Schiller, *Anal. Chem.*, **58**, 3017 (1986).
29. M. Nishioka, J. S. Bradshaw, M. L. Lee, Y. Tominaga, M. Tedjamulia and R. N. Castle, *Anal. Chem.*, **57**, 309 (1985).
30. R. L. Shearer, D. L. O'Neal, R. Rios and M. D. Baker, *J. Chromatogr. Sci.*, **28**, 25 (1990).
31. R. J. Skelton Jr., H.-C. K. Chang, P. B. Farnsworth, K. E. Markides and M. L. Lee, *Anal. Chem.*, **61**, 2292 (1989).
32. A. Ishihara, H. Tajima and T. Kabe, *Chem. Lett.*, 1992, 669.
33. T. Kabe, A. Ishihara and H. Tajima, *Ind. Eng. Chem. Res.*, **31**, 1577 (1992).
34. T. Kabe, A. Ishihara, Q. Zhang and H. Tsutsui, *J. Jpn. Petrol. Inst.*, **36** (6), 467 (1993).
35. H. Tajima, T. Kabe and A. Ishihara, *Bunseki Kagaku.*, **42**, 67 (1993) [in Japanese].
36. J. L. Buteyn and J. J. Kosman, *J. Chromatogr. Sci.*, **28**, 19 (1990).
37. A. Amorelli, Y. D. Amos, C. P. Halsig, J. J. Kosman, R. J. Jonker, M. De Wind and J. Vrieling, *Hydrocarbon Process.*, **71**, (6), 93 (1992).
38. I. Dzidic, M. D. Balicki, I. A. L. Rhodes and H. V. Hart, *J. Chromatogr. Sci.*, **26**, 236 (1988).
39. S. S. Brody and J. E. Chaney, *J. Gas Chromatogr.*, **4**, 42 (1966).
40. J. F. McGaughey and S. K. Gangwal, *Anal. Chem.*, **52**, 2079 (1980).
41. C. D. Pearson and W. J. Hines, *Anal. Chem.*, **49**, 123 (1977).
42. F. Berthou, Y. Dreano and P. Sandra, *HRC CC, J. High Resolut. Chromatogr. Chromatogr. Commun.*, **7**, 679 (1984).
43. E. Campaigne, Hewitt and L., Ashby, J., *J. Heterocyclic Chem.*, **6**, 553 (1969).
44. R. Gerdil and E. A. C. Lucken, *J. Am. Chem. Soc.*, **87**, 213 (1965).
45. E. G. G. Werner, *Recl. Trav. Chim.*, **68**, 509 (1949).
46. J. E. Banfield, W. Davies, N. W. Gamble and S. J. Middleton, *J. Chem. Soc.*, 4791 (1956).
47. C. Hansch and H. G. Lindwall, *J. Org. Chem.*, **10**, 381 (1945).
48. Y. Miki, Y. Sugimoto, S. Yamadaya and M. Oba, *Natl. Chem. Lab. Ind. Rep.*, **84** (12), 661 (1989) [in Japanese].
49. B. C. Gates, J. R. Katzer and G. C. A. Schuit, in:*Chemistry of Catalytic Processes*, p.390, McGraw-Hill: New York (1979).
50. M. L. Vrinat, *Appl. Catal.*, **6**, 137 (1983).
51. M. L. Vrinat, *Proc. of the NATO Advanced Study Institute on Surface Properties and Catalysis by Non-Metals: Oxides Sulfides and Other Transition Metal Compounds* (J. P. Bommelle, B. Delmon and E. Derouane

eds.), p.391, Reidel, Dordtecht (1983).

52. F. E. Massoth and G. Muralidhar, *Proc. 4th Int. Conf. Chemistry and Use of Molybdenum*, (H. F. Barry and P. C. H. Mitchell eds.), p.343, Climax Molybdenum Company Ltd. (1982).

53. M. J. Girgis and B. C. Gates, *Ind. Eng. Chem. Res.*, **30**, 2021 (1991).

54. T. Kabe, A. Ishihara, M. Nomura, T. Itoh and P. Qi, *Chem. Lett.*, 1991, 2233.

55. A. Ishihara, T. Itoh, T. Hino, M. Nomura, P. Qi and T. Kabe, *J. Catal.*, **140**, 184 (1993).

56. D. R. Kilakowski and B. C. Gates, *J. Catal.*, **62**, 70 (1980).

57. D. H. Broderick and B. C. Gates, *AIChE J.*, **27**, 663 (1981).

58. G. P. Singhal, R. L. Espino, J. E. Sobel and G. A. Huff, *J. Catal.*, **67**, 457 (1981).

59. M. L. Vrinat and L. de Mourgues, *J. Chim. Phys.*, **79-1**, 45 (1982).

60. D. R. Kilanowski, H. Teeuwen, V. H. J. de Beer, B. C. Gates, G. C. A. Schuit and H. Kwart, *J. Catal.*, **55**, 129 (1978).

61. M. Houalla, D. H. Broderick, A. V. Sapre, N. K. Nag, V. H. J. de Beer, B. C. Gates and H. Kwart, *J. Catal.*, **61**, 523 (1980).

62. C. G. Frye and J. F. Mosby, *Chem. Eng. Prog.*, **63** (9), 66, (1967).

63. P. Chakraborty and A. K. Kar, *Ind. Eng. Chem. Process Des. Dev.*, **17** (3), 252 (1978).

64. J. A. Mahoney, K. K. Robinson and E. C. Myers, *Chemtech*, 1978, 758.

65. K. Sakanishi, M. Ando, S. Abe and I. Mochida, *J. Jpn. Petrol. Inst.*, **34** (6), 553 (1991).

66. E. N. Givens and P. B. Venuto, *Prepr. Am. Chem. Soc. Div. Pet. Chem.*, **15** (4), A183 (1970).

67. Q. Zhang, A. Ishihara, H. Yashima, W. Qian, H. Tsutsui and T. Kabe, *J. Jpn. Petrol. Inst.*, **40** (1), 29 (1997).

68. Q. Zhang, A. Ishihara and T. Kabe, *J. Jpn. Petrol. Inst.*, **39** (6), 410 (1996).

69. S. S. Shih, S. Mizrahi, L. A. Green and M. S. Sarli, *Ind. Eng. Chem. Res.*, **31** (4), 1232 (1992).

70. K. Sakanishi, M. Ando and I. Mochida, *J. Jpn. Petrol. Inst.*, **35** (5), 403 (1992)[in Japanese].

71. J. A. Anabtawi, K. Alam, M. A. Ali, S. A. Ali and M. A. B. Siddiqui, *Fuel*, **74** (9), 1254 (1995).

72. X. Ma, K. Sakanishi and I. Mochida, *Ind. Eng. Chem. Res.*, **33** (2), 218 (1994).

73. X. Ma, K. Sakanishi, T. Isoda and I. Mochida, *Prepr.-Am. Chem. Soc., Div. Pet. Chem.*, **39** (4), 622 (1994).

74. H. Qabazard, F. Abu-Seedo, A. Stanislaus, M. Andari and M. Absi-Halabi, *Fuel Sci. Technol. Int.*, **13** (9), 1135 (1995).

75. O. Weisser and S, Landa, *Sulfide Catalysts, Their Properties and Applications*, Pergamon Press: Oxford (1973).

76. G. C. A. Schuit and B. C. Gates, *AIChE J.*, **19** (3), 417 (1973).

77. H. Topsøe, B. S. Clausen and F. E. Massoth, *Hydrotreating Catalysis*, p.111, Springer: Berlin (1996).

78. A. N. Startsev, *Catal. Rev.-Sci. Eng.*, **37** (3), 353 (1995).

79. B. Delmon, *Bull. Soc. Chim. Belg.*, **104** (4-5), 173 (1995).

80. J. G. Speight, *The Desulfurization of Heavy Oils and Residua*, Marcel Dekker: New York (1981).

81. R. D. Obolentsev and A. V. Mashkina, *Hydrogenolysis of Sulfur-containing Organic Compounds from Petroleum*, Gos. Nauch. Tekhn. Izd. Neft. i Gorno-Tiplivoi, Lit. Moscow (1961).

82. V. N. Perchenko and S. R. Sergienko, *Selective Catalytic Hydrogenation of Organic Sulfur Compounds*, Izd. A. N. Turkmenskoi S. S. R. Ashkhabad (1962).

83. J. J. Philippson, paper presented at *Am. Inst. Chem. Eng. Meet.*, Houston (1971).

84. N. K. Nag, A. V. Sapre, D. H. Broderick and B. C. Gates, *J. Catal.*, **57**, 509 (1979).

85. Y. Miki, Y. Sugimoto and S. Yamadaya, *J. Jpn. Petrol. Inst.*, **35** (4), 332 (1992) [in Japanese].

86. R. Papadopoulos and M. J. G. Wilson, *Chem. and Ind.*, 1965, 427.

87. S. Landa and A. Mrnkova, *Collect. Czech. Chem. Commun.*, **31**, 2202 (1966).

88. I. A. Van Parijs and G. F. Froment, *Ind. Eng. Chem. Prod. Res. Dev.*, **25**, 431 (1986).

89. I. A. Van Parijs, L. H. Hosten and G. F. Froment, *Ind. Eng. Chem. Prod. Res. Dev.*, **25**, 437 (1986).

90. R. Ramachandran and F. E. Massoth, *Chem. Eng. Commun.*, **18**, 239 (1982).

91. R. Bartsch and C. Tanielian, *J. Catal.*, **35**, 353 (1974).

92. L. D. Rollman, *J. Catal.*, **46**, 243 (1977).

93. P. J. Owens and C. H. Amberg, *Adv. Chem. Ser.*, **33**, 182 (1961).

94. P. Desikan and C. H. Amberg., *Can. J. Chem.*, **42**, 843 (1964).

95. S. Kolboe and C. H. Amberg, *Can. J. Chem.*, **44**, 2623 (1966).

96. A. E. Hargreaves and J. R. H. Ross, *J. Catal.*, **56**, 363 (1979).

97. K. F. McCarthy and G. L. Schrader, *J. Catal.*, **103**, 261 (1987).

98. P. C. H. Mitchell, *Catalysis* (C. Kembell ed.), **Vol.1**, p.204, The Chemical Society: London (1977).

99. M. R. Blake, M. Eyre, R. B. Moyes and P. B. Wells, *Bull. Soc. Chim. Belg.*, **90**, 1293 (1981).

100. P. Pokorny and M. Zdrazil, *Collect. Czech. Chem. Commun.*, **46**, 2185 (1981).

101. A. N. Startsev, V. A. Burmistrov and Yu. I. Hermakov, *Appl. Catal.*, **45**, 191 (1988).

102. C. N. Satterfield and G. W. Roberts, *AIChE. J.*, **14**, 159 (1968).

103. B. Ozimek and B. Radomyski, *Chem. Stosowana*, **20**, 245 (1976).

104. F. E. Massoth, *J. Catal.*, **47**, 316 (1977).

105. S. Morooka and C. E. Hamrin Jr., *Chem. Eng. Sci.,* **32**, 125 (1977).
106. H. C. Lee and J. B. Butt, *J. Catal.,* **49**, 320 (1977).
107. Y. Kawaguchi, J. G. Dalla Lana and F. D. Otto, Canad. *J. Chem. Eng.,* **56**, 65 (1978)
108. V. Vyskocil and M. Kraus, *Collect. Czech. Chem. Commun.,* **44**, 3676(1979).
109. P. Fott and P. Schneider, *Chem. Eng. Sci.,* **39**, 643 (1984).
110. I. A. Van Parijs and G. F. Froment, *Appl. Catal.,* **21**, 273 (1986).
111. B. Radomyski, J. Szczygiel and J. Trawczynski, *Appl. Catal.,* **39**, 25 (1988).
112. S-K. Ihm, S. J. Moon and H. J. Choy, *Ind. Eng. Chem. Res.,* **29**, 1147 (1990).
113. M. Kamil, S. S. Alam and S. K. Saraf, *Indian J. Chem. Technol.,* **1**, 319 (1994).
114. F. P. Daly, *J. Catal.,* **51**, 221 (1978).
115. P. Geneste, P. Amblard, M. Bonnet and P. Graffin, *J. Catal.,* **61**, 115 (1980).
116. E. Furimsky and C. H. Amberg, *Can. J. Chem.,* **54**, 1507 (1976).
117. M. Yamada, Y. L. Shi, T. Obara and K. Sakaguchi, *J. Jpn. Petrol. Inst.,* **33** (4), 227 (1990).
118. S. Kasahara, Y. L. Shi, J. Zou, K. Kawahara and M. Yamada, *J. Jpn. Petrol. Inst.,* **37** (2), 194 (1994).
119. D. R. Kilanowski and B. C. Gates, *J. Catal.,* **62**, 70 (1980).
120. M. Houalla, N. K. Nag, A. V. Sapre, D. H. Broderick and B. C. Gates, *AIChE J.,* **24**, 1015 (1978).
121. M. Nagai and T. Kabe, *J. Jpn. Petrol. Inst.,* **23**, 82 (1980).
122. T. Kabe, A. Ishihara and Q. Zhang, *Appl. Catal.,* **A97**, L1 (1993).
123. I. Dhainaut, C. Gachet and L. de Mourgues, *C. R. Hebd. Seances Acad. Sci. Ser. C,* **288**, 339 (1979).
124. G. Aguilar Rios, G. C. Gachet and L. de Mourgues, *J. Chim. Phys.,* **76**, 661 (1979).
125. W. S. O'Brien, J. W. Chen, R. V. Nayak and G. S. Carr, *Ind. Eng. Chem. Proc. Des. Dev.,* **25**, 221 (1986).
126. T. C. Ho and J. E. Sobel, *J. Catal.,* **128**, 581 (1991).
127. R. Edvinsson and S. Irandoust, *Ind. Eng. Chem. Res.,* **32**, 391 (1993).
128. Q. Zhang, W. Qian, A. Ishihara and T. Kabe, *J. Jpn. Petrol. Inst.,* **40** (3), 185 (1997).
129. M. J. Ledoux, C. P. Huu, Y. Segura and F. Luck, *J. Catal.,* **121**, 70 (1990).
130. S. Betteridge and R. Burch, *Appl. Catal.,* **23**, 413 (1986).
131. Q. Qusro and F. E. Massoth, *Appl. Catal.,* **29**, 375 (1987).
132. A. V. Sapre, D. H. Broderick, D. Fraenkel, N. K. Nag and B. C. Gates; *AIChE J.,* **26**, 690 (1980).
133. P. Desikan and C. H. Amberg, *Can. J. Chem.,* **41**, 1966 (1963).
134. M. Zdrazil, *Collect. Czech. Chem. Commun.,* **40**, 3491 (1975).
135. Y. Miki, Y. Sugimoto and S. Yamadaya, *J. Jpn. Petrol. Inst.,* **36** (1), 32 (1993) [in Japanese].
136. D. Levache, A. Guida and P. Geneste, *Bull. Soc. Chim. Belg.,* **90**, 1285 (1981).
137. S. S. Katti, D. W. B. Westerman, B. C. Gates, T. Youngless and L. Petrakis, *Ind. Eng. Chem. Proc. Des. Dev.,* **23**, 773 (1984).
138. T. Kabe, K. Akamatsu, A. Ishihara, S. Otsuki, M. Godo, Q. Zhang and W. Qian, *Ind. Eng. Chem. Res.,* **36** (12), 5146 (1997).
139. T. Isoda, X. Ma and I. Mochida, *Prepr.-Am. Chem. Soc., Div. Pet. Chem.,* **39** (4), 584 (1994).
140. V. Lamure-Meille, E. Schulz, M. Lemaire and M. Vrinat, *Appl. Catal.,* A, **131** (1), 143(1995).
141. V. Lamure-Meille, E. Schulz, M. Lemaire and M. Vrinat, *J. Catal.,* **170** (1), 29 (1997).
142. V. Vanrysselberghe, R. Le Gall and G. F. Froment, *Ind. Eng. Chem. Res.,* **37** (4), 1235 (1998).
143. H. Gilman and A. L. Jacoby, *J. Org. Chem.,* **4**, 108 (1938).
144. E. Campaigne, L. Hewitt and J. Ashby, *J. Heterocyclic Chem.,* **6**, 553 (1969).
145. R. Gerdil and E. A. C. Lucken, *J. Am. Chem. Soc.,* **87**, 213 (1965).
146. M. V. Landau, D. Berger and M. Herskowitz, *J. Catal.* **158**, 236 (1996).
147. J. M. J. G. Lipsch and G. C. A. Schuit, *J. Catal.,* **15**, 179 (1969).
148. H. Kwart, G. C. A. Schuit and B. C. Gates, *J. Catal.,* **61**, 128 (1980).
149. E. J. Markel, G. L. Schrader, N. N. Sauer and R. J. Angelici, *J. Catal.,* **116**, 11 (1989).
150. N. N. Sauer, E. J. Markel, G. L. Schrader and R. J. Angelici, *J. Catal.,* **117**, 295 (1989).
151. V. LaVopa and C. N. Satterfield, *J. Catal.,* **110**, 375 (1988).
152. M. Nagai and T. Kabe, *J. Catal.,* **81**, 440 (1983).
153. M. L. Vrinat, C. G. Gachet, C. Cavalletto and L. de Mourgues, *Appl. Catal.,* **3**, 57 (1982).
154. S. Irandoust and O. Gahne, *AIChE J.,* **36** (5), 746 (1990).
155. M. Nagai, *Bull. Chem. Soc. Jpn.,* **64** (10), 3210 (1991).
156. A. Ishihara and T. Kabe, *Ind. Eng. Chem. Res.,* **32** (4), 753 (1993).
157. T. Isoda, X. Ma and I. Mochida, *J. Jpn. Petrol. Inst.,* **37** (5), 506 (1994) [in Japanese].
158. M. Sohrabi and M. Zahedi, *Chem. Eng. Technol.,* **18** (6), 420 (1995).
159. T. B. Metcalfe, *Chim. Ind. Gen. Chim.,* **102**, 1300 (1969).
160. H. Gissy, R. Bartsch and C. Tanielian, *J. Catal.,* **65**, 158 (1980).
161. D. H. Broderick, G. C. A. Schuit and B. C. Gates, *J. Catal.,* **54**, 94 (1978).
162. J. Leglise, J. van Gestel and J.-C. Duchet, *Prepr.-Am. Chem. Soc., Div. Pet. Chem.,* **39**, 533 (1994).
163. L. A. Rankel, *Fuel Sci. Technol. Int.,* **9** (4), 435 (1991).

164. N. N. Bulgakov and A. N. Startsev, *Mendeleev Commun.*, 97 (1991).
165. R. C. Pille, C.-Y. Yu and G. F. Froment, *J. Mol. Catal.*, **94** (3), 369 (1994).
166. T. Isoda, X. Ma and I. Mochida, *J. Jpn. Petrol. Inst.*, **38** (1), 25 (1995) [in Japanese].
167. J. van Gestel, L. Finot, J. Leglise and J.-C. Duchet, *Bull. Soc. Chim. Belg.*, **104** (4-5), 189 (1995).
168. C. N. Satterfield, M. Modell and J. F. Mayer, *AIChE J.*, **21**, 1100 (1975).
169. C. N. Satterfield, M. Modell and J. A. Wilkens, *Ind. Eng. Chem. Proc. Des. Dev.*, **19**, 154 (1980).
170. C. H. Amberg, *J. Less-Common Met.,* **36**, 339 (1974).
171. S. W. Cowley and F. E. Massoth, *J. Catal.,* **51**, 291 (1978).
172. M. Nagai, T. Sato and A. Aiba, *J. Catal.,* **97**, 52 (1986).
173. J. Miciukiewicz, W. Zmierczak and F. E. Massoth, *Proc. 8th Int. Congr. Catal.*, Verlag Chemie, Weinheim, **Vol. 2**, p. 671 (1984).
174. J. Yang and F. E. Massoth, *Appl. Catal.*, **34**, 215 (1987).
175. L. C. Gutberlet and R. J. Bertolacini, *Ind. Eng. Chem. Prod. Res. Dev.*, **22**, 246 (1983).
176. H. Kwart, J. R. Katzer and J. Horgan, *J. Phys. Chem.*, **86**, 2641 (1982).
177. W. Kaernbach, W. Kisielow, L. Warzecha, K. Miga and R. Klecan, *Fuel*, **69** (2), 221 (1990).
178. C. Vladov, L. A. Petrov, A. D. Cao and B. Ytua, *Collect. Czech. Chem. Commun.*, **57** (12), 2524 (1992).
179. M. Nagai, *J. Jpn. Petrol. Inst.*, **38** (1), 52 (1995).
180. M. Nagai, *Ind. Eng. Chem., Prod. Res. Dev.*, **24**, 489 (1985).
181. M. Nagai, *Chem. Lett.*, 1023 (1987).
182. S. Boon-Long and J. Tscheikuna, *Asahi Garasu Zaidan Josei Kenkyu Seika Hokoku*, 653 (1994).
183. S. Chand, *Res. Ind.*, **39** (3), 194 (1994).
184. E. O. Odebunmi and D. F. Ollis, *J. Catal.,* **80**, 65 (1983).
185. C.-L. Lee and D. F. Ollis, *J. Catal.,* **87**, 332 (1984).
186. S. Krishnamurthy and Y. T. Shah, *Chem. Eng. Commun.*, **16**, 109 (1982).
187. K. W. Lee, M. J. Choi and S. B. Kim, *Korean J. Chem. Eng.*, **8** (3), 143 (1991).
188. G. R. Kocis and T. C. Ho, *Chem. Eng. Res. Des.*, **64**, 288 (1986).
189. H.-M. Lin, H. M. Sebastian and K.-C. Chao, *J. Chem. Eng. Data*, **25**, 252 (1980).
190. J. J. Simnick, C. C. Lawson, H. M. Lin and K. C. Chao, *AIChE. J.*, **23**, 469 (1977).
191. P. Wentreck and H. Wise, *J. Catal.,* **51**, 80 (1978).
192. H. Wise, *J. Less Comm. Met.*, **54**, 331 (1974).
193. K. N. Mathur, Z. Sarbak, N. Islam, H. Kwart and J. R. Katzer, *Tenth and Eleventh Quarterly Reports for the Period August 16, 1981 to February 15*, 1982; Prepared for Office of Fossil Energy, Department of Energy, Washington, DC (1982).
194. K. N. Mathur, M. D. Schrenk, H. Kwart and J. R. Katzer, *Final report for the Period September 15, 1978 to September 1981*; Prepared for Office of Fossil Energy, Department of Energy, Washington, DC (1982).
195. R. Prins, V. H. J. de Beer and G. A. Somorjai, *Catal. Rev.-Sci. Eng.*, **31** (1&2), 1(1989).
196. S. Kolboe, *Can. J. Chem.*, **47**, 352 (1969).
197. F. Ruette and E. V. Ludena, *J. Catal.,* **67**, 266 (1981).
198. G. V. Smith, C. C. Hinckly and F. Behbahany, *J. Catal.,* **30**, 218 (1973).
199. M. R. Blake, M. Eyre, R. B. Moyes and P. B. Wells, *Proc. 7th Int. Congr. on Congr. on Catal.*, (T. Seiyama, K. Tanabe eds.), **Part A**, p.591, Elsevier: Amsterdam (1980).
200. J. Joffre, D. A. Lerner and P. Geneste, *Bull. Soc. Chim. Belg.*, **93**, 831 (1984).
201. J. Joffre, P. Geneste and D, A. Lerner, *J. Catal.,* **97**, 543 (1986).
202. R. P. Diez and A. H. Jubert, *J. Mol. Catal.*, **83**, 219 (1993).
203. C. M. Friend and J. T. Roberts, *Acc Chem. Res.* **21**, (11), 394 (1988).
204. A. J. Gellman, M. H. Farias and G. A. Somorjai, *J. Catal.,* **88**, 546 (1984).
205. M. Zdrazil, *Collect. Czech. Chem. Commun.*, **42**, 1484 (1977).
206. N. V. Richardson and J. C. Campuzano, *Vacuum*, **31**, 449 (1981).
207. R. E. Preston and J. B. Benziger, *J. Phys. Chem.*, **89**, 5010 (1985).
208. G. R. Schoofs, R. E. Preston and J. B. Benziger, *Langmuir*, **1**, 313 (1985).
209. J. Stohr, J. L. Gland, E. B. Kollin, R. J. Koestner, A. L. Johnson, E. L. Muetterties and F. Sette, *Phys. Rev. Lett.*, **53**, 2161 (1984).
210. C. Rong and X. Qin, *Proc. 10th Int. Congr. Catal.* (L. Guczi *et al.* eds.), p.1919, Elsevier: Amsterdam (1992).
211. R. J. Mikovsky, A. J. Silvestri and H. Heinemann, *J. Catal.,* **34**, 324 (1974).
212. P. Kieran and C. J. Kemball, *J. Catal.,* **4**, 380 (1965).
213. J. Maternová and M. Zdrazil, *Collect. Czech. Chem. Commun.*, **45**, 2532 (1980).
214. A. J. Duben, *J. Phys. Chem.*, **82**, 348 (1978).
215. S. Harris and R. R. Chianelli, *J. Catal.,* **86**, 400 (1984).
216. M. C. Zonnevylle, R. Hoffmann and S. Harris, *Surf. Sci.*, **199**, 320 (1988).
217. M. J. Calhordo, R. Hoffmann and C. M. Friend, *J. Am. Chem. Soc.*, **112**, 50 (1990).
218. M. Neurock and R. A. van Santen, *J. Am. Chem. Soc.*, **116**, 4427 (1994).

219. P. Kieran and C. J. Kemball, *J. Catal.*, **4**, 394 (1965).
220. G. P. Singhal, R. L. Espino and J. E. Sobel, *J. Catal.*, **67**, 446 (1981).
221. M. Zdrazil, *Appl. Catal.*, **4**, 107 (1982).
222. H. Topsøe and B. S. Clausen, *Catal. Rev.-Sci. Eng.*, **26**, 395 (1984).
223. M. Zdrazil and J. Sedlacek, *Collect. Czech. Chem. Commun.*, **42**, 3133 (1977).
224. J. Joffre, P. Geneste, J.-P. Mensah and C. Moreau, *Bull. Soc. Chim. Belg.*, **108**, 865 (1991).
225. E. L. Muetterties and J. Stein, *Chem. Rev.*, **79**, 479 (1979).
226. A. Ishihara, Ph.D. Dissertation, Kyoto University, Kyoto (1989).
227. H. D. Kaesz, R. B. King, T. A. Manuel, L. D. Nichols and F. G. A. Stone, *J. Am. Chem. Soc.*, **82**, 4749 (1960).
228. R. B. King and F. G. A. Stone, *J. Am. Chem. Soc.*, **82**, 4557 (1960).
229. R. B. King, P. M. Treichel and F. G. A. Stone, *J. Am. Chem. Soc.*, **83**, 3600 (1961).
230. R. J. Angelici, *Acc. Chem. Res.*, **21**, 387 (1988).
231. R. J. Angelici, *Coord. Chem. Rev.*, **105**, 61 (1990).
232. R. J. Angelici, *Bull. Soc. Chim. Belg.*, **104** (4-5), 265 (1995).
233. T. B. Rauchfuss, *Prog. Inorg. Chem.*, **39**, 259 (1991).
234. B. C. Wiegand and C. M. Friend, *Chem. Rev.*, **92** (4), 491 (1992).
235. R. A. Sanchez-Delgado, *J. Mol. Cat.*, **86**, 287 (1994).
236. M. Draganjac, C. J. Ruffing and T. B. Rauchfuss, *Organometallics*, **4**, 1909 (1985).
237. M. Choi and R. J. Angelici, *Organometallics*, **10**, 2436 (1991).
238. M. Choi, M. Robertson and R. J. Angelici, *J. Am. Chem. Soc.*, **113**, 4005 (1991).
239. R. Cordone, W. D. Harman and H. Taube, *J. Am. Chem. Soc.*, **111**, 5969 (1989).
240. A. E. Ogilvy, A. E. Skaugset and T. B. Rauchfuss, *Organometallics*, **8**, 2739 (1989).
241. J. Chen and R. J. Angelici, *Organometallics*, **8**, 2277 (1989).
242. J. Chen and R. J. Angelici, *Organometallics*, **9**, 849 (1990).
243. A. E. Skaugset, T. B. Rauchfuss and C. L. Stern, *J. Am. Chem. Soc.*, **112**, 2432 (1990).
244. D. A. Lesch, J. W. Richardson Jr., R. A. Jacobson and R. J. Angelici, *J. Am. Chem. Soc.*, **106**, 2901 (1984).
245. R. A. Sanchez-Delgado, R. L. Marquez-Silva, J. Puga, A. Tiripicchio and M. Tiripicchio-Camellini, *J. Organomet. Chem.*, **316**, C35 (1986).
246. S. C. Huckett, N. N. Sauer and R. J. Angelici, *Organometallics*, **6**, 591 (1987).
247. N. N. Sauer and R. J. Angelici, *Organometallics*, **6**, 1146 (1987).
248. G. H. Spies and R. J. Angelici, *Organometallics*, **6**, 1897 (1987).
249. S. C. Huckett, L. L. Miller, R. A. Jacobson and R. J. Angelici, *Organometallics*, **7**, 686 (1988).
250. J. W. Hachgenei and R. J. Angelici, *J. Organomet. Chem.*, **355**, 359 (1988).
251. J. R. Lockemeyer, T. B. Rauchfuss, A. L. Rheingold and S. R. Wilson, *J. Am. Chem. Soc.*, **111**, 8828 (1989).
252. J. W. Hachgenei and R. J. Angelici, *Organometallics*, **8**, 14 (1989).
253. E. A. Ganja, T. B. Rauchfuss and C. L. Stern, *Organometallics*, **10**, 270 (1991).
254. M. G. Choi, L. M. Daniels and R. J. Angelici, *J. Am. Chem. Soc.*, **111**, 8753 (1989).
255. J. Chen, L. M. Daniels and R. J. Angelici, *J. Am. Chem. Soc.*, **112**, 199 (1990).
256. J. Chen, L. M. Daniels and R. J. Angelici, *J. Am. Chem. Soc.*, **113**, 2544 (1991).
257. J. Chen and R. J. Angelici, *Organometallics*, **9**, 879 (1990).
258. A. E. Skaugset, T. B. Rauchfuss and S. R. Wilson, *Organometallics*, **9**, 2875 (1990).
259. J. Chen and R. J. Angelici, *Appl. Organomet. Chem.*, **6**, 479 (1992).
260. J. Chen and R. J. Angelici, *Polyhedron*, **9**, 1883 (1990).
261. C. A. Dullaghan, S. Sun, G. B. Carpenter, B. Weldon and D. A. Sweigart, *Angew. Chem., Int. Ed. Engl.*, **35** (2), 212 (1996).
262. W. D. Jones and L. Dong, *J. Am. Chem. Soc.*, **113**, 559 (1991).
263. L. Dong, S. B. Duckett, K. F. Ohman and W. D. Jones, *J. Am. Chem. Soc.*, **114**, 151 (1992).
264. W. D. Jones, R. M. Chin and A. W. Meyer, *Prepr.-Am. Chem. Soc., Div. Pet. Chem.*, **38** (3), 650 (1993).
265. C. Bianchini, A. Meli, M. Peruzzini, F. Vizza, P. Frediani, V. Herrera and R. A. Sánchez-Delgado, *J. Am. Chem. Soc.*, **115**, 2731 (1993).
266. C. Bianchini, A. Meli, M. Peruzzini, F. Vizza, P. Frediani, V. Herrera, and R. A. Sanchez-Delgado, *J. Am. Chem. Soc.*, **115**, 7505 (1993).
267. C. Bianchini, A. Meli, M. Peruzzini, F. Vizza, S. Moneti, V. Herrera and R. A. Sanchez-Delgado, *J. Am. Chem. Soc.*, **116**, 4370 (1994).
268. J. S. Merola, A. Grieb, F. T. Ladipo and H. E. Selnau, *Prepr.-Am. Chem. Soc., Div. Pet. Chem.*, **38** (3), 674 (1993).
269. J. J. Garcia, B. E. Mann, H. Adams, N. A. Bailey and P. M. Maitlis, *J. Am. Chem. Soc.*, **117**, 2179 (1995).
270. M. H. Chisholm, S. T. Haubrich, J. C. Huffman and W. E. Streib, *J. Am. Chem. Soc.*, **119** (7), 1634 (1997).
271. C. Bianchini, P. Frediani, V. Herrera, M. V. Jimenez, A. Meli, L. Rincon, R. A. Sanchez-Delgado and F. Vizza, *J. Am. Chem. Soc.*, **117**, 4333 (1995).

272. C. Bianchini, M. V. Jimenez, A. Meli, S. Moneti, F. Vizza, V. Herrera and R. A. Sanchez-Delgado, *Organometallics*, **14**, 2342 (1995).
273. G. P. Rosini and W. D. Jones, *J. Am. Chem. Soc.*, **114**, 10767 (1992).
274. W. D. Jones and R. M. Chin, *J. Am. Chem. Soc.*, **116**, 198 (1994).
275. W. D. Jones and R. M. Chin, *Organometallics*, **11**, 2698 (1992).
276. S. Luo, A. E. Skaugset, T. B. Rauchfuss and S. C. Wilson, *J. Am. Chem. Soc.*, **114**, 1732 (1992).
277. A. E. Ogilvy, M. Draganjac, T. B. Rauchfuss and S. R. Wilson, *Organometallics*, **7**, 1171 (1988).
278. A. Arce, P. Arrojo, A. J. Deeming and Y. de Sanctis, *J. Chem. Soc.*, *Dalton Trans.*, **1992**, 2423.
279. U. Riaz, O. Curnow and M. D. Curtis, *J. Am. Chem. Soc.*, **113**, 1416 (1991).
280. U. Riaz, O. Curnow and M. D. Curtis, *J. Am. Chem. Soc.*, **116**, 4357 (1994).
281. K. Matsubara, R. Okamura, M. Tanaka and H. Suzuki, *J. Am. Chem. Soc.*, **120** (5), 1108 (1998).
282. M. Salmeron, G. A. Somorjai, A. Wold, R. R. Chianelli and K. S. Liang, *Chem. Phys. Lett.*, **90**, 105 (1982).
283. M. H. Farias, A. J. Gellman, G. A. Somorjai, R. R. Chianelli and K. S. Liang, *Surf. Sci.*, **140**, 181 (1984).
284. M. Salmeron and G. A. Somorjai, *Surf. Sci.*, **126**, 410 (1983).
285. A. J. Gelmann, M. H. Farias, M. Salmeron and G. A. Somorjai, *Surf. Sci.*, **136**, 217 (1984).
286. D. G. Kelly, M. Salmeron and G. A. Somorjai, *Surf. Sci.*, **175**, 465 (1986).
287. D. G. Kelly, J. A. Odriozola and G. A. Somorjai, *J. Phys. Chem.*, **91**, 5695 (1987).
288. F. Zaera, E. B. Kollin and J. L. Gland, *Surf. Sci.*, **184**, 75 (1987).
289. J. T. Roberts and C. M. Friend, *Surf. Sci.*, **186**, 201 (1987).
290. R. A. Cocco and B. J. Tatarchuk, *Surf. Sci.*, **218**, 127 (1989).
291. T. M. Gentle, *Energy Res. Abstr.*, **9**, 28229 (Abstract) (1984).
292. J. Stöhr, J. L. Gland, E. B. Kollin, R. J. Koestner, A. L. Johnson, E. L. Muetterties and F. Sette, *Phys. Rev. Lett.*, **53**, 2161 (1984).
293. J. F. Lang and R. I. Masel, *Surf. Sci.*, **183**, 44 (1987).
294. F. P. Netzer, E. Bertel and A. Goldmann, *Surf. Sci.*, **201**, 257 (1988).
295. G. R. Schoofs, R. E. Preston and J. B. Benziger, *Langmuir*, **1**, 313 (1985).
296. R. E. Preston and J. B. Benziger, *J. Phys. Chem.*, **89**, 5010 (1985).
297. B. A. Sexton, *Surf. Sci.*, **163**, 99 (1985).
298. A. C. Liu and C. M. Friend, *J. Am. Chem. Soc.*, **113**, 820 (1991).
299. F. Zaera, E. B. Kollin and J. L. Gland, *Langmuir*, **3**, 555 (1987).
300. J. Stöhr, E. B. Kollin, D. A. Fischer, J. B. Hastings, F. Zaera and F. Sette, *Phys. Rev. Lett.*, **55**, 1468 (1985).
301. N. V. Richardson and J. C. Campuzano, *Vacuum*, **31**, 449 (1981).
302. J. T. Roberts and C. M. Friend, *J. Am. Chem. Soc.*, **108**, 7204 (1986).
303. R. E. Preston and J. B. Benziger, *J. Phys. Chem.*, **89**, 5002 (1985).
304. A. J. Gellman, D. Neiman and G. A. Somorjai, *J. Catal.*, **107**, 92 (1987).
305. A. J. Gellman, D. Neiman and G. A. Somorjai, *J. Catal.*, **107**, 103 (1987).
306. M. E. Bussel and G. A. Somorjai, *J. Catal.*, **106**, 93 (1987).
307. J. R. Katzer and R. Sivasubramanian, *Catal. Rev.-Sci. Eng.*, **20**, 155, (1979).
308. M. J. Ledoux, *Catalysis*, A Specialist Report, Royal Society of Chemistry, Burlington House, London, **7**, 131 (1986).
309. H. Schulz, M. Schon and N. M. Rahman, *Studies in Surface Science and Catalysis, Vol. 27, Catalytic Hydrogenation, a Modern Approach* (L. Cerveny, ed.), p.204, Elsevier: Amsterdam (1986).
310. T. C. Ho, *Catal. Rev.-Sci. Eng.*, **30**, 117 (1988).
311. C. Moreau and P. Geneste, *Theoretical Aspects of Heterogeneous Catalysis* (J. B. Moffat ed.), p.256, Van Nostrand Reinhold catalysis series (1990).
312. G. Perot, *Catalysis Today*, **10**, 447 (1991).
313. S. Kasztelan, T. des Courieres and M. Breysse, *Catalysis Today*, **10**, 433 (1991).
314. J. K. Minderhoud and J. A. R. van Veen, *Fuel Process. Technol.*, **35**, 87 (1993).
315. S. H. Pine, J. B. Hendrickson, D. J. Cram and G. S. Hammond, in: *Organic Chemistry*, 4th edition, p.85, McGraw-Hill (1981).
316. J. F. Cocchetto and C. N. Satterfield, *Ind. Eng. Chem.*, *Process Des. Dev.*, **15**, 272 (1976).
317. J. F. Cocchetto and C. N. Satterfield, *Ind. Eng. Chem.*, *Process Des. Dev.*, **20**, 49 (1981).
318. C. N. Satterfield and J. F. Cocchetto, *AIChE. J.*, 21, 1107 (1975).
319. S. R. Shih, J. R. Katzer, H. Kwart and A. B. Stiles, *Prepr. Am. Chem. Soc., Div. Petrol. Chem.*, **22**, 919 (1977).
320. E. W. Stern, *J. Catal.*, **57**, 390 (1979).
321. J. Shabtai, N. K. Nag and F. E. Massoth, *10th N. Amer. Catal. Soc. Meeting*, San Diego (1987).
322. J. Shabtai, L. Veluswamy and A. G. Oblad, *Prepr. Pa.-Am. Chem. Soc., Div. Fuel Chem.*, **23**, 114 (1978).
323. Z. Sarbak, *Acta Chim. Hung.*, **127** (3), 371 (1990).
324. M. V. Bhinde and Ph.D. Dissertation, University of Delaware, Newark, 1979.
325. K. E. Cox and L. Berg, *Chem. Eng. Prog.*, **58**, 54 (1962).

326. M. Cerny, *Coll. Czechoslov. Chem. Commun.*, **44**, 85 (1979).
327. H. G. McIlvried, *Ind. Eng. Chem., Process Des. Dev.*, **10**, 125 (1971).
328. R. T. Hanlon, *Energy & Fuels*, **1**, 424 (1987).
329. J. Sonnemans, G. H. van den Berg and P. Mars, *J. Catal.*, **31**, 220 (1973).
330. C. D. Ajaka and R. S. Mann, *Ind. J. Technol.*, **31**, 131 (1993).
331. A. Kherbeche, R. Hubaut, J. P. Bonnelle and J. Grimblot, *J. Catal.*, **131**, 204 (1991).
332. C. N. Satterfield and S. H. Yang, *Ind. Eng. Chem., Process Des. Dev.*, **23**, 11 (1984).
333. E. O. Odebunmi and D. F. Ollis, *J. Catal.*, **80**, 76 (1983).
334. J. L. Olivé, S. Biyoko, C. Moulinas and P. Geneste, *Appl. Catal.*, **19**, 165 (1985).
335. C. Moreau, R. Durand, N. Zmimita and P. Geneste, *J. Catal.*, **112**, 411 (1988).
336. C. Moreau, L. Bekakra, J-L. Olive and P. Geneste, *Pro. 9th Int. Congr. Catal.* (M. J. Phillips, M. Ternan eds.), p.58 (1988).
337. S. H. Yang and C. N. Satterfield, *Ind. Eng. Chem., Process Des. Dev.*, **23**, 20 (1984).
338. S. Shih, E. Reiff, R. Zawadski and J. R. Katzer, *Prepr. Am. Chem. Soc., Div. Fuel Chem.*, **23**, 99 (1978).
339. C. N. Satterfield and J. F. Cocchetto, *Ind. Eng. Chem., Process Des. Dev.*, **20**, 53 (1981).
340. F. Gioia and V. Lee, *Ind. Eng. Chem., Process Des. Dev.*, **25**, 918 (1986).
341. J. T. Miller and M. F. Hineman, *J. Catal.*, **85**, 117 (1984).
342. M. Nagai, K. Sawahiraki and T. Kabe, *Nippon Kagakukaishi*, 1350 (1979) [in Japanese].
343. R. Zawadski, S. S. Shih, E. Reiff, J. R. Katzer and H. Kwart, *Tenth and Eleventh Quarterly Reports for the Period August 16, 1981 to February 15, 1982*; Prepared for Office of Fossil Energy, Department of Energy, Washington, DC (1982).
344. J. Shabtai, G. J. C. Yeh, C. Russell and A. G. Oblad, *Ind. Eng. Chem. Res.*, **28**, 139 (1989).
345. K. Malakani, P. Magnoux and G. Perot, *Appl. Catal.*, **30**, 371 (1987).
346. Y. Liu, F. E. Massoth and J. Shabtai, *Bull. Soc. Chim. Belg.*, **93**, 627 (1984).
347. A. K. Aboul-Gheit and I. K. Abdou, *J. Inst. Petrol.*, **59** (568), 188 (1973).
348. F. E. Massoth, K. Balsami and J. Shabtai, *J. Catal.*, **122**, 256 (1990).
349. V. Morávek, J.-C. Duchet and D. Cornet, *Appl. Catal.*, **66**, 257 (1990).
350. M. Nagai, K. Sawahiraki and T. Kabe, *Nippon Kagakukaishi*, **69** (1980) [in Japanese].
351. M. Nagai, T. Masunaga and N. Hana-oka, *Energy & Fuels*, **2**, 645 (1988).
352. Z. Sarbak, *Acta Chim. Hung.*, **127** (3), 359 (1990).
353. A. Finiels, P. Geneste, C. Moulinas and J.-L. Olive, *Appl. Catal.*, **22**, 257 (1986).
354. C. N. Satterfield and S. Gultekin, *Ind. Eng. Chem., Process Des. Dev.*, **20**, 62 (1981).
355. M. V. Bhinde, S. Shih, R. Zawadski, J. R. Katzer and H. Kwart, *Proc. 3rd Int. Conf. on Chemistry and Uses of Molybdenum* (H. F. Barry and P. C. H. Mitchell eds.), Climax Molybdenum Company, p.184, 1979.
356. T. C. Ho, A. A. Montagna and J. J. Steger, *Proc. 8th Int. Cong. Catal., Dechema*, 1984, II-257.
357. F. Goudriaan, H. Gierman and J. C. Vlugter, *J. Inst. Petrol*, **59**, 40 (1973).
358. K. Shanthi, C. N. Pillai and J. C. Kuriacose, *Appl. Catal.*, **46**, 241 (1989).
359. K. Shanthi, C. N. Pillai and J. C. Kuriacose, *Indian J. Chem.*, **30A**, 584 (1991).
360. J. van Gestel, J. Leglise and J.-C. Duchet, *Appl. Catal. A, Gen.* **92**, 143 (1992).
361. U. S. Ozkan, S. Ni, L. Zhang and E. Moctezuma, *Energy & Fuels*, **8**, 249 (1994).
362. U. S. Ozkan, S. Ni, E. Moctezuma and L. Zhang, *Prepr. Pap.-Am. Chem. Soc., Div. Petrol. Chem.*, **38**, 696 (1993).
363. H. S. Lo, Ph.D. Dissertation, University of Delaware, Newark, 1981.
364. M. Nagai, T. Masunaga and N. Hana-oka, *J. Catal.*, **101**, 284 (1986).
365. M. Nagai, *J. Jpn. Petrol. Inst.*, **36** (6), 502 (1993).
366. M. Nagai, *Bull. Chem. Soc. Jpn.*, **64** (1), 330 (1991).
367. J. P. Franck and J. F. Le Page, *Proc. 7th Int. Conf. Catal.*, (T. Seiyama, K. Tanabe eds.), Kodansha, Elsevier, Tokyo, p. 792, 1980.
368. M. J. Ledoux, G. Agostini, R. Benazouz and O. Michaux, *Bull. Soc. Chim. Belg.*, **93**, 635 (1984).
369. G. Perot, S. Brunet and N. Hamzé, *Proc. 9th Int. Conf. Catal.*, (M. J. Phillips, M. Ternan, eds.), Calgary **Vol. 1**, The Chemical Institute of Canada, Ottawa, p. 19, 1988.
370. M. J. Ledoux and B. Djellouli, *Appl. Catal.*, **67**, 81 (1990).
371. C. N. Satterfield and S. H. Yang, *J. Catal.*, **81**, 335 (1983).
372. C. N. Satterfield and C. M. Smith, *Ind. Eng. Chem., Process Des. Dev.*, **25**, 942 (1986).
373. C. N. Satterfield, C. M. Smith and M. Ingalls, *Ind. Eng. Chem., Process Des. Dev.*, **24**, 1000 (1985).
374. H. Toulhoat and R. Kessas, *Rev. Inst. Fr. du Petrole*, **41** (4), 511 (1986).
375. H. Toulhoat, R. Kessas, I. Ignatiadis, J. M. Schmitter, P. Arpino and G. Guiochon, *Proc. 9 th Ibero-American Symp. Catal.*, 657 (1984).
376. G. Perot, S. Brunet, C. Canaff and H. Toulhoat, *Bull. Soc. Chim. Belg.*, **96**, 865 (1987).
377. C. Moreau, L. Bekakra, R. Durand, N. Zmimita and P. Geneste, in: *Hydrotreating Catalysts*, (M. L. Occelli and R. G. Anthony esd.), **50**, p.115, Elsevier, Amsterdam, Studies Surface Science and Catalysis (1989).

378. L. Vivier, P. D'Araujo, S. Kasztelan and G. Perot, *Bull. Soc. Chim. Belg.*, **100**, 807 (1991).
379. L. Vivier, S. Kasztelan and G. Perot, *Bull. Soc. Chim. Belg.*, **100**, 801 (1991).
380. M. Callant, P. Grange, K. A. Holder and B. Delmon, *Bull. Soc. Chim. Belg.*, **100**, 823 (1991).
381. B. J. Bandy, D. R. Lloyd and N. V. Richardson, *Surf. Sci.*, **89**, 344 (1979).
382. K. Kishi and S. Ikeda, *J. Phys. Chem.*, **73**, 2559 (1969).
383. G. E. Calf, J. L. Garnett and V. A. Pickles, *Aust. J. Chem.*, **21**, 961 (1968).
384. J. L. Gland and G. A. Somorjai, *Adv. Coll. Interface Sc.*, **5**, 205 (1976).
385. G. R. Schoofs and J. B. Benziger, *J. Phys. Chem.*, **92**, 741 (1988).
386. A. K. Myers and J. B. Benziger, *Langmuir*, **5**, 1270 (1989).
387. S. X. Huang, D. A. Fischer and J. L. Gland, *Prepr. Pap.-Am. Chem. Soc., Div. Petrol. Chem.*, **38**, 692 (1993).
388. P. A. Armstrong, A. T. Bell and J. A. Reimer, *J. Phys. Chem.*, **97**, 1952 (1993).
389. M. J. Ledoux, P. E. Puges and G. Maire, *J. Catal.*, **76**, 285 (1982).
390. T. Fransen, O. van der Meer and P. Mars, *J. Phys. Chem.*, **80**, 2103 (1976).
391. J. A. Marzari, S. Rajagopal and R. Miranda, *J. Catal.*, **156**, 255 (1995).
392. S. H. Yang and C. N. Satterfield, *J. Catal.*, **81**, 168 (1983).
393. W. S. Millman, C. H. Bartholomew and R. L. Richardson, *J. Catal.*, **90**, 10 (1984).
394. G. C. Hadjiloizou, J. B. Butt and J. S. Dranoff, *Ind. Eng. Chem. Res.*, **31**, 2503 (1992).
395. J. L. Lemberton, N. Gnofam and G. Perot, *Appl. Catal.*, A **90**, 175 (1992).
396. U. S. Ozkan, L. Zhang, S. Ni and E. Moctezuma, *J. Catal.*, **148**, 181 (1994).
397. S. E. Lott, T. J. Gardner, L. I. McLaughlin and J. B. Oelfke, *Prepr. Pap.-Am. Chem. Soc., Div. Fuel Chem.*, **39**, 1073 (1994).
398. E. S. Olson and R. K. Sharma, *Prepr. Pap.-Am. Chem. Soc., Div. Fuel Chem.*, **39**, 1073 (1994).
399. N. Nelson and R. B. Levy, *J. Catal.*, **58**, 485 (1979).
400. J. L. Portefaix, M. Cattenot, M. Guerriche and M. Breysse, *Catal. Lett.*, **9**, 127 (1991).
401. J. L. Portefaix, M. Cattenot, M. Guerriche, J. Thivolle-Cazat and M. Breysse, *Catal. Today*, **10**, 473 (1991).
402. L. Vivier, V. Dominguez, G. Perot and S. Kasztelan, *J. Mol. Catal.*, **67**, 267 (1991).
403. G. Perot, P. d'Araujo, L. Vivier, S. Kasztelan and F. R. Ribeiro, *Prepr. Pap.-Am. Chem. Soc., Div. Petrol. Chem.*, **38**, 712 (1993).
404. R. M. Laine, *Catal. Rev.-Sci. Eng.*, **25**, 459 (1983).
405. S. Kasztelan, *Prepr. Pap.-Am. Chem. Soc., Div. Petrol. Chem.*, **38**, 642 (1993).
406. J. B. Bonanno, T. P. Henry, D. R. Neithamer, P. T. Wolczanski and E. B. Lobkovsky, *J. Am. Chem. Soc.*, **118**, 5132 (1996).
407. S. D. Gray, D. P. Smith, M. A. Bruck and D. E. Wigley, *J. Am. Chem. Soc.*, **114**, 5462 (1992).
408. S. D. Gray, K. J. Weller, M. A. Bruck, P. M. Briggs and D. E. Wigley, *J. Am. Chem. Soc.*, **117**, 10678 (1995).

3

Structure of Hydrodesulfurization and Hydrodenitrogenation Catalysts

Catalytic hydrodesulfurization (HDS) and hydrodenitrogenation (HDN) are processes in which organic nitrogen- and organic sulfur-containing compounds are removed from hydrocarbon feedstock to produce processible, stable, and environmentally acceptable liquid fuels and lube base stocks. This process is commonly referred to as "hydrotreatment," which is an integral part of oil refining. Although it has long been recognized that HDN is more difficult than HDS, the former has historically been of little concern to refiners because of the comparatively small quantities of nitrogen compounds present in conventional petroleum stocks. As a result, HDN was very similar to HDS either in the catalysts used or in the reaction conditions. The purpose of this chapter is to describe the structure of catalysts for hydrotreatment, with emphasis on HDS.

Conventional industrial hydrotreatment is generally carried out over sulfided molybdenum or tungsten-based catalyst promoted by a transient metal such as cobalt and nickel. Different combinations of these elements are used for different kinds of reactions. Co-Mo catalysts are principally used for HDS, Ni-Mo catalysts have better catalytic properties for HDN and aromatic saturation,[1] while the Ni-W system is usually employed for hydrocracking.[2,3] It is usually assumed that the properties of these catalysts are similar, if not identical, to the properties of their Mo analogues. It is true that W and Ni-W catalysts behave in roughly the same way as Mo and Co-Mo or Ni-Mo catalysts. In HDS, hydrogenation and HDN, the dependence of Ni-Mo and Ni-W catalysts on catalytic activity is similar.[4-8]

This chapter begins with the structure of molybdenum sulfide and tungsten sulfide. Following this is a discussion of the related effects of some factors such as sulfiding conditions, loadings of active metal species, and properties of support on the structure of catalysts. The current understanding on the structure of conventional catalysts is then summarized. A major portion of this part discusses the behavior of sulfur during hydrotreatment and the correlation between the hydrotreating activity and structure and other topics such as the effect of support, presulfiding state, etc.

3.1 Structure of Conventional Molybdenum- and Tungsten-based Catalysts

3.1.1 Structure of Oxide Catalysts

The structure of unpromoted molybdenum and tungsten catalysts has been widely studied.

One of main aims is to establish a basis for understanding the more complex and industrially more important promoted catalysts. In fact, the unpromoted catalysts are also important in their own right and many informative structure-activity relationships have been established.

Supported molybdenum-based catalysts are widely used in the hydrogenation, HDN and HDS of petroleum feedstocks. It is well known that the structures of metal sulfides and their relative proportion in sulfided molybdenum catalysts are very dependent on the content of the molybdenum. The content of molybdenum is consequently one of the more important parameters in optimizing the commercial molybdenum-based hydrotreating catalysts. Thus, their structure has been one of the most widely investigated and debated questions in heterogeneous catalysis. The introduction of various physical and chemical characterization approaches has resulted in significant progress in the understanding of the surface structures present in calcined molybdenum-based catalysts. Regarding the formation of the surface structures, it is commonly accepted that molybdenum interacts in the calcined state with the hydroxyl group on the alumina carrier surface, resulting in the formation of monolayer of structure.[9,10] An ESR study of MoO_3 on η- and γ-Al_2O_3 was reported by Dufaux et al.[11] If MoO_3 is calcined at temperatures around 750 K, it usually exhibits a weak signal at g = 1.93, which is assigned to Mo^{5+} in tetragonal square configuration. For γ-Al_2O_3, no Mo^{5+} signal was observed below 10 wt% MoO_3, suggesting that the MoO_3 interacted with the support. The authors ascribed the results to the formation of the monolayer of MoO_3 on the surface of the support. Later, Lim and Weller have confirmed the monolayer on alumina surface in a thermodynamic study.[12] A similar conclusion was also reached in a combined ^{1}H-NMR and low temperature chemisorption study,[13] and from several EXAFS studies.[14,15] Even for a Mo/Al_2O_3 of relatively high loading (15 wt% MoO_3) infrared spectroscopy could not detect bands due to a MoO_3 phase, indicating that the Mo is present in a highly dispersed phase.[16]

As noted above, in the oxide state there are many similarities between alumina-supported W and Mo catalysts. The species present at low loading are isolated tetrahedral WO_4^{2-} units. At higher loadings polytungstate or polymolybdate species are formed, and at still higher loadings bulk WO_3 or MoO_3 is formed.[17-19] In a series of XPS and Raman spectroscopy studies, Wachs and coworkers[20-23] showed, as in the case of molybdenum-alumina catalysts, that the formation and location of the surface tungsten-oxide species are controlled by the surface hydroxyl chemistry. The surface oxide species were found to be located in the outermost layer of the catalysts as an overlayer. By means of XANES[24,25] and Raman spectroscopy[24] it was revealed that polymeric tungsten-oxide structure form chains of WO_4 and WO_6 (or WO_5) units with a relative distribution depending on the tungsten loading. The nature of the tungsten-oxide species in W/Al_2O_3 catalysts has also been elucidated by TPR studies,[26-31] as well as by HREM, energy disperse X-ray spectroscopy, etc.[29] In contrast to the formation of $Al_2(MoO_4)_3$ at moderate calcination temperature, $Al_2(WO_4)_3$ are formed at a calcination temperature higher than 1000°C.[32,33] Thomas et al.[26,31] showed that the interaction of WO_3 with the alumina is stronger than that of MoO_3, leading to different sulfiding capacity of the Mo/Al_2O_3 and W/Al_2O_3 catalysts.[30] This was explained by differences in the polarization of the metal-oxygen bond.[30]

3.1.2 Structure of Sulfide Catalysts

A. Structure of MoS$_2$ or WS$_2$ and Related Thermodynamics

Hydrotreating catalysts are normally prepared by pore volume impregnation of alumina with aqueous solutions of $(NH_4)_6Mo_7O_{24}$, $(NH_4)_6W_{12}O_{39}$, $Co(NO_3)_2$, and $Ni(NO_3)_2$ with intermediate drying and calcination steps. The resulting oxide precursor is presulfided prior to the actual HDS and HDN reactions by a sulfiding procedure that may consist of reacting in a mixture of H_2S and H_2 or sulfur compounds easy to decompose at lower temperatures such as CS_2, dimethyldisulfides (DMDS) and alkyl polysulfides in an atmosphere of H_2, or in a liquid feed of sulfur-containing molecules and H_2. This treatment is typically carried out in the reactor itself since the metal sulfides are unstable under oxygen. The initial and the final states of the supported phases have been the subject of detailed characterizations by means of numerous physicochemical techniques. Today, most researchers agree that the resulting catalyst is (almost) completely sulfided, that is, MoO_3 is transformed into MoS_2 and the cobalt ions have passed from an oxide into a sulfide environment. During sulfidation as well as during actual HDS reaction, because the catalysts are highly reduced with H_2S always present, thermodynamics predict that molybdenum should be in the MoS_2 form, cobalt in the Co_9S_8, and nickel in the NiS or Ni_3S_2 form. Indeed, extended X-ray adsorption fine structure (EXAFS) studies of the Mo K-edge absorption spectra demonstrated that in sulfided Mo/Al$_2$O$_3$ catalysts the average Mo ion has the same environment as a Mo ion in MoS_2.[34-37] The only difference was that in the catalyst the N (Mo-Mo) coordination number for the second shell of (Mo) neighbors surrounding each Mo ion was less than 6, which was found in pure MoS_2. This indicates either that the size of MoS_2 particle slab on the alumina surface is small, or that the long-range order in these particles is not perfect. Arnoldy *et al.* also showed by temperature-programmed sulfiding of Mo/Al$_2$O$_3$ catalysts that sulfiding of a 4.5 atom-Mo/nm^2 catalyst can be completed at *ca.* 500 K, up to a S/Mo ratio of 1.9.[38]

Molybdenum sulfide are present mainly as molybdenum disulfide, which occurs in nature as the mineral molybdenite, in the viewpoint of thermodynamics. The crystal structure of molybdenite has been determined by Dickinson and Pauling.[39] The crystal lattice of molybdenum disulfide is built up of layers of the type shown in Fig. 3.1, within the layers six sulfur atoms at the vertices of a trigonal prism surround the metal atom.[40,41] Each layer is

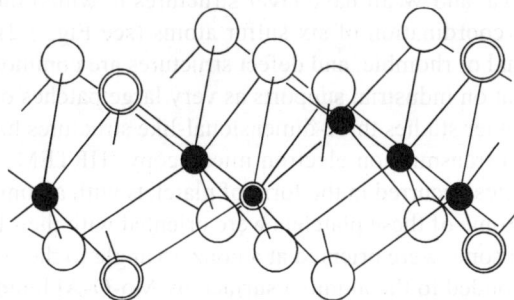

Fig. 3.1   Fragment of a layer of MoS$_2$. The metal atoms are indicated by small shaded circles, the sulfur atoms by open circles. The atoms drawn with double circles lie in one (110) plane.[41]
(From J. C. Wildervanck and F. Jellinek, *Z. anorg. allg. Chem.*, **328**, 310 (1964))

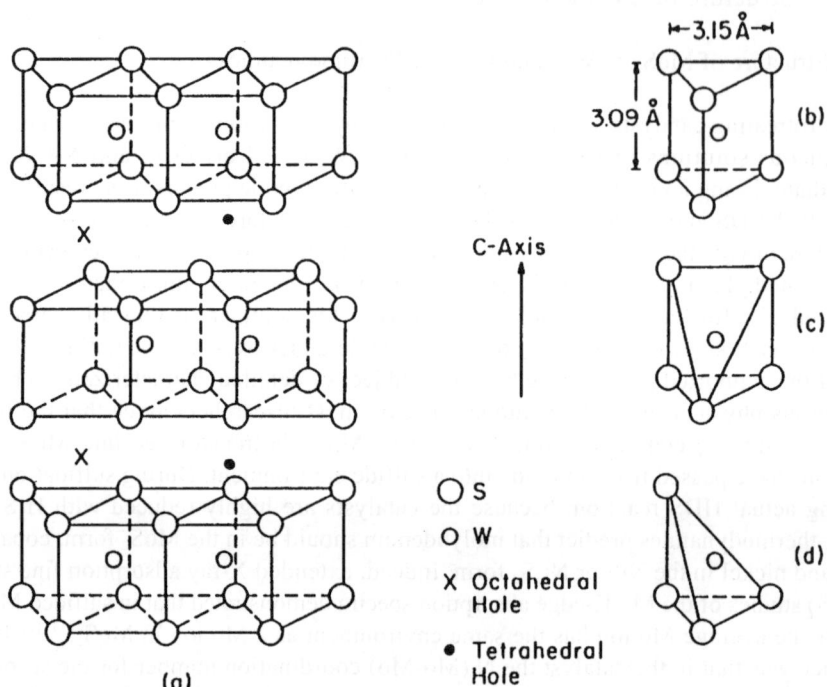

Fig. 3.2   Structure of WS$_2$: (a) stacking of layers illustrating the position of octahedral holes which may be partly occupied by Ni; (b) site symmetry of W^{3+} ions in the bulk; (c) in the side face; and (d) in an edge parallel to the c-axis.[42]
(From R. J. H. Voorhoeve and J. C. M. Stuiver, *J. Catal,* **23**, 246 (1971))

composed of sheets of Mo sandwiched between sheets of sulfur atoms. The bonding within a given layer is mainly covalent, whereas the bonding between layers is of the van der Waals type.[40,41] Crystals grow in the form of platelets with large dimensions parallel to the basal sulfur planes and a small dimension perpendicular to the basal plane. WS$_2$ as MoS$_2$ has a layer structure, discussed in detail by Voorhoeve and Stuiver,[42] and Farragher and Cossee.[43] Disulfides of Nb, Mo, Ta, and W all have layer structures in which the metal is surrounded by a trigonal prismatic coordination of six sulfur atoms (see Fig. 3.2). The stacking of the sulfur layers is hexagonal or rhombic, and defect structures are common. Topsøe *et al.* claim that MoS$_2$ can be present on industrial supports as very large patches of a wrinkle, one-slab-thick MoS$_2$ layer.[44] In other studies three-dimensional-like structures have been described. In a recent high resolution transmission electron microscopy (HRTEM) model study of HDS catalysts, MoS$_2$ crystallites occurred in the form of platelets with among the ratio of a height-to-width of 0.4–0.7.[45] Some of these platelets were oriented with their basal plane parallel to the alumina surface and some were oriented at a nonzero angle to the surface, suggesting that the MoS$_2$ platelets be bonded to the alumina surface by Mo-O-Al bonds.

Models of the active sites in Mo/Al$_2$O$_3$ catalysts have usually been developed taking into consideration the morphology of MoS$_2$. On the basis of bond energy considerations, Voorhoeve assumed as early as 1971 that the anions in the basal planes of MoS$_2$ are more

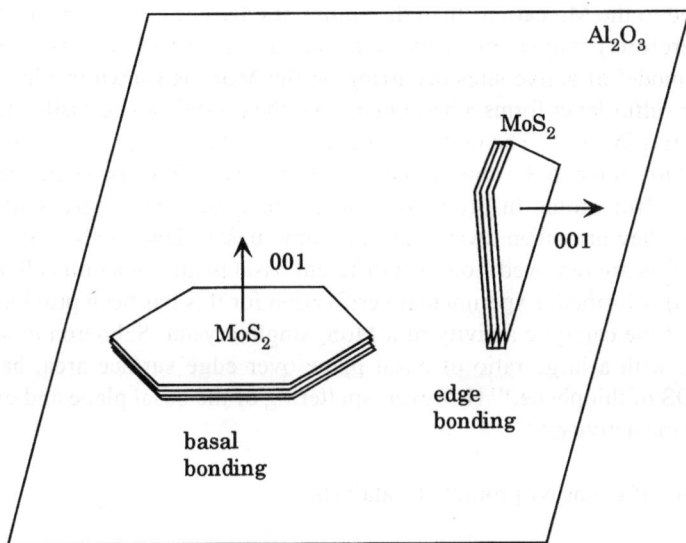

Fig. 3.3 Orientation of small MoS$_2$ crystallites on the surface of alumina.[45]
(From T. F. Hayden and J. A. Dumesic, *J. Catal*, **103**, 374 (1987))

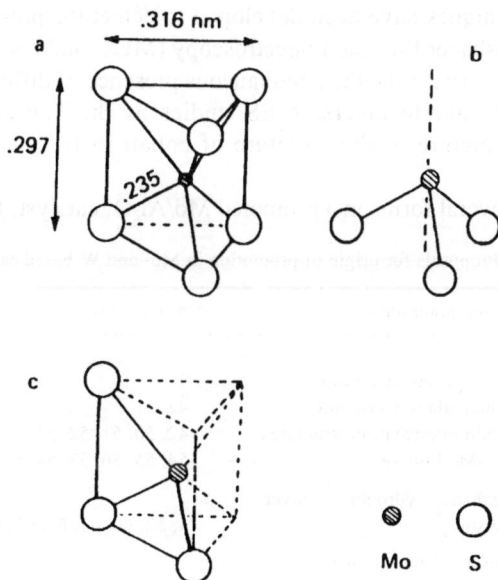

Fig. 3.4   (a) Unit cell of MoS$_2$ layer structure; (b) Model site 1 with its C$_{3v}$ axis as shown by the dashed line; (c) Model site 2.[47]
(From J. Joffre, P. Geneste and D. A. Lerner, *J. Catal*, **97**, 544 (1986))

strongly bonded to the Mo cations than the anions at edges or corners (as in the layer lattice of $TiCl_3$). Therefore catalysis most likely occurs at edges and corners, and not at basal planes.[46] The model of active sites occurring on the MoS_2 is shown in Fig. 3.3 because in MoS_2 the outer sulfur layer forms a basal plane and the crystal can be easily cleaved between the adjacent sulfur layers.[47] The unit cell of a layer is composed of a trigonal prismatic unit with a central Mo^{4+} ion and S^{2-} ions on the six summits (Fig. 3.4(a)). There are two ways-to-remove three-sulfur atoms in order to form a site with three degrees of coordinative unsaturation. If they are taken away from the same basal plane, a C_{3v} site (Fig. 3.4(b)) is obtained. If sulfurs are removed from two different basal planes of a unit cell, an asymmetric site (Fig. 3.4(c)) is formed. Experimental verification for this has been provided by a surface science study of the catalytic activity of a MoS_2 single crystal. Salmeron *et al.* showed that such a crystal, with a large ratio of basal plane over edge surface area, has a negligible activity for HDS of thiophene.[48] However, sputtering of the basal plane and exposure of Mo ions increased the activity.[49]

B. Structure of Co or Ni-promoted Catalysts

The structure of a catalyst containing a promoter such as cobalt and nickel is more complicated than a Mo or W/Al_2O_3 catalyst. With the addition of the promoter, the enhancement in catalytic activity of Mo $(W)/Al_2O_3$ is significant. In order to explain the promotional behaviors, numerous models and proposals have been suggested (Table 3.1). Many spectroscopic techniques have been developed to detect the presence of cobalt in one structure or another. Møssbauer Emission Spectroscopy (MES) and EXAFS technique can be employed to quantitatively determine the simultaneous presence of different cobalt structures. In recent years, primarily due to inverse MES studies by the Topsøe group and EXAFS research, a quantitative picture of the structure of cobalt in HDS catalysts has emerged out.[35,82]

Cobalt can exist in several forms in a promoted Mo/Al_2O_3 catalyst. In the oxide precursor

Table 3.1 Proposals for origin of promotion in Mo- and W-based catalysts

Role of Co and Ni promoter atoms	References
As a textural promoter	
formation of Co species in alumina	9
formation of intercalation structures	42
formation pseudo intercalation structures	42, 50, 51, 52, 53
formation of CoMoS phase	54, 55, 56, 57, 58, 59
As a promoter to enhance hydrogen spillover	
contact synergism	4, 52, 60, 61, 62, 63, 64
As a bonding modifier of the Mo or W site	
increased stability/dispersion of Mo phase	51, 65
change in sulfur-sulfur interaction	66
increasing number of sulfur vacancies	42, 56, 58, 67, 68, 69
increasing electron density of sulfur	70, 71, 72, 73
decrease in metal sulfur bond energy	67
enhancing the mobility of labile sulfur	74, 75, 76, 77, 78
As a producer of a new catalytic site	5, 79, 80, 81

form, cobalt ions interact strongly with the spinel type γ-Al_2O_3 lattice and occupy octahedral sites just below the Al_2O_3 surface or tetrahedral sites in the Al_2O_3 bulk.[83–85] At higher loading cobalt can also form Co_3O_4 crystallites on the surface of the support. It was also reported that cobalt aluminate was formed after undergoing a high temperature calcination.[86] In the sulfide form, cobalt may be present in three forms, as Co_9S_8 crystallites on the support, as cobalt ions adsorbed onto the surface of MoS_2 crystallites, and in tetrahedral sites in the Al_2O_3 lattice.[82] Depending on the relative concentrations of cobalt and molybdenum and on the pretreatment, a sulfided catalyst contains a relatively large amount of either Co_9S_8 or cobalt adsorbed on MoS_2 (so-called CoMoS phase). The structure of the catalyst in the sulfided state is predetermined by the structure of the oxide precursor. Therefore Co_3O_4 was found to transform into Co_9S_8, cobalt ions in octahedral support sites transformed into cobalt adsorbed on MoS_2 (CoMoS phase), and cobalt ions in tetrahedral support sites remained in those positions.[82]

By combining MES studies with catalytic activity studies, Topsøe et al. established that the promoter effect of cobalt is related to the cobalt ions adsorbed on MoS_2.[54] They have therefore confirmed suggestions made by Voorhoeve and Stuiver[46] and by Farragher and Cossee[43] that the promoter effect is not due to separate Co_9S_8 crystallites but to cobalt ions in contact with MoS_2. Originally the occurrence of this adsorption state was somewhat confusing, since thermodynamically the most stable phase of cobalt under sulfide conditions is Co_9S_8. Furthermore, solid-state chemistry studies have shown that $CoMo_2S_4$ is catalytically inactive[87] and that cobalt does not form ternary compounds with MoS_2, as it does with NbS_2 and TaS_2. Nevertheless, it can be adsorbed on the surface of MoS_2 crystallites. Farragher was the first to suggest that the cobalt is located at the edges of the MoS_2 platelets.[43,50] Proof for this suggestion was obtained by Chianelli et al.[88] in scanning Auger studies of cobalt-promoted single crystals of MoS_2. An EXAFS study of the Co K-edge demonstrated that the cobalt ion in a sulfided Co-Mo/Al_2O_3 catalyst was surrounded by sulfur ions. However, no second shell of neighboring ions can be determined, indicating that the cobalt ions are not present in a unique, well-ordered structure. Farragher suggested that the cobalt ions at the MoS_2 edges are located between subsequent MoS_2 layers, and he therefore called this a pseudo-intercalation structure, to differentiate it from real intercalation in which the cobalt ions would be randomly distributed between alternate MoS_2 layers.[43] Previously, the intercalation proper had been suggested by Voorhoeve to be the structure of cobalt.[46] On the other hand, Topsøe and Topsøe claimed that the cobalt ions were located at the edges of the molybdenum plane of a MoS_2 layer, and therefore extend the MoS_2 layer.[89] Convincing evidence for this model came from an infrared study on a series of sulfided Co-Mo/Al_2O_3 catalysts in which NO was used as probe molecule. The IR spectrum of NO molecules adsorbed on cobalt ions can be distinguished from that of NO molecules on molybdenum ions. By increasing the cobalt loading at fixed molybdenum loading it was demonstrated that the spectrum of NO adsorbed on Co sites increased in intensity, while that of NO adsorbed on Mo sites decreased in intensity. If the cobalt ions had been in the location proposed by Farragher, the intensity of the NO on Co spectrum should have increased, but the intensity of the NO on Mo spectrum should remain constant. Cobalt ions in the locations proposed by Topsøe and Topsøe, however, cover molybdenum ions and block adsorption of NO on these Mo ions. Therefore the observed behavior follows the Co location proposed by Topsøe and Topsøe.[89] Of course, the point that should be further studied is whether NO molecules actually chemisorb on the catalyst surface without disturbing its structure. If corrosive chemisorption were indeed to occur, the conclusions emerged from the IR adsorption study would need reinterpretation.

Farragher and Topsøe *et al.* describes the environment of the cobalt ions with octahedral and trigonal prismatic holes. On the basis of the solid-state NMR work, Ledoux has recently suggested that the cobalt ions may indeed be situated in tetrahedral positions at the MoS_2 edges.[79] Kasztelan *et al.* have quantitatively considered the solid-state structure of MoS_2 and the edge location of the cobalt promoter by calculating the number of edge and corner Mo and Co sites as a function of MoS_2 particle size.[90] The geometry for a single MoS_2 slab was considered. The reasonable fit between predictions and experimental results indicates that the assumptions underlying the model are not unrealistic.

In summary, the structure of a sulfided Co-Mo/Al_2O_3 catalyst consists of small MoS_2 crystallites that either lie on their basal planes parallel to the Al_2O_3 surface or are edge-bonded to the Al_2O_3 surface. The majority of the cobalt is present as cobalt ions adsorbed on the edges of the MoS_2 crystallites and as Co_9S_8 crystallites on the Al_2O_3 surface. A high Co/Mo ratio and a high sulfidation temperature of the oxide precursor favor Co_9S_8 formations. However, some cobalt is always present in tetrahedral sites in the Al_2O_3 lattice. A schematic picture of the resulting structure is presented in Fig. 3.5.[44]

Structure information on hydrotreatment catalysts has been interpreted in term of several models. The four main types of structural models, as illustrated in Fig. 3.6,[91] are the monolayer model, pseudo-intercalation model, contact synergy model and edge decoration model, or "CoMoS" phase model.

The monolayer model[92,93] is based on the common belief that MoO_3, a catalyst precursor, can be well dispersed as a monolayer on the Al_2O_3 support[51] through dehydration of the surface (Fig. 3.6(a)). During sulfiding, sulfide ions replace the terminal oxide ions. Since the sulfur anion is larger than the oxygen anion, sulfur uptake in the monolayer configuration can occur only to a limited extent, corresponding to a S/Mo ratio of no more than unity.[51]

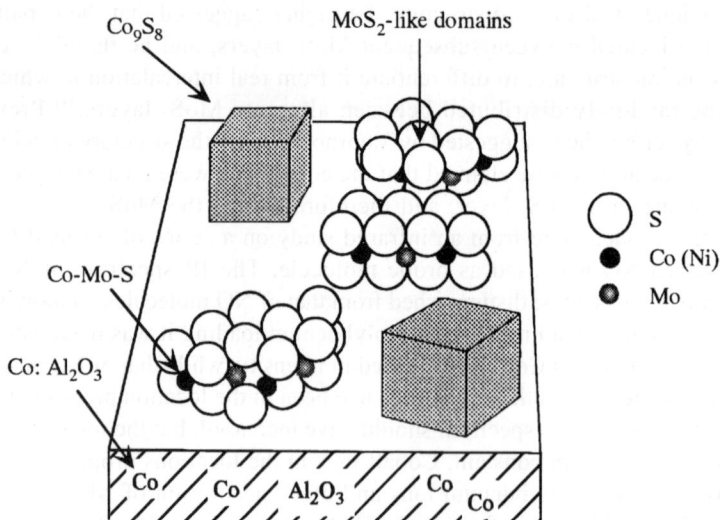

Fig. 3.5   Structure of the different forms in which cobalt ions can be present in Co-promoted MoS_2/Al_2O_3.[44]
(From H. Topsøe and B. S. Clausen, *Catal. Rev.-Sci. Eng.*, **26**, 401 (1984))

Monolayer Model   Contact Synergy Model

Intercalation Model   Edge Decoration Model

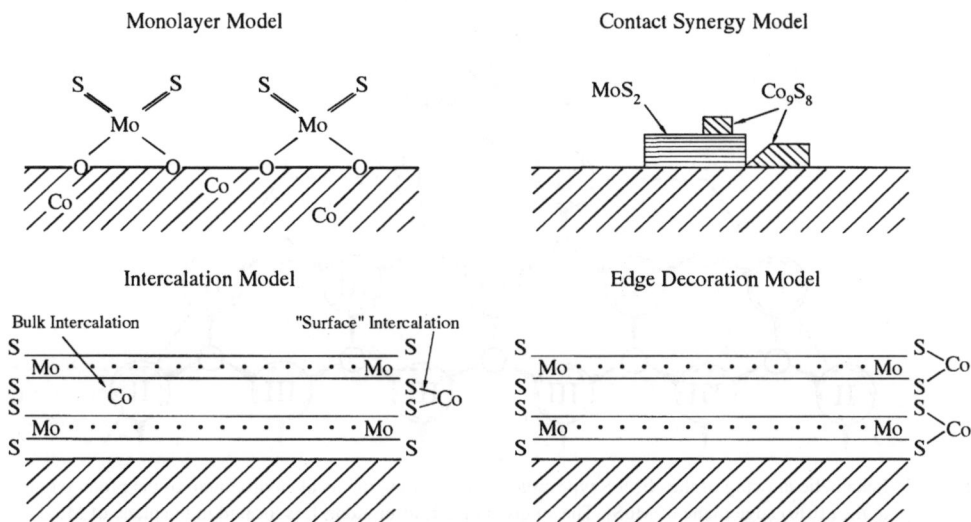

Fig. 3.6   Models for the promotion of MoS_2.[91] (a)Monolayer model.[9]; (b) Locations of the promoter atoms in the
MoS_2 structure in pseudo-intercalation models.[43,50]; (c) Contact synergy model.[60]; (d) Edge decoration
model.[56,93]
(From T. R. Halbert, T. C. Ho, E. I. Stieffel, R. R. Chainelli and M. Daage, *J. Catal.*, **130**, 117 (1991))

Massoth,[93] however, observed ratios as high as 2.75, which led him to propose a surface structure involving one-dimensional chains of MoS_2 instead of two-dimensional layers. The difficulties of this model are that the role of the promoter (such as Co or Ni) is uncertain, and that unsupported catalysts show catalytic behavior similar to the corresponding supported catalysts.

The pseudo-intercalation model[43,46] is based on the structure of bulk MoS_2 that has a prismatic arrangement of sulfur atoms around each Mo atom. The promoter Co or Ni is intercalated in the octahedral holes between S-Mo-S layers at the crystallite edge. The active sites are thus believed to be exposed Mo^{3+} ions located at the sulfur-deficient edge sites. It is not clear right now whether pseudo-intercalation is essential to the promotional effect. Karroua et al.[94], Zabala et al.[95] and Goetsch et al.[96] have demonstrated that Co_9S_8, when intimately blended with MoS_2, shows an appreciable promotional effect.

On the basis of his work on mixed bulk sulfide catalysts, Delmon[97] has advocated that the promotional role of Co occurs at the interface between the thermodynamically favored MoS_2 and Co_9S_8 phases, that is, a contact synergy model. An electron transfer at the interface explains this contact synergism between the two phases. Recently, Delmon et al. have proposed a modified contact synergy model,[94,98–101] according to which the contact is between CoMoS and Co_9S_8 instead of MoS_2 and Co_9S_8. It was suggested that spillover hydrogen produced by Co_9S_8 is able to activate, by remote control, the CoMoS species. This model earns some recognition. Iwamoto et al. proposed that the presence of Co promoter decreased the temperature of which the Mo-oxi-sulfide or MoS_3 phase was hydrogenated by the spillover of hydrogen from promoter to Mo species.[102]

The edge decoration model was initially proposed by Ratnasamy and Sivasanker[55] and was improved as "CoMoS" phase model by Topsøe and his coworkers.[44,103] The presence of

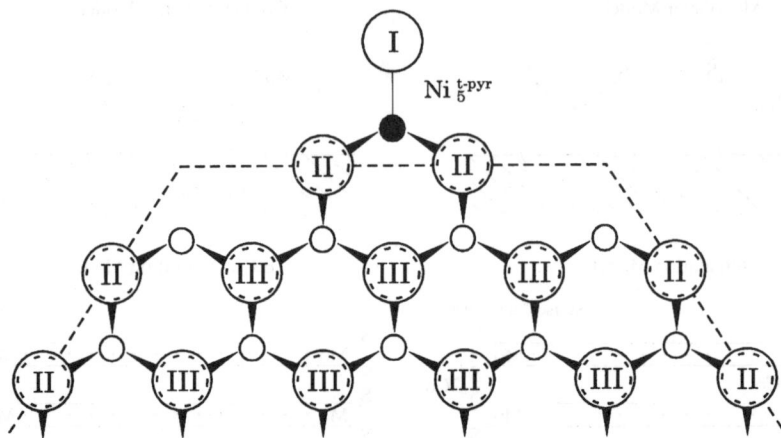

Fig. 3.7   Schematic drawing of the NiMoS type structure illustrating the local environment of the promoter atoms at the (10$\bar{1}$0) edge plane of MoS$_2$. The singly-bonded sulfur atom, I, is located in the plane of Mo. II and III denote doubly and triply bonded sulfur atoms which are present both in the top and bottom sulfur layers.[103]

(From H. Topøse, B. S. Clausen, N.-Y. Topøse and P. Zeuthen, *Sud. Surf. Sci. Catal,* **53**, 79 (1989))

an active phase, called the "CoMoS" phase, in supported and unsupported Co-Mo catalysts has recently been identified using Mössbauer spectroscopy. In alumina-supported catalysts it is suggested that the "CoMoS" phase is present as a single S-Mo-S sheet, with Co most likely present at molybdenum sites. In unsupported catalysts, the "CoMoS" phase likely consists of several layers with bulk MoS$_2$-like structures. The CoMoS phase was shown to be MoS$_2$-like structures with the promoter atoms located at the edges in five-fold coordinated sites (tetragonal pyramidal-like geometry) at the (10$\underline{1}$0) edge planes of MoS$_2$ (see Fig. 3.7). Later, the authors of other works[36,37,104–107] came to similar conclusions. At present the model of the "CoMoS" phase has gained the greatest recognition.

To summarize, each of these models has its merits as far as HDS is concerned. While the details of the catalyst structures are still unclear, for high activity it seems important to have the promoter metals intimately associated with a MoS$_2$-like species. Several such models of the active phase(s) have recently been hypothesized by Yermakov et al.[57,108] and Chianelli et al.[70] Models that are based on geometrical considerations have been proposed by Kasztelan et al.[90]

In addition to the models described above, other models have also attracted attention. For example, Wise and coworkers[71,109] found from electrical conductivity measurements on single crystals of MoS$_2$ that the conductivity of MoS$_2$ changes from n- to p-type upon addition of Co^{2+}, and these authors related the catalytic activity to the formation of electron-hole carriers. Based on surface segregation considerations, Phillips and Fote[110] proposed two empirical models for the surface structure of HDS catalysts, namely the "structure complex model" and the "boundary model". It was shown that both models could qualitatively explain the activity variations of HDS catalysts.

De Beer, Prins and coworkers[5,80,111] found that Co/C catalysts exhibit higher HDS activities than Mo/C catalysts. This led these authors to suggest that cobalt alone may be responsible for the activity in the Co-Mo catalyst system. In this Co only model, MoS$_2$ is

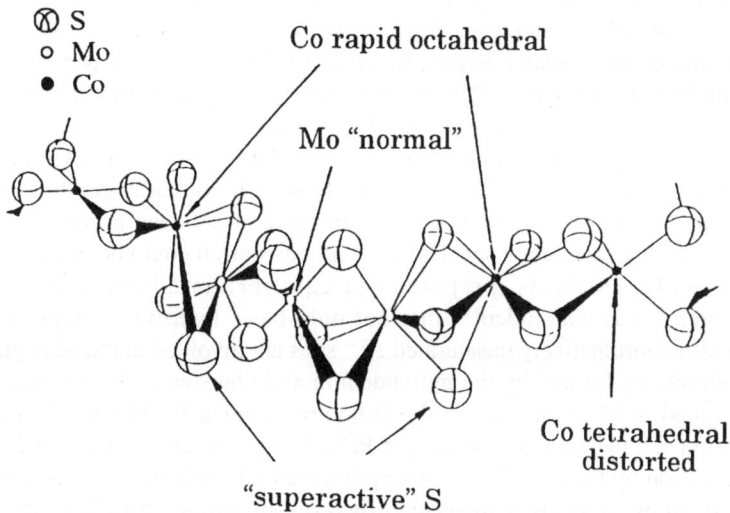

Fig. 3.8 A new model of the Co-Mo sulfided dispersed catalyst.[117]
(From M. J. Ledoux, *J. Chem. Soc., Faraday Trans. 1*, **83**, 2174 (1987))

regarded as an inert support for dispersing the cobalt sulfide. Van der Kraan *et al.*[112–115] found that the Mössbauer signal for Co/C after sulfiding at low temperature shows some resemblance to that CoMoS. This result was taken as support for the "Co only" model.

From ^{59}Co-NMR studies, Ledoux *et al.*[116] presented evidence for both tetrahedrally and octahedrally coordinated Co atoms in sulfided Co-Mo catalysts. The authors suggested that the active, distorted tetrahedral Co species are stabilized on the MoS$_2$ phase via a so-called "rapid octahedral" cobalt phase. Using the theoretical calculations of Harris and Chianelli,[72] an electron transfer was proposed to take place from the "rapid octahedral" cobalt to the molybdenum atoms, and this should result in an activation of the sulfur atoms ("super active" S) between the cobalt and the molybdenum atoms (Fig. 3.8).[117]

3.1.3 Effect of Preparation Parameters on the Structure of Catalysts

A. Effect of Impregnation Method and Calcination Conditions

The structure of sulfided catalysts depends essentially on the oxide precursors and the activation procedure. The preparation parameters include impregnation procedure, metal loading, calcination, sulfided conditions, etc. Hydrotreating catalysts are usually prepared by impregnation of the support. This aspect of the preparation has also been reviewed recently by Hall,[118] Knözinger,[119] Zdrazil [120] and Startsev.[121] Generally, the impregnation may be carried out in different ways, e.g., by pore filling (incipient wetness), adsorption of the metals by soaking the support in solutions containing one or more of the active metals, or a combination of these. Other more unconventional methods, like kneading or mixed-mull extrusion, may also be applied. Impregnation or soaking may be carried out in various solutions and the metals can be introduced by either co-impregnation or sequential impregnation (e.g., Mo first, then Co or vice versa). Further, the addition of each metal may be carried out all at once or stepwise. Most of the published literatures on the effect of the impregnation procedure is related to the properties of the calcined catalyst[122–134] and only in a few cases to the sulfided state.[54,56,135,136] The influence of preparation parameters, such as solute concentration, procedure, and pH of solution on the Co and Mo concentration profiles of alumina extrudates, has been studied in detail by Fierro et al.[130] and Goula et al.[137] By measurements of the pH in ammonium-heptamolybdate solutions after equilibrium adsorption on alumina, van Veen et al.[138] showed that a distinction between precipitation formation (polymerization of the molybdate species) and real adsorption could be made. From EXAFS study of Mo/Al$_2$O$_3$ catalysts, prepared via equilibrium adsorption of ammonium heptamolybdate, it was concluded[139] that not only basic hydroxyl groups, as commonly believed, but also coordinatively unsaturated Al^{3+} sites are involved in the adsorption reaction. The surface structures formed by the molybdenum and tungsten at the point of zero charge (PZC) were found to be analogous to the structures formed by Mo and W oxy-anions in aqueous solution at a pH equal to the sample PZC.[140] Specifically, pH values lower than six during impregnation by the equilibrium adsorption method result in the formation of tungsten monolayer structures, whereas multi-layer structures are formed for pH values higher than six.[141]

Recently, a lot of literature on the preparation has been reported. In these new preparation methods, common attempt is to obtain highly dispersed active metal species using metal carbonyl compounds[142] or other heteropolycompounds,[143] chelating agents (nitrilotric acetic acid, NTA).[144–147]

In a number of studies, Vordonis et al.[148–150] showed that an enhancement in the adsorption of negative (or positive) species could be brought about by an increase in the surface concentration of AlOH^{2+} (or AlO$^-$) species. Evidence for a better dispersion of the Mo phase using a Mo η^3-allyl complex instead of the normally used ammonium heptamolybdate solution has been given by Rodrigo et al.[127] These authors found a very homogeneous distribution of Mo throughout the alumina particles when using the allyl complex, whereas impregnation with heptamolybdate solutions seems to leave some droplets on the outer surface of the carrier, yielding an increased local Mo concentration which may favor the formation of MoO$_3$ crystallites. It has been reported[151] that preparation of the Type-II CoMoS structures with similar degrees of dispersion is possible on Al$_2$O$_3$, SiO$_2$, and C by use of a N-containing tetradentate organic ligand (nitrilotriacetic acid, NTA) to complex simultaneously Co and Mo. If NTA is used in the preparation of Ni-Mo/Al$_2$O$_3$ catalysts, Type-II NiMoS is also formed.[152]

Several researchers have pursued the influence of the calcination temperature on the sulfided state of the catalysts. Li and Hercules[153] measured XPS and ISS Mo/Al intensity ratios for Mo/Al$_2$O$_3$ catalysts calcined at temperatures between 300°C and 700°C and then sulfided at 400°C. These authors found that high calcination temperatures did not affect the degree of coverage of the support after sulfidation. Upon increasing the calcination temperature, there is substantial evidence from DRS[124,135] infrared measurements of NO adsorption[89,135], and MES[135,154] that the relative amount of tetrahedrally coordinated Co increases. This occurs at the expense of the amount of octahedrally coordinated Co.[89,124,135,154] Since this Co species is a precursor for CoMoS, the sulfided catalysts contain decreasing amounts of CoMoS with increasing calcination temperatures.[135] This behavior is depicted in Fig. 3.9. Infrared studies are also in agreement with this result, since increasing the calcination

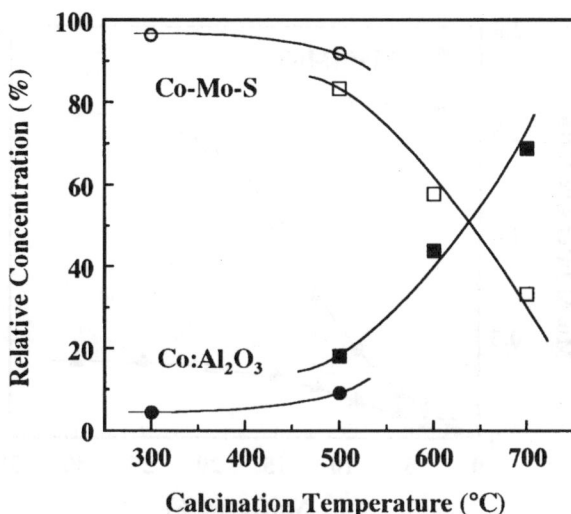

Fig. 3.9   Effect of calcination temperature on the phase distribution of cobalt in sulfided Co-Mo/Al$_2$O$_3$ catalysts. Circles and squares refer to catalysts with a Co/Mo ratio of 0.27 and 0.50, respectively.[135] (From R. Candia, N.-Y. Topøse, B. S. Clausen et al., *Proc. 4th Intern. Conf. on Chemistry and Uses of Molybdeum* 377 (1982))

temperature leads to a decrease in the number of Co atoms adsorbing NO in the sulfided catalyst.[135]

The sulfidation condition and properties of support influenced significantly the structure of sulfided catalysts. It is well known that the structures and their relative proportion in sulfided Mo- or W-based catalysts are very dependent on the loading of active metal.[155] One of most important parameters-active metal loadings is discussed below.

The loading used in industrial applications are usually governed by the desire to achieve as high an activity as possible with as small amount of expensive metals as possible. Thus, "wasting" metals in undesired low activity phases must be avoided. The molybdenum loading used is typically *ca.* 8 to 15 wt% Mo. This corresponds to about monolayer coverage (of the hydroxyl part of the surface) for an alumina support with a typical surface area of about 250 m^2/g. Above this Mo level or for low-surface-area alumina, separate MoO$_3$ entities are formed, resulting in bulk-like MoS$_2$ phases upon sulfidation. Johnson et al.[156] compared the surface areas measured with that calculated when assuming simple dilution of alumina with MoO$_3$. Results for both pore volume and surface area are consistent with a model in which the molybdenum are spread out in monolayer when a content of MoO$_3$ was less than 30 wt%. For instance, on a 289 m^2/g alumina support, a catalyst which incorporates 30% MoO$_3$ has an available area 202 m^2/g, or about 16 Å^2 per Mo atom. In crystalline MoO$_3$ (d=4.69 g/cm^3), 51 Å^3 is occupied by each Mo (with associated oxygen atoms), equivalent to an effective area of 13.7 Å^2. Hence, a value of 16 Å^2 for the monolayer cross section of an MoO$_x$ is reasonable. This agrees with recent FTIR study result obtained by Topsøe et al.[157] On the other hand, the promoter (Co or Ni) loading is optimized from experience and depending on the molybdenum

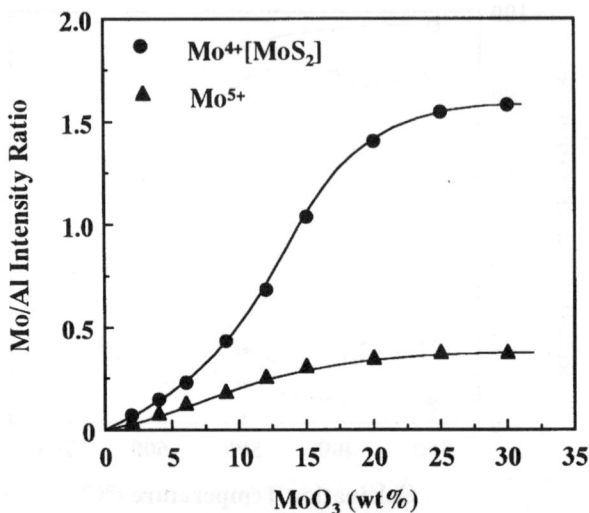

Fig. 3.10   Distribution of molybdenum oxidation states in sulfided Mo/Al$_2$O$_3$ catalysts as a function of MoO$_3$ content.[153]
(From C. P. Li and D. M. Hercules, *J. Phys. Chem.*, **88**, 461 (1984))

level and type of hydrotreating application, the promoter concentration usually amounts to around 1 to 5 wt%, giving a Co/Mo atomic ratio in the range 0.1 to 1.0.

It is generally found that low Mo-loading catalysts are more difficult to sulfide than high-loading catalysts. As discussed above, the MoS$_2$ species may be bonded to the alumina surface via a few oxygen bonds and it is possible that the relative proportion of such bonds increases with decreasing Mo loading. The X-ray photoelectron spectroscopy (XPS) studies of Li *et al.*[153] are in agreement with this suggestion. They indicate that in addition to Mo^{4+} (MoS$_2$), Mo^{5+} is also present but in a significantly lower concentration than Mo^{4+} (Fig. 3.10).

B. Effect of Support on the Structure of Catalyst

The support that is currently most used for industrial HDS catalysts is the γ-cubic type alumina, which typically has a surface area of the order of 2 to 3 $\times$ 10^2 m^2/g, a pore volume of 0.5 to 1.0 cm^3/g, and an average pore diameter of *ca.* 10 nm. Since the chemical nature of this support may be important to the surface chemistry of HDS, a simple discussion of the properties of alumina is presented here (see Refs. 9 and 158). The aluminas are invariably prepared by precipitation of a hydrated alumina formed by mixing solutions of alkali and salts of an aluminum compound (sulfate, nitrate, or occasionally chloride). The precipitate formed is gelatinous with a diffuse X-ray diagram resembling that of the mineral boehmite, AlO(OH), (gelatinous boehmite). The precipitate is usually aged by gentle heating in aqueous slurry, the aging conditions determining the properties of the final product. Aging at 313 K produced a solid that has an X-ray diagram more closely akin to that of bayerite Al(OH)$_3$. If subsequently aged at 353 K, the solid is converted into a product again similar in its X-ray diagram to boehmite, which now is a much more highly crystalline form (crystalline boehmite). Alternatively, aging in an alkaline solution leads to a different form of $Al(OH)_3$, gibbsite, which usually contains some alkali.

Subsequent to the wet stages in the synthesis crystalline boehmite is filtered and calcined at a temperature *ca.* 875 K, and γ- and δ-Al_2O_3 are produced. The calcination leads to the elimination of water, either that initially present as such in the precipitate, or that formed by reaction of OH^- group in a more advanced stage of the calcination:

$$2\ OH^- \rightarrow H_2O + O^{2-} + \square \tag{3.1}$$

where "$\square$" is an anion vacancy site. The reaction of two hydroxyl groups on the same surface results in the formation of "$\square$" on the surface. The reaction of two hydroxyl groups on different surfaces will result in the formation of bulk due to the combination of one O^{2-} with another "$\square$", producing a larger unit of alumina. In both reactions, as the process of water elimination takes place, the surface area starts to increase as pores are formed.

At calcining temperatures higher than 1375 K the product is invariably α-Al_2O_3, having low surface area and no longer containing either water or -OH^- groups (See Table 3.2). The most important feature of the structure of Al_2O_3 is that both in γ-Al_2O_3 and in η-Al_2O_3 the oxygen ions are cubic close packed and their structure are similar to that of the spinel like $MgAl_2O_4$. This packing is characterized by a slightly different type of stacking of the close packed oxygen layers (Fig. 3.11), that is, 1-2-3-1-2-3⋯. Further, there is a change in the Al^{3+} ion distribution, instead of being confined to the octahedral sites, hence they are tetrahedrally surrounded by O^{2-} ions. In spinel (Fig. 3.12) which has 32 O^{2-} ions, 32 octahedral sites (half-occupied by Al^{3+} ions), and 64 tetrahedral sites (8 occupied by Mg^{2+} ions), the Al^{3+} ions occur in octahedral sites and the Mg^{2+} ions in tetrahedral sites. Because of the similarity in structure of the γ-Al_2O_3 and η-Al_2O_3 and the spinel, their structures can be represented as $(H_{1/2}Al_{1/2})Al_2O_4$. This implies that the H^+ ions do not occupy the tetrahedral sites but exist as OH^- on the surface of alumina. Hence, one-eighth of all O^{2-} ions is present in the form of OH^- on the surface.

The interaction between Mo and alumina is known to influence the structure and activity of HDS catalyst-based Mo.[160–163] The strong interaction between Mo and alumina facilitates the formation of a monolayer rather than a multi-layer Mo structure, resulting in higher

Table 3.2 Preparation of aluminas [9]

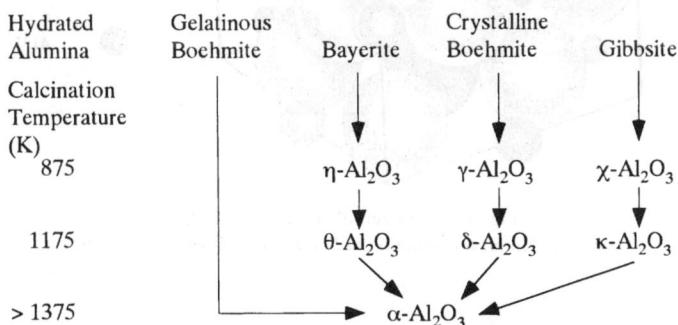

Hydrated Alumina	Gelatinous Boehmite	Bayerite	Crystalline Boehmite	Gibbsite
Calcination Temperature (K)				
875		η-Al_2O_3	γ-Al_2O_3	χ-Al_2O_3
1175		θ-Al_2O_3	δ-Al_2O_3	κ-Al_2O_3
> 1375		α-Al_2O_3		

(From G. C. A. Schuit and B. C. Gates, *AIChE J.*, **19**, 426 (1973))

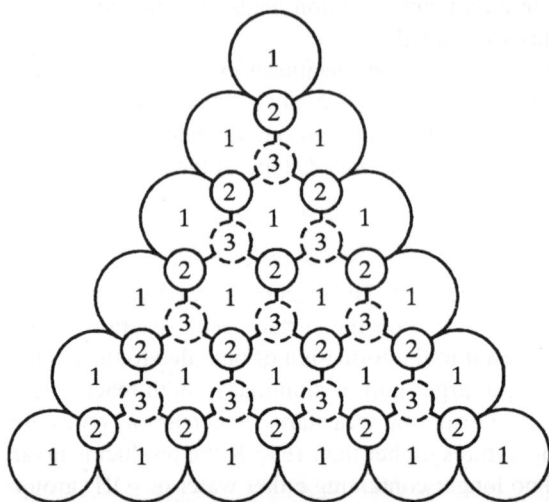

Fig. 3.11 Close packing structures: Hexagonal, sequence 1-2-1-2...; Cubic, sequence 1-2-3-1-2-3....[9]
(From G. C. A. Schuit and B. C. Gates, *AIChE J.*, **19**, 426 (1973))

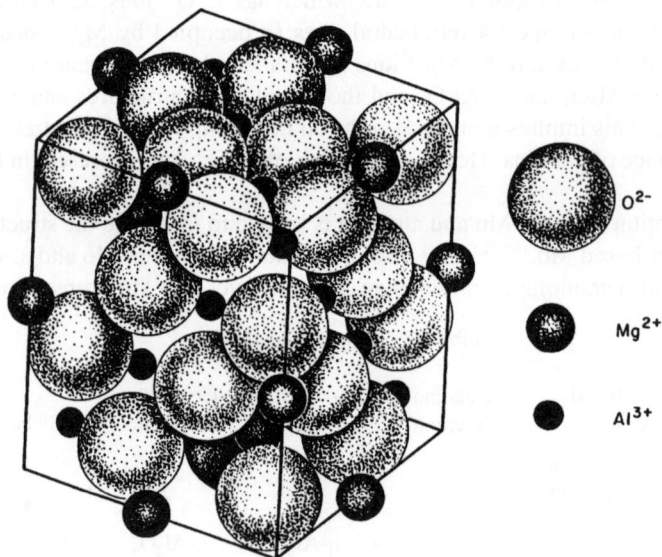

Fig. 3.12 Unit cell of spinel (MgAl$_2$O$_4$).[159]
(From C. Kittel, *Introduction to Solid State Physics*, p.550 (1971))

dispersion of Mo on the support.[161] The interaction between Mo and alumina may decrease the HDS activity because of the potential suppression of molybdate sulfidation.[160] On the other hand, Mo sulfide that interacts strongly with alumina has a lower intrinsic activity compared with weak interaction Mo sulfide.[162] For Co- and Ni-Mo/Al$_2$O$_3$, it was also recognized that alumina is not an inert carrier. The interaction between Mo and alumina is the result of a reaction between Mo anion and the basic (protonated) hydroxyl groups on the alumina surface during the catalyst preparation.[164,165] The molybdate adsorption on the alumina surface might occur by the decomposition of the adsorbing molecules/ion and physisorption on coordinatively unsaturated Al^{+3} sites.[164,165] On the other hand, the promoter ions, Co and Ni, can also react with the support and can occupy octahedral or tetrahedral sites in the external layers or even form CoAl$_2$O$_4$ (NiAl$_2$O$_4$) depending on the conditions of preparation. The origin of the exclusive use of alumina can be ascribed to its outstanded textural and mechanical properties and its relatively low cost. Another important factor is the ability to regenerate catalytic activities after intensive use under hydrotreating conditions.[166] Due to the necessity of developing hydrotreating catalysts with enhanced properties, other supports such as carbon,[167-169] silica,[170,171] zeolite,[172-174] and titania [175,176] have been studied. In several cases, it was claimed that activities higher than those of alumina-supported catalysts were obtained. In recent year, much effort was put to improve the properties of supports, which was decorated using P,[177] F,[178] B [179] etc. or using mixed oxide as a suppot.[180-185] The studies using other supports except for alumina are described in detail in Chapter 4.

C. Sulfidation Reaction

1. Studies Using Conventional Approaches

The structures and reactivities of sulfided Mo/Al$_2$O$_3$ catalysts have been widely investigated in HDS, because Mo sulfide supported on alumina is the most basic form among HDS catalysts.[186,187] It is believed that MoS$_2$ is present on alumina surface in sulfided molybdena-alumina catalysts. For example, it has been shown by means of EXAFS that molybdenum atoms in the sulfided catalysts are present in MoS$_2$-like structures and that these are ordered in very small domains.[104,105,188] Hayden and Dumesic reported that, in a transmission electron microscopy (TEM) study of HDS catalysts, MoS$_2$ crystallites were created on thin films of alumina.[189] Further, the amount of sulfur on sulfided molybdena-alumina has been directly measured by several methods. Arnoldy et al. showed by temperature-programmed sulfiding of Mo/Al$_2$O$_3$ catalysts that sulfiding of a 4.5 atoms-Mo/nm^2 catalyst can be completed at ca. 500 K, up to a S/Mo ratio of 1.9.[38] Kabe et al. also reported that, by the thermogravimetric analysis of sulfiding processes, total sulfur present in sulfided Mo/Al$_2$O$_3$ corresponded to MoO$_{1.5}$S$_{1.5}$.[190]

Several investigators have studied the detailed mechanism of the sulfiding reaction of unpromoted and promoted Mo/Al$_2$O$_3$ catalysts. From thermal gravimetric analysis (TGA) measurements, Massoth[93] found that the extent of sulfiding of a Mo/Al$_2$O$_3$ catalyst increased with temperatures between 300°C and 700°C. The predominant reaction was claimed to be the exchange of oxygen (associated with the molybdenum) by sulfur. Above 500°C, very little reactive oxygen was left and the sulfur level was found to be close to S/Mo = 2.0. This agrees qualitatively with an XPS study by Prada Silvy et al.,[191] who showed that below 400°C, incomplete sulfidation of the cobalt and molybdenum phase occurs. Li and Hercules[153] followed the sulfiding behavior at 350°C of a Mo/Al$_2$O$_3$ catalyst by XPS and observed that even after extended sulfiding times in addition to Mo^{4+}, a small amount of Mo^{5+} was still present. Moulijn and coworkers[38,160,192–194] showed by use of temperature-programmed sulfidation (TPS) that extensive sulfiding of unpromoted and promoted Mo- and W- based catalysts takes place even at temperatures as low as 200°C (Fig. 3.13). At these low temperatures, sulfiding was described to occur by simple O-S exchange on surface Mo^{6+} ions. At about 225°C, according to their scheme, elemental sulfur (from breaking of Mo-S bonds) is reduced to H_2S with concomitant reduction of molybdenum from +6 to +4. Above 225°C, the sulfiding is described mainly as O-S exchange on Mo^{4+} ions. In this temperature region, the S/Mo ratio approaches 2, which implies the formation of an almost stoichiometric MoS_2 phase.[38] From studies of wafers of $MoO_3/SiO_2/Si$ catalysts, de Jong et al.[195] found no evidence for the presence of elemental sulfur. These authors found that O-S exchange can begin below 100°C presumably forming Mo^{4+} oxysulfides on the surface and MoO_2 or H_xMoO_3 species in the interior. Sulfur was present partly as S^{2-}, partly as S_2^{2-} species. Above 125°C, the molybdenum oxides and oxysulfides convert to MoS_2, and the S_2^{2-} species disappears probably because of hydrogenation to H_2S. In accordance with this study, de Boer et al.[196] showed by use of EXAFS that MoO_xS_y species are formed by O-S exchange on a Mo/SiO_2 catalyst, even at room temperature. Increasing the temperature to 150°C appeared to result in a MoS_3-like structure, which was reported to transform to MoS_2 between 250 and 300°C. Low temperature sulfiding was also observed by Lojacono et al.[197,198] They observed that the relative intensity of a Mo^{5+} ESR signal initially increased to a maximum at about 100°C during sulfiding, then steeply decreased and essentially disappeared at around 500°C, presumably because of a reduction to Mo^{4+}. Essentially the same behavior was observed by Seshadri et al.[199] Recent TPR studies on oxide and sulfided alumina-supported W and NiW catalysts suggest that only parts of the metal oxides are sulfided at typical temperatures.[200] At the same time, it was suggested that the sulfidation of W-based catalysts is more difficult than Mo-based catalyst.

Parham and Merrill[201] using EXAFS previously studied the genesis of the active Mo phase during sulfiding of a $Co-Mo/Al_2O_3$. Below 100°C during sulfiding, a Mo-S but no Mo-Mo distance was found, indicating that MoS_2 had not yet formed. At sulfiding temperatures above 200°C, MoS_2 was formed, as evidenced by appearance of a distinct Mo-Mo bond distance. The size of the MoS_2 crystallites increased with increasing sulfiding temperature, being ca. 1.5 nm between 300°C and 400°C. With the recent development of the Quick scanning EXAFS (QEXAFS) technique, these structural changes can now be followed in much more detail.[202] In agreement with the above, the results show that already at room temperature extensive O-S exchange occurs and that with increasing sulfidation temperature, O peak decreases and S peak increases. Oxygen in the local environments of Mo was observed below 200°C, suggesting the presence of oxysulfides. A recent Mo 3d XPS study result of MoO_3 and its sulfidation products is shown in Fig. 3.14.[203] The spectra of the samples sulfided at intermediate temperature (25–300°C) can all be described in terms of the Mo^{6+} and Mo^{4+} doublets and one additional double, which was attributed to Mo^{5+} present in oxysulfide intermediate phases. The Mo 3d spectra of the sulfided sample contain contributions of these three states in varying percentages, reflecting the stepwise transition of Mo^{VI} in MoO_3 to Mo^{IV} in MoS_2 (Fig. 3.14). The most obvious trends are increasing formation of Mo^{5+} centers at sulfidation temperatures below 200°C and their reduction to the Mo^{4+} states at temperatures above 200°C. Some discrepancy exists between the different EXAFS studies on the values of the average sulfur coordination numbers around Mo. Values anywhere in the range between

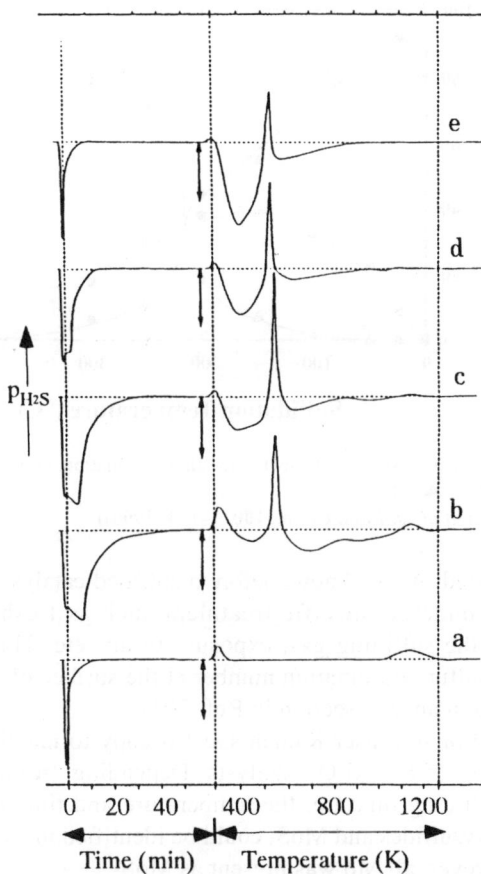

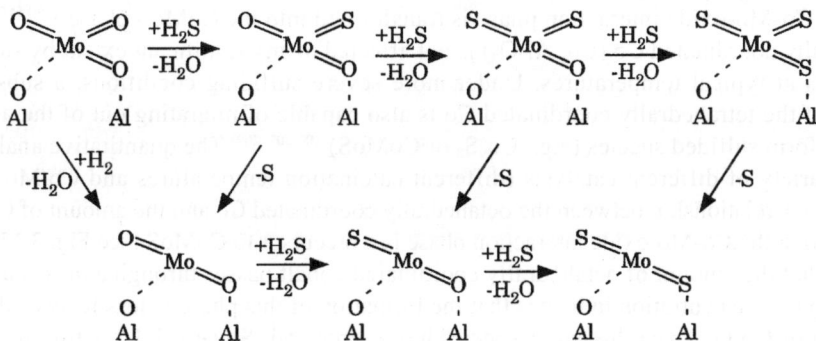

Fig. 3.13 (1) TPS patterns (H_2S) of MoO_3/Al_2O_3 with varying Mo content, pretreated in Ar at room temperature.
(a) 0.0 atoms/nm², (b) 0.5 atom/nm², (c) 1.0 atom/nm², (d) 2.2 atom/nm² (e) 4.5 atom/nm². The 50%
conversion level of H_2S is indicated by a double-headed arrow. The reaction scheme proposed for the
sulfidation of the Mo structure is also given.[38]
(From P. Arnold *et al.*, *J. Catal*, **92**, 42 (1985))

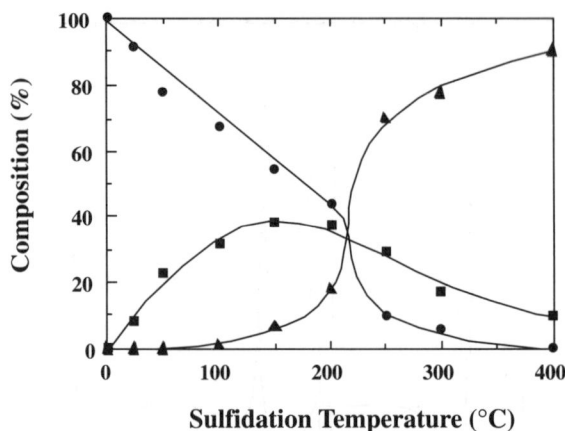

Fig. 3.14 Composition of the XPS spectra of MoO_3 after 3h of sulfidation, as a function of temperature.[203]
●: Mo^{6+}, ■: Mo^{5+}, ▲: Mo^{4+}.
(From Th. Weber et al., J. Phys. Chem., **100**, 14164 (1996))

3 and 6 have been reported. As we know, before a sulfided catalyst is subjected to EXAFS measurements, it will undergo specific treatment such as flushing with an inert gas, evacuation, cooling in the sulfiding gas, exposure to air, etc. Thus, such values are not surprising because the sulfur coordination number at the surface of MoS_2 is sensitive to the treatment (see the question/answer section in Ref. 204).

Payen et al.[205] used in situ laser Raman spectroscopy to identify intermediate species during stepwise sulfiding of Mo/Al_2O_3 catalysts. Depending on the hydration state of the oxide precursor, the sulfiding mixture, the temperature and time of sulfidation, different intermediates such as oxysulfides and MoS_3 could be identified in various proportions. In the fully sulfided state, however, all Mo was present as MoS_2.

From a quantitative phase analysis of typical catalysts before and after sulfiding, it was found that during sulfiding Co_3O_4 is transformed into Co_9S_8 and the octahedrally coordinated Co in the Co-Mo-oxide interaction phase is transformed into the CoMoS phase.[135,154,206] The tetrahedrally coordinated Co (Co: Al_2O_3) is not affected to any significant extent by sulfiding treatments at typical temperatures. Under more severe sulfiding conditions, a substantial fraction of the tetrahedrally coordinated Co is also capable of migrating out of the alumina lattice to form sulfided species (e.g., Co_9S_8 or CoMoS).[103,207–209] The quantitative analysis of a great variety of different catalysts (different calcination temperatures and Co/Mo ratios) showed a 1:1 relationship between the octahedrally coordinated Co and the amount of CoMoS, i.e., the Co in the Co-Mo-oxide interaction phase is a precursor to CoMoS (see Fig. 3.15).[135,154] The fact that the amount of octahedrally coordinated cobalt passes through a maximum with increasing Co concentration indicates that the formation of this phase is less favored than the formation of Co_3O_4, once this latter species has aggregated. Since a 1:1 relationship exists between the phases in the calcined and in the sulfided state, this behavior explains why the amount of CoMoS also goes through a maximum.[154]

The Ni-promoted Mo/Al_2O_3 catalyst is of the same character as the Co-Mo catalyst. Octahedrally coordinated nickel interacting with the molybdenum surface-phase is likewise

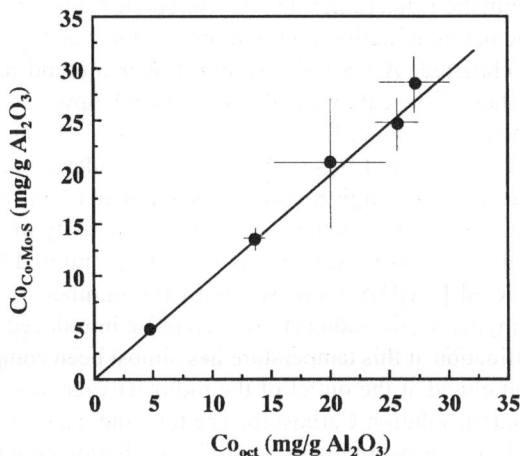

Fig. 3.15   Absolute amount of cobalt in CoMoS in the sulfided state of Co-Mo/Al$_2$O$_3$ catalysts with different Co/Mo ratios plotted as function of the amount of cobalt as Co$_{oct}$ in the calcined state.[154] (From C. Wivel et al., J. Catal, **87**, 510 (1984))

found to be a precursor to NiMoS.[89,211-214] The fact that the relative fraction of such species is typically larger for nickel than for cobalt for similar preparation conditions suggests that nickel is more efficient in forming the active phase. On the other hand, there are a few studies on the sulfided W-based catalysts.[210] Kim et al. reported that the structure of Ni-W/Al$_2$O$_3$ catalyst varied with sulfiding temperature.[210] The results of XRD and XPS indicated that similar to the Mo-based catalyst, tungsten sulfide was present in the WS$_2$-like phase and the sulfidability and crystallinity of WS$_2$ increased with increase in sulfidation temperature. The effect of pretreatment procedure on the structure and HDS activity of Mo/Al$_2$O$_3$ and Co-Mo/Al$_2$O$_3$ catalysts was also investigated using EXAFS.[215] In both catalysts, MoS$_2$-like structures were formed when the catalysts were initially sulfided, whereas no discrete structure was formed even after post-sulfation when the catalysts were initially reduced with H$_2$. It was suggested that the presence of Co strongly affects the sensitivity of the catalyst activity to the pretreatment, but does not affect the fine structure of the Mo site.

2. Studies Using Radioisotope Tracer Methods

The active sites in sulfided Mo/Al$_2$O$_3$ catalysts have been discussed and many reaction mechanisms have been proposed in recent years.[187,192,193] If it is assumed that the anions in the basal planes of MoS$_2$ are more strongly bonded to the Mo cations than the anions at edges or corners, sulfur vacancy would be created at edges and corners rather than at basal planes.[46] Therefore, the catalytic activity of HDS is expected to be much higher at edges and corners. This assumption has been experimentally verified by surface science studies of the catalytic activity of a single MoS$_2$ crystal.[48,49] When the HDS reaction begins with the adsorption of thiophenes to the sulfur vacancy, two mechanisms have been proposed: one is the one-point (end-on) mechanism where the reactant thiophene molecule is assumed to adsorb upright on the surface.[92] The other is the side-on mechanism where the adsorption of thiophene takes place via "multipoint" adsorption.[216] Recently, it has been shown in kinetic study of HDS of methyl-substituted DBTs that DBTs are adsorbed on the catalyst through the

π-electrons of their aromatic rings rather than unshared electrons of their sulfur.[217] Despite unremitting efforts, the nature of active sites and the behavior of surface sulfur species have not been completely clarified. A ^{35}S radioisotope tracer method has been developed to determine the sulfide state of the catalyst and is discussed below.

After being calcined in air at 430°C for *ca.* 20 h, a Mo/Al$_2$O$_3$ catalyst was sulfided with an about 20–50 vol% of ^{35}S-labeled H$_2$S ([^{35}S]H$_2$S) in H$_2$ under 30 kg/cm^2. The [^{35}S]H$_2$S in hydrogen was introduced with a high-pressure gas sampler (2.48 ml) in the manner of a pulse every 8 min. Fig. 3.16 [218] shows the change in radioactivity of eluted [^{35}S]H$_2$S during the sulfidation of catalyst at 200°C, which was directly monitored by radioanalyzer. The radioactivity of unreacted [^{35}S]H$_2$S increased with the number of introduced pulse and approached the constant, that is, the radioactivity in a pulse introduced. Therefore, it could be considered that the sulfidation at this temperature has almost been completed. In addition, the unreacted [^{35}S]H$_2$S that eluted at the outlet of the radioanalyzer was recovered by bubbling through a basic scintillation solution-Carbosorb. The total radioactivity of unreacted [^{35}S]H$_2$S trapped with Carbosorb was measured with a liquid scintillation counter after completion of sulfidation. This amount of radioactivity was consistent with that measured by the radioanalyzer. The numbers of the pulse incorporated into the catalyst (N_T) could be calculated from the balance of radioactivity introduced and released:

$$N_T = N_p - R_{trap}/R_p \qquad (3.2)$$

where N_p is the numbers of introduced pulse, R_p is the amount of radioactivity in a standard pulse, and R_{trap} is the amount of total radioactivity collected by the trap at the outlet of the reactor. The uptake amount of sulfur to the catalyst in the sulfidation (S_T) was obtained from:

$$S_T = N_T \times S_p \qquad (3.3)$$

where S_p is the amount of sulfur in a standard pulse. Sulfidation degree was used to evaluate

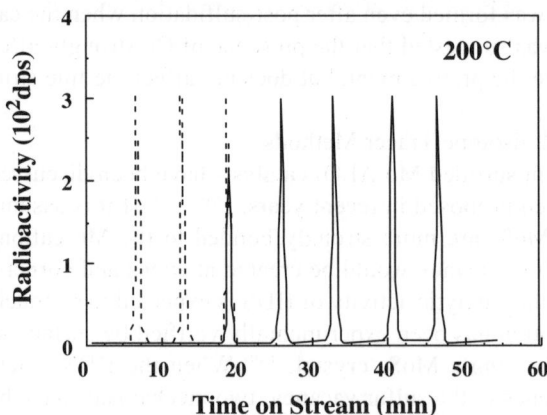

Fig. 3.16   The release of [^{35}S]H$_2$S in the sulfidation of Mo/Al$_2$O$_3$ with [^{35}S]H$_2$S pulse. The solid and dotted lines represent the radioactivities of [^{35}S]H$_2$S unreacted and that incorporated into the catalysts, respectively.[218] (From T. Kabe, W. Qian, A. Ishihara *et al., J. Chem. Soc., Faraday Trans.,* **93**, 3711 (1997))

the sulfidation extent of the catalyst. The sulfidation degree (R) was defined as the ratio of the uptake amount of sulfur in the sulfidation to total sulfur that was calculated when all molybdenum sulfides were present in MoS_2.

The results for Mo/Al_2O_3 catalysts are shown in Figs. 3.17 and 3.18, and Table 3.3. The amounts of sulfur accumulated on the Mo/Al_2O_3 catalysts increased with increasing temperature. The amounts of sulfur accumulated on Mo/Al_2O_3 with 12 wt% MoO_3 at 100°C, 200°C, 300°C and 400°C were 34.0, 38.5, 50.7 and 60.5 mg-S/g-cat, respectively and the value at 400°C corresponded to the sulfidation state of $MoS_{2.2}/Al_2O_3$. At the same temperature, the amount of sulfur accumulated on Mo/Al_2O_3 increased with increasing amount of molybdenum loaded as shown in Fig. 3.17. In Fig. 3.18, where the sulfidation degree was plotted against content of molybdenum oxide, the sulfidation degree at the same temperature did not change with increasing content of molybdenum oxide up to 16 wt% of MoO_3. These results show that the same kinds of sulfides are formed on the catalysts having different amounts of molybdenum at the same temperature and that molybdenum sulfide can be uniformly

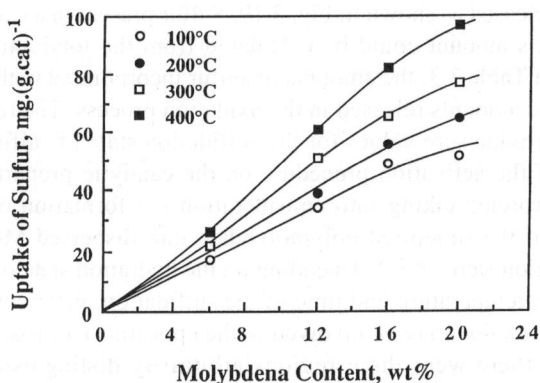

Fig. 3.17 Effect of Mo content on uptake amount of sulfur at various sulfiding temperatures.[218]
(From T. Kabe, W. Qian, A. Ishihara et al., J. Chem. Soc., Faraday Trans., **93**, 3711 (1997))

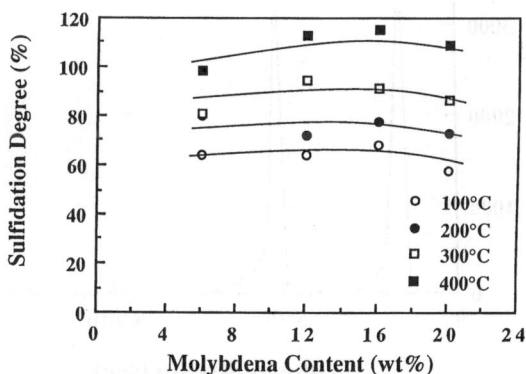

Fig. 3.18 Effect of Mo content on degree of sulfidation at various temperatures.[218]
(From T. Kabe, W. Qian, A. Ishihara et al., J. Chem. Soc., Faraday Trans., **93**, 3711 (1997))

Table 3.3 Sulfidation state of Mo/Al$_2$O$_3$ catalysts containing different content of MoO$_3$ [a) 218)]

Temperature(°C)	100	200	300	400
6% Mo/Al$_2$O$_3$ [b)]	17.1(17.9)	21.4(23.2)	21.5(23.8)	26.3(27.7)
12% Mo/Al$_2$O$_3$ [b)]	34.0(36.7)	38.5(39.2)	50.7(47.4)	60.5(56.0)
16% Mo/Al$_2$O$_3$ [b)]	48.6(40.2)	55.3(52.7)	65.0(67.4)	82.1(69.0)
20% Mo/Al$_2$O$_3$ [b)]	51.6(49.6)	64.7(68.3)	77.0(82.2)	96.7(92.5)

[a)] The amounts of sulfur accumulated during sulfidation with hydrogen sulfide, mg-sulfur/g-catalyst. Values in parentheses represent the amounts of sulfur released during oxidation of oxygen, mg-sulfur/g-catalyst.
[b)] The values before Mo/Al$_2$O$_3$ represent the loaded amount of MoO$_3$ in Mo/Al$_2$O$_3$.
(From T. Kabe, W. Qian, A. Ishihara et al., J.Chem. Soc., Faraday Trans., **93**, 3711 (1997))

supported on alumina up to 16 wt% of MoO$_3$. These are consistent with similar results from HDS of ^{35}S-labeled dibenzothiophene ([^{35}S]DBT) using sulfided molybdena-alumina catalysts having different amounts of molybdenum.[219)]

When the sulfided catalysts were oxidized by a pulse of oxygen gas at 430°C, radioactive oxidized sulfur was released as shown in Fig. 3.19. Sulfur present in a catalyst can be removed by this method and its amount could be calculated from the total amount of radioactivity released. As shown in Table 3.3, the amounts of sulfur incorporated in the sulfidation process are consistent with the amounts released in the oxidation process. The result shows that these facile methods can provide sure values for the sulfidation state of sulfided catalysts.

The influence of the activation procedure on the catalytic properties of the Mo/Al$_2$O$_3$ catalyst may be interpreted taking into consideration the formation of the metal sulfides. The transformation of the supported polymolybdate into dispersed MoS$_2$ was achieved by sulfidation and reduction steps.[220,221)] Depending on the hydration state of the oxide precursor, the sulfiding mixture, temperature and time of the sulfidation, different intermediates could be identified. The sulfidation may be involved in the appearance of trisulfide MoS$_3$. Payen et al.[205)] proposed that there were three regions arbitrarily distinguished according to the temperature of the treatment (Fig. 3.20). The successive O-S exchange to oxysulfides was

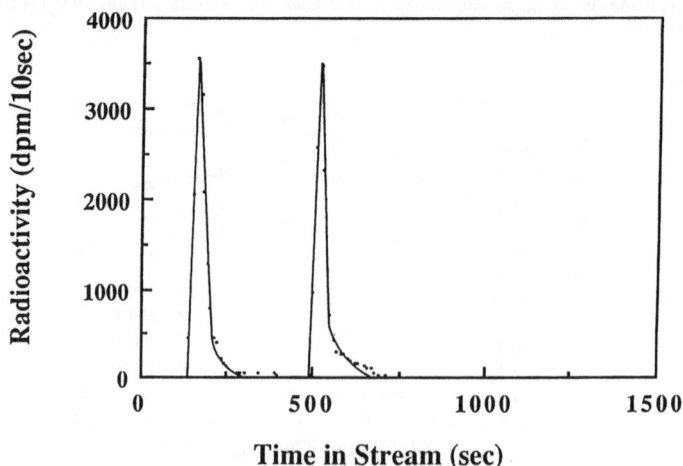

Fig. 3.19 The release of ^{35}S species in calcination of Mo/Al$_2$O$_3$ catalyst sulfided with O$_2$ pulse at 430°C.[218)]
(From T. Kabe, W. Qian, A. Ishihara et al., J. Chem. Soc., Faraday Trans., **93**, 3711 (1997))

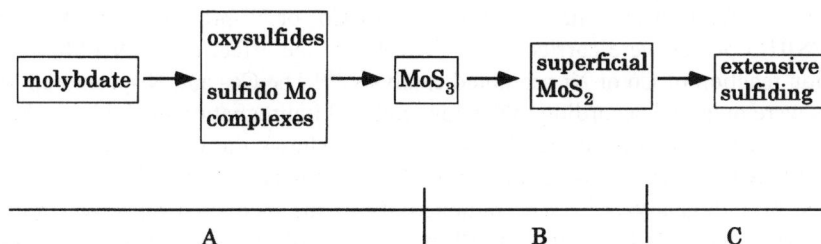

Fig. 3.20 Schematic illustration of the sulfidation process.[221]
(From J. van Gestel, J. J. Leglise and J. C. Duchet, *J. Catal.*, **145**, 434 (1994))

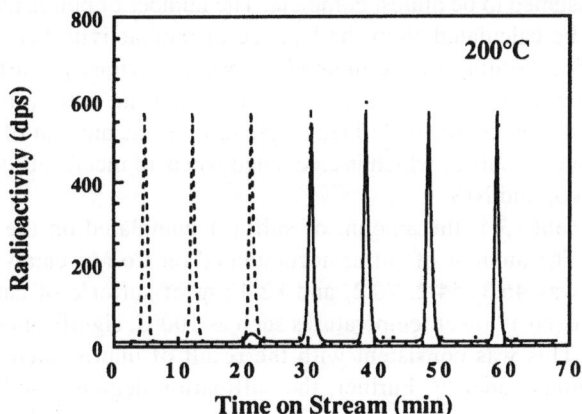

Fig. 3.21 Sulfidation of a Co-Mo/Al$_2$O$_3$ catalyst with [^{35}S]H$_2$S pulse at 200°C.[223] The solid line represents the
radioactivity of unreacted [^{35}S]H$_2$S. The dotted line represents the radioactivity of [^{35}S]H$_2$S incorporated
into the catalyst.
(From T. Kabe, A. Ishihara, W. Qian *et al.*, *Catal. Today*, **451**, 287 (1998))

readily observed in the first and second regions that are under mild conditions below 200°C.

The sulfided state of working catalyst MoO$_3$/Al$_2$O$_3$ was also directly investigated in
HDS of [^{35}S]DBT without presulfiding.[230] Despite the high conversion of DBT, [^{35}S]H$_2$S
was not detected during the initial step of HDS of [^{35}S]DBT. Thereafter, the radioactivity of
[^{35}S]H$_2$S increased slowly and approached a steady state. At this state, an amount of ^{35}S
supplied by the decomposition of [^{35}S]DBT into the catalyst was equal to that of ^{35}S released
from the catalyst as [^{35}S]H$_2$S. This indicated that the catalyst approached the saturation of the
sulfided state under this reaction condition. From the balance of radioactivities between
introduced and recovered ^{35}S-containing species, total radioactivity remaining on the catalyst
in the reaction was calculated:

$$S_R = S_T - S_H - S_D \tag{3.4}$$

where S_T is total radioactivity of [^{35}S]DBT in reactant solution supplied into the reactor, S_H
is total radioactivity of [^{35}S]H$_2$S released from the catalyst, S_D is the total radioactivity of
unreacted [^{35}S]DBT for this time interval. From this calculation, it can be concluded that Mo
on the catalyst exists as MoS$_{1.92}$.[230]

Similar sulfidation experiments of a series of Co- or Ni-promoted Mo/Al_2O_3 catalysts using [^{35}S]H_2S pulses were carried out to investigate the effect of Co(Ni)/Mo ratio on the sulfidation reaction of Co or Ni promoted catalysts.[222] The Co- and Ni-promoted Mo/Al_2O_3 catalysts were prepared according to the conventional impregnation method, in which nickel nitrate tetrahydrate was used. Similarly, Fig. 3.21 shows the change in radioactivity of recovered [^{35}S]H_2S during the sulfidation of Co-Mo catalyst at 200°C, which was directly monitored by radioanalyzer.[223] The radioactivity of the recovered pulse was not detected until the 3rd pulse was introduced at 20 min. This indicates that ^{35}S in these pulses was incorporated into the catalyst. Further, the radioactivity of the recovered [^{35}S]H_2S increased with the number of introduced pulses and approached the radioactivity of the introduced pulse after the 5th pulse was introduced at 37 min. Therefore, the sulfidation at this temperature was assumed to be almost complete. The number of pulses (N_T) incorporated into the catalyst could be calculated from the balance of radioactivity between introduced and recovered [^{35}S]H_2S according to the method reported previously. Further, the amount of sulfur uptake to the catalyst in the sulfidation (S_T) and the sulfidation degree (R) were obtained. The sulfidation degree is defined as the ratio of the amount of sulfur incorporated in the sulfidation to total sulfur, which is calculated when all metal sulfides are present in the forms of MoS_2, Co_9S_8 and NiS.

As shown in Table 3.4, the amount of sulfur accumulated on the catalysts increased with temperature. The amount of sulfur accumulated on Co-Mo catalyst at 100°C, 200°C, 300°C and 400°C was 46.3, 54.2, 70.2, and 82.4 mg of sulfur/g of catalyst, respectively. This indicates that even at lower temperatures such as 100°C, significant O/S exchange in the catalyst occurred. This was consistent with the result of unpromoted Mo/Al_2O_3 catalysts reported in a previous paper.[218] Further, the sulfidation degree at 400°C was 98.1% and approached 100%, corresponding to the sulfidation state of MoS_2 and CoS.[223]

Figure 3.22 shows the uptake amounts of sulfur on the Co-Mo or Ni-Mo catalysts sulfided at several sulfiding temperatures. The amount of sulfur (S_T) accumulated to the catalyst increased with sulfidation temperature for all catalysts. Moreover, the uptake amount of sulfur at 300°C and 400°C increased monotonically with increasing ratio of Ni/Mo while little difference in S_T was observed among the catalysts at 100°C and 200°C.

The changes in the degree of sulfidation (100% when Ni and Mo in Ni-Mo/Al_2O_3 are present in the forms of NiS, Co_9S_8, and MoS_2) of the catalysts at various temperatures are shown in Fig. 3.23. Comparing the sulfidation behavior of the Co-Mo/Al_2O_3 catalysts with that of the Ni-Mo/Al_2O_3 catalysts, striking similarities are found. Both nickel- and cobalt-promoted

Table 3.4   Uptake amount of sulfur measured in sulfidation and oxidation process of Co-Mo/Al_2O_3 and Mo/Al_2O_3 catalysts [a) 223]

Temperature (°C)	100	200	300	400
Mo(16)	48.6	55.3	65.0	77.7
Co(1)-Mo(16)	42.9	47.6	63.5	81.3
Co(2)-Mo(16)	47.6	52.7	68.8	79.8
Co(3)-Mo(16)	46.3	54.2	70.2	82.4
Co(5)-Mo(16)	44.7	56.4	73.5	94.3
Co(8)-Mo(16)	49.2	59.4	91.6	111.2

[a)] The amounts of sulfur accummulated during sulfidation with hydrogen sulfide, mg-sulfur/g-catalyst.

(From T. Kabe, A. Ishihara, W. Qian et al., *Catal. Today*, **45**, 287 (1998))

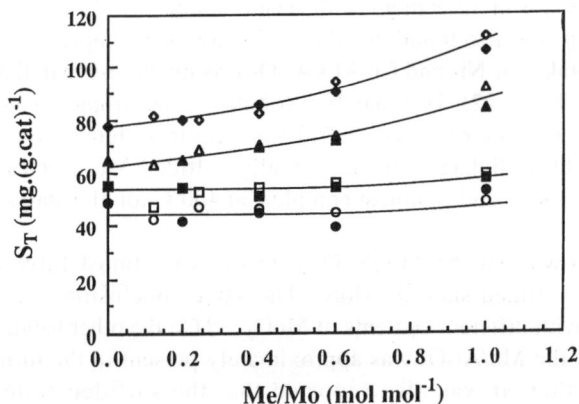

Fig. 3.22   Effect of Co (Ni)/Mo ratio on uptake amount of sulfur at various temperatures.[222] Solid symbols for Ni-Mo catalysts and open symbols for Co-Mo catalysts.
● ○: 100°C, ■ □: 200°C, ▲ △: 300°C, ◆ ◇: 400°C.

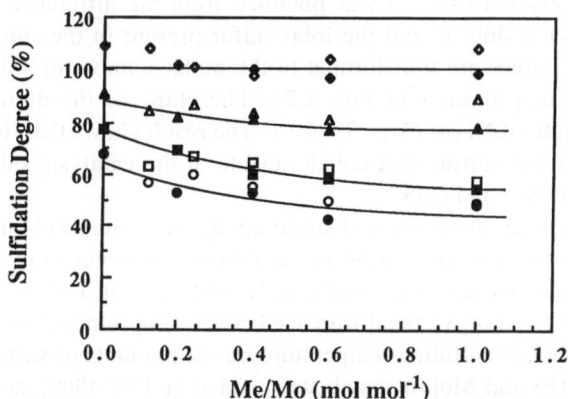

Fig. 3.23   Effect of Co (Ni)/Mo ratio on sulfidation degree at various temperatures.[222] Solid symbols for Ni-Mo catalysts and open symbols for Co-Mo catalysts.
● ○: 100°C, ■ □: 200°C, ▲ △: 300°C, ◆ ◇: 400°C.

molybdena catalysts show very similar features in the sulfidation with [³⁵S]H₂S. Sulfidation degrees below 300°C decreased slightly with increasing molar ratio of Ni/Mo whereas they remained approximately constant at 400°C, corresponding to 100%, despite the change in the molar ratio of Co (Ni)/Mo. This result suggests that at lower temperatures molybdenum oxide is sulfided preferentially whereas the sulfidation of cobalt or nickel oxide in the Co (Ni)-Mo/Al₂O₃ catalysts was very difficult, i.e., higher temperatures may be necessary for sulfiding nickel oxide.

Industrial HDS catalysts typically consist of a mixture of Mo, Co, and Ni sulfides on γ-alumina support. When Co or Ni is added to molybdenum sulfide catalysts, the HDS rate constant can be increased by a factor of 25.[74] This has been ascribed to the formation of new active phase Co(Ni)MoS.[54,224] Since the activity of the catalysts depends eventually on the activation procedure, the sulfide state of the catalysts during sulfidation has been widely investigated.[93,224,225] For unpromoted Mo/Al₂O₃ catalysts, the sulfidation has been noted above. These results are in agreement with recent studies.[203,226–228] On the other hand, for promoted catalysts, it was also found that the sulfidation is incomplete below 200°C since the amount of sulfur uptake for Ni- and Co-Mo/Al$_2$O$_3$ was much less than that of stoichiometric sulfur such as MoS$_3$ (Fig. 3.22). This may be attributed to the stronger interaction between Mo and alumina.[49] Hence, more extensive sulfidation requires higher temperatures in which a fraction of the Mo-O-Al linkages of the partially sulfided Mo species are sulfided. This process starts above 300°C and is almost complete at 400°C for all catalysts, as shown in Fig. 3.22.[222]

Figure 3.23 shows that the Mo/Al$_2$O$_3$ catalyst was almost fully sulfided at 400°C, corresponding to a sulfided state of MoS$_2$. The same conclusion was obtained for other Mo/Al$_2$O$_3$ catalysts with different contents of MoO$_3$.[218] On the other hand, it was reported that the sulfided state of the Mo/Al$_2$O$_3$ was approximately present in the form of MoS$_2$ in many studies.[36,37,229] Further, it was also reported that the sulfided state of the Mo/Al$_2$O$_3$ corresponded to MoS$_{1.92}$ under practical HDS conditions.[230] Therefore, it is reasonable to assume that all Mo species were fully sulfided at 400°C and still present in the form of MoS$_2$ on sulfided catalysts with promoter. The net amount of sulfur bonded to the promoter species on the promoted catalysts (S_T-S_{MoS2}) was obtained from the difference in the total uptake amount of sulfur (S_T) at 400 °C and the total sulfur present in the stoichiometric state as MoS$_2$ (S_{MoS2}). These values are transformed to the molar number of sulfur and are plotted against the content of promoter in Fig. 3.24. The slope of the dotted line is one and corresponds to the state of Ni (or Co) : S = 1 : 1. The result shows that, within experimental error, most of the nickel sulfide and cobalt sulfide in the catalysts sulfided at 400°C are present in the form of NiS and CoS.

As shown in Fig. 3.22, the amount of sulfur uptake on both nickel- and cobalt-promoted catalysts scarcely varied with increasing molar ratio of promoter to molybdenum at lower temperatures, whereas it increased significantly with the ratio of Ni (Co)/Mo at higher temperatures. In order to observe the difference between unpromoted and promoted catalyst more clearly, the effect of sulfiding temperature on the amounts of sulfur uptake on Ni(5)-Mo(16), Co(5)-Mo(16) and Mo(16) catalysts sulfided at 100–400°C are compared in Fig. 3.25.[218] At 100°C and 200°C, the amounts of sulfur uptake onto the three catalysts were almost the same. In contrast, on raising the temperature to 300°C and 400°C, the amount of

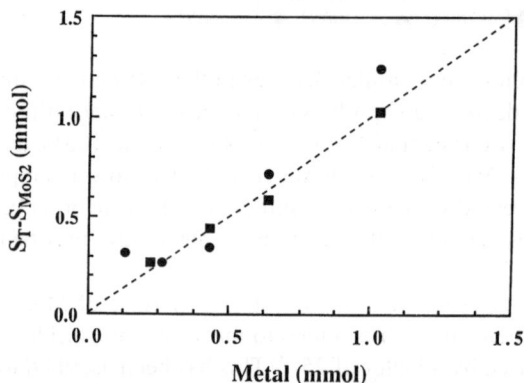

Fig. 3.24   Sulfidation of nickel and cobalt species in Ni-Mo/Al$_2$O$_3$ (■)and Co-Mo/Al$_2$O$_3$ (●)catalysts sulfided at 400°C. The dotted line corresponds to Ni (Co)/S = 1.0.[222]

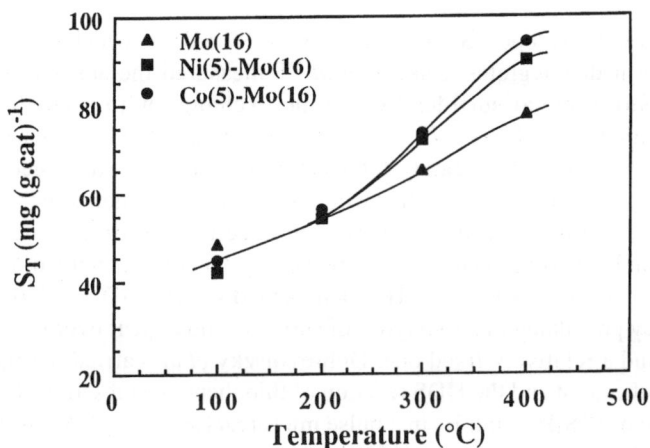

Fig. 3.25 Uptake amount of sulfur on unpromoted and promoted catalysts sulfided at various temperatures.[222]

total uptake of sulfur on the Ni-Mo or Co-Mo/Al$_2$O$_3$ catalyst was higher than that on the Mo/Al$_2$O$_3$ catalyst. This implies that molybdenum oxide is preferably sulfided at 100°C and 200°C since there is similar molybdenum content in three catalysts. That is, sulfidation of nickel oxide and cobalt oxide in the Ni or Co-Mo/Al$_2$O$_3$ catalyst is very difficult at lower temperatures and higher temperatures are necessary for sulfiding nickel and cobalt species. Korányi et al. reported similar results in an XPS study using Co-Mo catalysts sulfided by thiophene where cobalt was sulfided after the initial sulfidation of molybdenum.[231] In a recent sulfidation study for a model catalyst of Co-Mo/Al$_2$O$_3$,[232] Jong et al. proposed that the sulfidation Co oxide was more difficult than the sulfidation of Mo oxide, this being consistent with our results. Further, it was also observed that the differences in total uptake of sulfur between Co-Mo/Al$_2$O$_3$ and Mo/Al$_2$O$_3$ catalysts at 400°C was slightly larger than that between Ni-Mo/Al$_2$O$_3$ and Mo/Al$_2$O$_3$ catalysts, which were ca. 17.2 and 12.4 mg (g · cat)$^{-1}$, respectively. This means that the sulfidation of cobalt species proceeds more easily than the sulfidation of nickel species although the two catalysts show similar features of sulfidation.

3.2 Behavior of Sulfur on Sulfided Catalyst in Hydrotreating Reaction

3.2.1 Behavior of Sulfur in HDS Reaction

Despite extensive studies on the structure of sulfided Mo-based catalysts, some basic problems concerning the participation of surface sulfur species and the behavior of the adsorbed hydrogen in the HDS reaction remain to be resolved. Moreover, the process of HDS of one sulfur-containing compounds is a process in which the sulfur is transformed to hydrogen sulfide after cleavage of the C-S bond in the compound. Hence, it is very important to elucidate the behavior of sulfur on the "working catalyst" in HDS. For this purpose, a radioisotope tracer method has recently been developed. The method is a true in situ technique because it enables us to monitor changes in the process of introduced or labeled ^{35}S by tracing the changes in radioactivity of ^{35}S-containing species without removing the catalyst from the bed.

Sulfur has a radioisotope ^{35}S with a half-life of 87.5 days, which emits soft β-radiation (167 keV) when it disintegrates. Lukens et al.[233] measured the accessible surface area of supported transition metal sulfides by isotope exchange with a labeled H_2S in liquid scintillation solution. Kalechits et al.[234,235] have shown that in the hydrogenation of a mixture of benzene and [^{35}S]CS_2 on WS_2 catalyst, the catalyst sulfur was exchanged with radioactive sulfur of the feedstock. This labile sulfur would be a part of the non-stoichiometric sulfur of the catalyst that would be responsible for the acceleration of acid-catalyzed reaction (isomerization and cracking). Gachet et al. presulfided a commercial Co-Mo/Al_2O_3 catalyst with the gas mixture of [^{35}S]H_2S and H_2, then carried out the HDS of DBT at atmospheric pressure.[236] They postulated that two types of sulfur would appear over the sulfided catalyst: a labile sulfur and a relatively fixed one. Dobrovolszky et al. carried out the hydrogenation reaction of cyclohexanol and the HDS reaction of thiophene over a series of catalysts sulfided by a gas mixture of [^{35}S]H_2S and H_2 in a pulse microreactor.[161,237–239] A pulse-micro-catalytic system was used for sulfidation (at 400°C) and for catalytic reaction of cyclohexanol (at 200°C). The first cyclohexanol pulse was introduced to an oxide catalyst, thereafter alternating [^{35}S]H_2S and cyclohexanol pulse followed, each at its respective temperature in hydrogen carrier gas. One fraction of the interacting H_2S molecules became reversibly adsorbed, and was removed by the carrier gas flow or by the cyclohexanol pulse, while another bonded

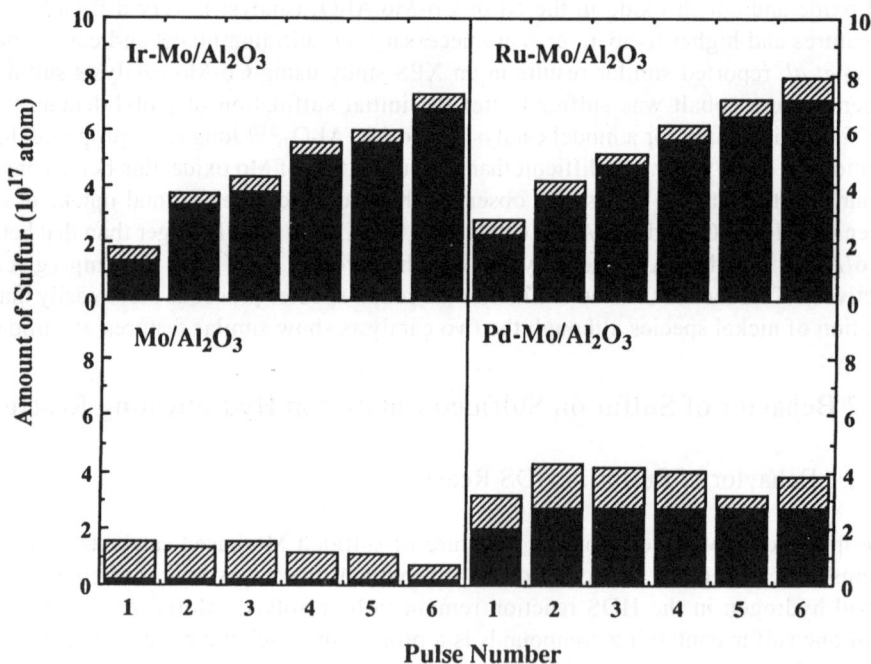

Fig. 3.26 Amount of reversibly (shaded column) and irreversibly (solid column) retained sulfer as a function of the pulse number introduced onto the catalysts. The amount of sulfur in a introduced pulse is 4.8×10^{17} atom.[161]

(From M. Dobrovolszky, Z. Paál and P. Tétényi, Catal. Today, 9, 115 (1991))

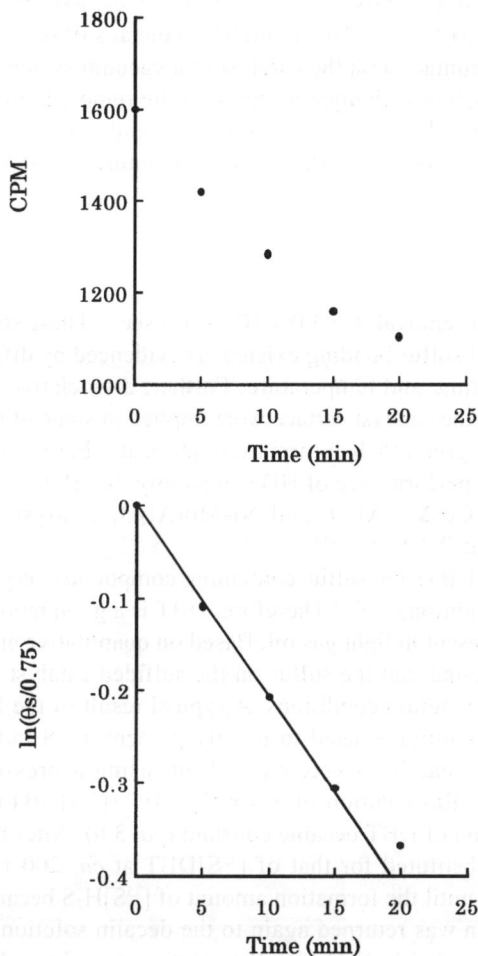

Fig. 3.27 (a) Decrease in ^{35}S signal with time during reduction in H_2.[243] Initial coverage was $\theta_s = 0.75$, $T = 525°C$,
$P(H_2) = 780$ Torr; (b) Plot of $\ln(\theta_s/0.75)$ vs. t shows first-order kinetics.
(From A. J. Gellman, M. Z. Bussell and G. A. Somorjai, *J. Catal.*, **107**, 108 (1987))

irreversibly, and was not removed. As shown in Fig. 3.26, sulfur uptake by the Mo/Al$_2$O$_3$
catalyst was small, and the promoter appeared to enhance the sulfur uptake onto catalysts.
Isagulyants *et al.* made an effort to estimate quantitatively the amount of sulfur held on Co-
Mo catalysts sulfided by ^{35}S elements and ^{35}S-labeled thiophene.[240] By tracing changes in the
^{35}S in produced [^{35}S]H$_2$S during the reaction of [^{32}S]thiophene, it was suggested that two
types of sulfur appeared during the HDS reaction: a labile sulfur and a fixed sulfur. The
amount of labile sulfur that was progressively replaced during the reaction was *ca.* 20 wt%
of the sulfur of the catalyst in its stationary state. It was deduced that a fixed sulfur that did
not directly participate in the catalytic action would be located on the molybdenum and
cobalt sulfide lattices. It was also found that H$_2$S was not formed directly from the sulfur of
DBT but from the sulfur on the catalyst.

Unsupported MoS_2, WS_2 catalysts were used to investigate the behavior of sulfur in sulfur exchange with [^{35}S]H$_2$S.[241,242] The unpromoted metal sulfide picked up radioactive ^{35}S from [^{35}S]H$_2$S during its contact with the catalyst in a vacuum system. Gellman *et al.* used a radiotracer ([^{35}S]CS$_2$) labeling technique to measure the removal rate of sulfur adsorbed on the Mo (100) surface in the HDS reaction of thiophene and in the reduction with H_2.[243] Fig. 3.27 shows the loss of ^{35}S from the surface at a temperature of 535°C in H_2. A fit to a first-order kinetic equation:

$$r = d\theta_s/dt = k\theta_s \tag{3.5}$$

yields the rate constant for removal, $k = 3.0 \times 10^{-4}$ site^{-1} sec^{-1}. These studies generally showed that more than one type of sulfur bonding existed, as evidenced by different rates and extents of sulfur exchange with time and temperature. Further, dissociative adsorption of H_2S and presence of SH groups on the catalyst surface were implied in some of these studies. However, these methods could not give much information about the behavior of sulfur on working catalysts during practical performance of HDS. Recently, the HDS reaction of [^{35}S] DBT on the sulfided Mo/Al$_2$O$_3$, Co-Mo/Al$_2$O$_3$ and Ni-Mo/Al$_2$O$_3$ catalysts under practical HDS conditions was carried out.[74,75,76,219,223,230]

It is well known that DBTs are sulfur-containing compounds very difficult to desulfurize even under deep HDS conditions.[244,245] Therefore, DBT is a good model compound of sulfur-containing compounds present in light gas oil. Based on quantitative analysis of the formation rate of [^{35}S]H$_2$S, it was found that the sulfur on the sulfided catalyst was labile, the amount of which varied with the reaction conditions. A typical result of the HDS of [^{35}S]DBT over a sulfided Mo/Al$_2$O$_3$ was initially used to simply present a ^{35}S radioisotope pulse tracer method (Fig. 3.28). The reactions were carried out using a pressurized fixed bed flow reactor.[230] Initially, a decalin solution of 1.0 wt% DBT ([^{32}S]DBT) was pumped into the reactor until the conversion of DBT became constant (*ca.* 3 h). After that, decalin solution of 1 wt% [^{35}S]DBT was substituted for that of [^{32}S]DBT at *ca.* 200 min. The reaction with [^{35}S]DBT was performed until the formation amount of [^{35}S]H$_2$S became constant (*ca.* 3.5 h). Then the reactant solution was returned again to the decalin solution of 1 wt% [^{32}S]DBT at *ca.* 390 min and reacted for 4–5 h.[230] During the reaction period (*ca.* 11 h), the conversion of DBT was held almost constant at *ca.* 38%. Major products were biphenyl (BP) and cyclohexylbenzene (CHB). After the decalin solution of [^{32}S]DBT was replaced by that of [^{35}S]DBT (2159 dpm/ml), the radioactivities of unreacted [^{35}S]DBT in liquid products increased and reached a steady state (1336 dpm/ml) immediately. The conversion (38.1%) calculated from the radioactivity of the steady state was in good agreement with the conversion (37.5%) estimated from GC analysis. In the case of produced [^{35}S]H$_2$S, however, *ca.* 135 min was needed to reach the steady state in released radioactivity. When the solution of [^{35}S]DBT was replaced with that of [^{32}S]DBT at 400 min, the radioactivities of unreacted [^{35}S]DBT also decreased immediately from the steady state (1336 dpm/ml) to normal state (25 dpm/ml). In contrast, the time delay for produced [^{35}S]H$_2$S from its steady state (823 dpm/ml) to normal state (41 dpm/ml) was *ca.* 135 min.

When BP (0.5 wt%) was added to the reactant solution of [^{35}S]DBT (DBT = 1.0 wt%), the changes in radioactivities of unreacted [^{35}S]DBT and produced [^{35}S]H$_2$S are shown in Fig. 3.29, and the changes in the concentration of produced BP and CHB with the reaction time plotted. With the increase in radioactivities of unreacted [^{35}S]DBT, the concentration of BP

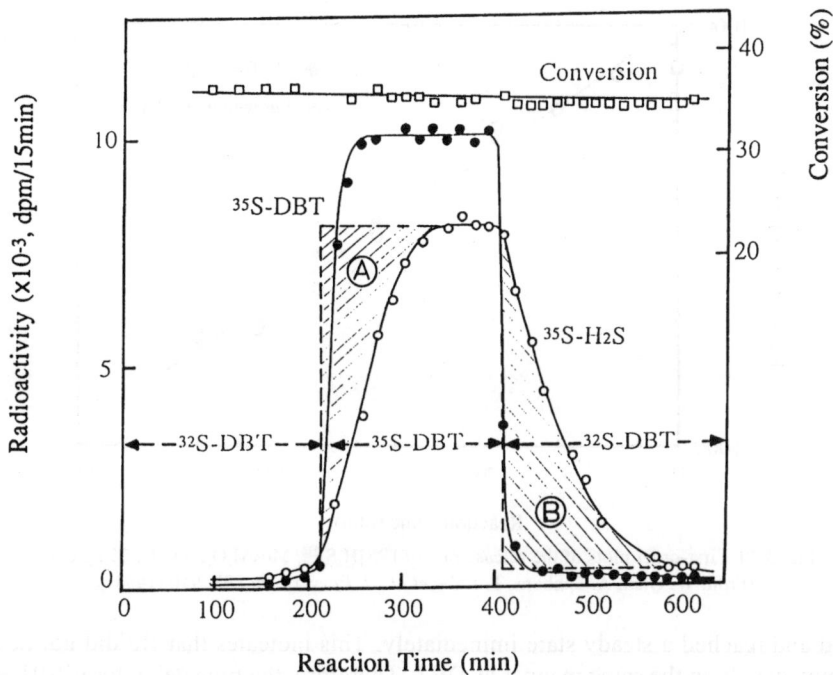

Fig. 3.28   Operation procedure in HDS of [^{35}S]DBT.[230] Mo/Al$_2$O$_3$, temperature 360°C, pressure 25 kg/cm^2;
●: [^{35}S]DBT; ○: [^{35}S]H$_2$S; □: Conversion.
(From W. Qian, A. Ishihara, T. Kabe et al., J. Phys. Chem., **98**, 908 (1994))

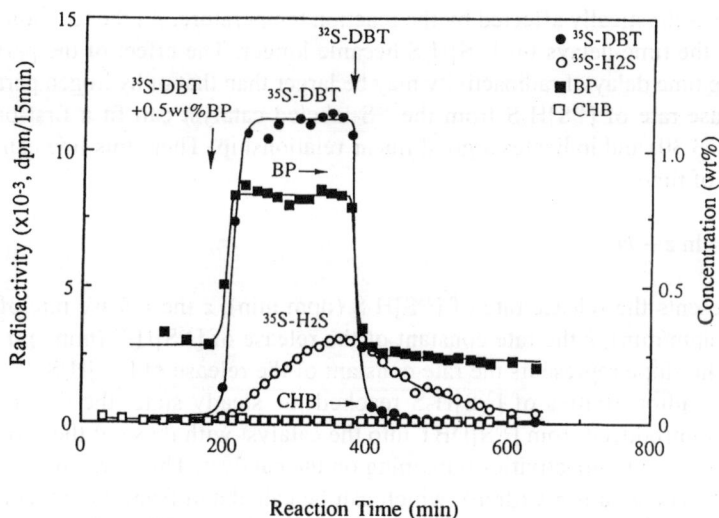

Fig. 3.29   Changes in radioactivities of unreacted [^{35}S]DBT and formed [^{35}S]H$_2$S with reaction time and
concentrations of produced biphenyl and cyclohexylbenzene with reaction time.[211] Mo/Al$_2$O$_3$, 360°C,
10 kg/cm^2, DBT 1 wt%, BP 0.5 wt%.
(From W. Qian, A. Ishihara, T. Kabe et al., J. Phys. Chem., **98**, 909 (1994))

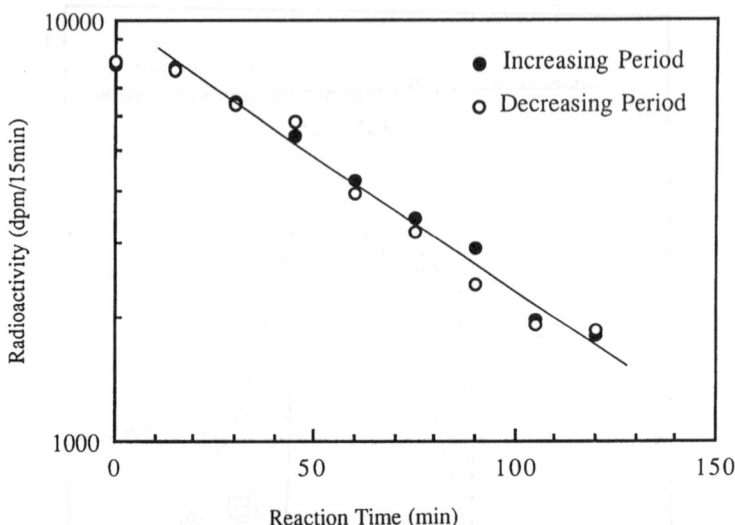

Fig. 3.30 First-order plots of the release rate of [^{35}S]H$_2$S.[230] Mo/Al$_2$O$_3$, 360°C, 25 kg/cm^2.
(From W. Qian, A. Ishihara, T. Kabe *et al.*, *J. Phys. Chem.*, **98**, 910 (1994))

increased and reached a steady state immediately. This indicates that BP did not delay but eluted immediately in the same manner as DBT. Therefore, the time delay for [^{35}S]H$_2$S could not be due to the reaction system. From these results, it is suggested that the sulfur in DBT is not directly released as H$_2$S but accommodated on the catalyst. It was observed that DBT concentrations and temperature influence the release of [^{35}S]H$_2$S.[230] The time delay to reach the steady state of the radioactivities for [^{35}S]H$_2$S increased with decrease in the concentration of DBT and was drastically affected by the reaction temperature. As the reaction temperature was lowered, the time delays for [^{35}S]H$_2$S became longer. The effect of the partial pressure of DBT on the time delay of radioactivity may be larger than that of hydrogen partial pressure.

The release rate of [^{35}S]H$_2$S from the ^{35}S-labeled catalyst can fit a first-order plot, as shown in Fig. 3.30, and indicates a good linear relationship. Then, this line can be revealed as a function of time:

$$\ln y = \ln z - kt \tag{3.6}$$

where y represents the release rate of [^{35}S]H$_2$S (dpm/min), z the release rate of [^{35}S]H$_2$S at steady state (dpm/min), k the rate constant of the release of [^{35}S]H$_2$S (min^{-1}), t the reaction time (min). The slope represents the rate constant of the release of [^{35}S]H$_2$S.

After the radioactivities of [^{35}S]H$_2$S reached the steady state, the difference in total radioactivities introduced from [^{35}S]DBT into the catalyst with those of the formed [^{35}S]H$_2$S is equivalent to total radioactivities remaining on the catalyst. This corresponds to area A or B in Fig. 3.28. The area is z/k (dpm), which can be calculated from the integral (t: 0–∞) of Eq.(3.6). Since all ^{35}S on the catalyst originated from the HDS of [^{35}S]DBT, the concentration of ^{35}S in sulfur introduced to the catalyst at the steady state should be equal to that in sulfur of [^{35}S]DBT because the isotope effect between ^{35}S and ^{32}S was thought to be negligible. The concentration of ^{35}S in sulfur of [^{35}S]DBT could be defined as $^{35}S_{DBT}/S_{DBT}$ (dpm/g), where

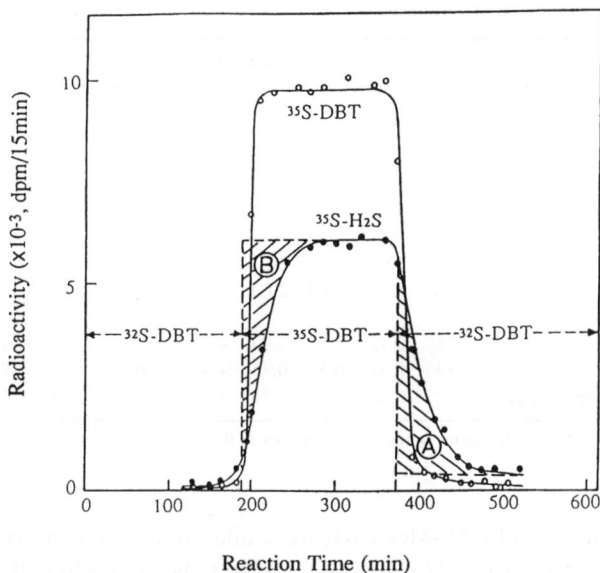

Fig. 3.31   Change in radioactivities of unreacted [^{35}S]DBT and formed [^{35}S]H$_2$S with reaction time.[75)] Mo/Al$_2$O$_3$, 250°C, 50 kg/cm^2, ○: Unreacted [^{35}S]DBT, ●: Formed [^{35}S]H$_2$S.
(From T. Kabe, W. Qian, A. Ishihara et al., J. Catal., 143, 241 (1993))

$^{35}S_{DBT}$ is radioactivities in 1 mol of DBT (dpm/mol) and S_{DBT} is the amount of sulfur in 1 mol DBT (g/mol). According to this, the amount of labile sulfur on the catalyst (S_0) can be presented by $(z/k)/(^{35}S_{DBT}/S_{DBT})$. If the sulfur on the catalyst was assumed to exist in MoS$_2$ under the reaction condition,[51)] the total amount of sulfur on the catalyst is 55.7 m/g of catalyst. At 25 kg/cm2 and 360 °C the amount of labile sulfur was 21.4 mg/g of catalyst, corresponding to 38.4% of total sulfur in the catalyst.

The behavior of sulfur on sulfided Co- and Ni-Mo/Al$_2$O$_3$ catalysts under practical conditions of HDS was also investigated using the ^{35}S RPTM.[75,76)] The changes in radioactivities of the unreacted [^{35}S]DBT and the produced [^{35}S]H$_2$S with the reaction time when a solution of 1 wt% [^{35}S]DBT was reacted on sulfided Co-Mo/Al$_2$O$_3$ at 250°C and 50 kg/cm^2 are shown in Fig. 3.31. After [^{35}S]DBT was substituted for [^{32}S]DBT, the radioactivities of the unreacted [^{35}S]DBT in the liquid product increased and approached a steady state immediately. Similar to the case of Mo/Al$_2$O$_3$ (MoO$_3$ 12.5%) described above, in the case of the produced [^{35}S]H$_2$S, however, ca. 145 min was needed to approach the steady state in released radioactivities. When the solution of [^{35}S]DBT returned to that of [^{32}S]DBT at 370 min, the radioactivities of the unreacted [^{35}S]DBT also decreased immediately from the steady state to the normal state. However, the time delay for the produced [^{35}S]H$_2$S from its steady state to the normal state was still ca. 100 min, as shown in Fig. 3.31. As described above, these results indicate that the sulfur in DBT is not directly released as H$_2$S but is accommodated on the catalyst. The amount of labile sulfur participating in HDS reactions at various temperatures and the release rate constants of [^{35}S]H$_2$S were determined and listed in Table 3.5.[246)]

When the HDS reaction of [^{35}S]DBT was performed with the sulfided Ni-Mo/Al$_2$O$_3$, it can

Table 3.5 Kinetic parameters at various HDS conditions on Co-Mo/Al$_2$O$_3$ and Ni-Mo/Al$_2$O$_3$ catalysts[246]

catalysts	Co-Mo/Al$_2$O$_3$						Ni-Mo/Al$_2$O$_3$				
Reaction pressure, kg/cm^2	50	50	50	50	50	10	50	50	50	50	50
Reaction temperature, °C	230	250	260	280	280	280	210	230	240	260	280
Concentration of DBT, wt%	1.0	1.0	1.0	1.0	3.0	1.0	1.0	1.0	1.0	1.0	1.0
Conversion from GC analysis, %	23.4	39.4	62.6	92.4	53.9	59.8	12.7	29.1	43.3	64.5	99.5
Conversion from radioactivity of ^{35}S, %	23.7	37.8	59.7	90.1	53.0	57.2	11.8	28.9	42.1	63.0	97.1
Labile sulfur, S_0, mg/g-cat.	5.8	12.6	19.9	26.1	30.3	6.6	29.5	11.2	14.7	16.2	18.4
S_0/S_t [a], %	8.27	18.0	28.4	37.2	43.2	42.0	8.2	13.9	18.2	20.1	22.8
Rate constant of formed H$_2$S, k, $\times 10^{-2}$ min^{-1}	2.26	2.45	2.71	2.92	4.65	3.34	1.57	2.13	2.40	3.26	4.40
$S_0 \times k$, mg/min · g-cat.	0.13	0.31	0.51	0.76	1.41	0.99	0.10	0.24	0.35	0.52	0.81
Rate of DBT HDS, mg/min · g-cat.	0.19	0.32	0.51	0.76	1.32	0.98	0.10	0.24	0.35	0.53	0.81

[a] S_t is defined as the amount of total sulfur when metal sulfides in the sulfided catalysts were present as MoS$_2$, Co$_9$S$_8$, and NiS.

be observed that the result for Ni-Mo/Al$_2$O$_3$ are similar to those for Co-Mo/Al$_2$O$_3$ (see Table 3.5). For example, reactivity of Ni-Mo/Al$_2$O$_3$ at 260°C is similar to that at 260°C with Co-Mo/Al$_2$O$_3$. The radioactivities of the formed [^{35}S]H$_2$S with reaction time at 210°C, 230°C, 240°C, 260°C and 280°C are shown in Fig. 3.32. Compared with the case of [^{35}S]DBT, the time delays observed for [^{35}S]H$_2$S were also significantly affected by the reaction temperature. As in the case of Co-Mo/Al$_2$O$_3$, the time delay for [^{35}S]H$_2$S became longer with decreasing reaction temperature.

In order to understand more clearly promotion effect of Co or Ni on the molybdenum catalyst, the HDS reactions of [^{35}S]DBT on the sulfided Co/Al$_2$O$_3$ and Ni/Al$_2$O$_3$ were also performed. The reactivity at 360°C is only one-seventh of the reactivity in the case of Mo/Al$_2$O$_3$ under the same reaction conditions. Similar to Mo/Al$_2$O$_3$, the steady state for the radioactivity of the unreacted [^{35}S]DBT was always immediately achieved at every temperature, while the time delay for the produced [^{35}S]H$_2$S to approach the steady state was ca. 160 and 130 min at 360°C and 400°C, respectively. The time delay for Co/Al$_2$O$_3$ at 360°C was longer than that (ca. 115 min) for Mo/Al$_2$O$_3$ at 360°C. This result indicates that the time delay for [^{35}S]H$_2$S elution is not due to the adsorption/desorption of H$_2$S on alumina support, but to the sulfur exchange between the sulfur in DBT and sulfur on the catalyst. In the case of Ni/Al$_2$O$_3$ catalyst, results similar to those for Co/Al$_2$O$_3$ were obtained. Ni/Al$_2$O$_3$ shows reactivity similar to Co/Al$_2$O$_3$ under the same reaction conditions. The time delay for the produced [^{35}S]H$_2$S to approach the steady state was ca. 150, 120 and 90 min at 360°C, 380°C and 400°C, respectively.

The formation rate of [^{35}S]H$_2$S from all catalysts could be treated as a first-order reaction (see Fig. 3.33). The release rate constants of [^{35}S]H$_2$S for all catalysts were obtained. The apparent activation energies of HDS for DBT on all catalysts calculated from the Arrhenius plots of the rates of HDS were about 20 ± 1 kcal/mol for Mo/Al$_2$O$_3$, Co/Al$_2$O$_3$, Co-Mo/Al$_2$O$_3$, Ni/Al$_2$O$_3$, and Ni-Mo/Al$_2$O$_3$.[76] This was consistent with the results reported in Ref. 217. It implies that the same reaction process occurs on sulfided Mo/Al$_2$O$_3$, Co-Mo/Al$_2$O$_3$, and Ni-Mo/Al$_2$O$_3$.

For the sulfided Co/Al$_2$O$_3$, the stable sulfide is Co$_9$S$_8$.[83–85,247] Results of EXAFS have also

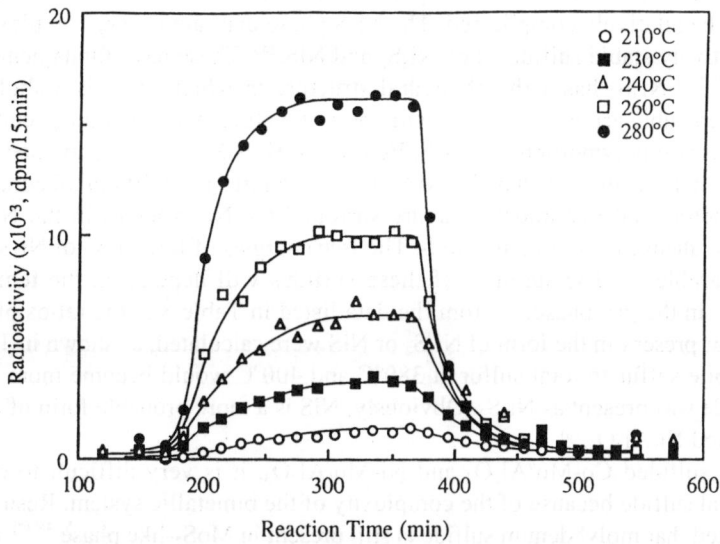

Fig. 3.32 Changes in radioactivities of formed [^{35}S]H$_2$S with reaction time at various temperatures.[76] Ni-Mo/Al$_2$O$_3$, 210–280°C, 50 kg/cm^2.
(From T. Kabe, W. Qian, A. Ishihara et al., J. Catal., **149**, 174 (1994))

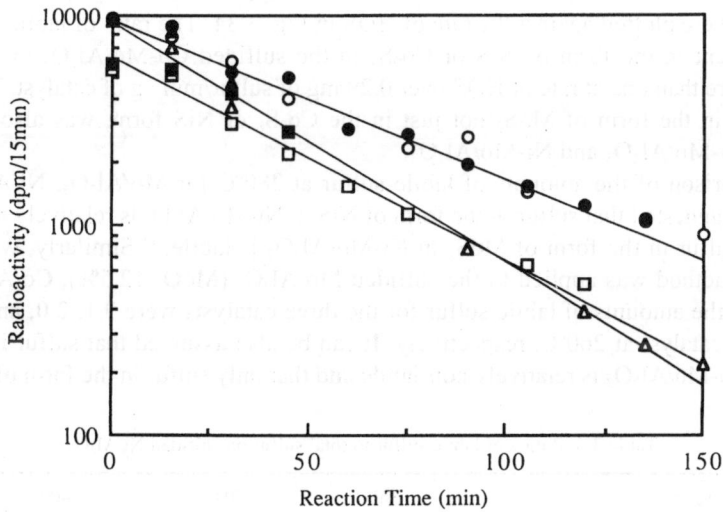

Fig. 3.33 First-order plots of release rate of [^{35}S]H$_2$S.[76] Open symbols for the decreasing period and solid symboles for the increasing period. ○ ●: Mo/Al$_2$O$_3$ (340°C, 50 kg/cm^2); □ ■ : Co-Mo/Al$_2$O$_3$ (260°C); △ ▲ : Ni-Mo/Al$_2$O$_3$ (260°C).
(From T. Kabe, W. Qian, A. Ishihara et al., J. Catal., **149**, 175 (1994))

indicated the existence of a Co_9S_8-like phase. For the sulfided Ni/Al_2O_3, the forms of nickel sulfide may be relatively complicated. The Ni-S phase diagram is very complex [248] but there are two relatively stable sulfides, i.e., Ni_3S_2 and NiS.[249] These have fundamentally different structures.[250,251] Ni_3S_2 has a rhombohedral structure in which there is a slightly distorted body-centered cubic arrangement of S with the metal atom at some of the pseudo-tetrahedral sites. There is two polymorphism of NiS. Below 620 K, NiS has the rhombohedral symmetry, but at higher temperatures it has the NiAs-type structure. In the rhombohedral form five S atoms in a tetragonal pyramidal structure surround the Ni, whereas in the NiAs structure each Ni is octahedrally coordinated to S. The free energies of formation of Ni_3S_2 and NiS are quite comparable.[249] The stability of these sulfides will depend on the temperature and H_2/H_2S ratio in the gas phase.[81] From the data listed in Table 3.5, the ratios of labile sulfur to total sulfur present in the form of Ni_3S_2 or NiS were calculated, as shown in Table 3.6. The ratios of labile sulfur to total sulfur at 380°C and 400°C would become more than 1, if the nickel sulfide was present as Ni_3S_2. Obviously, NiS is a more probable form of nickel sulfide in the sulfided Ni/Al_2O_3.[76]

For the sulfided $Co-Mo/Al_2O_3$ and $Ni-Mo/Al_2O_3$, it is very difficult to determine the form of metal sulfide because of the complexity of the bimetallic system. Results of EXAFS have indicated that molybdenum sulfide is still present in MoS_2-like phase.[36,37] Therefore, we could assume that cobalt or nickel sulfide and molybdenum sulfide in the sulfided Co-Mo/Al_2O_3 and $Ni-Mo/Al_2O_3$ were still present in the form of Co_9S_8 or NiS and MoS_2, respectively.

Based on this assumption, it was further assumed that only sulfur present in the Co_9S_8 or NiS phase in the sulfided $Co-Mo/Al_2O_3$ or $Ni-Mo/Al_2O_3$ was labile, and that sulfur present in the form of MoS_2 was non-labile. The ratios of the amount of labile sulfur to total amount of sulfur present in the Co_9S_8 or NiS phase for Co/Al_2O_3, $Co-Mo/Al_2O_3$, Ni/Al_2O_3 and $Ni-Mo/Al_2O_3$ were plotted against the rate of HDS in Fig. 3.34. The ratio of labile sulfur to total sulfur present in the form of NiS or Co_9S_8 in the sulfided $Co-Mo/Al_2O_3$ or $Ni-Mo/Al_2O_3$ became more than one at rate of HDS over 0.29 mg of sulfur/min · g of catalyst. This indicates that sulfur in the form of MoS_2 not just in the Co_9S_8 or NiS form, was also labile in the sulfided $Co-Mo/Al_2O_3$ and $Ni-Mo/Al_2O_3$.

Comparison of the amounts of labile sulfur at 280°C for Mo/Al_2O_3, Ni/Al_2O_3 and Ni-Mo/Al_2O_3 suggested that sulfur in the form of NiS in $Ni-Mo/Al_2O_3$ is relatively non-labile and that only sulfur in the form of MoS_2 in $Ni-Mo/Al_2O_3$ is labile.[76] Similarly, when the same treatment method was applied to the sulfided Mo/Al_2O_3 (MoO_3 12.5%), Co/Al_2O_3 and Co-Mo/Al_2O_3, the amounts of labile sulfur for the three catalysts were 9.1, 2.0, and 19.9 mg of sulfur/g of catalyst at 260°C, respectively. It can be also assumed that sulfur in the form of Co_9S_8 in $Co-Mo/Al_2O_3$ is relatively non-labile and that only sulfur in the form of MoS_2 in Co-

Table 3.6 Ratios of labile sulfur to total sulfur on sulfided Ni/Al_2O_3 [76]

Temperature (°C)	360	380	400
Labile Sulfur (mg-sulfur/g-catalyst)	7.2	9.9	12.8
Ratio I [a] (%)	84.5	115.2	149.4
Ratio II [b] (%)	55.8	76.7	99.2

It was assumed that all sulfur were present as [a] Ni_3S_2; [b] NiS.

(From W. Qian, A. Ishihara, T. Kabe et al., J. Catal., **149**, 176 (1994))

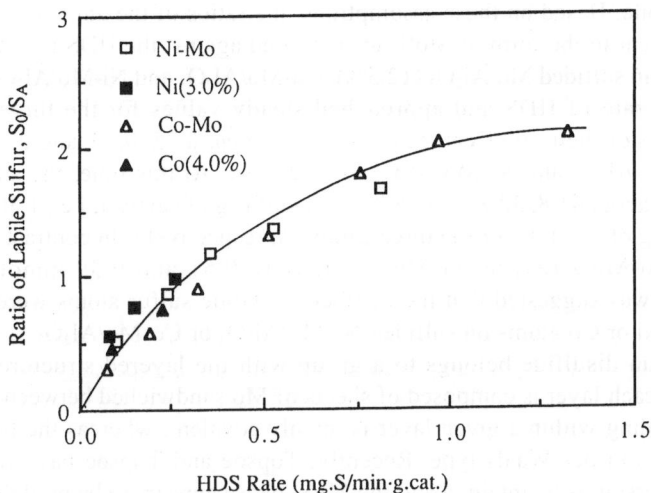

Fig. 3.34   Plots of ratio of labile sulfur vs. rate of DBT HDS.[76]
Open symbols belong to Co-Mo/Al$_2$O$_3$ ($\triangle$) and Ni-Mo/Al$_2$O$_3$ ($\square$). It was assumed that the sulfur in MoS$_2$ phase was not labile but only the sulfur in Co$_9$S$_8$ or NiS phase was labile in the sulfided Ni-Mo/Al$_2$O$_3$ and Co-Mo/Al$_2$O$_3$. The solid symbols belong to Co/Al$_2$O$_3$ ($\blacktriangle$) and Ni/Al$_2$O$_3$ ($\blacksquare$). The ratios of labile sulfur on the sulfided catalysts were estimated from S_0/S_A (S_A: total amount of sulfur present in the form of Co$_9$S$_8$ or NiS in the sulfided Co/Al$_2$O$_3$, Ni/Al$_2$O$_3$, Co-Mo/Al$_2$O$_3$, and Ni-Mo/Al$_2$O$_3$).[76]
(From T. Kabe, W. Qian, A. Ishihara et al., J. Catal., **149**, 176 (1994))

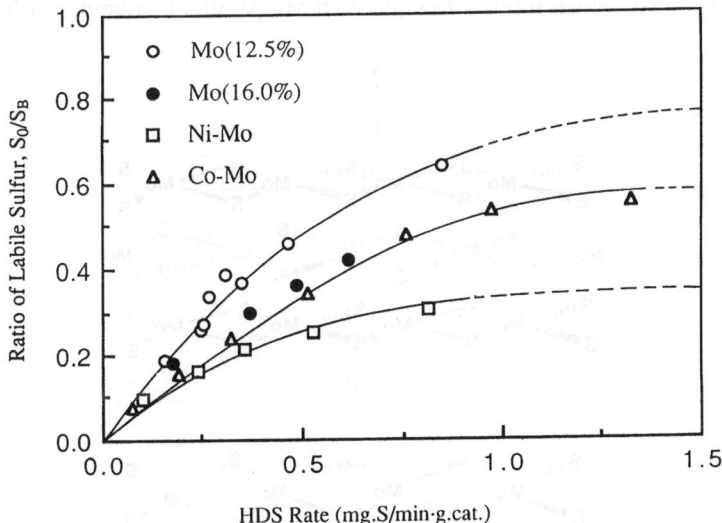

Fig. 3.35   Plots of ratio of labile sulfur vs. rate of DBT HDS.[76]
$\bigcirc$: Mo/Al$_2$O$_3$ (12.5 %); $\bullet$: Mo/Al$_2$O$_3$ (16 %); $\triangle$: Co-Mo/Al$_2$O$_3$; $\square$: Ni-Mo/Al$_2$O$_3$. It was assumed that the sulfur in Co$_9$S$_8$ and NiS phase was not labile but only the sulfur in MoS$_2$ phase was labile in the sulfided Co-Mo/Al$_2$O$_3$ and Ni-Mo/Al$_2$O$_3$. The ratios of labile sulfur on the sulfided catalysts were estimated from S_0/S_B (S_B: total amount of sulfur present in the form of MoS$_2$ in the sulfided Mo/Al$_2$O$_3$, Co-Mo/Al$_2$O$_3$ and Ni-Mo/Al$_2$O$_3$).
(From T. Kabe, W. Qian, A. Ishihara et al., J. Catal., **149**, 177 (1994))

Mo/Al_2O_3 is labile. Based on these assumptions, the ratios of the amount of labile sulfur to total sulfur present in the form of MoS_2 were plotted against the HDS rate as shown in Fig. 3.35. The ratios in sulfided Mo/Al_2O_3 (12.5%), $Co-Mo/Al_2O_3$ and $Ni-Mo/Al_2O_3$ increased with increase in the rate of HDS and approached steady values for the three catalysts. The maximum values of their ratios can be deduced to be *ca.* 0.75, 0.59, and 0.37 for Mo/Al_2O_3 (12.5%), $Co-Mo/Al_2O_3$ and $Ni-Mo/Al_2O_3$, respectively. At this time, the amounts of labile sulfur would be about 41.8, 32.3, and 25.2 mg of sulfur/g of catalyst, i.e., 1.30, 1.01, and 0.79 mmol of sulfur/g of catalyst for the three catalysts, respectively. In contrast, the contents of Co or Ni in $Co-Mo/Al_2O_3$ or $Ni-Mo/Al_2O_3$ were 0.51 and 0.39 mmol/g of catalyst, respectively. It was suggested that the numbers of labile sulfur atoms were approximately twice those of Ni or Co atoms on sulfided $Ni-Mo/Al_2O_3$ or $Co-Mo/Al_2O_3$.

Molybdenum disulfide belongs to a group with the layered structure shown in Fig. 3.36(a),[252] and each layer is composed of sheets of Mo sandwiched between sheets of sulfur atoms. The bonding within a given layer is mainly covalent, whereas the bonding between layers is of the van der Waals type. Recently, Topsøe and Topsøe have reported that the monolayer dispersion is maintained and the MoS_2 phase appears to be predominantly present as a two-dimensional single-slab structure oriented flat-wise on the alumina support (c-axis perpendicular to alumina surface) for Mo/Al_2O_3 up to 12% Mo.[157,253] Thus, it is an acceptable hypothesis that the MoS_2 phase is present as the single slab structure flat on the surface of alumina as shown in Fig. 3.36(b). However, since the locations of sulfurs on the surface of alumina were different from each other, the labile capacities of sulfurs would also be different. The sulfur between the molybdenum layer and alumina surface (S_b) may be the most difficult to move, and the sulfur over molybdenum layer (S_a) may be the most mobile. The sulfur in other sites (S_c or S_d), which forms a triangle with Mo parallel to alumina surface, may have

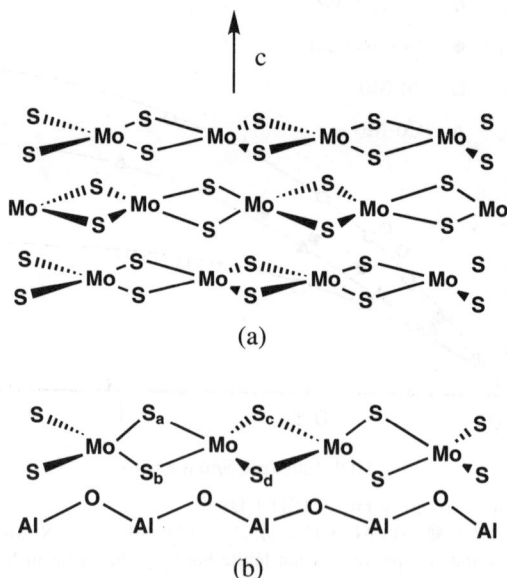

(a)

(b)

Fig. 3.36 Structure of MoS_2: (a) Crystal structure of molybdenum disulfide; (b) Tetrahedral structure of MoS_2 phase on alumina. Mobile capacity of sulfur: $S_a > S_c$ (S_d) $> S_b$.[76)]
(From T. Kabe, W. Qian, A. Ishihara *et al.*, *J. Catal.*, **149**, 177 (1994))

intermediate mobile capacity. This explains why the amount of labile sulfur changes depending on the reaction conditions. If the sulfur between molybdenum and alumina surface, S_b, was non-labile, the amount of labile sulfur in the sulfided Mo/Al_2O_3 would be 75% of the total sulfur. This is good agreement with results that the maximum amount of labile sulfur is 75% of the total sulfur, as shown in Fig. 3.35.

For the sulfided $Co-Mo/Al_2O_3$ and $Ni-Mo/Al_2O_3$, the structure of MoS_2-like phase located in the edge may be rearranged because of the presence of Co or Ni atoms, and a square pyramidal model may be an acceptable model. This idea originated from Ratnasamy and Sivasanker[55] and later in a more detailed model was proposed by Topsøe's group.[254,255] In recent works, Bouwens et al. and Louwers and Prins have given further evidence about this model by the use of EXAFS (Fig. 3.37).[36,37] The square pyramidal coordination of the Co or Ni atoms resembles that of the millerite structure. Co or Ni atoms are connected to the MoS_2 crystallite by four sulfur atoms. An additional sulfur atom is attached in front of the Ni atom

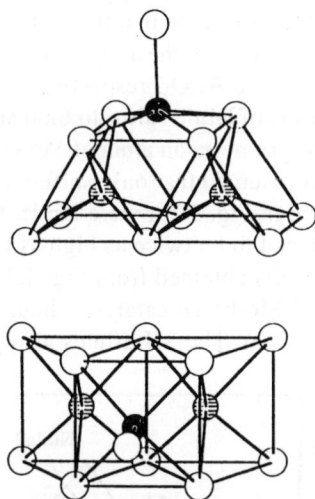

Fig. 3.37 Top and side view of a square-pyramidal "CoMoS" structure.[36]
(From S. M. A. M. Bouwens et al., J. Phys. Chem., **95**, 133 (1991))

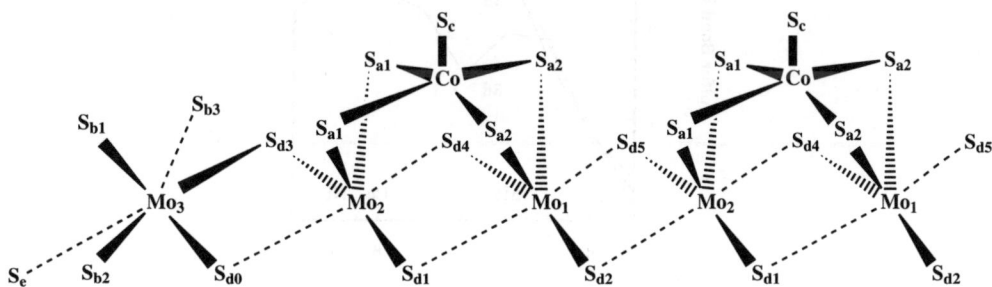

Fig. 3.38 A deformed tetrahedral structure of MoS_2 and structure of Mo-S-Co (Ni).[74]
(From W. Qian, A. Ishihara, T. Kabe et al., J. Chem. Soc., Faraday Trans., **93**, 4399 (1997))

as shown in Fig. 3.38.[74] Even in this model, one can still assume that the structure of MoS_2 is a deformed tetrahedral structure and only the locations of other two weak Mo-S bond within the layers are changed, as shown in Fig. 3.38. The Mo_1-S_{c0} or Mo_2-S_{c1}, and Mo_1-S_{b2} or Mo_2-S_{b3} bonds were considered as two other weak bonds assigned to the van der Waals type in this structure. As noted above, the sulfur in the Co-Mo/Al_2O_3 or Ni-Mo/Al_2O_3 is the most labile among sulfided Co/Al_2O_3 or Ni/Al_2O_3, Mo/Al_2O_3, and Co-Mo/Al_2O_3 or Ni-Mo/Al_2O_3. This is consistent with the bond energies of metal sulfide calculated by Nørskov et al.[67] and Topsøe et al.[58,256] Based on *ab initio* calculations, a model was developed to determine variations in the metal-sulfur bond energies for all the transition metal sulfides. It was found that the trends in HDS activities follow quite closely the trends in the calculated bond energies (Fig. 3.39). Specifically, it was found that the largest activities are observed for the systems that have the weakest bound sulfur atoms. They proposed that the bond energies of metal sulfide varied as follows: nickel or cobalt sulfide > molybdenum sulfide > CoMoS or NiMoS. Taking into account the fact that the bond energy of Co-S or Ni-S is higher than that of CoMoS or NiMoS, it is reasonable to assume that the sulfur attached to only Co or Ni atom is more difficult to move. On the other hand, the atomic ratios of Co/Mo and Ni/Mo were 0.59 and 0.37 for Co-Mo/Al_2O_3 and Ni-Mo/Al_2O_3, respectively. These values are in very good agreement with the maximum ratios of labile sulfur to total sulfur obtained from Fig. 3.35. It indicates that an atom of Co or Ni promotes an atom of Mo or two atoms of sulfur in adjacent MoS_2 phase. Furthermore, it was assumed that only sulfurs in MoS_2 phase adjacent to Co or Ni atoms, i.e., S_{a1} or S_{a2} as shown in Fig. 3.38, were labile, the numbers of labile sulfurs in Mo-S-Co (Ni) phase can be deduced to be twice as high as that of Co (Ni) atoms. This is in very good agreement with the results obtained from Fig. 3.35.

The HDS activity of sulfided Mo-based catalysts should be relative to the existence of sulfur vacancies (uncoordinated sites). The SH groups also played an important role in the

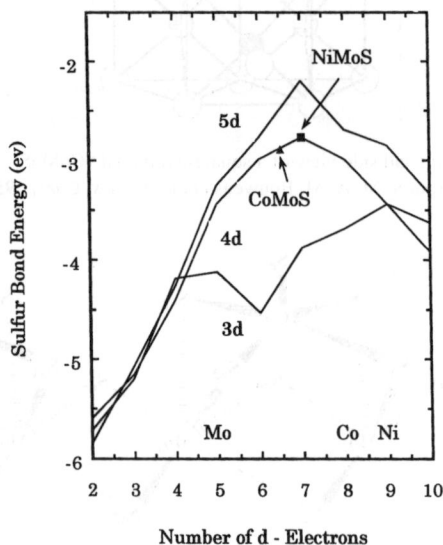

Fig. 3.39 Calculated variation in the sulfur bond energy of different transition metal sulfides.[242]
(From H. Topsøe et al., Bull. Soc. Chim. Belg., **104**, 284 (1995))

HDS reaction.[257,258] Regarding the evidence for the presence of SH groups, studies of deuterium exchange,[93] chemical titration by silver ions[260] and Raman spectroscopy[261] have provided such evidence. More recently, Topsøe and Topsøe postulated that SH groups existed at the edges of MoS_2, and found by FTIR study on sulfided Mo-based catalysts that SH groups and vacancies could interconvert and coexist in close proximity.[157] As mentioned above, H_2S was not formed directly from the sulfur in DBT, but from the sulfur on the catalyst. The absence of DBT did not generate H_2S, while the incorporation of sulfur in DBT onto catalyst generated H_2S. If the vacancies were the sites for the coordination with the heteroatoms of reactants, the mechanism of DBT HDS would be illustrated more simply as shown in Fig. 3.40. It is also assumed that only sulfur bonded to both Co (or Ni) and Mo in the form of MoS_2 is labile in the scheme. When ^{35}S in DBT occupied the vacancy and the carbon-sulfur bonds were cleaved,^{35}S remained on the catalyst as Mo-S species. The generation of H_2S will form a new vacancy on the catalyst. Thus, a shift of vacancy on the catalyst surface would occur. Ruette and Ludena have proposed the possibility that the vacancy position may be easily shifted in the molecular orbital calculations of the desulfurization reaction of thiophene over a Co-Mo catalyst.[262] For this mechanism of HDS, it should be noted that only after the sulfur in DBT was incorporated into the catalyst was sulfur on the catalyst surface released as H_2S. This means that after an anion vacancy is occupied by a sulfur atom removed from DBT, a new anion vacancy appears on the catalyst surface.

It has been suggested that the amount of labile sulfur is equal to the total sulfur that can be converted into vacancies under a certain reaction condition for a catalyst. Although this amount corresponds closely to the reactivity of the catalyst, it is definitely not equal to the number of active sites. The reactivity of a catalyst will also depend upon the conversion rate between labile sulfur and the vacancy-regeneration rate of active sites, as well as the amount of labile sulfur. Thus, the regeneration rate of active sites as well as the amount of labile sulfur varies with reaction conditions. Furthermore, the variation in the regeneration rate of active sites is more than in the amount of labile sulfur. This could explain the correspondence between the amounts of labile sulfur and the HDS rates shown in Fig. 3.35, where the

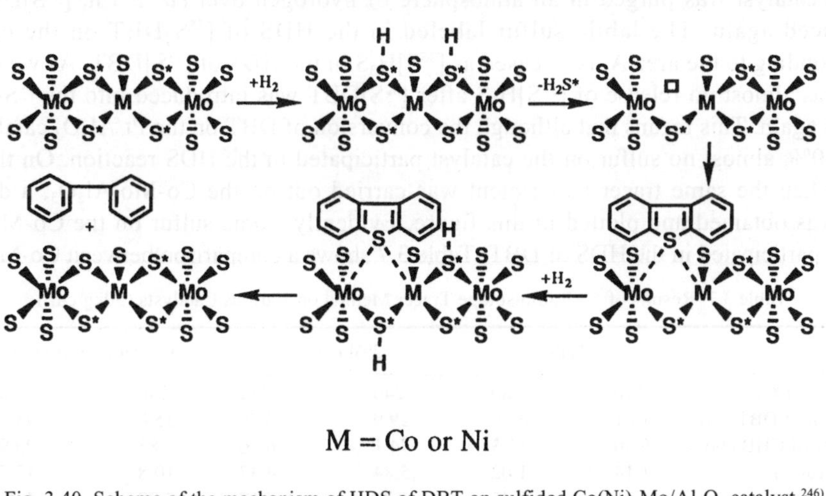

M = Co or Ni

Fig. 3.40 Scheme of the mechanism of HDS of DBT on sulfided Co(Ni)-Mo/Al_2O_3 catalyst.[246]

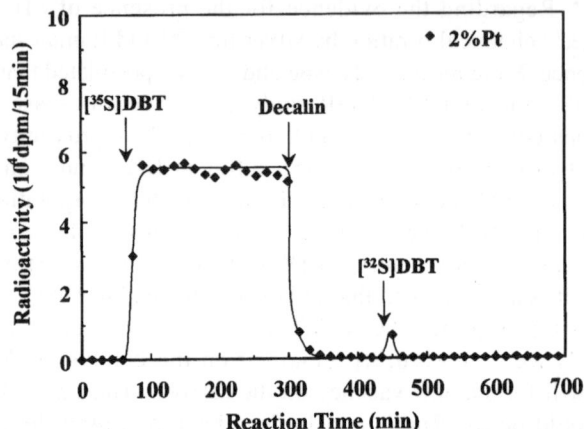

Fig. 3.41   Change in radioactivities of formed [35S]H₂S with reaction time. 2%Pt/Al₂O₃, 280°C, 50 kg/cm², DBT 1.0 wt%.

reactivity of a catalyst varied significantly even though the amount of labile sulfur for the catalyst remained more or less constant.

In order to satisfy the requirement for the reduction of aromatics in gasoline and light oil and hydrodenitrogenation, it is necessary to develop a new catalyst with high activity of hydrogenation. Therefore, hydrotreating catalysts using novel metals such as Pt, Pd, etc. have been developed.[263–265] When the 35S radioisotope tracer method was used to investigate the behavior of sulfur on noble metal (e.g., Pt, Pd, etc.) catalysts, different results were obtained. Fig. 3.41 shows the result of a 35S tracer experiment using a Pt/Al₂O₃ with 2 wt% Pt at 280°C, 50 kg/cm² and WHSV 28h⁻¹. In order to estimate the amount of labile sulfur more accurately, the operation procedure was improved and was slightly different from the typical operation procedure described above. Instead of the replacement of [35S]DBT solution for the [32S]DBT solution, the [35S]DBT solution was replaced by decalin solvent. After the 35S-labeled catalyst was purged in an atmosphere of hydrogen over *ca.* 2.5 h, [32S]DBT was introduced again. The labile sulfur labeled in the HDS of [35S]DBT on the catalyst, corresponding to the area A, is released as [35S]H₂S in the HDS of [32S]DBT. As we can see, there was almost no release of [35S]H₂S after [32S]DBT was introduced into the 35S-labeled catalyst again. This means that although the conversion of DBT on the Pt/Al₂O₃ catalyst was about 50%, almost no sulfur on the catalyst participated in the HDS reaction. On the other hand, when the same tracer experiment was carried out on the Co-Mo/Al₂O₃, a different result was obtained and plotted in this figure. Evidently, some sulfur on the Co-Mo/Al₂O₃ catalyst participated in the HDS of DBT. Table 3.7 shows a comparison between Co-Mo/Al₂O₃

Table 3.7 Results of 35S Radioisotope Tracer Method on Various Catalysts (50kg/cm²)

Catalyst	2%Pt		2%Pt–10%Pd		Co-Mo(Co/Mo:0.6)	
Temperature (°C)	240	280	240	280	240	280
Conversion of DBT (%)	16.1	60.0	29.9	8.46	35.8	94.7
Selectivity of CHB (%)	5.50	11.3	74.1	63.6	8.85	9.69
S_0 (mg/g-cat.⁻¹)	1.14	1.02	5.84	4.37	10.8	17.5
k_{RE} (10⁻²min⁻¹)	4.86	7.45	4.70	5.67	2.97	4.95

and Pt/Al_2O_3 catalysts. The amount of labile sulfur on the Co-Mo increases with increasing reaction temperature but the amount of labile sulfur on the Pt catalyst was only *ca.* 1mg/g·Cat., similar to that obtained in a blank experiment.

3.2.2 Comparison of Sulfur Behavior in HDS, HDO and HDN

In the above section, the behavior of sulfur on sulfided Mo-based catalysts under the operating conditions of HDS was discussed. It was found that some sulfur on the catalyst in the HDS reaction is mobile and that the amount of labile sulfur increased with change in reaction conditions. However, it is not yet clear whether sulfur on the catalyst can be released in other hydrotreating process such as HDO and HDN. It has been reported that sulfur, oxygen and nitrogen compounds can be adsorbed more strongly on the catalyst than hydrocarbons in HDS, HDN and HDO.[217, 266]

In order to more clearly observe the release of $[^{35}S]H_2S$ during various reactions, labeling of the catalyst with ^{35}S is necessary. Similar to the experiment mentioned above, the catalyst was labeled with ^{35}S in the HDS reaction of $[^{35}S]DBT$ (Fig. 3.42). First, the reaction was carried out in the same manner as mentioned above (operation procedure 1). Of course, the same result was obtained. Then, instead of replacing the $[^{35}S]DBT$ solution for the $[^{32}S]DBT$ solution, the $[^{35}S]DBT$ solution was replaced by decalin solvent (operation procedure 2). The change in radioactivities of formed $[^{35}S]H_2S$ with the reaction time is also shown in Fig. 3.42. It is observed that a portion of ^{35}S, which is represented by the shaded area A in Fig. 3.42, remained on the catalyst when only decalin solvent was substituted for the reactant solution of $[^{35}S]DBT$. Even though the catalyst was reduced in an atmosphere of high pressure

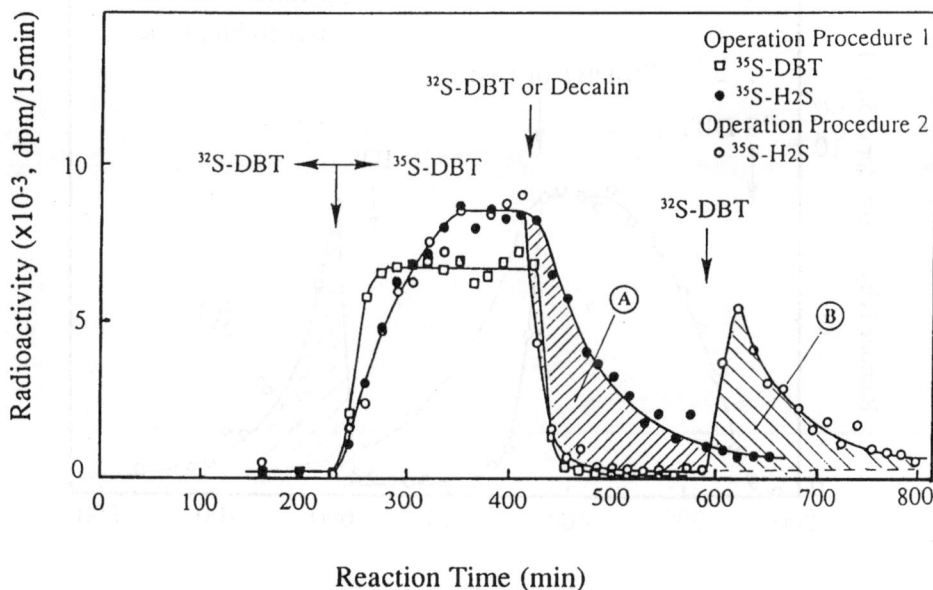

Fig. 3.42   Change in radioactivities of formed $[^{35}S]H_2S$ with reaction time.[267] Mo/Al_2O_3, 360°C, 50 kg/cm², DBT 1.0 wt%.
(From T. Kabe, W. Qian, A. Ishihara *et al.*, *J. Phys. Chem.*, **98**, 913 (1994))

Table 3.8   Radioactivities of [^{35}S]H$_2$S during operating steps (d) in various hydrotreating reaction carried out according to operation procedure 2 [267]

Reaction	HDS					HDO		HDN		
Reactant	DBT	BT	T	T	TP	DBF	CBL		QNL	
Total radioactivities of formed [^{35}S]H$_2$S (10^3 dpm)	1.92	2.11	2.07	1.95	1.92	0.500	1.50 [a]	0	0	1.09[a]
Reference area	(B)	(C)	(D)	(E)		(F)	(G)			(H)
	Fig.3.42	Fig.3.43	Fig.3.44	Fig.3.44	——	Fig.3.45	Fig.3.45	Fig.3.46	Fig.3.46	Fig.3.46

[a] Radioactivities of formed [^{35}S]H$_2$S during another added operatiion step (e) (See text). Area A in Fig. 3.50 is 1.92 × 10^3 dpm.
(From T. Kabe et al., J. Phys. Chem, **98**, 914 (1997))

of hydrogen for ca. 3h, [^{35}S]H$_2$S was hardly produced. This indicates that the sulfur accommodated on the catalyst was not eluted without the supply of sulfur by HDS of DBT. That is,^{35}S remaining on the catalyst did not adsorb as [^{35}S]H$_2$S on the catalyst but was exchanged with the labile sulfur on the catalyst. When the reactant solution was replaced with [^{32}S]DBT at ca. 590 min, this portion of ^{35}S was released again as [^{35}S]H$_2$S, as shown in Fig. 3.42. Almost all ^{35}S on the catalyst could be replaced by ^{32}S derived from HDS of [^{32}S]DBT. This can be verified by the fact that the shaded area B is approximately equal to the shaded area A as shown in Table 3.8. At the same time, a ^{35}S-labeled catalyst under practical HDS conditions can be obtained according to this method.

To survey the effect of sulfur compounds on the substitution rate of sulfur on the catalyst,

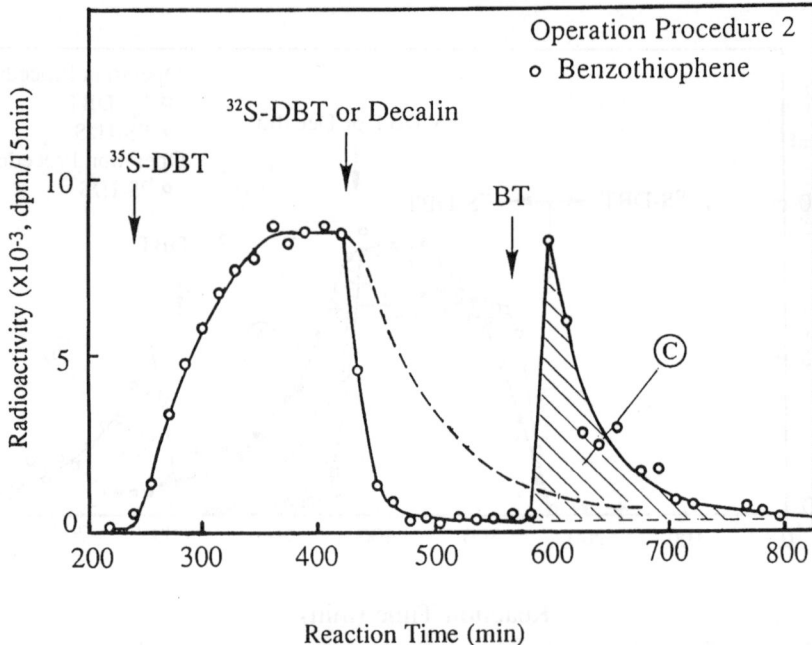

Reaction Time (min)

Fig. 3.43   Change in radioactivities of formed [^{35}S]H$_2$S with reaction time.[267] Mo/Al$_2$O$_3$, 360°C, 50 kg/cm^2, BT 0.73 wt%.
(From T. Kabe, W. Qian, A. Ishihara et al., J. Phys. Chem., **98**, 914 (1994))

Table 3.9 Conversions and pseudo-first-order rate constants of the HDS reaction for various heteroatom compounds [267]

Compound	Concentration in decalin		Conversion %	Rate Constant $\times 10^6$, L/s $\cdot$ g-catalyst
	wt %	$\times 10^2$, mol/L		
Dibenzothiophene (DBT)	1.0	4.77	59.3	8.0
Benzothiophene (BP)	0.73	4.77	100	—
Thiophene (T)	0.46	4.77	100	—
	1.00	10.4	100	—
Thiophenol (TP)	1.20	9.54	100	—
Dibenzofuran (DBF)	0.91	4.77	7.3 [a]	0.67
Carbazole (CRL)	0.91	4.77	0.0	0.0
Quinoline (QNL)	0.70	4.78	90.0 [b]	20.5

[a] Products were cyclohexylbenzene and biphenyl. [b] Products were mainly 1,2,3,4-tetrahydroquinoline and 3,4,5,6-tetrahydroquinoline.
(From T. Kabe, W. Qian and A. Ishihara, *J. Phys. Chem.*, **98**, 914 (1994))

a decalin solution of 0.73 wt% benzothiophene (BT) containing the same molar concentration of sulfur as that of DBT was introduced into the ^{35}S-labeled catalyst. The change in radioactivity of formed [^{35}S]H$_2$S with reaction time is shown in Fig. 3.43. It can be observed that the formation curve of [^{35}S]H$_2$S is the same as in the case of [^{32}S]DBT, until operation step (c) during the reaction according to operation procedure 2. However, when the reactant solution was changed from decalin to the decalin solution of BT in operation step (d), [^{35}S]H$_2$S formation rate, i.e., the rate for which ^{35}S on the catalyst was replaced by ^{32}S in BT, was more rapid than that for the case of [^{32}S]DBT. This may be attributed to an increase in ^{32}S incorporated from BT into the catalyst since the HDS rate of BT was more rapid (conversion = 100%) than that of DBT (conversion = 59.3%; see Table 3.9). Similar results were obtained in the case of thiophene (see Fig. 3.44 and Table 3.9). These results also indicate that the replacement rate of sulfur is independent of the kind of sulfur compounds. Using a solution containing higher thiophene concentration caused an increase in the removal rate of sulfur, as shown in Fig. 3.44. Since the conversion of thiophene was still 100% (see Table 3.9), the amount of sulfur incorporated into the catalyst is about three times as much as that in the case of 1 wt% DBT. In addition, when a decalin solution of 1 wt% thiophenol, containing twice as much sulfur as that of 1 wt% DBT was used, a formation curve of [^{35}S]H$_2$S similar to that for 1 wt% thiophene was obtained, because the conversion of thiophenol was also 100% and the rate of ^{32}S incorporation was similar to that of 1 wt% thiophene (see Table 3.9). This further indicates that the removal rate of sulfur is independent of the kind of sulfur compounds and depends only upon the rate of sulfur incorporated into the catalyst. Values of areas representing the amount of sulfur released during the various operation steps shown in Figs. 3.42–3.44 are listed in Table 3.8. Since all areas were approximately the same, all ^{35}S on the catalyst that was incorporated into the catalyst in operation step (b) was almost replaced by ^{32}S during the operation step (d), even though the replacement rates of sulfur were different from each other.

The behavior of sulfur in other hydrotreating reactions differs from that in HDS. Fig. 3.45 shows the changes in radioactivities of formed [^{35}S]H$_2$S when dibenzofuran (DBF) was introduced into the ^{35}S-labeled catalyst.[267] As reported by Lavopa et al.,[268] the major products were still BP and CHB. It can be observed that the formation curve of [^{35}S]H$_2$S is very different from those curves in the case of sulfur compounds. During operation step (d), only

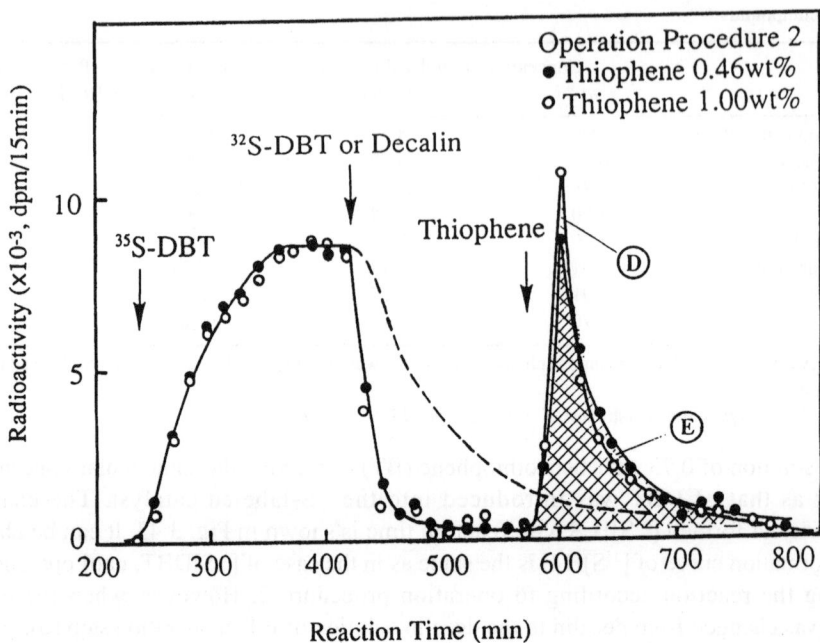

Fig. 3.44 Change in radioactivities of formed [^{35}S]H$_2$S with reaction time.[267] Mo/Al$_2$O$_3$, 360°C, 50 kg/cm^2, T: 0.46 wt% and 1.00 wt%.
(From T. Kabe, W. Qian and A. Ishihara, *J. Phys. Chem.*, **98**, 914 (1994))

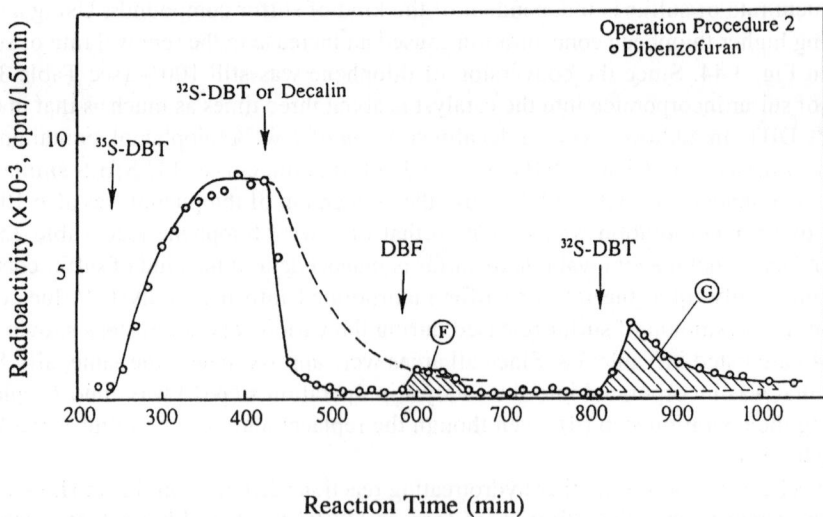

Fig. 3.45 Change in radioactivities of formed [^{35}S]H$_2$S with reaction time.[267] Mo/Al$_2$O$_3$, 360°C, 50 kg/cm^2, DBF: 0.91 wt%.
(From T. Kabe, W. Qian and A. Ishihara, *J. Phys. Chem.*, **98**, 914 (1994))

a small portion of ^{35}S on the catalyst was eluted and released as [^{35}S]H$_2$S during the HDO reaction of DBF. This may be due to the low conversion of DBF (7.3%). To verify whether ^{35}S, which was not eluted during the HDO reaction of DBF, was still present on the catalyst, another operation step (e) was added. After HDO reaction of DBF was conducted for *ca.* 4h, the solution of [^{32}S]DBT was substituted for that of DBF, and then reacted for *ca.* 4 h. As shown in Fig. 3.45, [^{35}S]H$_2$S was produced again, and the sum of area F and area G, which represented respectively radioactivities of [^{35}S]H$_2$S released during operation steps (d) and (e), is approximately equal to the area A (see Table 3.8). These findings indicate that the portion of ^{35}S remaining on the catalyst could be eluted and released as [^{35}S]H$_2$S if ^{32}S was supplied.

When a decalin solution of a non-basic nitrogen-containing compound (carbazole) was used in operation step (d) of operation procedure 2, [^{35}S]H$_2$S was scarcely detected during the operation step (d) (see Fig. 3.46).[267] This is because the conversion of carbazole is nearly zero. As reported by Bhinde,[269] when a decalin solution of quinoline was used in operation steps (d) of operation procedure 2, quinoline was hydrogenated rapidly to produce 1,2,3,4-tetrahydroquinoline and 5,6,7,8-tetrahydroquinoline, but HDN of quinoline to give hydrocarbons hardly occurred (see Table 3.9). At the same time, radioactivities of [^{35}S]H$_2$S could hardly be detected (see Fig. 3.46). After HDN reaction of quinoline was conducted for *ca.* 4 h, operation step (e) was also added, i.e., the solution of [^{32}S]DBT was substituted for that of quinoline and reacted for *ca.* 4 h. Although [^{35}S]H$_2$S was produced again as shown in Fig. 3.46, the total radioactivities of released [^{35}S]H$_2$S (area H in Fig. 3.46) were smaller than the total radioactivities of ^{35}S remaining on the catalyst (area A, see Table 3.8). This may be due to the fact that the catalyst was poisoned by quinoline since the reactivity of DBT HDS decreased from 59.3% to 33.3%. This makes it possible for a portion of ^{35}S to still be held on the catalyst because the amount of labile sulfur on the catalyst decreased with decrease in the

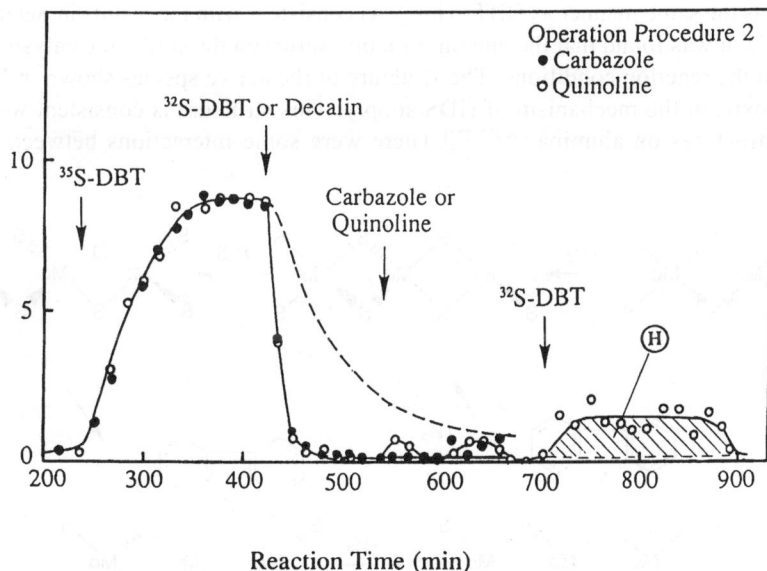

Fig. 3.46 Change in radioactivities of formed [^{35}S]H$_2$S with reaction time.[267] Mo/Al$_2$O$_3$, 360°C, 50 kg/cm^2, CBL: 0.91 wt%, QNL: 0.70 wt%.
(From T. Kabe, W. Qian and A. Ishihara, *J. Phys. Chem.*, **98**, 914 (1994))

HDS rate.[230)]

As shown in Fig. 3.42, H$_2$S was not formed in the absence of DBT while the incorporation of sulfur in DBT onto catalyst generated H$_2$S. The release of H$_2$S formed anion vacancies on the catalyst. It seems that the catalyst tends to keep a constant amount of anion vacancies under each reaction condition. Although we cannot estimate the amount of anion vacancies, this may be related to the amount of labile sulfur. To understand why there is a time delay for [^{35}S]H$_2$S to achieve a steady state of radioactivities, why the sulfur in DBT is not directly released as H$_2$S and why H$_2$S is formed from the sulfur on the catalyst, the following tentative reaction scheme (Fig. 3.47) was considered to account for these phenomena. According to this scheme, the reaction would proceed with the following steps: (a) The sulfur compound is adsorbed on an anion vacancy on the catalyst. (b) After the hydrogenolysis of C-S bonds, the hydrocarbon species is released into the gaseous phase, whereas the sulfur atom remains on the catalyst. Most hydrogen probably originates from the SH groups. (c) The sulfur remaining on the catalyst is hydrogenated and forms a new SH group. (d) At the same time, the release of hydrogen sulfide generates a new anion vacancy. Therefore, a shift of active sites on the catalyst surface would occur.

For this mechanism of HDS, only after the sulfur in DBT was incorporated into the catalyst was the sulfur on the catalyst surface released as H$_2$S. When an anion vacancy was occupied by sulfur removed from DBT, a new anion vacancy appeared on the catalyst surface. At this time, the probability of sulfur being released as H$_2$S for all labile sulfur may be the same. Therefore, after [^{32}S]DBT was substituted for [^{35}S]DBT in operation procedure 1 (see Fig. 3.42), the decreasing curve of formed [^{35}S]H$_2$S can be revealed as an exponential function of time. In contrast to this, after the solution of [^{35}S]DBT was substituted for that of [^{32}S]DBT, the increasing curve of formed [^{35}S]H$_2$S can be revealed as a logarithmic function of time. According to this mechanism, the product of HDS, biphenyl, was not delayed but eluted immediately as the same manner as DBT. This was consistent with the results in Section 3.2.1 (see Fig. 3.29). It was found that the amount of labile sulfur on the sulfided catalysts changes depending on the reaction conditions. The structure of the active species shown in Fig. 3.36 was used to explain the mechanism of HDS simply. This structure is consistent with MoS$_2$ single-slab structures on alumina.[256,270,271)] There were some interactions between Mo and

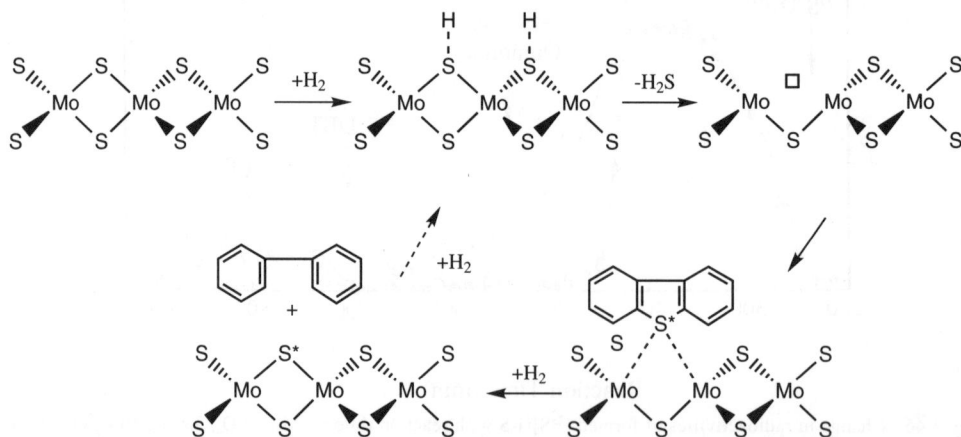

Fig. 3.47 Mechanism of DBT HDS on sulfided Mo/Al$_2$O$_3$. S*: ^{35}S, □: Anion vacancy.[246)]

alumina. This would cause differences among the environments of each sulfur as shown in Fig. 3.36. Thus, the mobile capacity of each sulfur would be different from each other as discussed in Section 3.2.1.

When the HDO reaction of DBF was carried out on ^{35}S-labeled catalyst , only a small portion of ^{35}S was replaced with oxygen atoms and released as $[^{35}S]H_2S$ (see Fig. 3.45) because of the low conversion of HDO of DBF. In addition, for the case of HDN of quinoline, although the hydrogenation rate of quinoline was very rapid, the HDN reaction of quinoline hardly occurred (see Table 3.9). Thus,^{35}S was scarcely replaced by nitrogen compounds (see Fig. 3.46). On the contrary, for the HDS reaction of sulfur compounds such as thiophene which can be desulfurized more easily than DBT,^{35}S remaining on the catalyst was replaced at a more rapid rate than that in the case of DBT. These results indicated that ^{35}S remaining on the catalyst could not be removed and released as $[^{35}S]H_2S$ until the HDN, HDO, or HDS reaction has proceeded. The hydrogenation reaction could not cause the formation of H_2S. And the rate of $[^{35}S]H_2S$ formation will increase with increase in the rate of heteroatoms incorporated into the catalyst. The sulfur exchange reaction between the sulfur on the catalyst and the sulfur in the sulfur compounds may not be the rate-determining step for HDS reaction but a fast reaction. This is consistent with the result that the formation rate of H_2S from the catalyst depended only upon the rate of sulfur incorporation into the catalyst.

3.3 Correlation between Structure and Catalytic Activity

The possibility of establishing a meaningful correlation between the catalytic activities and the physical structural or chemical properties of the catalyst is one of the most important consequences of the significant progress achieved in the characterization of the active sulfided state of hydrotreating catalysts. Specifically, the activity correlation have provided important information on topics such as the origin of the activity, the promoting effect, the nature of active precursors and the influence of preparation parameters, among others.

3.3.1 Unpromoted Mo and W Catalysts

A. Studies Using Conventional Approaches

The activity behavior of unpromoted Mo and W catalysts has been the subject of numerous studies. Some of these have aimed at establishing a basis for understanding the more complicated and industrially more important promoted catalyst, but the unpromoted catalyst itself is also important in their own right and many informative structure-activity relationships have been obtained.

Unpromoted catalysts can be prepared with or without a support. From a structural point of view, the supported catalysts may be more complex than the unsupported model systems. Nevertheless, due to the high dispersion of the active phases in supported catalysts, such systems provide special opportunities for establishing structure-activity correlations. In both unsupported and supported Mo and W catalysts, the metals in the active phase represent similar MoS_2- or WS_2-like phases. Therefore, it is not surprising that supported and unsupported catalysts often exhibit quite similar types of structure-activity correlation and both systems are discussed in this section.

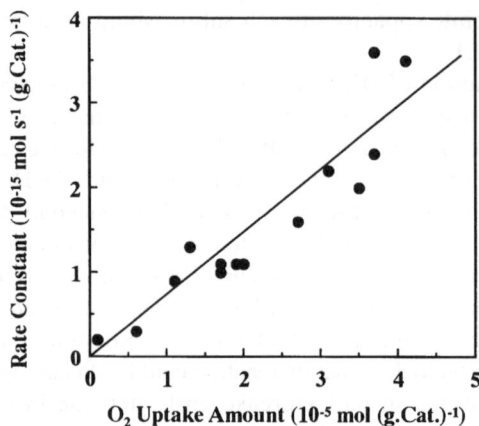

Fig. 3.48 HDS activity as a function of adsorption capacity of oxygen for unsupported MoS$_2$ catalyst.[274)]
(From S. J. Tauster, T. A. Pecoraro and R. R. Chianelli, *J. Catal*, **63**, 518 (1980))

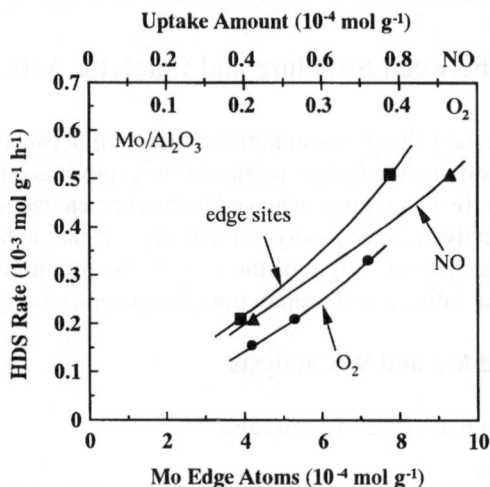

Fig. 3.49 HDS activity as a function of adsorption capacity of NO and number of Mo edge atoms estimated from EXAFS.[374)]
(From H. Topsøse, B. S. Clause and F. E. Massoth, *Hydrotreating Catalysis*, p.157 (1996))

Stevens and Edmonds[272)] concluded in an early study, from examinations of polycrystalline MoS$_2$ samples with different relative concentrations of edge and basal planes, that the active sites for hydrogenation are located at the edges, whereas HDS occurs on the basal planes. Although Suvasanker et al.[273)] at that time also reached similar conclusions for the HDS reaction, most subsequent studies have indicated that the active sites for HDS are located at the edge planes of MoS$_2$. Tauster et al.[274)] used oxygen chemisorption and found that it correlated with the HDS activity (Fig. 3.48). These authors concluded, based on the results of Bahl et al.[275)] who showed that the oxidation of MoS$_2$ starts at the edges of MoS$_2$, that the HDS reaction is favored at the edge planes. Topsøe et al.[59)] also found this conclusion to hold for supported Mo/Al_2O_3 catalysts and correlation were observed between the HDS activity and the number of MoS_2 edge sites measured *in situ* by EXAFS (Fig. 3.49).

. Silbernagel *et al.*[276] found a good correlation between HDS activity and the intensity of an ESR signal assigned to a Mo^{5+} species. They assigned these species to bulk defects, but recent combined ESR and NO chemisorption experiments by Derouane *et al.*[277] have shown the Mo^{5+} species to be located at the MoS_2 edges. Thus, the ESR results also confirm that the HDS sites are located at the MoS_2 edges. This is further substantiated by the work of Roxlo *et al.*,[278] who observed a proportionality between the HDS activity and number of surface defects at the edge planes of MoS_2. Furthermore, microscopy investigations of unsupported MoS_2 prepared by solid state synthesis also showed a high concentration of edge defects in the more active catalysts.[279]

Chemisorption studies may provide more detailed information, and better activity correlations have been obtained by the use of probe molecules which chemisorb on specific sites. In the case of unpromoted catalysts, O_2 chemisorption has been the most widely used probe molecule. Following the early O_2 chemisorption studies of Tauster *et al.*,[274] Chung and Massoth,[280] Bachelier *et al.*[281] and Okamoto *et al.*[52] many such studies of both unsupported and supported sulfide Mo and W catalysts have appeared.[59,85,279,283–291] Other probe molecules, such as CO,[281,286] NO,[59,225,292–296] H_2,[297] pyridine,[292,298] and toluene,[299] have also been used to establish activity correlations for different hydrotreating reactions. Figs. 3.48–3.50 show that using CO and NO chemisorption, the correlations obtained may be quite similar to those obtained using O_2 chemisorption.

The detailed nature of the chemisorption sites differs for different probe molecules so the chemisorption stoichiometry will also vary. The limitations of titration studies are further illustrated in Figs. 3.48 and 3.50, which show that for certain catalysts, the rates may change without a corresponding change in the O_2 or CO chemisorption uptakes.[281,282,294] Since O_2 chemisorption reflects the general state of dispersion of the unpromoted and promoted catalyst rather than specific sites, at present O_2 chemisorption cannot be used for the quantitative

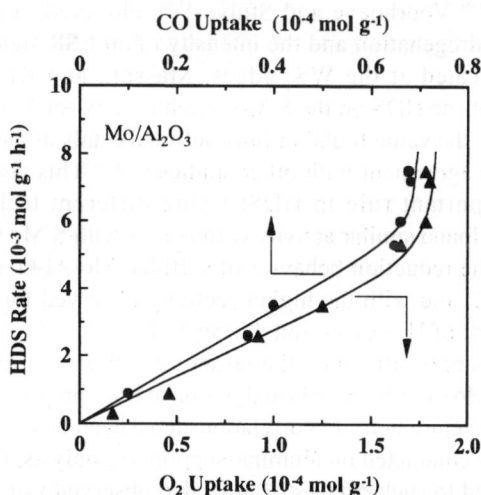

Fig. 3.50   HDS activity as a function of adsorption capacities of oxygen and CO for unpromoted Mo/Al_2O_3 catalyst.[281] (From J. Bachelier, J. C. Duchet and D. Cornet, *Bull. Soc. Chim. Belg.*, **90**, 1307 (1981))

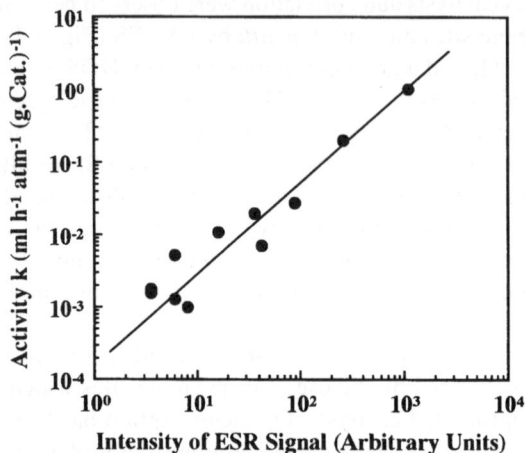

Fig. 3.51 Benzene hydrogenation activity vs. intensity of a tungsten ESR signal.[68]
(From R. J. H. Voorhoeve, *J. Catal.*, **23**, 241 (1971))

determination of active sites.[285] From XPS studies of the adsorption of O_2 and NO on polycrystalline MoS_2, Shuxian et al.[300] conclude that NO adsorbs dissociatively above 130 K and that the resulting oxygen atoms lead to surface oxidation above 200 K. Therefore, one should be careful in interpreting IR spectra of adsorbed NO.

It is fair to conclude from the above-mentioned studies that most hydrotreating reactions involve sites associated with the edges of MoS_2. For nonacid catalyzed reactions, most studies indicate that the edge sites are coordinatively unsaturated Mo edge sites or sulfur vacancies. This is also in accord with the previously discussed chemisorption studies, the activity studies on "cut" single crystals and the bond energy model (BEM). Few studies have suggested other types of sites.[71,109] Voorhoeve and Stuiver[42,68] observed a correlation between the activity for benzene hydrogenation and the intensity of an ESR signal (Fig. 3.51) attributed to sulfur vacancies located at the WS_2 edges. Massoth and Kibby[301] investigated the dependence of the thiophene HDS on the S/Mo stoichiometry for Mo/Al_2O_3. Upon decreasing the stoichiometry below the value found in fully sulfided catalysts (S/M $\approx$ 2), the activity was observed to increase in agreement with other studies.[302,303] This also strongly supports that vacancies play an important role in HDS. Using different techniques, Okamoto and coworkers[298,304,305] later found similar activity correlations with S/Mo stoichiometries. Scheffer et al.[160,306] found that the reduction behavior of sulfided Mo/Al_2O_3 catalysts correlates with HDS activity (Fig. 3.52) and with the highest activity observed for catalysts exhibiting the lowest onset temperature of H_2S evolution during TPR.

Support effects strongly influence the structure-activity correlations for unpromoted catalysts. Unpromoted Mo- and W-based catalysts have been prepared on a variety of supports and many interesting structure-activity correlations have been observed.[307–314] Although most of the research has been conducted on alumina-supported catalysts, the promotional behavior is certainly not restricted to such catalysts but is also observed using other supports such as carbon, silica-alumina, etc. (for reviews see Refs. 171 and 315 and Sections 3.1.3B and 4.1.1).

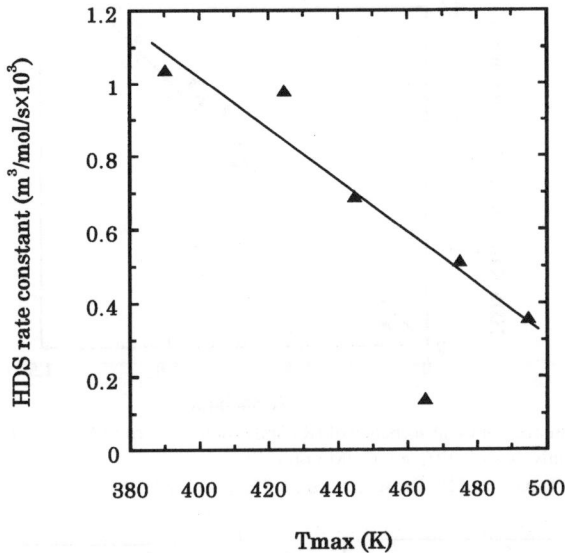

Fig. 3.52   Relation between the temperature of peak maximum for H_2 consumption during TPR-S of Mo/Al$_2$O$_3$ catalyst and thiophene HDS activity.[306]
(From B. Scheffer et al., J. Catal., **121**, 44 (1990))

The metal loading used in the preparation of a catalyst is one of the most important parameters used in optimizing catalyst performance, and numerous studies have dealt with the effect of this parameter on the activity.[316,317] With the recent insight available regarding the nature of the active phases, one can now establish some useful elementary criteria for choosing the appropriate metal loading.[85,103,271] For low Mo-loading catalysts, increasing the Mo loading will increase the total concentration of MoS$_2$ edges sites.[281] However, after reaching a certain Mo loading, which corresponds to about a monolayer coverage, bulk phases will begin to form and a further increase in Mo loading will decrease the MoS$_2$ edge dispersion and hence the activity.[209,318] Thus, there will generally be an optimum metal loading for Mo (W) which will depend on the support surface area. Also for the promoter atoms, there will typically be an optimum concentration-optimum Co/Mo or Ni/Mo ratio. At high promoter concentrations, there will be a tendency to form separate promoter phases, such as Co$_3$O$_4$ (Fig. 3.53). The choice of optimal metal loading is in practice more complex than outlined above, and the effects of many other preparation variables are also important.[118-120]

The chemisorption of probe molecules and infrared spectroscopy are used in combination to estimate the dispersion of the active metal and to correlate the activity and the amount of chemisorbed probe molecules. In the many instances where good activity correlations have been observed, it is likely that the active sites are related to the adsorption sites. Oxygen chemisorption has been one of the most frequently used methods to study promotional behaviors. Many of the early studies dealt mainly with reduced catalysts,[319-324] but more recent studies have also been carried out directly on sulfided catalysts.[285-288,318,325-332] Although good correlations between the HDS activity and the total oxygen adsorption are typically observed for unpromoted sulfided Mo catalysts, many studies have shown that these types of correlations are not generally valid for promoted catalysts. For example, Topsøe and

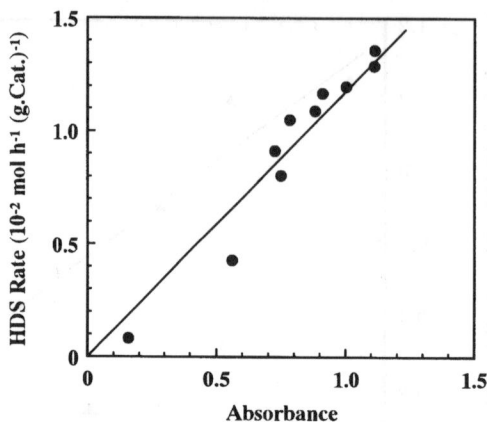

Fig. 3.53   HDS rate of thiophene vs. absorbence of IR band (1800 cm^{-1}) from NO adsorbed on Co in calcinated Co-Mo/Al$_2$O$_3$ catalysts with different Co/Mo ratio.[206)]
(From N.-Y. Topsøe and H. Topsøe, *J. Catal*, **77**, 294 (1982))

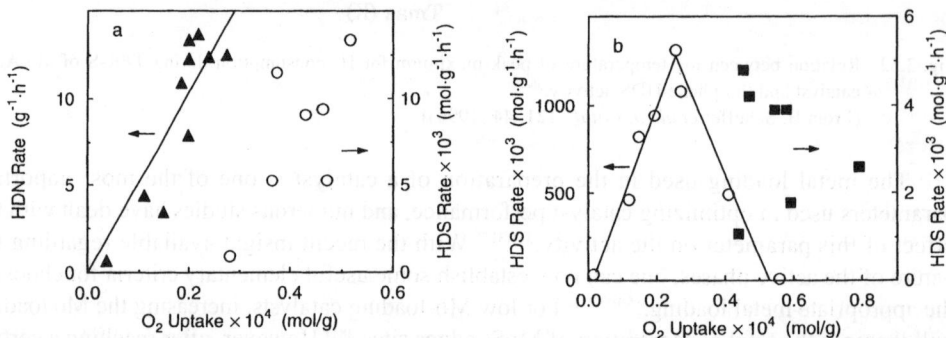

Fig. 3.54   Correlations between HDS and HDN activities and total oxygen uptakes for different supported Co-Mo (a) and Ni-Mo (b) catalysts. ○: thiophene HDS [325)]; ▲: indole hydrogenolysis [333)]; ○: thiophene HDS [332)]; ■: thiophene HDS.[286)]
(From H. Topsøe *et al.*, *Hydrotreating Catalysis*, p.170 (1996))

coworkers[325)] and Reddy *et al.*[287)] observed large differences in activity between different Co-promoted and unpromoted catalysts but only small variations in the oxygen uptake (Fig. 3.54(a)). NO is another probe molecule which has been used to establish correlations with the activity of promoted catalysts.[89,124,125,325,334)] The total uptake of NO does not generally correlate with the activity since sites of very different intrinsic activities are being titrated rather nonselectively.[89)] Other probe molecules are CO, H$_2$, H$_2$S and a few basic molecules.[286,332,334–338)] However, one cannot in general draw firm conclusions regarding the detailed nature and coordination geometry of the active sites based only on such information since the sites for adsorption and catalysis may not necessarily be identical. For example, the number of sites involved in the catalysis will depend on reaction conditions such as temperature, type of reaction, partial pressure of reactants, etc., whereas the number and type of sites involved in the chemisorption depend on the experimental conditions.

Although the sulfidation condition affects the structure of the catalyst, as described in

Section 3.1.3, studies on the relationship between presulfiding and activity of catalysts have not been pursued in detail. Gestel et al.[221] used alkylpolysulfides as a presulfiding agent and reported differences in selectivity and activation energies. Recently, Kasahara et al. investigated the effect of different pretreatments on the activity and structure of catalysts.[215] Leliveld et al. also studied the sulfidation reaction in detail by the use of EXAFS.[339] In all cases, the detailed activity correlations were not given.

B. Studies Using ^{35}S Radioisotope Tracer Methods

To correlate the sulfided state of the catalyst with the behavior of a catalyst during the HDS reaction, several groups have developed a radioisotope ^{35}S tracer method to trace the behavior of sulfur in HDS.[161,236,237,241] The catalyst was labeled by sulfiding the calcined catalyst with the radioisotope ^{35}S or ^{35}S-labeled H_2S, thiophene and CS_2, and was then used in the HDS and HYD of model compounds. By tracing the release of $[^{35}S]H_2S$ from the catalyst, it was pointed out in some of these studies that ca. 20–30% of sulfur on the catalysts is exchangeable and that more than one type of sulfur bonding exists since the rates and the extents of sulfur exchange change with time and temperature. However, these methods could not give information about the behavior of sulfur on working catalysts.

Recently, the ^{35}S RPTM method has been noted.[74,219] The effect of molybdenum content on catalyst properties was examined by studying a series of sulfided molybdena-alumina containing 6–20 wt% molybdena.[219] Fig. 3.55 shows results using several catalysts with different molybdenum content at 360°C. From Fig. 3.55, the values of labile sulfur (S_0) corresponding to Area C and the release rate constant of $[^{35}S]H_2S$ were determined and are given in Table 3.10. It is noteworthy that k_{ER} scarcely varied between catalysts at the same temperature, as shown in Fig. 3.56. Within experimental error, at a given temperature, the rate constant of sulfur exchange did not vary with the content of molybdenum for all the catalysts. This indicates a similarity in the nature of the active sites within this series of catalysts. The activation energy of the sulfur exchange reaction was obtained from Fig. 3.56 and was

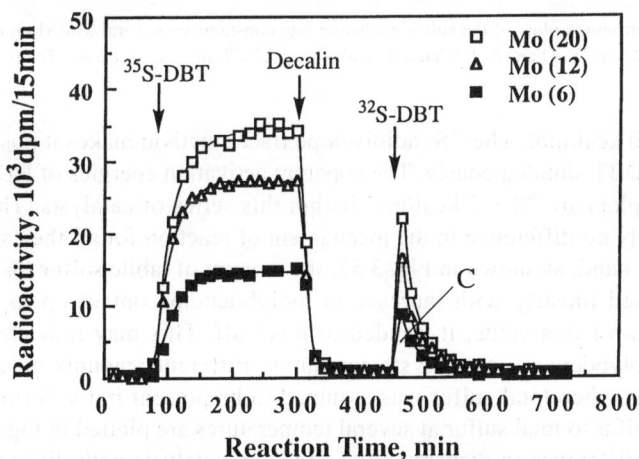

Fig. 3.55 Effect of molybdenum content on the release of $[^{35}S]H_2S$.[219] 50 kg/cm², 360°C.
(From W. Qian, A. Ishihara, T. Kabe et al., J. Chem. Soc., Faraday Trans., **93**, 1823 (1997))

Table 3.10 Results in the HDS reactions and the exchange reactions on the sulfided Mo/Al$_2$O$_3$ catalyst [219]

Catalyst	Conversion (%)	S_0 [a] (atom/nm^2)	k_{RE} [b] (10^{-4}/s)	S_0/S_t [c] (%)	r_{HDS} [d] (10^{-4}atom/s/nm^2)
γ-Al$_2$O$_3$	2.5	0.034	——	——	0.28
6% MoO$_3$	36.9	0.435	7.83	21.7	3.83
12%MoO$_3$	66.9	0.898	7.70	21.4	7.01
16%MoO$_3$	75.4	1.200	7.82	21.4	8.64
20%MoO$_3$	83.7	1.344	7.32	17.4	9.74

[a] Amount of labile sulfur.
[b] Release rate constant of [^{35}S]H$_2$S.
[c] Ratio of labile sulfur to total sulfur.
[d] Hydrodesulfurization reaction rate.
(From W. Qian, A. Ishihara, T. Kabe *et al.*, *J. Chem. Soc., Faraday Trans.*, **93**, 1823 (1997))

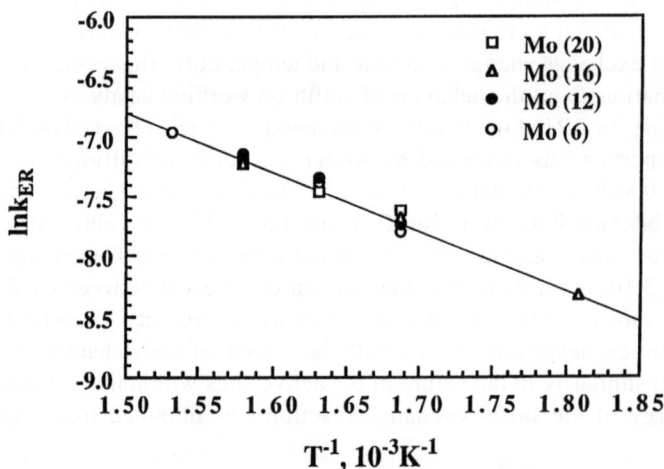

Fig. 3.56 Arrhenius plots of the sulfur exchange rate constants on several Mo/Al$_2$O$_3$ catalysts.[219]
(From W. Qian, A. Ishihara, T. Kabe *et al.*, *J. Chem. Soc., Faraday Trans.*, **93**, 1824 (1997))

approximately 10 kcal/mol. The ^{35}S radioisotope tracer method makes it possible to evaluate the HDS rate of DBT simultaneously. The apparent activation energies of HDS were obtained from Arrhenius plots are 20 ± 2 kcal/mol within this series of catalysts. This indicates that there is essentially no difference in the mechanism of reaction for all the catalysts.

On the other hand, as shown in Fig. 3.57, the amount of labile sulfur (S_0) on the series of catalysts increased linearly with increase in molybdenum content up to 2.89 atoms/nm^2 (Mo(16)), but above this value, it trended to level off. This may indicate a difference in dispersion of molybdenum on catalysts containing different amounts of molybdenum. As mentioned above, when total sulfur was assumed to be present in the form of MoS$_2$ phase, ratios of labile sulfur to total sulfur at several temperatures are plotted in Fig. 3.58. This more clearly shows a difference in dispersions of MoS$_2$ on catalysts with different molybdenum content. A monolayer dispersion was maintained up to 2.89 atom/nm^2 (Mo(16)) but occurrence of polylayers of MoS$_2$ was observed above this value with formation of crystallites of MoS$_2$

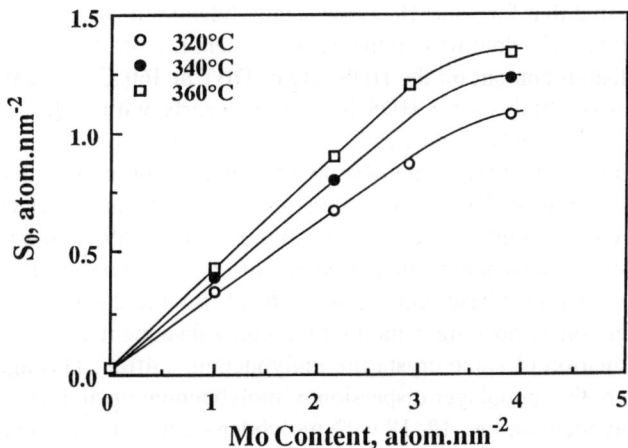

Fig. 3.57 Effect of molybdenum content on the amount of labile sulfur.[219]
(From W. Qian, A. Ishihara, T. Kabe *et al.*, *J. Chem. Soc., Faraday Trans.*, **93**, 1824 (1997))

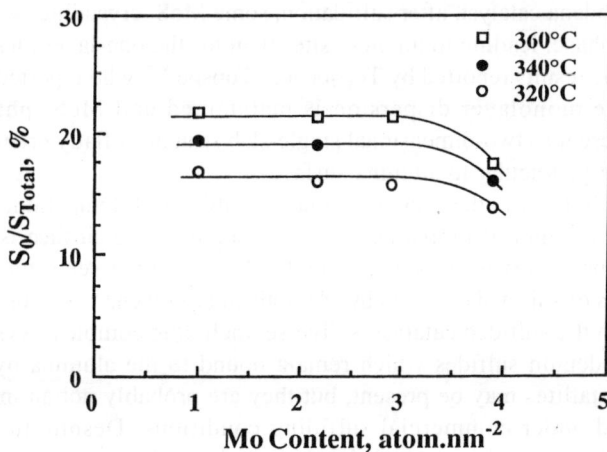

Fig. 3.58 Effect of molybdenum content on the ratio of labile sulfur to total sulfur.[219]
(From W. Qian, A. Ishihara, T. Kabe *et al.*, *J. Chem. Soc., Faraday Trans.*, **93**, 1824 (1997))

on Mo(20). Several studies on coverage of alumina as a function of molybdenum content in sulfided catalysts have been performed. Most of the structural studies presented evidence that the majority of molybdenum atoms are present as MoS_2, with perhaps a minor fraction forming oxysulfide species, the amount of which depends on the sulfiding conditions. From IR studies of the surface hydroxyl groups in calcined and sulfided Mo/Al_2O_3 catalysts, Topsøe concluded that the fraction of uncovered alumina is approximately the same in calcined and in sulfided states of the catalysts.[259] Thus, it was suggested that the MoS_2 crystallites in the sulfided state are present as a monolayer. It is generally found that coverage of alumina by molybdenum species increases linearly with the molybdenum content up to a certain limit, above which it levels off. This indicates the formation of bulk-like MoS_2 structures, the size of which is reported to increase somewhat with increasing molybdenum content.[105,205,340,341]

It was also suggested that for < *ca.* 10–12 wt% molybdenum in a typical catalyst, monolayer single slab structures will dominate in the whole content region.[157,281] Fig. 3.59(a) shows the effect of molybdenum content on the HDS rate of DBT at 360°C. The catalytic activities of the present series of catalysts for HDS increased linearly with added molybdenum up to about 2.89 atom/nm^2 (Mo(16)) and was kept constant until molybdenum was added beyond this value. This implies that the effect of molybdenum content on the activity comprises two distinct ranges due to the difference in dispersion of molybdenum on alumina. These results suggest that molybdenum interacted with the alumina up to 2.89 atom/nm^2 (Mo(16)) during impregnation and that subsequent calcination or sulfidation did not result in the formation of a crystallite molybdenum phase. On the other hand, for the 20 wt% catalyst (Mo(20)), a monolayer dispersion is no longer maintained and subsequent calcination or sulfidation results in the formation of some crystallite molybdenum sulfide. This agrees with reported results,[157,281] where the monolayer dispersion of molybdenum on alumina was maintained up to 8–12 wt% molybdenum, i.e., 12–18 wt% molybdena. Further, as shown in Figs. 3.55 and 3.56, the activation energies of HDS reaction and the exchange rate of sulfur for this series of catalysts are very similar to each other. Therefore, it can be assumed that there is no difference in the mechanism of HDS and the nature of the active sites within these catalysts. For 20 wt% molybdena catalyst, after sulfidation, some MoS$_2$ crystallites were present in other than a one-layer phase, leading to no more sites than for the one-layer phase. This is in good agreement with the results reported by Topsøe and Topsøe,[157] who reported that up to 12 wt% molybdenum, the monolayer dispersion is maintained and MoS$_2$ phase appears to be predominantly present as two-dimensional single-slab structures flatly oriented on the alumina support (c-axis perpendicular to alumina surface).

As shown in Fig. 3.57, the increase in labile sulfur with temperature for each catalyst suggests that the strength of molybdenum sulfur bonds in the surface is not uniform. The exchange reactions observed on sulfided molybdena-alumina seem more consistent with recent descriptions of sulfided catalysts by Massoth and Zeuthen,[342] i.e., the local environment of molybdenum in the sulfided catalyst is diverse, including complex oxysulfide species and disordered molybdenum sulfides which remain bound to the alumina by Mo-O-Al bonds. "True" MoS$_2$ crystallites may be present, but they are probably not an important feature of catalysts sulfided under commercial sulfiding conditions. Despite this complexity, the exchangeable sulfur was still correlated closely with dispersion of the active phase. As shown in Fig. 3.57, in sulfided Mo/Al$_2$O$_3$ catalysts, the number of exchangeable sulfur atoms was constant up to about 2.89 molybdenum atom/nm^2 (Mo(16)).

The amount of labile sulfur represents the number of active sites and the release rate constant represents the mobility of active sites, i.e., the turnover frequency.[76] It was observed that at the same temperature, the sulfur exchange rate constant did not vary with the molybdenum content but that the amount of labile sulfur varied directly with the rate of HDS from Fig. 3.59(a) and (b). Therefore, it can be concluded that at a given temperature, the increase in HDS activity with the content of molybdenum for this series of catalysts is attributed to the increase in total numbers of active sites. At the same time, the intrinsic structure and morphology of the catalysts of molybdenum-alumina catalyst were maintained with molybdenum content up to 2.89 atom/nm^2. On the other hand, for higher molybdenum contents (3.89 atom/nm^2), because a polylayer of molybdenum sulfide is now formed, the number of active sites no longer increased. This can be verified by the fact that the amount of labile sulfur also no longer increased. It is noteworthy that although polylayers of MoS$_2$

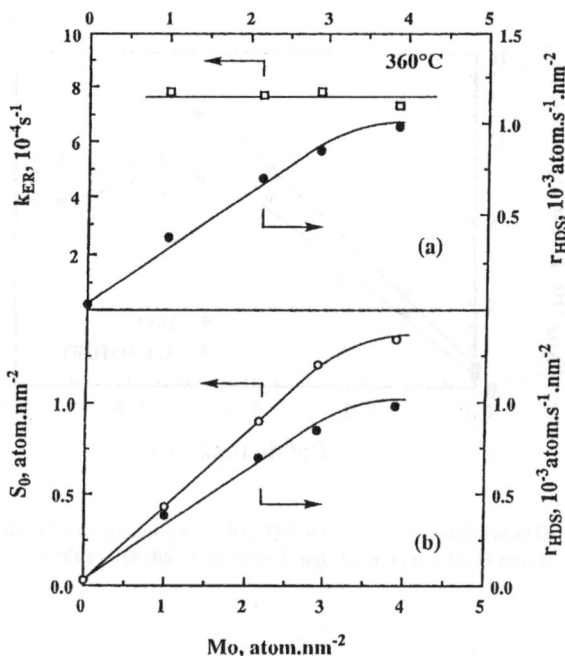

Fig. 3.59 Relationship within the sulfur exchange rate constant, the amount of labile sulfur and HDS rate.[219]
(From W. Qian, A. Ishihara, T. Kabe *et al.*, *J. Chem. Soc., Faraday Trans.*, **93**, 1824 (1997))

occurred on the alumina surface at higher molybdenum contents, the MoS_2 in outer layer may have the same characteristics of MoS_2 in the monolayer phase.

3.3.2 Promoted Mo and W Catalysts

One conventional method for enhancing the catalytic activity is to increase the content of active metal species, including the molybdenum and promoter cobalt or nickel. Therefore, the dispersion state of molybdenum on the surface of alumina support for the base catalyst-Mo/Al_2O_3 and the promotion effect of cobalt on molybdena-alumina catalyst must be clarified. As a result of continuous efforts, both structural and catalytic aspects are now relatively well understood. Generally, with increasing promoter concentration, the activity may increase significantly and typically passes through a maximum at a Co/Mo or Ni/Mo molar ratio of 0.3–1.0 (see Refs. 58, 107, 343, 344 and Figs. 3.60, and 3.61). As shown in Fig. 3.60, the activities of DBT and 4,6-DMDBT for HDS increased with increasing amount of cobalt added at lower Co/Mo molar ratios (below 0.5). At higher Co/Mo molar ratios (above *ca.* 0.5), however, the promoting effect of cobalt for DBT increased only slightly. Compared with the case of DBT, the promoting effect of cobalt for HDS of 4,6-DMDBT decreased with increasing Co/Mo molar ratio, the maximum effect was attained when the ratio was 0.5. At the same molar ratio, the rate constants of the formation of biphenyls were approximately 20 times that of unpromoted Mo/Al_2O_3, while the rate constants of the formation of CHBs were approximately 4 times. At the same time, cobalt enhanced the activity of HDS more than that of hydrogenation. Further, the mode of formation of decalin and cyclohexylbenzenes was

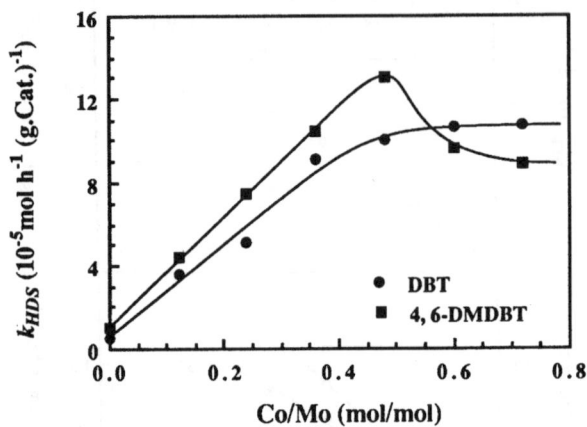

Fig. 3.60 HDS activities of DBT and 4,6-DMDBT as a function of Co/Mo ratio.[344]
(From Q. Zhang *et al.*, *J. Jpn. Petrol. Inst.*, **40**, 410 (1997))

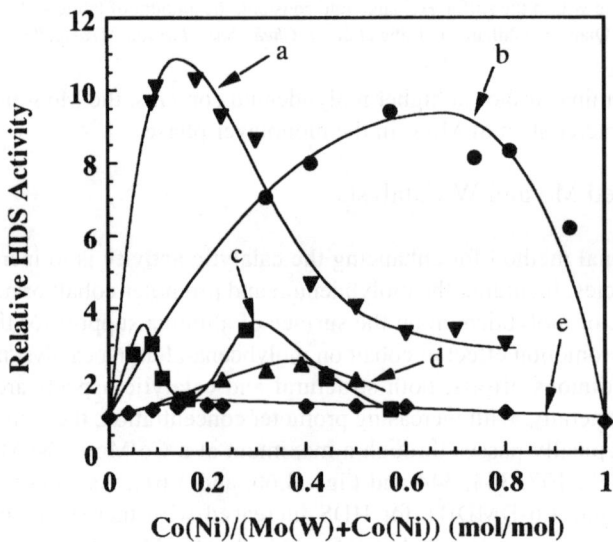

Fig. 3.61 Illustration of a variety of HDS promotional behaviors encountered in Co(Ni)-Mo(W) catalysts.[374]
(From H. Topsøe, B. S. Clausen and F. E. Massoth, *Hydrotreating Catalysis*, p.164 (1996))

nearly the same and it was suggested that hydrogenation of tetralin and DBTs occurred on the same active sites.

To understand the promoting effect of cobalt on HDS, several structural models have been proposed[43,46,94–97,345,346] and are reviewed.[347–351] In the classification given by Delmon,[352] 12 models of the active component structure are proposed (the "geometric"[90,353] models as well as the model of Ledoux *et al.*[354] are not taken into consideration here). Their appearance in the literature reflects objectively the approach of our knowledge to an understanding of the structure of the active component of HDS catalysts. At present the model of the "CoMoS phase" developed by Topsøe *et al.* has gained the greatest recognition. This model was first based on the results of catalyst studies by MES, and later confirmed by EXAFS. These models were explained qualitatively rather than quantitatively; therefore, it is not clear at the present time whether the intercalation model or the contact synergy model is essential to the promoting effect. Recently, many studies have shown that the activity of promoted catalysts is related to the presence of highly active CoMoS type structures. Topsøe *et al.* proposed the CoMoS model for Co-Mo/Al$_2$O$_3$ catalysts.[44,54,56,82,355,356] The reducible Co-oxide species in the

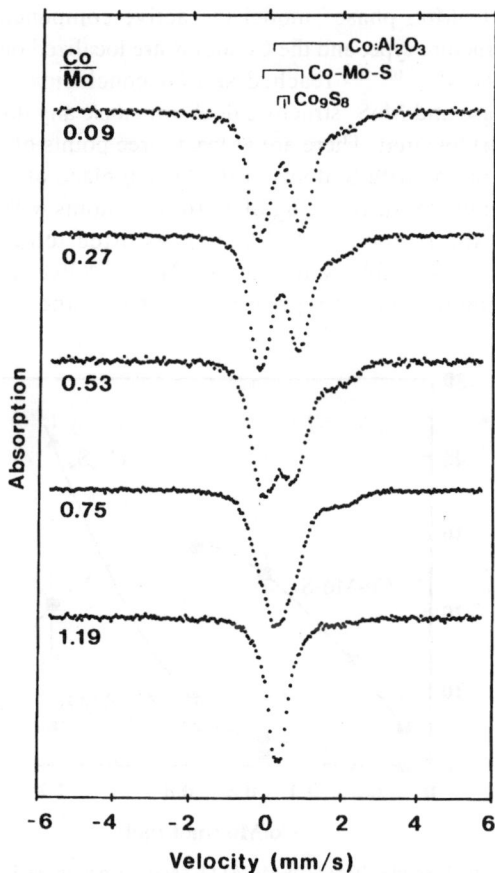

Fig. 3.62 Mössbauer emission spectroscopy of sulfided Co-Mo/Al$_2$O$_3$ catalysts with different Co/Mo atomic ratios.[54]

Co-Mo/Al$_2$O$_3$ catalyst were well distributed and associated as Co-Mo-O phase, which was considered to be the precursor of the highly active CoMoS phase. The existence of the CoMoS phase in the sulfided Co-Mo/Al$_2$O$_3$ catalyst was also verified by the XPS study of Lee et al.[357] and the EXAFS study of Bouwens et al.[36]

The MES results (Fig. 3.62) of Wivel et al.[54] showed that upon increasing the Co concentration in Co-Mo/Al$_2$O$_3$ catalysts with constant Mo loading, the amount of CoMoS increases initially. However, at high Co/Mo ratios, Co$_9$S$_8$ is also formed and for very high ratios this Co species comes to dominate at the expense of the amount of CoMoS (Fig. 3.63). The presence of Co$_9$S$_8$ in catalysts with high Co concentrations is a general feature[60,62,270,271,358–360] and may be related to the amount of Co$_3$O$_4$ present before sulfiding. The feature of hydrotreating catalysts which has attracted most attention is probably the remarkable increase in catalytic activity with the addition of Co or Ni promoter atoms to Mo- or W-based catalysts. Upon increasing the concentration of the promoter atoms, the activity may increase significantly and typically pass through a maximum. However, the detailed promotional behaviors depend on the hydrotreating reaction and may vary significantly from study to study, as illustrated in Fig. 3.60.

According to the "CoMoS phase" model the active component of HDS catalysts is crystallized in a MoS$_2$ structure type, and the Co atoms are localized on the MoS$_2$ edges. Later the authors of other works[36,37,104–107] reached similar conclusions. However, even if the similarity of CoMoS phase and MoS$_2$ structure does not cause any doubt, there is no unified opinion on the Co (or Ni) location. There are at least three points of view: (i) square-planar surrounding of Ni (Co) atoms with S atoms at the (1010̄) plane of a MoS$_2$ (or WS$_2$) single slab[361,362]; (ii) square-pyramidal surrounding of Ni (or Co) atoms with five sulfur atom at the edge plane of a MoS$_2$ single slab[36,37,362,363]; (iii) distorted tetrahedral and octahedral surrounding with sulfur atoms, as followed from ^{59}Co NMR.[354] Recently, some new "models" have been used to understand the promoting effect and the structure-activity

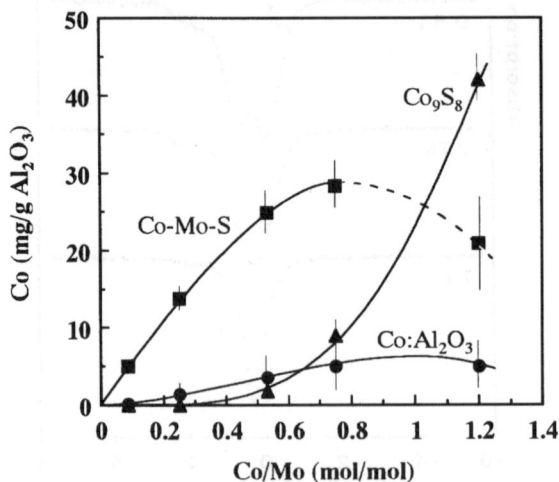

Fig. 3.63 Distribution of cobalt phase in sulfided Co-Mo/Al$_2$O$_3$ catalysts determined from the MES spectra in Fig. 3.62.[54]

(From C. Wivel, R. Candia, B. S. Clausen, S. Mørp and H. Topsøe, J. Catal. **68**, 459 (1981))

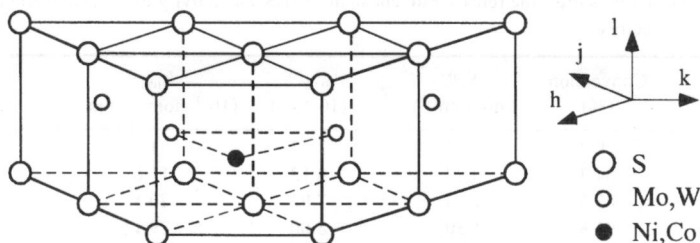

O   S
o   Mo,W
●   Ni,Co

Fig. 3.64 The structure of sulfide bimetallic species.[121]
(From A. N. Startsev, *Catal. Rev.-Sci. Eng.*, **37**, 383 (1995))

relationship.[57,108,256] Topsøe *et al.* proposed a bond energy model (BEM) on the basis of the *ab initio* calculations.[58,67] It has been shown that the BEM model may also form a basis for a semi-quantitative model explaining the change in promotion observed as a function of the Co/Mo (or Ni/Mo) ratio. Startsev recently proposed a improved "CoMoS phase" model, which is called the "sulfide bimetallic species"-SBMS model.[57,108,121] The structure of the sulfide bimetallic species, is shown in Fig. 3.64.[121] According to this model the active component of HDS catalysts represents a MoS₂ (WS₂) single slab, in the side plane (10$\bar{1}$0) of which Ni (Co) atoms are localized. In SBMS Mo (W) atoms are at the center of a triangular prism surrounded by six S atoms at distances of Mo-S = 2.41 Å and W-S =2.46 Å. The prisms are joined by the edge and form a "sandwich" with distances of Mo-Mo =3.17 Å and W-W =3.24 Å. All these distances are typical of bulk MoS₂ and WS₂.

As reported above, it was reported that a good correlation between the amount of labile sulfur and HDS activity was obtained using the ^{35}S radioisotope pulse tracer method (^{35}S RPTM) for a series of sulfided molybdena-alumina containing 6% to 20 wt% of molybdena.[219] ^{35}S RPTM was also used to investigate the promotion of Co to Mo/Al₂O₃.[74] Similar to the case

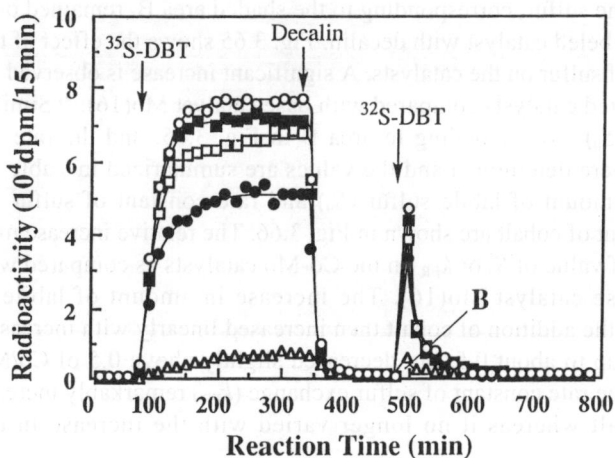

Fig. 3.65 Effect of ratio of Co/Mo on sulfur exchange.[74] 50 kg/cm² 280°C.
△: Mo(16), ●: Co(1)-Mo(16), □: Co(2)-Mo(16), ■: Co(3)-Mo(16), ○: Co(5)-Mo(16); the values in parentheses indicate the content of metal oxide.
(From W. Qian, A. Ishihara, T. Kabe *et al.*, *J. Chem. Soc., Faraday Trans.*, **93**, 4397 (1997))

Table 3.11 Amount of labile sulfur, the release rate constant of H_2S and activity of HYD and HDS on sulfided Co-Mo/Al_2O_3 catalyst [74]

Catalyst	Conversion (%)	S_0 [a] (atom nm^{-2})	k_{RE} [b] (10^{-4} s^{-1})	k_{HDS} [c] (10^{-4} atom s^{-1} nm^{-2})	k_{HYD} [d] (10^{-4} atom s^{-1} nm^{-2})
Mo(16)	9.4	0.27	2.43	1.1	0.17
Co(1)-Mo(16)	64.3	0.87	8.12	11.6	0.66
Co(2)-Mo(16)	83.5	1.13	7.92	20.5	1.08
Co(3)-Mo(16)	91.8	1.30	8.00	29.1	0.94
Co(5)-Mo(16)	94.7	1.38	8.25	34.9	1.14

For explanations of [a], and [b] see Table 3.10; [c] Rate constant of pseudo-first-order reaction for HDS; [d] Rate constant of pseudo-first-order reaction for HYD.
(From W. Qian, A. Ishihara, T. Kabe et al., J. Chem. Soc., Faraday Trans., **93**, 4397 (1997))

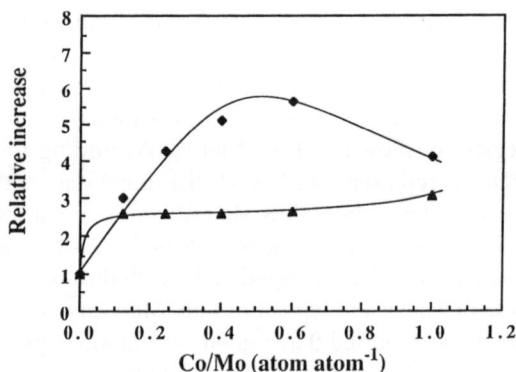

Fig. 3.66   Effect of Co/Mo on the increase in S_0 (◆) and k_{ER} (▲) relative to those of Mo/Al_2O_3 catalyst at 260°C.[74]
(From W. Qian, A. Ishihara, T. Kabe et al., J. Chem. Soc., Faraday Trans., **93**, 4397 (1997))

of Mo/Al_2O_3, some sulfur, corresponding to the shaded area B, remained on the catalyst after purging the ^{35}S-labeled catalyst with decalin. Fig. 3.65 shows the effect of the ratio of Co/Mo on the behavior of sulfur on the catalysts. A significant increase is observed in area B at 280°C for cobalt-promoted catalysts compared with base catalyst Mo(16).[74] Similarly, the amounts of labile sulfur (S_0), corresponding to area B in Fig. 3.65, and the rate constant of sulfur exchange (k_{ER}) were determined and the values are summarized in Table 3.11. The relative increase in the amount of labile sulfur (S_0) and rate constant of sulfur exchange (k_{ER}) at 260°C with content of cobalt are shown in Fig. 3.66. The relative increase means the enhanced magnifications of value of S_0 or k_{ER} on the Co-Mo catalysts as compared with the value of S_0 or k_{ER} of the base catalyst Mo(16). The increase in amount of labile sulfur increased remarkably with the addition of cobalt then increased linearly with increasing ratio of cobalt to molybdenum up to about 0.5, but decreased slightly above 0.5 of Co/Mo molar ratio. In contrast to this, the rate constant of sulfur exchange (k_{ER}) remarkably increased once with the addition of cobalt whereas it no longer varied with the increase in ratio of cobalt to molybdenum.

Fig. 3.67 shows the effect of Co/Mo molar ratio on the conversion of DBT, and the yield of CHB, on the series of Co-Mo catalysts at 260°C. The total conversion of DBT and the yield of CHB increased markedly at lower Co/Mo molar ratio then decreased slightly at higher Co/Mo molar ratio.

k_{HDS} and k_{HYD} were determined and listed in Table 3.11.[74] k_{HDS}/k_{HDS0} is plotted vs. Co/Mo for reaction at 260°C in Fig. 3.67, k_{HDS}[and k_{HDS0} being the rate constants of HDS on Co-Mo catalyst and Mo(16) catalyst, respectively. Similarly, k_{HYD}/k_{HYD0} is plotted vs. Co/Mo in Fig. 3.68, k_{HYD} and k_{HYD0} being the rate constants of HYD on Co-Mo catalyst and Mo(16), respectively. A significant promotion effect of cobalt on the molybdena catalyst for both HDS and HYD activity was observed. k_{HDS} and k_{HYD} increased linearly with increasing Co content at lower Co content. The maximum increases in the HDS and HYD activities occurred on the catalyst with Co/Mo = 0.5 where the values of k_{HDS}/k_{HDS0} and k_{HYD}/k_{HYD0} were approximately 25 and 8, respectively. The activities for HDS and HYD decreased slightly with further increase in the cobalt content.

The apparent activation energies of HDS and HYD for all catalysts were obtained from Arrhenius plots of the reaction rate constants, respectively. There is no significant difference in the apparent activation energies: 23 ± 2 and 25 ± 2 kcal/mol for HDS reaction and HYD reaction, respectively. This indicates that there is essentially no difference in the mechanism of reaction on addition of cobalt. The enhancement in k_{HDS} and k_{HYD} on cobalt addition are

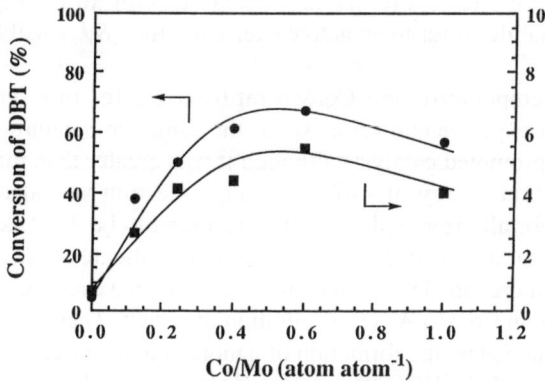

Fig. 3.67   Effect of ratio of Co/Mo on the conversion of DBT and the yield of CHB at 260°C.[74]
(From W. Qian, A. Ishihara, T. Kabe et al., *J. Chem. Soc., Faraday Trans.*, **93**, 4398 (1997))

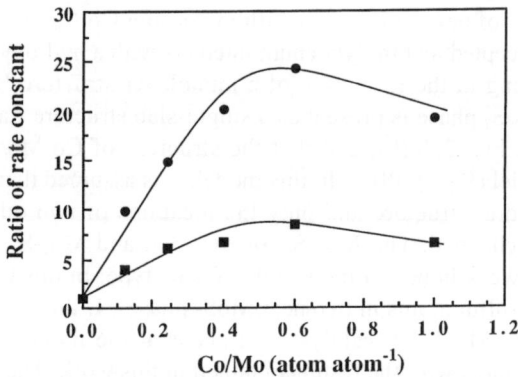

Fig. 3.68   Effect of Co/Mo on k_{HDS} and k_{HYD} relative to those of Mo/Al$_2$O$_3$ catalyst at 260°C. [74] ●: k_{HDS}/k_{HDS0}; ■: k_{HYD}/k_{HYD0}.
(From W. Qian, A. Ishihara, T. Kabe et al., *J. Chem. Soc., Faraday Trans.*, **93**, 4398 (1997))

10–25 fold and 4–8 fold, respectively, but the enhancement in k_{HDS} and k_{HYD} did not significantly increase with increasing Co content. The maximum enhancement was only a factor of *ca.* 2 when the Co/Mo ratio increased from 0.12 to 0.6. Hence, the promoting effect of cobalt is considered to arise from the formation of a more active phase and the increase in the catalytic activity to be due to the increase in the number of the same active sites.

For the alumina-supported Mo-based catalysts, coordinately unsaturated sites (CUS) or anion vacancy sites of sulfided catalysts are generally accepted to be the active sites of HDS and HYD.[283–285,335,365–367] The recently developed RPTM enables one to determine the amount of labile sulfur on the working catalyst, which is considered to represent the amount of sulfur that can be transformed to the vacancies, and the rate constant of sulfur exchange, which is considered to represent the ease of this transformation process.[240] In order to investigate the correlation of the HDS rate of DBT with S_0 and k_{ER}, the product of S_0 and k at several temperatures on the series of catalysts was plotted vs. the HDS reaction rate (r_{HDS}). A good straight line plot through the origin was obtained. This result can be explained as follows. Under a given reaction condition, S_0 is correlated with all sulfur which can be transformed to the vacancies (active sites), and k_{ER} is correlated with an average frequency of the transformation of a labile sulfur to an active site. Therefore, $S_0 k_{ER}$ will be correlated with the HDS rate.

The effect of temperature and Co/Mo ratio on k_{ER} for this series of catalysts was investigated. Within experimental error, k_{ER} at the same temperature did not vary with the Co/Mo ratio for all promoted catalysts although it was greater than that for the unpromoted catalyst. The activation energy of sulfur exchange reaction was also determined from an Arrhenius plot and for all promoted catalysts was found to be 7 ± 1 kcal/mol, less than that of the unpromoted catalyst (10 ± 1 kcal/mol). This indicates that the sulfur on the Co-promoted catalyst is more mobile than that on the unpromoted catalyst. This implies that a new active site is formed on Co-Mo/Al$_2$O$_3$ on addition of cobalt. Therefore, the promoting effect of cobalt can be explained by the formation of a more active phase, i.e., the so-called CoMoS phase. On the other hand, the HDS rate increased almost linearly with increasing S_0 while k_{ER} was virtually the same for all promoted catalysts at the same temperature. This indicates that the new active phase derived from the addition of cobalt is the same for all the promoted catalysts. Therefore, the increase in the catalytic activity with Co/Mo ratio is due to the increase in the number of new active sites with increasing Co/Mo.

It is commonly accepted that molybdenum interacts with a hydroxyl group on the alumina carrier surface, resulting in the formation of a monolayer structure.[9,10,355] It is an accepted hypothesis that the MoS$_2$ phase is present as a single-slab structure flat on the surface of the alumina, as shown in Fig. 3.36(b), and that the structure of Co-Mo/Al$_2$O$_3$ is an improved square-pyramidal model (Fig. 3.38).[76] In this model, it is assumed that the structure of MoS$_2$ is a deformed tetrahedral structure and only the locations of two other weak Mo-S bonds within the layers are changed. The Mo$_1$-S$_{d1}$ or Mo$_2$-S$_{d0}$, and Mo$_1$-S$_{d5}$ or Mo$_2$-S$_{d4}$ bonds are considered to be two weak bonds of the van der Waals type. In this model, one cobalt atom will promote the four sulfur atoms in two near MoS$_2$ phases. If this is the case, the promotion effect of cobalt on the Mo/Al$_2$O$_3$ catalyst will occur at the molar ratio of Co/Mo of 0.5, which is in good agreement with the results obtained in this work. The rates of both HDS and HYD were promoted with increasing Co/Mo molar ratio up to *ca.* 0.5, suggesting that cobalt may be well dispersed on Mo/Al$_2$O$_3$ and forms a CoMoS phase with the two neighboring MoS$_2$ phases at Co/Mo < 0.5. When the cobalt content is higher (Co/Mo > 0.5), the surplus cobalt may cover the highly active CoMoS sites.[344] In this case, the sulfur compounds with methyl groups which cause steric hindrance, such as 4,6-DMDBT cannot approach the highly active sites, Thus, HDS of 4,6-DMDBT and HYD of aromatic rings in 4,6-DMDBT will be inhibited. As a result, at higher Co/Mo molar ratios (Co/Mo > 0.5), k/k_0 values for both 3,3'-DMBP and 3,3'-DMCHB decreased with increasing Co/Mo molar ratio, as shown in Fig. 3.60. However, in the case of DBT, although the surplus cobalt may cover the highly active sites, the DBT can still approach the highly active CoMoS sites because it does not have a methyl group. That is, if large masses of Co_9S_8 which are formed from surplus cobalt on alumina exist near the CoMoS sites as in Fig. 3.60, the approach of 4,6-DMDBT to the CoMoS sites would be more difficult than that of DBT. According to the literature, the HDS rate of thiophene approached the maximum in the range of Co/Mo ratio from 0.5 to 1.0, but in the range of Co/Mo higher than 1.0, the rate decreased or was not affected.[54,89,107,206,368] The rate of HDS of DBT was constant and the rate of HDS of 4,6-DMDBT decreased at Co/Mo ratios above 0.5. It seems that the HDS of 3-ring 4,6-DMDBT as well as DBT is more sensitive to the change in the catalyst structure brought about by the addition of cobalt than HDS of thiophene.[344]

3.4 Genesis of Active Sites and Reaction Mechanism of HDS

Much hydrotreating research has focused on obtaining insight on the genesis of the active sites and the reaction mechanism to obtain guidelines for developing novel hydrotreating catalysts. There are many debated issues. Several reviews can be cited.[44,103,120,121,369,370] Detailed information on these issues is not straightforward. For example, unambiguous conclusions regarding the nature of the sites from either structure-activity correlations or from kinetic analysis alone are difficult to obtain. Indeed, several different types of sites may satisfy a given observed structure-activity correlation. Furthermore, knowledge of the elementary steps and their energetics is usually not available. Despite the complexities, insight at the atomic level has recently provided additional information on the nature of active sites. These results will be discussed in this section. It will be shown that the formation of coordinative unsaturated sites (CUS), e.g., sulfur vacancies, plays an essential role in most rections. SH groups appear to be involved in both the supply of hydrogen and in providing Brønsted acidity for acid-catalyzed reactions including, in certain cases, the HDN reaction.

3.4.1 Unpromoted Mo and W Catalysts

The structure-activity correlations for unpromoted catalysts have shown that HDS and other hydrotreating reactions are structure sensitive with the active sites residing along the MoS_2 (WS_2) edges. Regarding the nature of these active sites, there is now ample evidence that the active sites involve sulfur vacancies or CUS sites. During hydrotreating it is also essential to have hydrogen available in the vicinity of the adsorbed sulfur-containing molecule. In view of this, Lipsch and Schuit[92] proposed a mechanism (Fig. 2.20) involving a vacancy site and a neighboring source of hydrogen from an OH (or SH) group. Although this proposal was for a monolayer model, the basic ideas can be adapted to the edge of MoS_2 (WS_2). Different proposals for the dual function nature of hydrotreating catalysts have been offered by other authors.[56,256,272,296,371,372] It has been debated whether the dissociation of hydrogen occurs homolytically or heterolytically.[372,373] In the heterolytic dissociation of hydrogen over MoS_2:

$$H_2 \rightarrow H^- + H^+ \tag{3.7}$$

the proton will most likely react with a sulfide ion leading to the formation of SH groups. Alternatively, it is possible for the hydride ion to be oxidized to a proton by the molybdenum leading to the formation of an additional SH group.[372] The net result of the heterolytic dissociation of hydrogen can in this case be formally written as Fig. 3.69.[374] Consequently, the HDS reaction may involve the donation of protons and electrons and the semiconducting nature of MoS_2 may facilitate this.

Solid-state chemistry models for sulfidic HDS catalysts on the possible CUS sites along the edges and corners of MoS_2 were first proposed by Voorhoeve[46] and by Farragher and Cossee.[43,50] They explained their experimental results for the hydrogenation of benzene and cyclohexene by assuming that the catalytic active site was a Mo ion located the edge of a MoS_2 crystallite. In this model the desulfurization starts by adsorption of the sulfur atom of the reactant molecule on the sulfur vacancy. A four-electron reduction process[69,375] then leads to the formation of H_2S and butadiene from thiophene:

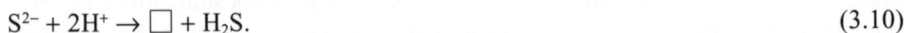

$$C_4H_4S + 2H^+ + 4e \rightarrow C_4H_6 + S^{2-} \tag{3.8}$$

$$2H_2 \rightarrow 4H^+ + 4e \tag{3.9}$$

$$S^{2-} + 2H^+ \rightarrow \square + H_2S. \tag{3.10}$$

The four electrons are delivered by a redox couple like:

$$4Mo^{n+} \rightarrow 4Mo^{(n+1)+} + 4e$$

$$\text{or} \quad 4Mo^{n+} \rightarrow 2Mo^{(n+2)+} + 4e. \tag{3.11}$$

Looking at these reactions in a different way,

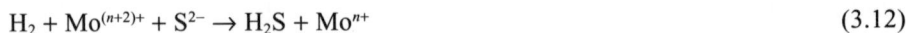

$$H_2 + Mo^{(n+2)+} + S^{2-} \rightarrow H_2S + Mo^{n+} \tag{3.12}$$

where hydrogen is used in a reductive addition reaction on sulfidic surfaces, as on some oxidic surfaces.[376] In this way, the HDS reaction shows much clearly in common with the Birch reduction of functional groups in organic molecules in a solution of Na or K in liquid

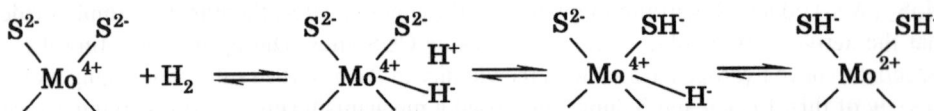

Fig. 3.69 Schematic illustration of heterolytic dissociation of hydrogen on Mo active site.[374]
(From H. Topsøe, B. S. Clausen and F. E. Massoth, *Hydrotreating Catalysis*, p.219 (1996))

NH3, in which the reduction is not performed by hydrogen atoms, but by electrons and protons. Under the high H_2 pressure conditions during HDS, transition metal sulfides become sulfur-deficient and good electron conductors. Protons are always available at the surface of transition metal sulfides in the form of SH groups. The analogy with the Birch reduction explains the observations made by Pecoraro and Chianelli,[377] Vissers et al.[282] and Ledoux et al.[354] that most transition metal sulfides are capable of catalyzing the HDS reaction.

When the reactant thiophene molecule is assumed to be adsorbed upright on the surface, the so-called one-point (end-on) mechanism for the HDS of thiophene is obtained, as shown in Fig. 2.20.

One objection to the end-on mode of adsorption has been that the basicity of the sulfur atom in thiophene is low, due to the delocalization of the lone pair electrons on sulfur onto the butadiene fragment in thiophene.[216] To overcome the problem of low basicity on the sulfur atom, Kuart et al. proposed that adsorption of thiophene takes place via "multipoint" (side-on) adsorption.[216] In this mechanism, the C_1-C_2 bond of thiophene is coordinated at a sulfur anion vacancy, with the adjacent sulfur atom interacting with a neighboring sulfur on at the catalyst surface (Fig. 2.21(a)). The assistance of an adjacent sulfur atom in the side-on mechanism leads to sulfur-sulfur bond formation and suggests that persulfide S_2^{-2} anions may be important in HDS. As a consequence of lower coordination, the Madelung energy of surface ions is lower than that of bulk ions and the oxidation state will therefore be lower. The surface of MoS_2 thus contains S_2^{2-} anions.[378] The involvement of such persulfide anions in HDS has already been proposed.[5,379] It is noteworthy that RuS_2 and OsS_2, which have the pyrite crystal structure with S_2^{2-} units, are among the very best HDS catalysts, and that Co, Ni, and Fe, which are known promoters for MoS_2, can also occur in the pyrite structure.

3.4.2 Periodic Trends and Promotion Effect

Studies of the periodic trends in activity variations of transition metal sulfides have provided interesting opportunities to obtain insight into the role of different types of sites. It has been known for a long time that many transition metal sulfides are active catalysts in hydrotreating reactions.[2,80,282,354,377,380-383] Studies of the activity variations as a function of the position of the metal in the periodic table are considered a useful approach for gaining insight into the catalysis by transition metals. Pecoraro and Chianelli[377] were among the first to extend such studies to sulfides of the transition metals. They studied the HDS of DBT at high pressures over unsupported sulfides and observed a characteristic volcano-type dependence of the activity on the periodic position for both the second row (4d) and third row (5d) transition metal sulfides (Fig. 4.16). The first row (3d) transition metal sulfides are much less active than sulfides in the second and third rows and the there appears to be a minimum in activity around manganese. The strong dependence of the activity on the position in the periodic table shows that the electronic structure of the sulfide plays an important role in determining the catalytic activity and in recent years, significant effects have been devoted to obtaining a physical basis for understanding the origin of the variations in HDS activity. Chianelli and coworkers [73,186,377] performed SCF-Xα scattered wave calculations on hypothetical $(MS_6)_n^{n-}$ clusters in which metal cations are octahedrally surrounded by sulfur. They observed a correlation between the HDS activity and the product of two calculated parameters, as shown in Fig. 3.70.[73] One factor is the ability of the transition metal to bond covalently, in both a σ and π fashion, to 3p orbital of sulfur, which measures the covalent

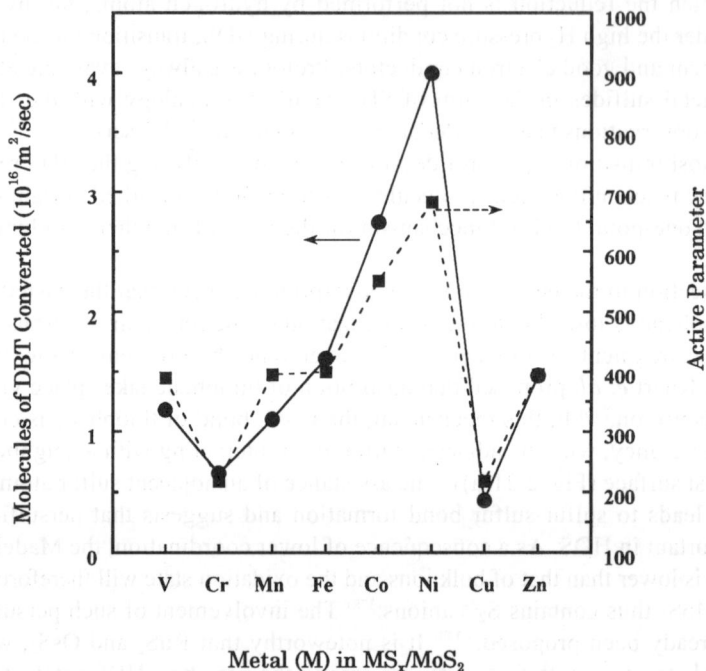

Fig. 3.70   Calculated activity parameter for each mixed metal sulfide system (right-hand scale).[73] Shown for
comparison, using the left-hand scale, are the measured HDS activities.
(From S. Harris and R. R. Chianelli, *J. Catal.,* **98**, 28 (1986))

contribution to the metal-sulfur bond strength. Another is the number of "d" electrons in the
highest occupied molecular orbital (HOMO). In the presence of Co or Ni, Mo is reduced
relative to the Mo present in MoS_2. That is, the formal Mo 4d orbital occupation increases
from 2 in MoS_2 to 3 and 4 in the presence of Co and Ni, respectively. In the presence of Cu,
on the other hand, Mo is formally oxidized so that the formal 4d occupation decreases to 1.
These effective reductions and oxidation of Mo correlate with an increase and decrease in
catalytic activity. Thus, the promotion effect is ascribed to the increase in the number of 4d
orbital electrons due to the ability of the effective promoters such as Co to donate electrons
to Mo. Likewise, a poison such as Cu has the ability to remove electrons from Mo and thus
decrease the number of 4d electrons. On the other hand, metals such as V, Cr, Mn, and Fe do
not serve as very effective promoters or poisons because they do not donate or withdraw the
electron to or from Mo. Although this approach has been questioned by several authors and
it remains unclear why the particular product of the two parameters should be correlated to
the activity, this indicates the importance of metal-sulfur bond strength for understanding the
promotion effect.

   The interaction between active metal species and support affects significantly the activity
of transition metals.[384,385] This is further discussed in Chapter 4.

   As mentioned in Chapter 2, the hydrotreating reaction is affected by the H_2S. Fig. 2.34
shows the effect of the partial pressure of H_2S on HDS rates of DBT and 4,6-DMDBT.[386] Both
the HDS rates of DBT and of 4,6-DMDBT decreased with increasing the partial pressure of
H_2S. The HDS rate of DBT at the H_2S partial pressure of 0.19 atm decreased to about 9% and

12% of those without addition of H_2S at 240°C and 260°C, respectively. The HDS rate of 4,6-DMDBT at the H_2S partial pressure of 0.19 atm decreased to about 22% and 29% of those without addition of H_2S at 240°C and 260°C, respectively. This result indicated that the HDS rate of DBT was more strongly inhibited by H_2S as compared with 4,6-DMDBT. On the other hand, it can be observed that the retarding effects of H_2S on HDS of DBT and 4,6-DMDBT decrease with rising temperature. The inhibition of H_2S on the HDS rate of DBT over a Mo/Al_2O_3 (MoO_3: 12 wt%) catalyst at 240°C was compared with that over the $Co-Mo/Al_2O_3$ catalyst in Fig. 3.71. The inhibition of H_2S on Mo/Al_2O_3 was much stronger than that on $Co-Mo/Al_2O_3$. Similar results were also obtained in the HDS reaction of 4,6-DMDBT. This implies that the presence of Co results in a decrease in the inhibition of H_2S. Furthermore, from kinetic analytic results, the activation energies of HDS reactions of DBTs and the heats of adsorption of DBTs and H_2S on the catalyst were estimated from Arrhenius (Fig. 3.72) and van't Hoff plots (Fig. 3.73).[77] The activation energies of HDS of DBT or 4,6-DMDBT on Mo/Al_2O_3 and $Co-Mo/Al_2O_3$ catalysts were almost the same, being 24 ± 2 and 33 ± 2 kcal/mol, respectively. Further, the heats of adsorption of DBT or 4,6-DMDBT on the two catalysts were approximately the same, 10 ± 1 and 14 ± 2 kcal/mol, respectively. In contrast, the heats of adsorption of H_2S on Mo/Al_2O_3 and $Co-Mo/Al_2O_3$ were different from each other, 21 ± 1 and 17 ± 1 kcal/mol, respectively. The heats of adsorption of DBT, 4,6-DMDBT and H_2S over each catalyst increased in the order DBT < 4,6-DMDBT < H_2S. Moreover, the adsorption equilibrium constants over each catalyst also increased in the same order. This means that H_2S was adsorbed more strongly than DBTs and thus inhibited the HDS of DBTs through competitive adsorption. This was verified by the fact that the inhibition of H_2S on the HDS of 4,6-DMDBT was less than that of DBT because 4,6-DMDBT more strongly is adsorbed on the catalyst than DBT. Compared with the Mo/Al_2O_3 catalyst, all the adsorption equilibrium constants on the $Co-Mo/Al_2O_3$ catalyst decreased because of the addition of Co. On the other

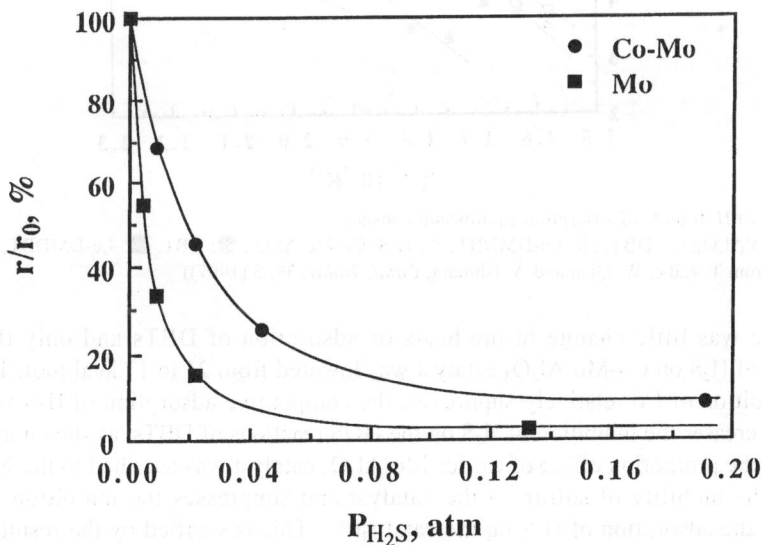

Fig. 3.71 Effect of partial pressure of H_2S on HDS reactivity of DBT.[77] 50 kg/cm², 240°C.
(From T. Kabe, W. Qian and A. Ishihara, *Catal. Taday*, **39**, 5 (1997))

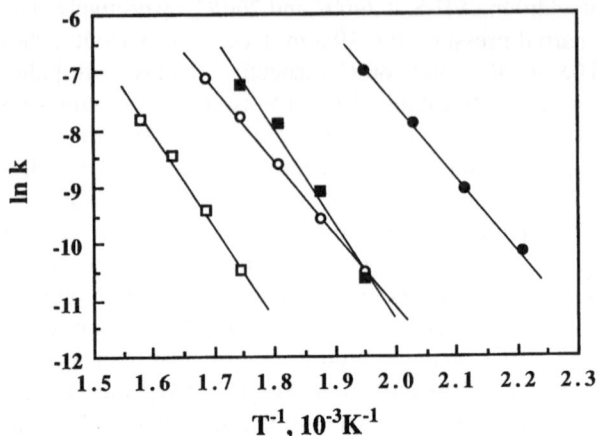

Fig. 3.72   Arrhenius plots of HDS reaction rate constants.[77]
Mo/Al$_2$O$_3$: ○: DBT, □: 4,6-DMDBT; Co-Mo/Al$_2$O$_3$: ●: DBT, ■: 4,6-DMDBT.
(From T. Kabe, W. Qian and A. Ishihara, *Catal. Taday*, **39**, 5 (1997))

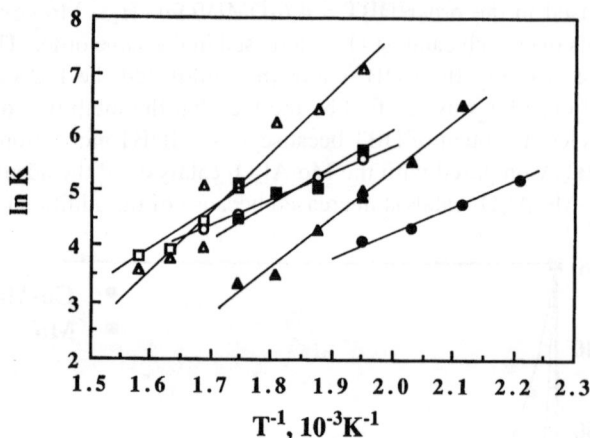

Fig. 3.73   van't Hoff plots of adsorption equilibrium constants. [77]
Mo/Al$_2$O$_3$: ○: DBT, □: 4,6-DMDBT, △: H$_2$S; Co-Mo/Al$_2$O$_3$: ●: DBT, ■: 4,6-DMDBT, ▲: H$_2$S.
(From T. Kabe, W. Qian and A. Ishihara, *Catal. Taday*, **39**, 5 (1997))

hand, there was little change in the heats of adsorption of DBTs and only the heat of adsorption of H$_2$S on Co-Mo/Al$_2$O$_3$ catalyst was lowered from 21 to 17 kcal/mol. This means that the addition of Co relatively suppresses the competitive adsorption of H$_2$S with DBTs, and then decreases the inhibition of H$_2$S on the HDS reactions of DBTs, as shown in Fig. 3.71. Therefore, the promotion effect of Co for Mo/Al$_2$O$_3$ catalyst was ascribed to the fact that Co enhances the mobility of sulfur on the catalyst and suppresses the inhibition of H$_2$S by weakening the adsorption of H$_2$S on the catalyst.[386] This is verified by the results obtained from the ^{35}S RPTM.[75,76,78]

As in the case of unpromoted Mo/Al$_2$O$_3$ catalyst, part of the sulfur on the promoted catalyst in the HDS is labile (See Section 3.2). Similarly, the release rate of H$_2$S from the

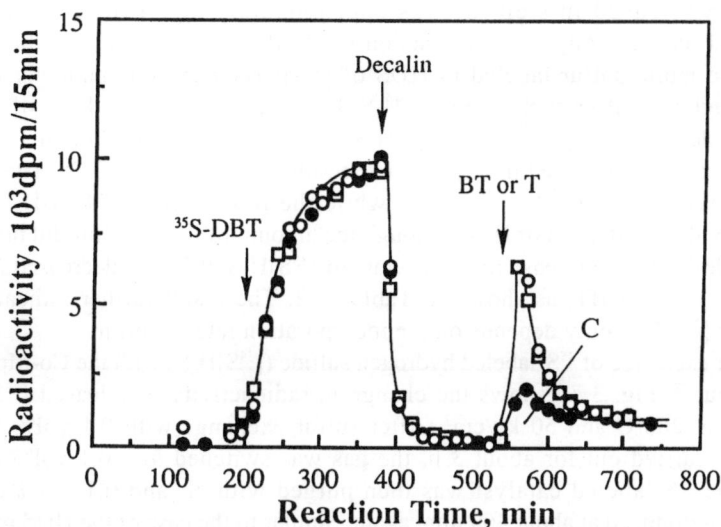

Fig. 3.74 Effects of sulfur compounds on the release rate of [^{35}S]H$_2$S.[78] ○: Thiophene (0.46 wt%, 260°C), ●: Thiophene (0.46 wt%, 200°C), □: Benzothiophene (0.73 wt%, 260°C). (From W. Qian, A. Ishihara, T. Kabe et al., J. Catal., **170**, 290 (1997))

Table 3.12 Results in the HDS reactions and the exchange reactions on the sulfided Co-Mo/Al$_2$O$_3$ catalyst [78]

Temperature (°C)	240	260	280	200	260	260	260	260	260	400
Reactant	DBT [a,d]			T [a,e]		BT [a,f]	DBT+H$_2$S [a,g]	H$_2$S [a]	H$_2$S [b]	
Conversion (%)	29.5	62.3	90.9	30.9	100	100	21.1	—	—	—
Concentration of formed H$_2$S (vol%)	0.040	0.084	0.125	0.042	0.137	0.137	0.029	0.300	0.10	0.10
Rate constant, k (10^{-2}min/g-cat.)	1.90	3.20	3.72	2.31	5.02	4.95	8.21	8.14	4.87	—
Labile sulfur, S_0 (mg/g-cat.)	13.1	15.6	21.8	10.4	16.1	16.5	15.8	16.1	16.0	30.0
S_0/S_t [c] (%)	18.9	22.5	31.4	15.0	23.2	23.8	22.8	23.2	23.1	43.3
$S_0 \times k$ (mg/min·g-cat.)	0.249	0.499	0.811	0.240	0.808	0.817	1.30	—	—	—
r_{HDS} (g/min·g-cat.)	0.240	0.508	0.741	0.252	0.815	0.815	0.391	—	—	—

[a] The catalyst was labeled by ^{35}S in the HDS reaction of [^{35}S]-DBT;
[b] The catalyst was labeled by ^{35}S in the presulfiding with 3.4 vol% [^{35}S] H$_2$S;
[c] S_t is defined as the amount of total sulfur when metal sulfides were present as MoS$_2$ and Co$_9$S$_8$;
[d] Dibenzothiophene; [e] Thiophene; [f] Benzothiophene; [g] Dibenzothiophene + 0.3 vol% H$_2$S.
(From W. Qian, A. Ishihara, T. Kabe et al., J. Catal., **170**, 290 (1997))

catalyst in the HDS reaction depends on the incorporation rate of sulfur but is independent of the kinds of sulfur-containing compounds (Fig. 3.74, Table 3.12).[78] This implies that the mobility of sulfur depends on the concentration and formation rate of H$_2$S. The sulfur exchange between the catalyst and H$_2$S was carried out using a tracer method.[78] It was found that sulfur on the catalyst could be exchanged by the sulfur in H$_2$S. After the catalyst was labeled with ^{35}S during HDS of [^{35}S]DBT, as shown in Fig. 3.74,[78] ^{35}S remaining on the catalyst was also exchanged by ^{32}S in [^{32}S]H$_2$S and released as [^{35}S]H$_2$S again. Although the release rate of [^{35}S]H$_2$S from the catalyst was more rapid than that in the case of DBT, the total amount of ^{35}S released in step (d) was approximately equal to the amount of ^{35}S accommodated on the catalyst in the reaction of [^{35}S]DBT, as shown in Table 3.12. This shows that the labile sulfur labeled in HDS of [^{35}S]DBT can be completely released as [^{35}S]H_2S through the sulfur exchange with H_2S. In order to investigate HDS of DBT in the presence of added H_2S, the gas of 0.3 vol% H_2S in H_2 and decalin solution of 1.0 wt% DBT were simultaneously introduced in step (d). The result is shown in Fig. 3.75. The conversion of DBT decreased from 62.3% to 21.1% while the release rate of [^{35}S]H_2S increased significantly with the addition of H_2S, although the amount of labile sulfur did not change as shown in Table 3.12. The release rate constants of [^{35}S]H_2S at 260°C decreased in the order H_2S + DBT > H_2S > DBT, as shown in Table 3.12. The result further indicates that the release rate of [^{35}S]H_2S only depends on the incorporation rate of sulfur.

The sulfur exchange of ^{35}S-labeled hydrogen sulfide ([^{35}S]H_2S) with the Co-Mo/Al_2O_3 was also carried out.[78] Fig. 3.76 shows the change in radioactivity of released [^{35}S]H_2S with reaction time at 260°C and 50 kg/cm². After sulfur exchange with 0.1 vol% [[32]S]H_2S in hydrogen was carried out for about 3 h, the gas was switched to a 0.1 vol% [^{35}S]H_2S in hydrogen. The ^{35}S-labeled catalyst was then purged with N_2 and 0.1 vol% [[32]S]H_2S in hydrogen was introduced at about 460 min, again. Similar to the case of the HDS of [^{35}S]DBT, the radioactivity of released [^{35}S]H_2S slowly increased and approached a steady state after [^{35}S]H_2S was introduced for about 120 min. When N_2 was introduced, the release of [^{35}S]H_2S decreased immediately, indicating that a portion of ^{35}S, corresponding to the shaded area A, was accommodated to the catalyst. The portion of ^{35}S remaining on the catalyst was released as [^{35}S]H_2S after N_2 was subsequently replaced by [[32]S]H_2S at about 460 min, as shown in Fig. 3.76 (area B). This portion of ^{35}S represented the total amount of sulfur exchanged on the catalyst under this reaction condition. The amount of labile sulfur (S_0) can be calculated from the total radioactivity corresponding to area B and is listed in Table 3.13. Similar to the case of the HDS reaction, a first-order plot of the release rate of [^{35}S]H_2S could be drawn and

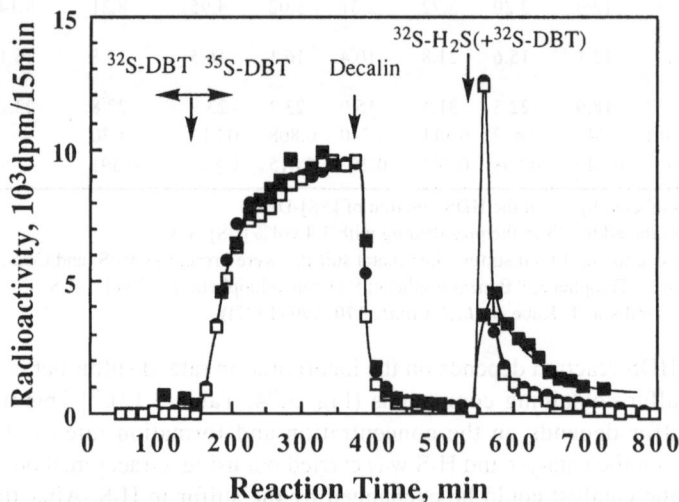

Fig. 3.75 Release of [^{35}S]H_2S in the sulfur exchange reaction with [[32]S]H_2S in the step (d) with or without DBT.[78]
●: 0.3 vol% H_2S + 1.0 wt% DBT; □: 0.3 vol% H_2S; ■: 1.0 wt% DBT.
(From W. Qian, A. Ishihara, T. Kabe et al., J. Catal., **170**, 291 (1997))

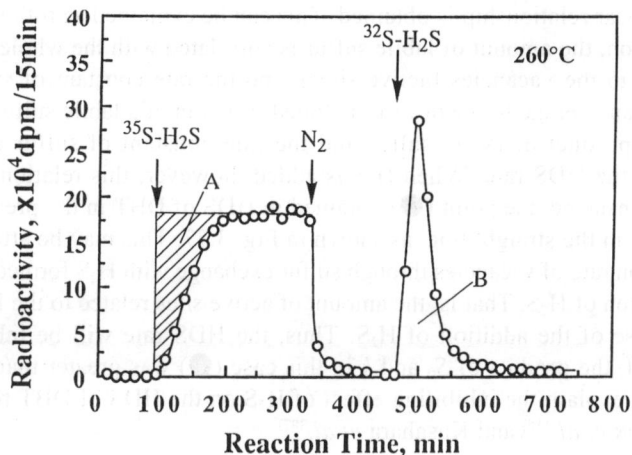

Fig. 3.76 Change in the radioactivity of released [^{35}S]H$_2$S with reaction time.[78] 50 kg/cm^2, 0.1 vol% H$_2$S, 80 ml/min.
(From W. Qian, A. Ishihara, T. Kabe *et al.*, *J. Catal.*, **170**, 289 (1997))

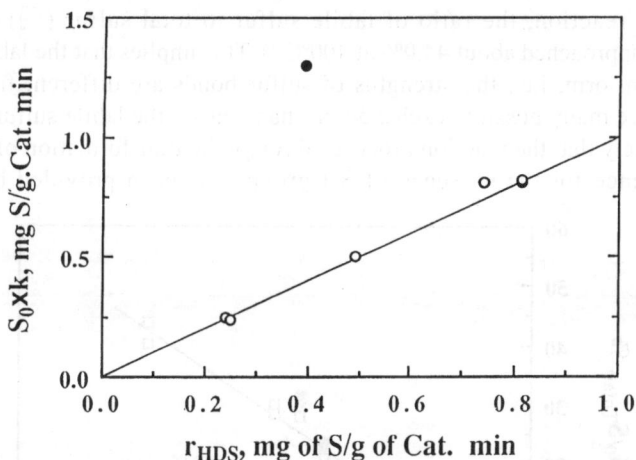

Fig. 3.77 Relation among the amount of labile sulfur, the exchange rate constant and the HDS rate.[78]
(From W. Qian, A. Ishihara, T. Kabe *et al.*, *J. Catal.*, **170**, 293 (1997))

a good linear relationship was also obtained.[78] The exchange rate constant (k) was also determined and is presented in Table 3.13. As shown in the table, the partial pressure of H$_2$S did not affect the amounts of labile sulfur and the release rate constants of [^{35}S]H$_2$S, but the concentration of H$_2$S and the flow rate of H$_2$S (i.e., the supply rate of H$_2$S) influenced significantly the release rate of [^{35}S]H$_2$S even though the amount of labile sulfur remained constant.

In order to investigate the relationship between the amount of labile sulfur and the activity of the catalyst, the product of the amount of labile sulfur (S_0) and the rate constant (k)

of sulfur exchange in the HDS reaction is plotted against the HDS reaction rate (r_{HDS}) in Fig. 3.77. A good linear relationship is obtained. This can be explained as follows. Under a given reaction condition, the amount of labile sulfur is correlated with the whole sulfur which can be transformed to the vacancies (active sites), and the rate constant of sulfur exchange is correlated with an average frequency of the transformation of a labile sulfur to an active site. Therefore, the product of labile sulfur and the rate constant of sulfur exchange will be correlated with the HDS rate. When H_2S is added, however, this relationship will become untenable. For instance, the point (●) obtained in HDS of DBT in the presence of 0.3 vol% H_2S deviated from the straight line, as shown in Fig. 3.77. This may be attributed to the fact that the migration rate of vacancies through sulfur exchange with H_2S formed will be enhanced due to the addition of H_2S. That is, the amount of active sites related to the HDS reaction will decrease because of the addition of H_2S. Thus, the HDS rate will be inhibited by adding H_2S. As a result, the product of S_0 and k in this case (●) was greater than the HDS rate of DBT. This may explain the inhibiting effect of H_2S on the HDS of DBT reported by Zhang et al.,[386] Ledoux et al.[116] and Kasahara et al.[387]

3.4.3 Amount of Labile Sulfur and Mechanism of Sulfur Exchange

In order to understand the role of labile sulfur in the HDS reaction, the behavior of sulfur in the HDS reaction and in the sulfur exchange reaction is compared in Fig. 3.78. In the sulfur exchange reaction, the ratio of labile sulfur to total sulfur (□) increased with temperature and approached about 44.0% at 400°C.[78] This implies that the labile sulfur on the catalyst is not uniform, i.e., the strengths of sulfur bonds are different from each other. Although there are many possible exchange mechanisms on the labile sulfur on the catalyst with H_2S, it is likely that the reaction proceeds through the transformation of SH groups and vacancies; evidence for the presence of SH groups has been provided by a number of

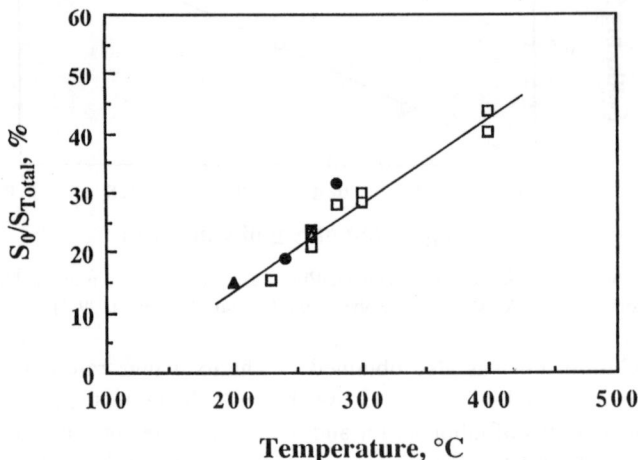

Fig. 3.78 Comparison of amount of labile sulfur obtained in various reactions.[78] □: Data obtained in sulfur exchange with [^{35}S]H_2S; ●: Data obtained in DBT HDS; ■: Data obtained in HDS of BT; ▲: Data obtained in HDS of T; ○: Data obtained in sulfur exchange with 0.3% H_2S; △: Data obtained in HDS of DBT in the presence of 0.3% H_2S.

(From W. Qian, A. Ishihara, T. Kabe et al., J. Catal., 170, 291 (1997))

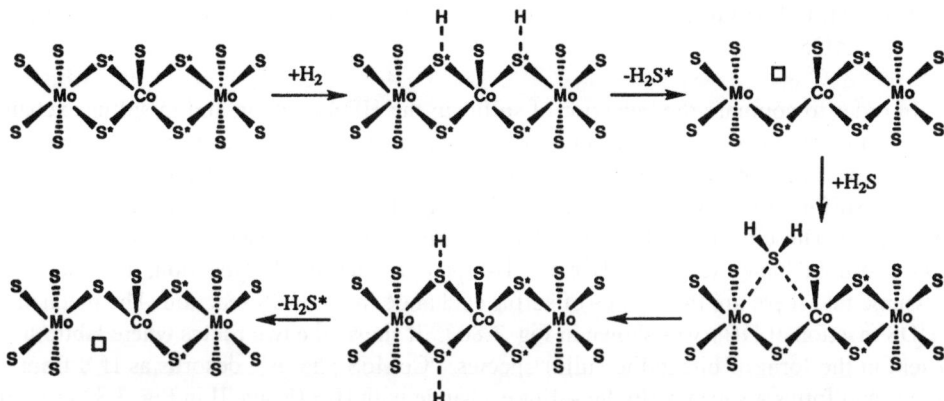

Fig. 3.79 Exchange of the labile sulfur on the sulfided Co-Mo/Al$_2$O$_3$ with H$_2$S.[78] S*: ^{35}S, □: Anion vacancy.
(From W. Qian, A. Ishihara, T. Kabe et al., J. Catal., **170**, 292 (1997))

studies.[93,260,261,388] More recently, in an FTIR study on sulfided Mo-based catalysts, it was postulated that SH groups existed at the edges of MoS$_2$ and that SH groups and vacancies could interconvert and coexist in close proximity.[157,253] Further, an electronic density analysis presented theoretically the possibility of the formation of SH groups.[389]

Therefore, a hypothetical release route of [^{35}S]H$_2$S was postulated as shown in Fig. 3.79.[78] As noted above, portion of the sulfur bonded with Co and Mo atoms was considered to be more labile and present as SH groups in H$_2$ atmosphere below 400°C. When H$_2$S was formed and subsequently desorbed from the catalyst, an anion vacancy occurred and the adjacent Mo^{4+} would be partially reduced to Mo^{3+} at the same time. On the other hand, if it is assumed that reduction of Mo^{4+} to Mo^{3+} is difficult as reported by McGarvey and Kasztelan,[390] the amount of active sites generated on the catalyst will actually be finite and this process perhaps may not continue only in H$_2$ under the present reaction conditions. This was supported indirectly by the fact that [^{35}S]H$_2$S was hardly detected even though the catalyst was reduced in H$_2$ as shown in Figs. 3.75 and 3.76. When H$_2$S was introduced, the adsorbed H$_2$S dissociated and formed new SH groups with an adjacent labile sulfur due to the high mobility of the hydrogen atom. The old anion vacancy disappears and a new anion vacancy will occur after the desorption of H$_2$S formed from the labile sulfur. It is noteworthy that the extent of sulfur exchange was only about 44.0% of total sulfur at 400°C. It is well known that HDS reactions are generally carried out below 400°C. Therefore, this portion of the labile sulfur is considered to be more closely related to the HDS reaction.

Startsev et al. reported a result where the sulfur exchange rates with H$_2$S on a series of Ni-Mo (W) /SiO$_2$ catalysts were one to two orders of magnitude greater than the rates of hydrogenation of butadiene in the HDS reaction of thiophene.[391] Massoth and Zeuthen also postulated qualitatively a similar result in the sulfur exchange reaction with H$_2$S on a sulfided Mo/Al$_2$O$_3$.[342] Therefore, it is reasonable to assume that the sulfur exchange rate is very rapid under typical HDS reaction conditions.

3.4.4 Transformation between Labile Sulfur and Vacancies and Mechanism of HDS

In order to compare the behavior of sulfur in the HDS reaction and in sulfur exchange reaction, the amounts of labile sulfur participating in various HDS reactions are also plotted against temperature in Fig. 3.78.[78] Essentially, there is no significant difference in the amounts of labile sulfur in the HDS reaction and those in the sulfur exchange reactions at every temperature. These results indicate that labile sulfur in the HDS reaction is equivalent to that in the sulfur exchange with H_2S. Because H_2S is a product of HDS reaction, it is reasonable to assume that there are two routes of sulfur exchange in the HDS reaction. The mechanism of HDS reaction of DBT was shown in Fig. 3.80.[78] It shows the two routes where labile sulfur present in the form of bimetallic sulfur species, "CoMoS phases", desorbs as H_2S from the catalyst and forms a vacancy. In the sulfur exchange with H_2S (Route II in Fig. 3.80 or in Fig. 3.79), when one vacancy (active site) is occupied by sulfur in H_2S formed in the HDS reaction, a labile sulfur in another site is released as H_2S to form another new vacancy. In HDS (Route I), when a sulfur compound adsorbs on a vacancy, the C-S bond is subsequently cleaved, and the sulfur remains on the catalyst. Simultaneously, another labile sulfur is released as H_2S and a new active site is formed. In the two routes, the migration of vacancies on the catalyst always occurs due to the transformation between labile sulfur and vacancies

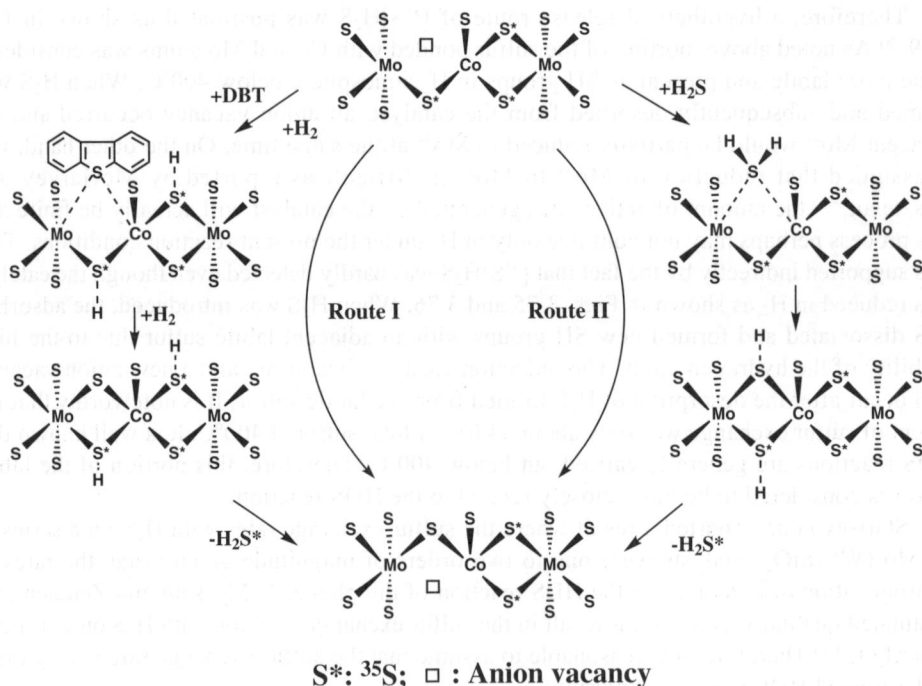

S*: ^{35}S; □ : Anion vacancy

Route I: Hydrodesulfurization; Route II: Sulfur Exchange

Fig. 3.80 Transformation between labile sulfur and vacancies on the sulfided Co-Mo/Al$_2$O$_3$ in HDS.[78]
(From W. Qian, A. Ishihara, T. Kabe et al., J. Catal., **170**, 293 (1997))

on the catalyst surface. In the sulfur exchange with [^{34}S]H$_2$S on a sulfided Mo/Al$_2$O$_3$ Massoth and Zeuthen also postulated such a migration of vacancies on the catalyst due to high mobility of labile sulfur.[342] This is consistent with results obtained by Leliveld et al.,[392] who investigated the genesis of sulfur vacancies on sulfided (Co)Mo/Al$_2$O$_3$ catalysts with EXAFS at 673 K in a H$_2$ and H$_2$/thiophene gas atmosphere. For Mo, no significant changes in sulfur coordination were observed for either the Co-promoted or the unpromoted sample. In Co-promoted Mo/Al$_2$O$_3$, vacancies are primarily formed on the Co atom. Table 3.12[78] shows that the release rate constant becomes greater with an increase in the concentration of H$_2$S formed in the HDS reaction of sulfur compounds such as DBT, BT, and thiophene at 260°C even though the amount of labile sulfur scarcely changes. Moreover, these release rate constants are almost the same as those in the sulfur exchange reactions where the concentrations of H$_2$S used are the same as those in the HDS reaction. Therefore, it can be assumed that a rapid adsorption/desorption of H$_2$S will always take place in the presence of H$_2$S under typical hydrotreating process conditions, leading to a rapid interconversion of the active site and labile sulfur. Thus, the vacancies under the reaction conditions will not be fixed, but mobile.

References

1. D. C. McCulloch in: *Applied Industrial Catalysis* (B. E. Leach ed.), **Vol. 1**, p.69, Academic Press, New York (1983).
2. O. Weisser and S. Landa, *Sulphide Catalysts, Their Properties and Application*. Pergamon Press, Oxford (1973).
3. P. B. Kwant and W. C. van Zijll Langhout, *Proces Technol.*, **10**, 11 (1986).
4. P. Grange, *Catal. Rev.-Sci. Eng.*, **21**, 135 (1980).
5. J. C. Duchet, E. M. van Oers, V. H. J. de Beer and R. Prins, *J. Catal.*, **80**, 386 (1983).
6. C. Gachet, M. Breysse, M. Cattenot, T. Decamp, R. Frety, M. Lacroix, L. de Mourgues, J. L. Protefaix, M. Vrinat, J. C. Duchet, S. Housni, M. Lakhdar, M. J. Tilliette, J. Bachelier, D. Cornet, P. Engelhard, C. Gueguen and H. Toulhoat, *Catal. Today*, **4**, 6 (1988).
7. G. Perot, G. Brunet and N. Hamze in: *Proceedings, 9th International Congress on Catalysis* (M. J. Phillips and M. Ternan eds.), p.19, Calgary (1988).
8. I. Mazzarino, G. Baldi and A. Iannibello, *Bull. Soc. Chim. Belg.*, **199**, 849 (1991).
9. G. C. A. Schuit and B. C. Gates, *AIChE J.*, **19**, 417 (1973).
10. F. E. Massoth in: *Advances in Catalysis and Related Subjects* (D. D. Eley, H. Pines and P. B. Weisz eds.), **Vol.27**, p.265, Academic Press, New York, (1987).
11. M. Dufaux, M. Che and C. Naccache, *J. Chim. Phys.*, **67**, 527 (1970).
12. T. Lim and S. W. Weller, *J. Catal.*, **108**, 175 (1987).
13. B. M. Reddy and V. M. Mastikhin, *Proceedings, 9th International Congress on Catalysis* (M. J. Phillips and M. Ternan eds.), p.82 (1988).
14. B. S. Clausen, H. Topsøe, R. Candia, J. Villadsen, B. Lengeler, J. Als-Nielsen and F. Christensen, *J. Phys. Chem.*, **85**, 3868 (1982).
15. N.-S. Chiu, S. H. Bauer and M. F. L. Johnson, *J. Catal.*, **89**, 226 (1984).
16. C. Li, Q. Xin, K-L. Wang and X. Guo, *Appl. Spectrosc.*, **5**, 874 (1991).
17. D. Ouafi, F. Mauge, J. C. Lavalley, E. Payen, S. Kasztelan, M. Jouari, J. Grimblot and J. O. Bonnelle, *Catal. Today*, **4**, 24 (1988).
18. D. S. Zingg, L. E. Makovsky, R. E. Tischer, F. R. Brown and D. M. Hercules, *J. Phys. Chem.*, **84**, 2898 (1980).
19. L. Salvati Jr., L. E. Makovsky, J. M. Stencel, F. R. Brown and D. M. Hercules, *J. Phys. Chem.*, **85**, 3700 (1981).
20. I. E. Wachs, C. C. Chersich and J. H. Hardenbergh, *Appl. Catal.*, **13**, 335 (1985).
21. I. E. Wachs, F. D. Hardcastle and S. S. Chan, *Spectroscopy*, **1**, 30 (1986).
22. I. E. Wachs and F. D. Hardcastle, *Proc. 9th Intern. Congr. Catal.* (M. J. Phillips and M. Terman eds.), p.1449 (1988).
23. I. E. Wachs, G. Deo, D. S. Kim, M. A.Vuurnam and H. Hu, *Proc. 10th Int. Congr. Catal.*, (L. Guczi et al. eds), p.30, Elsevier, Amsterdam (1992).
24. J. A. Horsley, I. E. Wachs, J. M. Brown, G. H. Via and F. D. Hardcastle, *J. Phys. Chem.*, **91**, 4014 (1987).
25. F. Hilbrig, H. E. Gobel, H. Knözinger, H. Schmeiz and B. Lengeler, *J. Phys. Chem.*, **95**, 6973 (1991).

26. R. Thomas, E. M. van Oers, V. H. J. de Beer, J. Medema and J. A. Moulijn, *J. Catal.*, **76**, 241 (1982).
27. B. Scheffer, P. Molhoek and J. A. Moulijn, *Appl. Catal.*, **46**, 11 (1989).
28. Y. Yan, Q. X. S. Jiang and X. Guo, *J. Catal.*, **131**, 234 (1991).
29. R. Zhang, J. Jagiello, J. F. Hu, Z.-Q. Huang and J. A. Schwarz, *Appl. Catal. General.*, **84**, 123 (1992).
30. P. J. Mangnus, B. Scheffer and J. A. Moulijn, *Prepr. Am. Chem. Soc. Pet. Div.*, **32**, 329 (1987).
31. R. Thomas, M. C. Mietelmeijer-Hazeleger, F. P. J. M. Kerkhof, J. A. Moulijn, J. Medema and V. H. J. de Beer, *Proc. 3rd, Int. Conf. on Chemistry and Uses of Molybdenum* (H. G. Barry and P. C. H. Mitchell eds), p.85, Climax Molybdenum Company (1979).
32. R. Thomas, F. P. J. M. Kerkhof, J. A. Moulijn, J. Medema and V. H. J. de Beer, *J. Catal.*, **61**, 559 (1980).
33. S. S. Chan, I. E. Wachs, L. L. Murrell and N. C. Dispenziere Jr., *J. Catal.*, **92**, 1 (1985).
34. N.-S. Chiu, S. H. Bauer and M. F. L. Johnson, *J. Catal.*, **98**, 32 (1986).
35. S. M. A. M. Bouwens, R. Prins, V. H. J. de Beer and D. C. Koningsberger, *J. Phys. Chem.*, **94**, 3711 (1990).
36. S. M. A. M. Bouwens, J. A. R. van Veen, D. C. Koningsberger, V. H. J. de Beer and R. Prins, *J. Phys. Chem.*, **95**, 123 (1991).
37. S. P. A. Louwere and R. Prins, *J. Catal.*, **133**, 94 (1992).
38. P. Arnoldy, J. M. A. van den Heijkant, G. D. de Bok and J. A. Moulijn, *J. Catal.*, **92**, 35 (1985).
39. R. G. Dickinson and L. Pauling, *J. Amer. Chem. Soc.*, **45**, 1466 (1923).
40. R. H. Williams and A. J. Mcevoy, *J. Phys., D: Appl. Phys.*, **4**, 456 (1971).
41. J. C. Wildervanck and F. Jellinek, *Zeitschrift fur Anorganische und allgemeine Chemie Band*, **328**, 309 (1964).
42. R. J. H. Voorhoeve and J. C. M. Stuiver, *J. Catal.*, **23**, 243 (1971).
43. A. L. Farragher and P. Cossee in: *Proceedings, 5th International Congress on Catalysis*, Palm Beach, 1972 (J. W. Hightower ed.), p.1301, North-Holland, Amsterdam (1973).
44. H. Topsøe and B. S. Clausen, *Catal. Rev. - Sci. Eng.*, **26**, 395 (1984).
45. T. F. Hayden and J. A. Dumesic, *J. Catal.*, **103**, 366 (1987).
46. R. J. H. Voorhoeve and J. C. M. Stuiver, *J. Catal.*, **23**, 228 (1971).
47. J. Joffre, P. Geneste and D. A. Lerner, *J. Catal.*, **97**, 543 (1986).
48. M. Salmeron, G. A. Somorjai, A. Wold, R. R. Chianelli and K. S. Liang, *Chem. Phys. Lett.*, **90**, 105 (1982).
49. M. H. Farias, A. J. Gellman, G. A. Somorjai, R. R. Chianelli and K. S. Liang, *Surf. Sci.*, **140**, 181 (1984).
50. A. L. Farragher, *Adv. Colloid. Interface Sci.*, **11**, 3 (1979).
51. V. H. J. de Beer, T. H. M. van Sint Fiet, G. H. A. M. van der Steen, A. C. Zwaga and G. C. A. Schuit, *J. Catal.*, **35**, 297 (1974).
52. Y. Okamoto, H. Tomioka, T. Imanaka and S. Teranishi, *Proc. 7th Int. Congr. Catal.* (T. Seiyama and K. Tanabe eds.), p.616, Elsevier, Amsterdam (1980).
53. A. J. A. Konings, W. L. J. Brentjens, D. C. Koningsberger and de V. H. J. Beer, *J. Catal.*, **67**, 145 (1981).
54. C. Wivel, R. Candia, B. S. Clausen, S. Mørup and H. Topsøe, *J. Catal.*, **68**, 453 (1981).
55. P. Ratnasamy and S. Sivasanker, *Catal. Rev. Sci. Eng.*, **22**, 401 (1980).
56. H. Topsøe, B. S. Clausen, R. Candia, C. Wivel and S. Mørup, *J. Catal.*, **68**, 433 (1981).
57. A. N. Startsev, V. A. Burmistrov and Yu. I. Yermakov, *Appl. Catal.*, **45**, 191 (1988).
58. H. Topsøe, B. S. Clausen, N.-Y. Topsøe, J. Hyldtoft and J. K. Nørskov, *Prepr. Am. Chem. Soc. Div. Pet. Chem.*, **38** (3), 638 (1993).
59. H. Topsøe, R. Candia, N.-Y. Topsøe and B. S. Clausen, *Bull. Soc.Chim. Belg.*, **93**, 783 (1984).
60. B. Delmon in: *Proceedings, Climax 3rd International Conference on Chemistry and Uses of Molybdenum* (H. F. Barry and P.C. H. Mitchell eds.), p.73, Climax Molybdenum Co., Ann Arbor, Michigan (1979).
61. B. Delmon, *Bull. Soc. Chim. Belg.*, **88**, 979 (1979).
62. F. Delannay, P. Gajardo, P. Grange and B. Delmon, *J. C. S. Faraday Trans. I*, **76**, 988 (1980).
63. B. Delmon, *Catal. Lett.*, **22**, 1 (1993).
64. S. Eijsbouts, V. H. J. de Beer and R. Prins, *J. Catal.*, **109**, 217 (1988).
65. Y. Okamoto, H. Nakamo, T. Shimokawa, T. Imanaka and S. Teranishi, *J. Catal.*, **50**, 447 (1977).
66. J. K. Burdett and J. T. Chung, *Surf. Sci. Lett.*, **236**, L353 (1990).
67. J. K. Nørskov, B. S. Clausen and H. Topsøe, *Catal. Lett.*, **13**, 1 (1992).
68. R. J. H. Voorhoeve, *J. Catal.*, **23**, 236 (1971).
69. V.H. J. de Beer and G. C. A. Schuit in: *Preparation of Catalysts* (B. Delmon, P. A. Jacobs and G. Poncelet eds.), p.343, Elsevier, Amsterdam (1976).
70. R. R. Chianelli, T. A. Pecoraro, T. R. Halbert, W. -H. Pan and E. I. Stiefel, *J. Catal.*, **86**, 226 (1984).
71. P. R. Wentrcek and H. Wise, *J. Catal.*, **51**, 80 (1978).
72. S. Harris and R. R. Chianelli, *J. Catal.*, **86**, 400 (1984).
73. S. Harris and R. R. Chianelli, *J. Catal.*, **98**, 17 (1986).
74. W. Qian, A. Ishihara, Y. Okoshi, M. Godo and T. Kabe, *J. Chem. Soc., Faraday Trans.*, **93** (24), 4395 (1997).
75. T. Kabe, W. Qian, S. Ogawa and A. Ishihara, *J. Catal.*, **143**, 239 (1993).
76. T. Kabe, W. Qian and A. Ishihara, *J. Catal.*, **149**, 171 (1994).
77. T. Kabe, W. Qian and A. Ishihara, *Catal. Today*, **39**, 3 (1997).

78. W. Qian, A. Ishihara, G. Wang, T. Tsuzuki, M. Godo and T. Kabe, *J. Catal.*, **170**, 286 (1997).
79. M. J. Ledoux, O. Michaux, G. Agostini and P. Panissod, *J. Catal.*, **96**, 189 (1985).
80. V. H. J. de Beer, J. C. Duchet and R. Prins, *J. Catal.*, **72**, 369 (1981).
81. R. Burch and A. Collins, *J. Catal.*, **97**, 385 (1986).
82. H. Topsøe, B. S. Clausen, N.-Y. Topsøe, N. and E. Pedersen, *Ind. Eng. Chem. Fundam.*, **25**, 25 (1986).
83. F. Delannay, E. Haeussler and B. Delmon, *J. Catal.*, **66**, 469 (1980).
84. S. Kasztelan, J. Grimblot and J. P. Bonnelle, *J. Phys. Chem.*, **91**, 1503 (1987).
85. H. Topsøe and B. S. Clausen, *Appl. Catal.*, **25**, 273 (1986).
86. P. H. Bolt, F. H. P. M. Habraken and J. W. Geus, *Catal. Lett.*, **36**, 183 (1996).
87. G. Hagenbach, Ph. Courty and B. Delmon, *J. Catal.*, **23**, 295 (1971).
88. R. R. Chianelli, A. F. Ruppert, S. K. Behal, B. H. Kear, A. Wold and R. Kershae, *J. Catal.*, **92**, 56 (1985).
89. N.-Y. Topsøe and H. Topsøe, *J. Catal.*, **84**, 386 (1983).
90. S. Kasztelan, H. Toulhoat, J. Grimblot and J. P. Bonnelle, *Appl. Catal.*, **13**, 127 (1984).
91. T. R. Halbert, T. C. Ho, E. I. Stiefel, R. R. Chianelli and M. Daage, *J. Catal.*, **130**, 116 (1991).
92. J. M. J. G. Lipsch and G. C. A. Schuit, *J. Catal.*, **15**, 179 (1969).
93. F. E. Massoth, *J. Catal.*, **36**, 164 (1975).
94. M. Karroua, H. K. Matralis, P. Grange and B. Delmon, *J. Catal.*, **139**, 371 (1993).
95. J. M. Zabala, M. Mainil, P. Grange and B. Delmon, *React. Kinetic Catal. Lett.*, **3**, 285 (1975).
96. D. A. Goetsch, J. C. Carver, R. L. Seiver and W. H. Sawyer, *8th North American Meeting of the Catalysis Society*, Philadelphia, May 1-4 (1983).
97. B. Delmon, *ACS. Div. Pet. Chem. Prepr.*, **22**, 503 (1977).
98. M. Karroua, P. Grange and B. Delmon, *Appl. Catal.*, **50**, L3 (1989).
99. M. Karroua, A. Centeno, H. K. Matralis, P. Grange and B. Delmon, *J. Catal.*, **51**, L21 (1989).
100. Y. W. Li and B. Delmon, *Stud. Surf. Sci. Catal.*, **112**, 349 (1997).
101. Y. W. Li and B. Delmon, *J. Mol. Catal. A: Chem.*, **127** (1-3), 163 (1997).
102. R. Iwamoto, K. Inamura, T. Nozaki and A. Iino, *Appl. Catal., A*, **163** (1-2), 217 (1997).
103. H. Topsøe, B. S. Clause, N.-Y. Topsøe and P. Zeuthen, *Stud. Surf. Sci. Catal.*, **53**, 77 (1989).
104. T. G. Parham and R. P. Merrill, *J. Catal.*, **85**, 295 (1984).
105. N.-S. Chiu, M. F. L. Johnson and S. H. Bauer, *J. Catal.*, **113**, 281 (1988).
106. S. M. A. M. Bouwens, D. C. Koningsberger, V. H. J. de Beer and R. Prins, *Catal. Lett.*, **1**, 55 (1988).
107. M. Boudart, J. S. Arrieta and R. Dalla Betta, *J. Ame. Chem. Soc.*, **105**, 6501 (1983).
108. Yu. I. Yermakov, A. N. Startsev, V. A. Burmistrov, O. N. Shumilo and N. N. Bulgakov, *Appl. Catal.*, **18**, 33 (1985).
109. H. Wise, *Proc. 2nd Int. Conf. Chem. Uses of Molybdenum* (P. C. H. Mitchell and A. Seaman eds.), p.160, Climax Molybdenum Company (1976).
110. R. W. Phillips and A. A. Fote, *J. Catal.*, **41**, 168 (1976).
111. J. P. R. Vissers, B. Scheffer, V. H. J. de Beer, J. A. Moulijn, J. A. and R. Prins, *J. Catal.*, **105**, 277 (1987).
112. A. M. van der Kraan, M. W. J. Craje, E. Gerkema, W. L. T. M. Ramselaar and V. H. J. de Beer, *Appl Catal.*, **39**, 7 (1988).
113. A. M. van der Kraan, *Hyperfine Interactions*, **40**, 211 (1988).
114. A. M. van der Kraan, M. W. J. Craje, E. Gerkema, W. L. T. M. Ramselaar and V. H. J. de Beer, *Hyperfine Interactions*, **46**, 567 (1989).
115. M. W. J. Craje, S. P. A. Louwers, V. H. J. de Beer, R. Prins and A. M. van der Kraan, *J. Phys. Chem.*, **96**, 5445 (1992).
116. M. J. Ledoux, C. P. Huu, Y. Segura and F. Luck, *J. Catal.*, **121**, 70 (1990).
117. M. J. Ledoux, *J. Chem. Soc., Faraday Trans. 1*, General Discussion, **83**, 2169 (1987).
118. W. K. Hall, *Proc. 4th Int. Conf. on the Chem. and Uses of Molybdenum*, p.224, Barry, Climmax Molybdenum Company Ltd (1982).
119. H. Knözinger, *Proc. 9th Int. Congr. Catal.*, (M. J. Phillips and M. Ternan eds.), p.20 (1988).
120. M. Zdrazil, *Catal. Today*, **3**, 269 (1988).
121. A. N. Startsev, *Catal. Rev-Sci. Eng.*, **37**, 353 (1995).
122. L. Karakonstantis, H. Matralis, Ch. Kordulis and A. Lycourghiotis, *J. Catal.*, **162** (2), 306 (1996).
123. F. J. Gil-Llambias and A. L. Agudo, *J. Mat. Soc. A.*, **17**, 936 (1982).
124. C. V. Caceres and J. L. G. Fierro, *Appl. Catal.*, **10**, 333 (1984).
125. G. M. Ismail, M. I. Zaki, G. C. Bond and R. Shukri, *Appl. Catal.*, **72**, L1 (1991).
126. M. A. Apecetche, M. Houalla and B. Delmon, *Surface and Interface Analysis*, **3**, 90 (1981).
127. L. Rodrigo, K. Marcinkowska, A. Adnot, P. C. Roberge, S. Kaliaguine, J. M. Stencel, L. E. Makovsky and J. R. Diehl, *J. Phys. Chem.*, **90**, 2690 (1986).
128. M. N. Blanco, C. V. Caceres, J. L. G. Fierro and H. J. Thomas, *Appl. Catal.*, **33**, 231 (1987).
129. W. C. Cheng and C. J. Pereira, *Appl. Catal.*, **33**, 331 (1987).
130. J. L. G. Fierro, J. M. Asua, P. Grange and B. Delmon, *Am. Chem. Soc., Pet. Div. Prepr.*, **32**, 271 (1987).

131. G. L. M. Souza, A. C. B. Santos, D. A. Lovate and A. C. Faro Jr., *Catal. Today*, **5**, 451 (1989).
132. A. M. Turek and I. E. Wachs, *J. Phys. Chem.*, **96**, 5000 (1992).
133. M. de Boer, R. G. Leliveld, A. J. van Dillen and J. W. Geus, *Appl. Catal. A General*, **102**, 35 (1993).
134. E. Hillerova and M. Zdrazil, *Appl. Catal.*, **A**, **138** (1), 13 (1996).
135. R. Candia, N.-Y. Topsøe, B. S. Clausen, C. Wivel, R. Nevald, S. Mørup and H. Topsøe in: *Proc. 4th Int. Conf. on the Chemistry and Uses of Molybdenum* (H. F. Barry and P. C. H. Mitchell eds), p.374, Climax Molybdenum Co Ltd. (1982).
136. P. K. Rao, V. V. D. N. Prasad, K. S. Rao and K. V. R. Chary, *J. Catal.*, **142**, 121 (1992).
137. M. A. Goula, Ch. Kordulis, A. Lycourghiotis and J. L. G. Fierro, *J. Catal.*, **137**, 285 (1992).
138. J. A. R. van Veen, H. de Wit, C. A. Emeis and P. A. J. M. Hendriks, *J. Catal.*, **107**, 579 (1987).
139. C. T. J. Mensch, J. A. R. van Veen, B. van Winderden and M. P. van Dijk, *J. Phys. Chem.*, **92**, 4961 (1988).
140. S. D. Kohler, J. G. Ekerdt, D. S. Kim and I. E. Wachs, *Catal. Lett.*, **16**, 231 (1992).
141. F. J. Gil-Llambias, J. Salvatierra, L. Bouyssieres, M. Escudey and R. Cid, *J. Catal.*, **59**, 185 (1990).
142. Y. Okamoto, *Stud. Surf. Sci. Catal.*, **100**, 77 (1996).
143. A. Griboval, P. Blanchard, E. Payen, M. Fournier and J. L. Dubois, *Stud. Surf. Sci. Catal.*, **106**, 181 (1997).
144. M. Adachi, C. Contescu and J. A. Schwarz, *J. Catal.*, **162** (1), 66 (1996).
145. T. Shimizu, T. Kiyohara, K. Hiroshima and M. Yamada, *Sekiyu Gakkaishi*, **39** (2), 158 (1996) [in Japanese].
146. L. Medici and R. Prins, *J. Catal.*, **163** (1), 38 (1996).
147. K. Hiroshima, T. Mochizuki, T. Honma, T. Shimizu and M. Yamada, *Appl. Surf. Sci.*, **121/122**, 433 (1997).
148. L. Vordonis, P. G. Koutsoukos and A. Lycoughiotis, *Langmuir*, **2**, 281 (1986).
149. L. Vordonis, P. G. Koutsoukos and A. Lycoughiotis, *J. Catal.*, **101**, 186 (1986).
150. L. Vordonis, A. Akratopulu, P. G. Koutsoukos and A. Lycoughiotis, *Preparation of Catalysts VI* (G. Poncelet, P. Grange and P. A. Jacobs eds.), p.309, Elsevier, Amsterdam (1987).
151. J. A. R. van Veen, E. Gerkema, A. M. van der Kraan and A. Knoester, *J. Chem. Soc. Commun.*, 1684 (1987).
152. J. A. R. van Veen, H. A. Colijn, P. A. J. M. Hendriks and A. J. van Welsenes, *Fuel Proc. Tech.*, **35**, 137 (1993).
153. C. P. Li and D. M. Hercules, *J. Phys. Chem.*, **88**, 456 (1984).
154. C. Wivel, B. S. Clausen, R. Candia and H. Topsøe, *J. Catal.*, **87**, 497 (1984).
155. T. Kameoka, T. Sato, Y. Yoshimura, H. Shimada, N. Matsubayashi, M. Imamura and A. Nishijima, *Sekiyu Gakkaishi*, **39** (2), 87 (1996) [in Japanese].
156. M. F. L. Johnson, P. V. Andrew, S. H. Bauer and N.-S. Chiu, *J. Catal.*, **98**, 51 (1986).
157. N.-Y. Topsøe and H. Topsøe, *J. Catal.*, **139**, 631 (1993).
158. J. W. Newsome, H. W. Heiser, A. S. Russell and H. C. Stumpf, *Alumina Properties*, Alumina Co. of Am., Pittsburgh (1960).
159. C. Kittel, *Introduction to Solid State Physics*, John Wiley and Sons, New York, p. 550 (1971).
160. B. Scheffer, P. Arnoldy and J. A. Moulijn, *J. Catal.*, **112**, 516 (1988).
161. M. Dobrovolszky, Z. Paál and P. Tétényi, *Catal. Today*, **9**, 113 (1991).
162. R. Candia, H. Topsøe and B. S. Clausen in: *Proceedings of the 9th Iberonoamerican Symposium on Catalysis*, Lisbon, Portugal, p.211 (1984).
163. P. Faye, E. Payen and D. Bougeard, *Stud. Surf. Sci. Catal.*, **106**, 281 (1997).
164. J. A. R. van Veen, P. A. J. M. Hendriks, E. J. G. M. Romes and R. R. Andrea, *J. Phys. Chem.*, **94**, 5275 (1990).
165. C. T. J. Mensch, J. A. R. van Veen, B. van Wingerden and M. P. van Dijk, *J. Phys. Chem.*, **92**, 4961 (1988).
166. Y. Okamoto and H. Katsuyama, *Stud. Surf. Sci. Catal.*, **101**, 503 (1996).
167. G. C. Stevens and T. Edmonds in: *Preparation of Catalysts II*, (B. Delmon, P. Grange, P. Jacobs and G. Poncelet eds.), p.507, Elsevier, Amsterdam (1979).
168. A. J. Bridgewater, R. Burch and P. C. H. Michell, *Appl. Catal.*, **4**, 267 (1982).
169. J. T. Trawczynski, J. Walendziewski and M. Kulazynski in *"Proceedings ICCS '97"*, **Vol. 3**, 1815 (1997).
170. G. Muralidhar, F. E. Massoth and J. Shabtai, *J. Catal.*, **85**, 44 (1984).
171. M. Breysse, J. L. Portefaix and M. Vrinat, *Catal. Today*, **10**, 489 (1991).
172. Y. Okamoto and H. Katsuyama, *Ind. Eng. Chem. Res.*, **35** (6), 1834 (1996).
173. H. Shimada, S. Yoshitomi, T. Sato, N. Matsubayashi, M. Imamura, Y. Yoshimura and A. Nishijima, *Stud. Surf. Sci. Catal.*, **106**, 115 (1997).
174. W. J. J. Welters, G. Vorbeck, H. W. Zandbergen, L. J. M. van de Ven, E. M. van Oers, J. W. de Haan, V. H. J. de Beer and R. A. van Santen, *J. Catal.*, **161** (2), 819 (1996).
175. I. Wang and R. C. Chang, *J. Catal.*, **117**, 266 (1989).
176. F. P. Daly, *J. Catal.*, **116**, 600 (1989).
177. P. Atanasova, T. Tabakova, Ch. Vladov, T. Halachev and A. Lopez Agudo, *Appl. Catal.*, *A*, **161** (1-2), 105 (1997).
178. A. Benitez, J. Ramirez, J. L. G. Fierro and A. A. Lopez Agudo, *Appl. Catal.*, *A*, **144** (1-2), 343 (1996).
179. D. Li, T. Sato, M. Imamura, H. Shimada and A. Nishijima, *J. Catal.*, **170**, 357 (1997).
180. A. Benitez, J. Ramirez, A. Vazquez, D. Acosta and A. Lopez Agudo, *Appl. Catal.*, *A*, **133** (1), 103 (1995).
181. M. J. Ledoux, A. Peter, E. A. Blekkan and F. Luck, *Appl. Catal.*, *A*, **133** (2), 321 (1995).

182. K. Segawa, M. Katsuta and F. Kameda, *Catal. Today*, **29** (1-4), 215 (1996).
183. A. Gutierrez-Alejandre, J. Ramirez and G. Busca, *Langmuir*, **14** (3), 630 (1998).
184. C. Pophal, F. Kameda, K. Hoshino, S. Yoshinaka and K. Segawa, *Catal. Today*, **39** (1-2), 21 (1997).
185. E. Olguin, M. Vrinat, L. Cedeno, J. Ramirez, M. Borque and A. Lopez Agudo, *Appl. Catal.*, *A*, **165** (1-2), 1 (1997).
186. R. R. Chianelli, *Catal. Rev.-Sci. Eng.*, **26** (3 & 4), 361 (1984).
187. R. Prins, V. H. J. de Beer and G. A. Somorjai, *Catal. Rev.-Sci. Eng.*, **31** (1 & 2), 1 (1989).
188. B. S. Clausen, H. Topsøe, R. Candia, J. Villadsen, B. Lengeler, J. Als-Nielsen and F. Christensen, *J. Phys. Chem.*, **85**, 3868 (1982).
189. T. F. Hayden and J. A. Dumesic, *J. Catal.*, **103**, 366 (1987).
190. T. Kabe, S. Yamadaya, M. Oba and Y. Miki, *Int. Chem. Eng.*, **12**, 366 (1972).
191. R. Prada Silvy, J. M. Beuken, P. Bertrand, B. K. Hodnett, F. Delannay and B. Delmon, *Bull. Soc. Chim. Belg.*, **93**, 775 (1984).
192. B. Scheffer, J. C. M. de Jonge, P. Arnoldy and J. A. Moulijn, *Bull. Soc. Chim. Belg.*, **93**, 751 (1984).
193. B. Scheffer, E. M. van Oers, P. Arnoldy, V. H. J. de Beer and J. A. Moulijn, *Appl. Catal.*, **25**, 303 (1986).
194. H. R. Reinoudt, A. D. van Langeveld, R. Mariscal, V. H. J. de Beer, J. A. R. van Veen, S. T. Sie and J. A. Moulijn, *Stud. Surf. Sci. Catal.*, **106**, 263 (1997).
195. A. M. de Jong, H. J. Borg, L. J. van Ijzendoorn, V. G. F. M. Soudant, V. H. J. de Beer, J. A. R. van Veen and J. W. Niemantsverdriet, *J. Phys. Chem.*, **97**, 6477 (1993).
196. M. de Boer, A. J. van Dillen, D. C. Koningsberger, J. W. Geus, *Jpn. Appl. Phys.*, Vol. 32, Suppl., **32** (2), 460 (1992).
197. M. Lojacono, J. L. Verbeek, G. C. A. Schuit in: *Proc. 5th Int. Congr. Catal.* (J. W. Hightower ed.), **2**, p.1409, North-Holland, Amsterdam (1972).
198. M. Lojacono, J. L. Verbeek and G. C. A. Schuit, *J. Catal.*, **29**, 463 (1973).
199. K. S. Seshadri, F. E. Massoth and L. Petrakis, *J. Catal.*, **124**, 416 (1970).
200. P. J. Mangnus, A. Bos and J. A. Moulijn, *J. Catal.*, **146**, 437 (1994).
201. T. G. Parham and R. P. Merrill, *J. Catal.*, **85**, 295 (1984).
202. B. S. Clausen, G. Steffensen, T. B. Zunic and H. Topsøe, HASYLAB, Annual Report (1994).
203. Th. Weber, J. C. Muijsers, J. H. M. C. van Wolput, C. P. J. Verhagen and J. W. Niemantsverdriet, *J. Phys. Chem.*, **100**, 14144 (1996).
204. R. Candia, B. S. Clausen, J. Barholdy, N.-Y. Topsøe, B. Lengeler and H. Topsøe, *Proc. 8th Int. Congr. Catal.*, Berlag. Chemie, Weinheim, Vol. II, 375 (1984).
205. E. Payen, S. Kasztelan, S. Houssenbay, R. Szymanski and J. Grimblot, *J. Phys. Chem.*, **93**, 6501 (1989).
206. N.-Y. Topsøe and H. Topsøe, *J. Catal.*, **77**, 293 (1982).
207. R. Candia, B. S. Clausen and H. Topsøe, Proc. 9th Iberoamerican Symposium on Catalysis, Lisbon, Portugal, 211 (1984).
208. R. Prada Silvy, P. Grange and B. Delmon, *Stud. Surf. Sci. Catal.*, **53**, 233 (1990).
209. J. A. R. van Veen, E. Berkema, A. M. van der Kraan, P. A. J. M. Hendriks and H. Beens, *J. Catal.*, **133**, 112 (1992).
210. C. H. Kim, W. L.Yoon, I. C. Lee and S. I. Woo, *Appl. Catal.*, *A*, **144** (1-2), 159 (1996).
211. F. Mauge, J. C. Duchet, J. C. Lavalley, S. Houssenbay, E. Payen and J. Grimblot, *Catal. Today*, **10**, 561 (1991).
212. S. Kasztelan, E. Payen, H. Toulhoat, J. Grimblot and J. P. Bonnelle, *Polyhedron.*, **5**, 157 (1986).
213. A. Morales, M. M. Ramirez and M. M. R. de Agudelo, *Appl. Catal.*, **23**, 23 (1986).
214. D. Chadwick, D. W. Aitchison, R. Badilla-Ohlbaum and L. Josefsson in: *Preparation of Catalysts III*, (G. Poncelet, P. Grange and P. A. Jacobs eds.), p.323, Elsevier, Amsterdam (1982).
215. S. Kasahara, Y. Udagawa and M. Yamada, *Appl. Catal. B*, **12** (2-3), 225 (1997).
216. H. Kwart, G. C. A. Schuit and B. C. Gates, *J. Catal.*, **61**, 128 (1980).
217. T. Kabe, A. Ishihara and Q. Zhang, *Appl. Catal. A*, **97**, L1 (1993).
218. T. Kabe, W. Qian, H. Tanihata, A. Ishihara and M. Godo, *J. Chem. Soc., Faraday Trans.*, **93** (20), 3709 (1997).
219. W. Qian, Q. Zhang, Y. Okoshi, A. Ishihara and T. Kabe, *J. Chem. Soc., Faraday Trans.*, **93** (9), 1821 (1997).
220. H. Hallie, *Oil Gas J.*, **80**, 69 (1982).
221. J. van Gestel, J. Leglise and J.-C. Duchet, *J. Catal.*, **145**, 429 (1994).
222. T. Kabe, A. Ishihara, W. Qian and M. Godo, *J. Catal.*, [to be published].
223. T. Kabe, A. Ishihara, W. Qian and M. Godo, *Catal. Today*, **45**, 285 (1998).
224. C. A. Henriques, A. M. P. Bentes Jr., R. Frety and M. Schmal, *Catal. Today*, **5**, 443 (1989).
225. J. Miciukiewicz and F. E. Massoth, *J. Catal.*, **119**, 531 (1989).
226. J. C. Muijsers, Th. Weber, R. M. van Hardeveld, H. W. Zandbergen and J. W. Niemantsverdriet, *J. Catal.*, **157**, 698 (1995).
227. A. M. de Jong, V. H. J. de Beer, J. A. R. van Veen and J. W. Niemantsverdriet, *J. Phys. Chem.*, **100**, 17722

(1996).
228. M. de Boer, A. J. van Dillen, D. C. Koningsberger and J. W. Geus, *J. Phys. Chem.*, **98**, 7862 (1994).
229. F. Delannay, *Appl. Catal.*, **16**, 135 (1985).
230. W. Qian, A. Ishihara, S. Ogawa and T. Kabe, *J. Phys. Chem.*, **98**, 907 (1994).
231. T. I. Korányi, I. Manninger and Z. Paál, *Solid State Ionics*, **32/33**, 1012 (1989).
232. A. M. de Jong, V. H. J. de Beer, J. A. R. van Veen and J. W. Niemantsverdriet, *J. Vac. Sci. Technol. A*, **15** (3), 1592 (1997).
233. H. R. Lukens, R. G. Meisenheimer and J. N. Wilson, *J. Phys. Chem.*, **66**, 496 (1962).
234. K. A. Pavlova, B. D. Panteleea, E. N. Deryagina and I. V. Kalechits, *Kinet. Katal.*, **6** (3), 493 (1965).
235. I. V. Kalechits and E. N. Deryagina, *Kinet. Katal.*, **8** (3), 604 (1969).
236. C. G. Gachet, E. Dhainaut, L. de Mourgnes, J. P. Candy and P. Fouillous, *Bull. Soc. Chim. Belg.*, **90**, 1279 (1981).
237. M. Dobrovolszky, Z. Paál and P. Tétényi, *Appl. Catal. A*, **142**, 159 (1996).
238. M. Dobrovolszky, P. Tétényi and Z. Paál, *Chem. Eng. Commun.*, **83**, 1 (1989).
239. M. Dobrovolszky, K. Matusek, Z. Paál and P. Tétényi, *J. Chem. Soc., Faraday Trans.*, **89** (16), 3137 (1993).
240. G. V. Isagulyants, A. A. Greish and V. M. Kogan, *Proc. 9th Internal. Congr. on Catal.*, (M. J. Phillips and M. Ternan eds.), **Vol. 1**, p.35, Calgary (1988).
241. K. C. Campbell, M. L. Mirza, S. J. Thomson and G. Webb, *J. Chem. Soc., Faraday Trans. I*, **80**, 1689 (1984).
242. J. Freel, J. G. Larson and J. F. Adams, *J. Catal.*, **96**, 544 (1985).
243. A. J. Gellman, M. E. Bussell and G. A. Somorjai, *J. Catal.*, **107**, 103 (1987).
244. A. Ishihara, T. Itoh, T. Hino, P. Qi and T. Kabe, *J. Catal.*, **140**, 184 (1993).
245. T. Kabe, A. Ishihara, Q. Zhang, H. Tsutsui and H. Tajima, *J. Jpn., Petrol. Inst.*, **36** (6), 467 (1993).
246. W.Qian in: *Ph. D. Dissertation, Tokyo University of Agriculture and Technology*, Tokyo (1995).
247. V. H. J. de Beer, T. H. M. van Sint Fiet, G. H. A. M. van der Steen, A. C. Zwaga and G. C. A. Schuit, *J. Catal.*, **35**, 297 (1974).
248. G. Kullerud and R. A. Yund, *J. Petrol.*, **3**, 126 (1962).
249. C. B. Alcock in: *Principles of Pyrometallurgy*, Academic Press, London/New York (1976).
250. A. Wold, *Adv. Chem. Ser.*, **98**, 17 (1971).
251. T. Sparks and T. Komoto, *Rev. Modern Phys.*, **40**, 752 (1968).
252. R. H. Williams and A. J. Mcevoy, *J. Phys., D: Appl. Phys.*, **4**, 456 (1971).
253. N.-Y. Topsøe and H. Topsøe, *J. Catal.*, **139**, 641 (1993).
254. B. S. Clausen, B. Lengeler, R. Candia, J. Als-Nielsen and H. Topsøe, *Bull. Soc. Chim. Belg.*, **90**, 1249 (1981).
255. H. Topsøe, B. S. Clausen, N.-Y. Topsøe, E. Pedersen, W. Niemann, A. Müller, H. Bögge and B. Lengeler, *J. Chem. Soc., Faraday Trans. I*, **83**, 2157 (1987).
256. H. Topsøe, B. S. Clausen, N.-Y. Topsøe, J. K. Nørskov, C. V. Ovesen and C. J. H. Jacobsen, *Bull. Soc. Chim. Belg.*, **104**, 283 (1995).
257. S. Eijsbouts, V. H. J. de Beer and R. Prins, *J. Catal.*, **127**, 619 (1991).
258. H. Topsøe in: *Proceedings of the NATO Advanced Study Institute on Surface Properties and Catalysis by Non-Metals: Oxides, Sulfides and Other Transition Metal Compounds*, 1982 (J. P. Bonnelle, B. Delmon and E. Derouane eds.), p.329, Reidel, Dordrecht (1983).
259. N.-Y. Topsøe, *J. Catal.*, **64**, 235 (1980).
260. J. Maternova, *Appl. Catal.*, **3**, 3 (1982).
261. E. Payen, S. Kasztelan and J. Grimblot, *J. Mol. Struct*, **174**, 71 (1988).
262. F. Ruette and E. V. Ludena, *J. Catal.*, **67**, 266 (1981).
263. M. Sugioka, F. Sado, Y. Matsumoto and N. Maesaki, *Catal. Today*, **29** (1-4), 255 (1996).
264. C. Jan, T. Lin and J. Chang, *Ind. Eng. Chem. Res.*, **35** (11), 3893 (1996).
265. S. A. De Leon, P. Grange and B. Delmon, *Catal. Lett.*, **47** (1), 51 (1997).
266. M. J. Girgis and B. C. Gates, *Ind. Eng. Chem. Res.*, **30**, 2021 (1991).
267. T. Kabe, W. Qian and A. Ishihara, *J. Phys. Chem.*, **98**, 912 (1994).
268. V. Lavopa and C. N. Scatterfield, *Energy & Fuels*, **1**, 323 (1987).
269. M. V. Bhinde, Ph. D. Dissertation, University of Delaware, Newark (1979).
270. J. C. Grimblot, P. Dufresne, L. Gengembre and J. P. Bonnelle, *Bull. Soc. Chim. Belg.*, **90**, 1261 (1981).
271. S. S. Pollack, J. V. Sanders and R. E. Tischer, *Appl. Catal.*, **8**, 383 (1983).
272. G. C. Stevens and T. Edmonds, *Proc. 2nd Int. Conf. Chem. Uses of Molybdenum* (P. C. H. Mitchell and A. Seaman eds.), p.155, Climax Molybdenum Company (1976).
273. S. Suvasanker, A. V. Ramaswamy and P. Ratnasamy, *Proc. 3rd Int. Conf. Chem. Uses of Molybdenum"* (P. C. H. Mitchell and A. Seaman eds.), p.98, Climax Molybdenum Company (1979).
274. S. J. Tauster, T. A. Pecoraro and R. R. Chianelli, *J. Catal.*, **63**, 515 (1980).
275. O. P. Bahl, E. L. Evans and J. M. Thomas, *Proc. Royal Soc. London Ser. A*, **386**, 53 (1968).
276. B. G. Silbernagel, T. A. Pecoraro and R. R. Chianelli, *J. Catal.*, **78**, 380 (1982).
277. E. G. Derouane, E. Pedersen, B. S. Clausen, Z. Gabelica and H. Topsøe, *J. Catal.*, **107**, 587 (1987).

278. C. B. Roxlo, M. Daage, A. F. Ruppert and R. R. Chianelli, *J. Catal.*, **108**, 176 (1986).
279. A. Sachdev, J. Lindner, Schwank and J. M. A. V. Garcia, *J. Solid State Chem.*, **87**, 378 (1990).
280. K. S. Chung and F. E. Massoth, *J. Catal.*, **64**, 332 (1980).
281. J. Bachelier, J. C. Duchet and D. Cornet, *Bull. Soc. Chim. Belg.*, **90**, 1301 (1981).
282. J. P. R. Vissers, C. K. Groot, E. M. van Oers, V. H. J. de Beer and R. Prins, *Bull. Soc. Chim. Belg.*, **93**, 813 (1984).
283. N. K. Nag, K. S. P. Rao, K. V. R. Chary, B. R. Rao and V. S. Subrahmanyam, *Appl. Catal.*, **41**, 165 (1988).
284. K. Ramanathan and S. W. Weller, *J. Catal.*, **95**, 249 (1985).
285. W. Zmierczak, G. Muralidhar and F. E. Massoth, *J. Catal.*, **77**, 432 (1982).
286. R. Burch and A. Collins, *Appl. Catal.*, **17**, 273 (1985).
287. B. N. Reddy and V. S. Subrahmanyam, *Appl. Catal.*, **27**, 1 (1986).
288. T. A. Bodrero and C. H. Bartholomew, *J. Catal.*, **84**, 145 (1983).
289. J. L. G. Fierro, L. G. Tejuca, A. L. Agudo and S. W. Weller, *J. Catal.*, **89**, 111 (1984).
290. J. Lindner, A. Sachdev, J. Schwank and M. Villa-Garcia, *J. Catal.*, **135**, 427 (1992).
291. J. Lindner, A. Sachdev, J. Schwank and M. Villa-Garcia, *J. Catal.*, **137**, 333 (1992).
292. J. Valyon, R. L. Schneider and W. K. Hall, *J. Catal.*, **85**, 277 (1984).
293. Y. Okamoto, Y. Katoh, Y. Mori, T. Imanaka and S. Teranishi, *J. Catal.*, **70**, 445 (1981).
294. S.-J. Moo and S.-K. Ihm, *Appl. Catal.*, **42**, 307 (1988).
295. J. Miciukiewicz, Q. Qaderm and F. E. Massoth, *Appl. Catal.*, **49**, 247 (1989).
296. J. Miciukiewicz, W. Zmierczak and F. E. Massoth, *Bull. Soc. Chim. Belg.*, **96**, 915 (1987).
297. V. I. Yerofeyev and I. V. Kaletchits, *J. Catal.*, **86**, 55 (1984).
298. Y. Okamoto, H. Tomioka, T. Imanaka and S. Teranishi, *Chem. Lett.*, **381** (1979).
299. G. M. K. Abotsi, A. W. Scaroni and F. J. Derbyshire, *Appl. Catal.*, **37**, 93 (1988).
300. A. Shuxian, W. K. Hall, G. Ertl and H. Knözinger, *J. Catal.*, **108**, 167 (1986).
301. F. E. Massoth and C. L. Kibby, *J. Catal.*, **47**, 316 (1977).
302. Y. Okamoto, T. Imanaka and S. Teranishi, *J. Phys. chem.*, **85**, 3798 (1981).
303. V. H. J. de Beer, C. Bevelander, T. H. M. van Sint Fiet, P. G. A. J. Werter and C. H. Amberg, *J. Catal.*, **43**, 68 (1976).
304. Y. Okamoto, H. Tomioka, T. Imanaka and S. Teranishi, *J. Catal.*, **57**, 153 (1980).
305. Y. Okamoto, H. Tomioka, Y. Katoh, T. Imanaka and S. Teranishi, *J. Phys. Chem.*, **84**, 1833 (1980).
306. B. Scheffer, N. J. J. Dekker, P. J. Mangnus and J. A. Moulijn, *J. Catal.*, **121**, 31 (1990).
307. J. Ramirez and A. Gutierrez-Alejandre, *J. Catal.*, **170**, 108 (1997).
308. K. C. Pratt, J. V. Sanders and V. Cristov, *J. Catal.*, **124**, 416 (1990).
309. J. M. Solar, F. J. Derbyshire, V. H. J. de Beer and L. R. Radovic, *J. Catal.*, **129**, 330 (1991).
310. J. Ramirez, L. Ruiz-Ramirez, L. Cedeno, V. Harle, M. Vrinat and M. Breysse, *Appl. Catal. A.*, **93**, 163 (1993).
311. D. Hamon, M. Vrinat, M. Breysse, B. Durand, F. Beauchesne and T. des Courieres, *Bull. Soc. Chim. Belg.*, **108**, 933 (1991).
312. J. C. Duchet, M. J. Tilliette and D. Cornet, *Catal. Today*, **10**, 579 (1991).
313. D. Hamon, M. Vrinat, M. Breysse, B. Durand, M. Jebrouni, M. Roubin, P. Magnoux and T. des Courieres, *Catal. Today*, **10**, 613 (1991).
314. K. S. Rao, H. Ramakrishna and G. M. Dhar, *J. Catal.*, **133**, 146 (1992).
315. F. Luck, *Bull. Soc. Chim. Belg.*, **108**, 781 (1991).
316. S. M. A. M. Bouwens, N. Barthe-Zahir, V. H. J. de Beer and R. Prins, *J. Catal.*, **131**, 326 (1991).
317. P. Da Silva, N. Marchal and S. Kasztelan, *Stud. Surf. Sci. Catal.*, **106**, 353 (1997).
318. J. W. Gosselink, H. Schaper, J. P. de Jonge and W. H. J. Stork, *Appl. Catal. A*, **32**, 337 (1987).
319. B. S. Parekh and S. W. Weller, *J. Catal.*, **47**, 100 (1977).
320. B. S. Parekh and S. W. Weller, *J. Catal.*, **55**, 58 (1978).
321. W. S. Millman and W. K. Hall, *J. Catal.*, **59**, 311 (1979).
322. H.-C. Liu and S. W. Weller, *J. Catal.*, **66**, 65 (1980).
323. A. L. Agudo, F. J. Gil-Llambias, P. Reyes and J. L. G. Fierro, *Appl. Catal.*, **1**, 59 (1981).
324. J. Uchtil, L. Berranek, L. Zakradnikova and M. Kraus, *Appl. Catal.*, **4**, 233 (1982).
325. N.-Y. Topsøe, H. Topsøe, O. Sørensen, B. S. Clausen and R. Candia, *Bull. Soc. Chim. Belg.*, **93**, 727 (1984).
326. A. L. Agudeo, F. J. Gil-Llambias, J. M. D. Tascon and J. L. G. Fierro, *Bull. Soc. Chim. Belg.*, **93**, 719 (1984).
327. R. Prada Silvy, F. Delannay, P. Grange and B. Delmon, *Polyhedron.*, **5**, 195 (1986).
328. R. Burch and A. Collins in: *Proc. 4th Int. Conf. on the Chemistry and Uses of Molybdenum* (H. F. Barry, P. C. H. Mitchell eds.), p.374, Climax Molybdenum Co Ltd. (1982).
329. W. S. Millman, C. H. Bartholomew and R. L. Richardson, *J. Catal.*, **90**, 10 (1984).
330. S. J. Tauster and K. L. Riley, *J. Catal.*, **67**, 250 (1981).
331. S. J. Tauster and K. L. Riley, *J. Catal.*, **70**, 230 (1981).
332. T. A. Bodrero, C. H. Bartholomew and L. C. Pratt, *J. Catal.*, **78**, 253 (1982).
333. Y. Liu, F. E. Massoth and J. Shabtai, *Bull. Soc. Chim. Belg.*, **93**, 627 (1984).

334. J. Laine, F. Severino, C. V. Caceres, J. L. G. Fierro and A. L. Agudo, *J. Catal.,* **103**, 228 (1987).
335. J. Bachelier, J. C. Duchet and D. Cornet, *J. Catal.,* **87**, 283 (1984).
336. B. I. Parsons, M. Ternam in: *Proc. 6th Int. Congr. Catal.* (G. C. Bond, P. B. Wells and F. C. Tompkins eds.), p.965, London (1977).
337. C. W. Colling, J.-G. Choi and I. T. Thompson, *J. Catal.,* **160**, 35 (1996).
338. S. I. Kim and S. I. Woo, *Appl. Catal.,* **74**, 109 (1991).
339. B. R. G. Leliveld, J. A. J. van Dillen, J. W. Geus, D. C. Koningsberger and M. de Boer, *J. Phys. Chem. B,* **101** (51), 11160 (1997).
340. A. Nishijima, S. Yoshitomi, H. Shimada, Y. Yoshimura, T. Sato and N. Matsubayashi, *Proc. 9th Int. Congr. Catal.* (M. J. Phillips and M. Ternay eds.), p.173 The Chem. Inst. of Canada, Ontario (1988).
341. G. L. Schrader and C. P. Cheng, *J. Catal.,* **49**, 247 (1983).
342. F. E. Massoth and P. Zeuthen, *J. Catal.,* **145**, 216 (1994).
343. R. Candia, B. S. Clausen and H. Topsøe, *J. Catal.,* **77**, 564 (1982).
344. Q. Zhang, W. Qian, S. Oshima, A. Ishiahara and T. Kabe, *J. Jpn., Petrol. Inst.,* **40** (5), 408 (1997).
345. B. C. Gates and H. Topsøe, *Polyhedron,* **16**, 3213 (1997).
346. P. T. Vasudevan and J. L. G. Fierro, *Catal. Rev. -Sci. Eng.,* **38** (2), 161 (1996).
347. J. A. Rodriguez, *Polyhedron,* **16** (18), 3177 (1997).
348. C. M. Friend and D. A. Chen, *Polyhedron,* **16** (18), 3165 (1997).
349. K. J. Weller, P. A. Fox, S. D. Gray and D. E. Wigley, *Polyhedron,* **16** (18), 3139 (1997).
350. R. J. Angelici, *Polyhedron,* **16** (18), 3073 (1997).
351. P. Grange and X. Vanhaeren, *Catal. Today,* **36** (4), 375 (1997).
352. B. Delmon, *Surf. Interface Anal.,* **9**, 195 (1986).
353. S. Kasztelan, *Langmuir,* **6**, 590 (1990).
354. M. J. Ledoux, O. Michaux and G. Agostini, *J. Catal.,* **102**, 275 (1986).
355. H. Topsøe, B. S. Clausen, R. Candia, C. Wivel and S. Mørup, *Bull. Soc. Chim. Belg.,* **90**, 1189 (1981).
356. H. Topsøe, N.-Y. Topsøe, O. Sørensen, R. Candia, B. S. Clausen, S. Kallesøe, E. Pedersen and R. Nevald, *ACS Symp. Series No 279*, p.235 (1985).
357. D. K. Lee, H. T. Lee, I. C. Lee, S. K. Park, S. Y. Bse, C. H. Kim and S. I. Woo, *J. Catal.,* **159**, 219 (1996).
358. S. H. Bauer, N.-S. Chiu and M. F. L. Johnson, *J. Phys. Chem.,* **90**, 4888 (1986).
359. P. Gajardo, A. Mathieux, P. Grange and B. Delmon, *Appl. Catal.,* **3**, 347 (1982).
360. R. L. Chin and D. M. Hercules, *J. Phys. Chem.,* **86**, 360 (1982).
361. D. I. Kochubey, M. A. Kozlov, K. I. Zamarsev, V. A. Burmistrov, A. N. Startsev and Yu. I. Yermakov, *Appl. Catal.,* **14**, 1 (1985).
362. W. Nieman, B. S. Clausen and H. Topsøe, *Catal. Lett.,* **41**, 355 (1990).
363. A. N. Startsev, *Prep ACS, (208th) Symp.,* p.434 (1994).
364. Yu. I. Yermakov, A. N. Startsev and V. A. Burmistrov, *Appl. Catal.,* **11**, 1 (1984).
365. B. M. Reddy, K. V. R. Chary, V. S. Subrahmanyam and N. K. Nag, *J. Chem. Soc., Faraday Trans. 1*, p. 1655 (1985).
366. H.C. Zonnevylle, R. Hoffmann and S. Harris, *Surf. Sci.,* **199**, 320 (1988).
367. J. Valyon and W. K. Hall, *J. Catal.,* **84**, 216 (1983).
368. E. G. Derouane, E. Pedersen, B. S. Clausen, Z. Gabelica, R. Candia and H. Topsøe, *J. Catal.,* **99**, 253 (1986).
369. F. E. Massoth and G. Muralidhar, *Proc. 4th Int. Conf. on the Chemistry and Uses of Molybdenum* (H. F. Barry and P. C. H. Mitchell eds.), p.343, Climax Molybdenum Co Ltd. (1982).
370. B. Delmon, *Lat. Am. Appl. Res.,* **26** (2), 87 (1997).
371. E. Furimsky, *Catal. Rev.-Sci. Eng.,* **22**, 371 (1980).
372. G. C. A. Schuit, *Int. Journal of Quantum Chemistry, XII, Suppl.,* **2**, 43 (1977).
373. A. B. Anderson, Z. Y. Al-Saigh and W. K. Hall, *J. Phys. Chem.,* **92**, 803 (1988).
374. H. Topsøe, B. S. Clausen and F. E. Massoth, *Hydrotreating Catalysis (Catalysis - Science and Technology)* (J. R. Anderson and M. Boudart eds.), **Vol.11**, Springer, New York (1996).
375. B. C. Gates, J. R. Katzer and G. C. A. Schuit in: *Chemistry of Catalytic Processes*, McGraw-Hill, New York (1979).
376. G. C. A. Schuit, *Int. J. Quantum.Chem.,* **12**, 43 (1977).
377. T. A. Pecoraro and R. R. Chianelli, *J. Catal.,* **67**, 430 (1981).
378. R. G. Leliveld, A. J. van Dillen, J. W. Geus and D. C. Koningsberger, *J. Catal.,* **171**, 115 (1997).
379. J. B. Goodenough, in *4th Int. Conf. on the Chemistry and Uses of Molybdenum* (H. F. Barry and P. C. H. Mitchell eds), p.1, Climax Molybdenum Co Ltd. (1982).
380. D. H. Broderick and B. C. Gates, *AIChE J.,* **27**, 663 (1981).
381. R. Navarro, B. Pawelec, J. L. G. Fierro, P. T. Vasudevan, J. F. Cambra and P. L. Arias, *Appl. Catal. A,* **137** (2), 269 (1996).
382. C. Bianchini, M. V. Jimenez, A. Meli, S. Moneti and F. Vizza, *J. Organomet. Chem.,* **504** (1-2), 27 (1995).
383. R. Navarro, B. Pawelec, J. L. G. Fierro and P. T. Vasudevan, *Appl. Catal. A,* **148** (1), 23 (1996).

384. J. Frimmel and M. Zdrazil, *J. Catal.,* **167** (1), 286 (1997).
385. H. Toulhoat, P. Raybaud, S. Kasztelan, G. Kresse and J. Hafner, *Prepr. - Am. Chem. Soc., Div. Pet. Chem.*, **42** (1), 114 (1997).
386. Q. Zhang, W. Qian, A. Ishihara and T. Kabe, *J. Jpn., Petrol. Inst.*, **40** (3), 185 (1997).
387. S. Kasahara, T. Shimizu and M. Yamada, *Catal. Today*, **35** (1-2), 59 (1997).
388. C. J. Wright, C. Sampson, D. Fraser, R. B. Moyes, P. B. Wells and C. Riekel, *J. Chem. Soc., Faraday Trans. 1*, **76**, 1585 (1980).
389. A. Sierraalta and F. Ruette, *J. Mol. Catal. A: Chem.*, **109** (3), 227 (1996).
390. G. B. McGarvey and S. Kasztelan, *J. Catal.,* **148**, 149 (1994).
391. A. N. Startsev, E. V. Artamonov and Yu. I. Yermakov, *Appl. Catal.*, **45**, 183 (1988).
392. B. R. G. Leliveld, J. A. J. van Dillen, J. W. Geus and D. C. Koningsberger, *J. Catal.,* **165**, 184 (1997).

4

Development of Novel Hydrodesulfurization and Hydrodenitrogenation Catalysts

Hydrotreatment processes which include hydrodesulfurization (HDS), hydrodenitrogenation (HDN), hydrodeoxygenation (HDO), hydrodemetallation (HDM) etc. remove sulfur, nitrogen, oxygen, and nickel and vanadium metals from petroleum and produce various petroleum products. Recent air pollution caused by diesel exhaust gas has extended over large urban areas and much attention has been focused on technology for clean fuel production such as deep desulfurization of light gas oil (LGO). Although alumina-supported Co-Mo and Ni-Mo have been conventional catalysts of petroleum hydrotreatment for a long time, the development of novel catalysts with high catalytic activity is required to achieve clean fuel production. For this purpose, it is important to remove alkyl-substituted dibenzothiophenes which are often less reactive (see Chapter 2). Therefore, novel catalysts may have to have multi-functions including not only desulfurization but also dealkylation, isomerization of alkyl groups and hydrogenation of aromatic rings.

On the other hand, the demand for white oil is increasing year by year while raw materials become heavier and heavier. Heavier fractions such as vacuum residue must be transformed to lighter fractions by a cracking process. In recent oil processing, the cracking processes of the heavier fraction are preceded by the hydrotreating processes because sulfur and nitrogen compounds retard the activity of cracking catalysts.[1] Sulfur compounds inhibit hydrogenation activity when catalysts include noble metals. Nitrogen compounds deactivate acidic catalysts such as zeolites. The hydrotreating process is also used as a thermal cracking reactor. Therefore, with respect to the combination of hydrotreatment with catalytic cracking and thermal cracking, the development of highly active hydrotreating catalysts is needed.

To prepare novel active catalysts, the relationship among the preparation methods and the structure and the reactivity of catalysts must be elucidated. Recent development of various analytical methods has brought about remarkable progress in the clarification of the relationship between the surface structure and the reactivity of the catalysts. However, there are few examples which deal with working catalysts. Further, few examples attach much importance to the preparation methods of catalysts.

In this chapter, we focus attention on the preparation methods and reactivities of catalysts. Conventional approaches using molybdenum as a major active element, and cobalt and nickel as promoters are described, followed by approaches using noble metals. Other approaches with the use of metals other than molybdenum and noble metals or composite catalysts are also discussed. Most discussion is limited to hydrodesulfurization with some mention of hydrodenitrogenation. There are recent reviews on this subject.[2,3]

4.1 Conventional Approaches

Molybdenum is a major active element for hydrotreatment, and cobalt and nickel are the most important promoters. In this section, the effects of supports and additives on the HDS reactivities of molybdenum-based catalysts and cobalt- or nickel-promoted molybdenum-based catalysts and the HDS reactivities of new materials using molybdenum compounds are described.

4.1.1 Effects of Supports on the HDS Reactivities of Molybdenum-based Catalysts and Cobalt- or Nickel-promoted Molybdenum-based Catalysts

Alumina has a relatively high surface area and mechanical strength, and it is cheap. Further, the reaction of alumina with ammonium heptamolybdate forms a monolayer of MoO_3 on alumina surface.[4] Alumina-supported catalysts also have the ability to regenerate the hydrotreating activities of used catalysts by repeating successive oxidation and sulfidation. Therefore, studies on the active species, the structure and the reactivity of hydrotreating catalysts have focused exclusively an alumina-supported catalysts and studies using supports other than alumina are very few. However, in order to develop novel hydrotreating catalysts with excellent abilities, it is very important to study supports other than alumina. Recent approaches using supports other than alumina, i.e., carbon, oxides, zeolites, etc., are described here and compared with approaches using alumina.

A. Carbon

Carbon-supported Co-Mo and Ni-Mo catalysts have been studied extensively over the past twenty years because these catalysts are usually more active in HDS than conventional alumina-supported ones.[5-58] Further, since the interaction of carbon support with the sulfide phase is relatively weaker and the use of this support may give the active sulfide forms of various transition metals easily in the presulfiding step, it was expected that the use of carbon support benefited the elucidation of the synergistic effect derived from cobalt or nickel and molybdenum in the HDS reaction. Recently, HDN catalyzed by carbon-supported catalysts has also been reported.[59-66]

Topsøe et al. reported that a so-called CoMoS structure in catalysts is related to the activity in the HDS reaction.[13-16] The CoMoS structure has been observed in alumina- and carbon-supported catalysts [12-16] as well as in unsupported ones.[17] In the CoMoS structure, cobalt is located on the edge site of MoS_2 crystallite. When carbon is used, the dispersion of the CoMoS phase increases compared with other catalysts. Further, two types of CoMoS phases (type I and type II) were reported for alumina-supported catalysts.[18] Type II CoMoS formed at a higher sulfiding temperature was more active than type I formed at a lower normal sulfiding temperature (Fig. 4.1). Type I CoMoS leaves the Mo-O-Al linkage while type II is fully sulfided and the interaction between CoMoS phase and the support was of the Van der Waals type. Topsøe et al. suggested that the CoMoS phase on carbon supports is similar to the type II CoMoS. Van Veen et al.[21,22] used nitrilotriacetic acid with cobalt nitrate and heptamolybdate to prepare fully sulfided type II CoMoS on carbon, alumina and silica, so that the difference in the precursor support interaction should be the least for all supports.

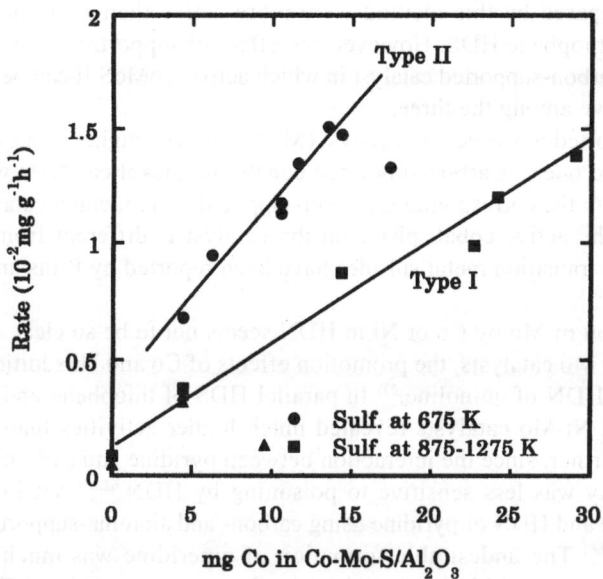

Fig. 4.1 HDS activity as a function of the amount of Co in CoMoS for catalysts sulfided at different temperatures.
(From H. Topsoe *et al.*, *Ind. Eng. Chem. Fundam.*, **25**, 33 (1986))

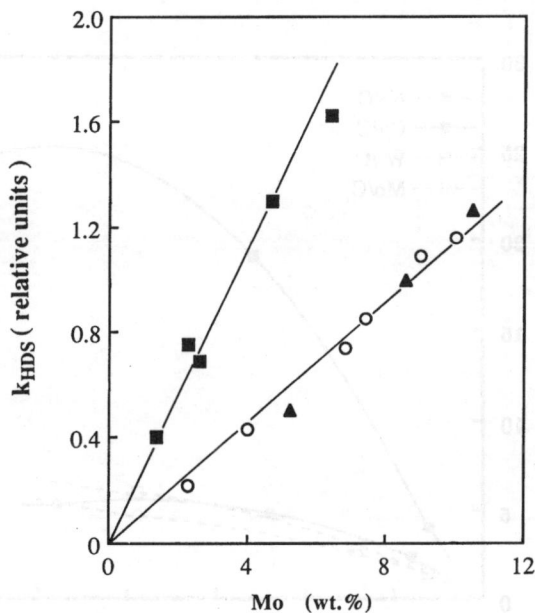

Fig. 4.2 Thiophene HDS activity of CoMoS II vs. Mo loading.
Co/M = 0.33 atom/atom, Supports: ○: γ-Al₂O₃; ▲: SiO₂; ■: C.
Catalysts are presulfided at 350 °C in 1/9 v/v H₂S/H₂ for 1h; test conditions, 0.2–0.3 g catalyst, H₂ flow
rate 55 ml min⁻¹, [thiophene] = 6 vol%, *T* = 350°C, *P* = 1 bar.
(From J. A. R. Van Veen *et al.*, *J. Chem. Soc., Chem. Commun.*, 1685 (1987))

The catalysts prepared by this method were more active than those prepared from usual impregnation in thiophene HDS. However, the effect of supports was observed as shown in Fig. 4.2 and the carbon-supported catalyst in which active CoMoS II can be formed selectively was the most active among the three.

Carbon-supported molybdenum catalyst (Mo/C) showed higher catalytic activity than the alumina-supported one.[28] Carbon-supported cobalt and nickel catalysts were also shown to be more active than the corresponding carbon-suppoted molybdenum catalyst (Fig. 4.3).[8,9] It is believed that the active cobalt phase on the catalyst is different from Co_9S_8.[15,30] Other carbon-supported transition metal sulfides have been reported by Prins and coworkers[26,59,60] and Ledoux et al.[61]

The promotion of Mo by Co or Ni in HDN seems not to be so clear as in HDS.[67-69] On carbon-supported Mo catalysts, the promotion effects of Co and Ni addition on activity have been reported in HDN of quinoline.[23] In parallel HDS of thiophene and HDN of pyridine, carbon-supported Ni-Mo catalysts revealed much higher activities than a commercial Ni-Mo/Al_2O_3.[51,63] Further, since the interaction between pyridine and carbon support is weaker, their HDS activity was less sensitive to poisoning by HDN.[62,63] Vít investigated parallel HDS of thiophene and HDN of pyridine using carbon- and alumina-supported Mo, Co-Mo and Ni-Mo catalysts.[66] The undesirable formation of piperidine was much lower on carbon-supported and unpromoted Mo catalyst than on alumina-supported and Co- or Ni-promoted Mo catalyst. Irrespective of the type of support, the presence of Co or Ni accelerated mainly HDS. Therefore, the unpromoted Mo catalysts had better HDN/HDS selectivity than Co-Mo and Ni-Mo catalysts.

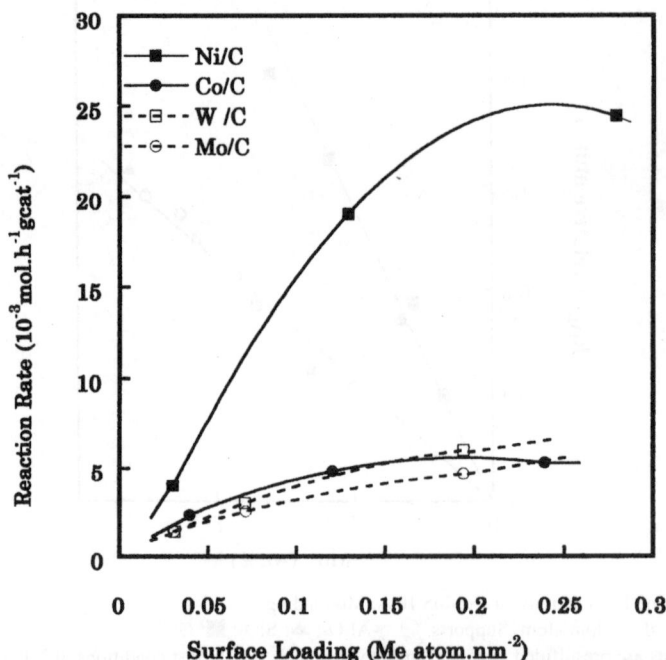

Fig. 4.3   Reaction rate of thiophene HDS as a function of surface metal loading.
(From V. H. L. de Beer et al., J. Catal., **72**, 371 (1981))

An important drawback in the utilization of carbon materials for hydrotreating reaction is their extensive microporosity. For catalytic reactions involving large molecules the micropores are of little use, since part of the transition metals will be deposited in these pores and be wasted. Most mesoporous carbons, on the other hand, have poor crushing strength, low bulk density or too low a surface area.

B. Oxides

In the studies using oxides as supports, there are many examples using titania.[70–85] Segawa et al. prepared highly dispersed molybdena-titania catalyst by an equilibrium adsorption method.[70] XRD, Raman, and XPS data of the calcined samples showed that monolayer coverage of molybdenum oxide over-layer possessed a highly distorted square pyramidal structure. A MoS_2-like structure was formed by sulfiding with H_2S, and hydrogenolysis of thiophene required an ensemble of two doubly and triply coordinatively unsaturated sites. Vrinat et al.[71] reported that, up to a molybdenum loading corresponding to 2.8 atoms/nm^2 of molybdenum, Mo/TiO$_2$ catalysts are about five times more active per molybdenum atom than the corresponding Mo/Al$_2$O$_3$ samples in the hydrogenation of biphenyl and the HDS of thiophene. Similar results were reported by other groups.[84,85] Ramirez et al.[72] reported that, in an electron microscopic study, the length of MoS_2 crystallites as well as the number of stacked layers for Mo/TiO$_2$ distributed toward the smaller particle range in comparison with Mo/Al$_2$O$_3$ (Fig. 4.4). The differences in the catalytic activity between these two catalysts was explained in terms of different activities between the smaller and larger particles. They proposed that the role of TiO$_2$ is to promote the formation of smaller molybdenum sulfide crystallites than these formed on alumina. These MoS_2 particles might be oriented edge-up on titania due to interaction between the MoS_2 edge planes and some planes of TiO$_2$. It was reported that the synergetic effect of cobalt was lower on TiO$_2$ than on

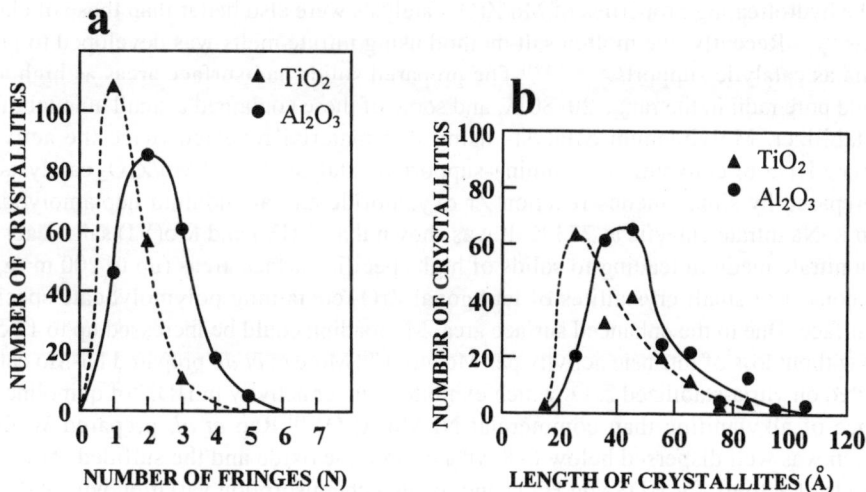

Fig. 4.4   Distribution of the number of fringes (a) and of the length of crystallites (b) for TiO$_2$ and Al$_2$O$_3$ molybdenum sulfide supported catalysts.
(From J. Ramirez et al., Appl. Catal., **52**, 219,220 (1989))

a

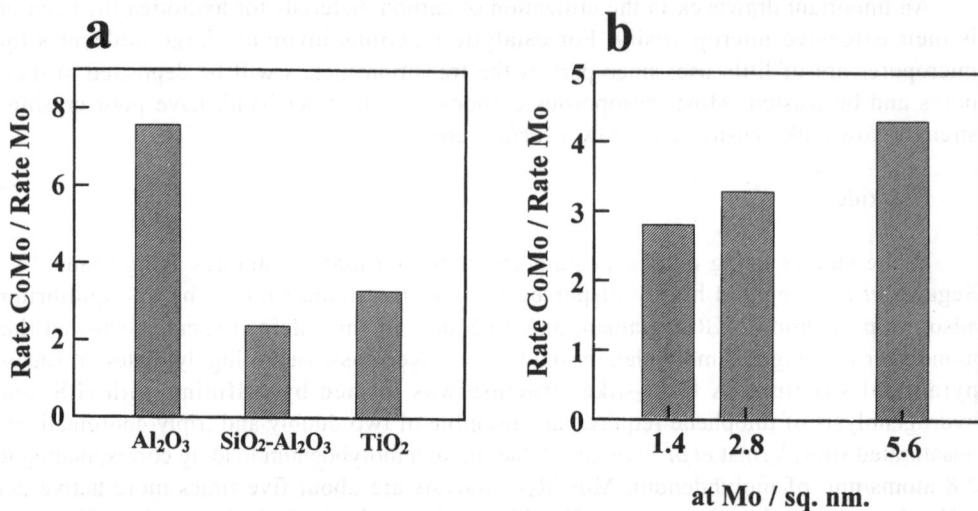

b

Fig.4.5   (a) Synergetic effect of cobalt for various supported molybdenum catalysts; (b) Variation of the synergy
with metal loading for Co-Mo/TiO$_2$ catalysts at constant Co/(Co + Mo) ratio.
(From G. Muralidhar *et al.*, *J. Catal.*, **85**, 46,47 (1984))

Al$_2$O$_3$ (Fig. 4.5).[72,73] This lower promoting effect was ascribed to the formation of the edge-up MoS$_2$ particles more difficult to promote or to a greater interaction of cobalt with the support, leading to a lower percentage of the promoter involved in the formation of the active CoMoS phase. However, this promoting effect increased with increasing molybdenum loading for catalysts with the same ratio of Co to Mo. This was ascribed to the preferential formation of the edge-up particles difficult to promote at low Mo contents.

The hydrotreating properties of Mo/ZrO$_2$ catalysts were also better than those of classical systems.[86–93] Recently, the molten salt method using nitrate melts was developed to prepare zirconia as catalytic supports.[87–89,92,93] The prepared solids had surface areas as high as 130 m^2/g and pore radii in the range 20–80 Å, and some of these contained a small amount of Y$_2$O$_3$ as a stabilizer. Molybdenum catalysts using this material revealed twice the activity in thiophene HDS of conventional alumina-supported catalysts.[88–90,92] Mo/ZrO$_2$ catalysts were also prepared by simultaneous reaction Zr oxychloride and ammonium heptamolybdate in molten K-Na nitrate eutectic at 773 K. It was shown that Zr(IV) and Mo(VI) salts react in the molten nitrate medium leading to solids of high specific surface areas (up to 200 m^2/g). The solids consist of small crystallites of tetragonal ZrO$_2$ containing polymolybdate species on their surface. Due to the enhanced surface area, Mo loading could be increased up to 12.5 wt% of Mo without loss of intrinsic activity per Mo atom.[93] Moreau *et al.* prepared Ni-Mo catalysts supported on yttria-stabilized ZrO$_2$ which exhibited higher activity in HDN of quinoline in the presence of alkylaniline than commercial Ni-Mo/Al$_2$O$_3$.[94] Rao *et al.* prepared W/ZrO$_2$.[95] Tungsten was well dispersed below 6–8 wt% in both the oxide and the sulfided states. From the trends of activity for HDS and HYD and oxygen chemisorption variation with W loading, they concluded that oxygen chemisorption was not specific to either HDS or HYD. Specific activities indicated that WS$_2$/ZrO$_2$ was as active as WS$_2$/Al$_2$O$_3$. The trends in variation of intrinsic activities on WS$_2$/ZrO$_2$ and WS$_2$/Al$_2$O$_3$ suggested that HDS and HYD originated from different sites on the catalyst surface.

Titania is unsuitable for industrial application due to its relatively low surface area and the low stability of the active anatase structure at high temperatures. Therefore, some binary and ternary oxides containing titania have been used as supports in a catalyst for hydrotreating reactions.[94,96–110] It was shown that binary oxides such as TiO_2-Al_2O_3 and TiO_2-ZrO_2 revealed higher HDN properties than simple oxides for upgrading of coal-derived liquids.[96] Ramírez *et al.* reported that the number of acid sites on the TiO_2-Al_2O_3, measured by pyridine adsorption, affects the dispersion and reactivity of the MoS_2 sites.[103] Spojakina *et al.* studied thiophene HDS over TiO_2-Al_2O_3-supported molybdenum catalysts. They suggested that the role of titania is to promote the formation of polymolybdenum compounds with octahedrally coordinated molybdenum, leading to an increase in the catalytic activity.[104] Thiophene HDS catalyzed by Ni-Mo/TiO_2-Al_2O_3 was also reported and it was shown that the increase in the HDS activity with titania loading is consistent with the better reducibility of molybdenum on titania and that nickel influences the reducibility of molybdenum, as revealed by the decreased temperature of reduction.[105] Further, Martin *et al.* reported that a correlation exists between the number of surface acid sites and the catalytic activity of Ni-Mo/TiO_2-Al_2O_3 in thiophene HDS, and that the activity and surface concentration of Brönsted acid sites increase with titania content in the TiO_2-Al_2O_3 supports.[106] Segawa and coworkers reported that Mo/TiO_2-Al_2O_3 and Ni-Mo/TiO_2-Al_2O_3 catalysts had much higher activities in HDS of DBT, 4-MDBT and 4,6-DMDBT than corresponding alumina-supported catalysts.[107,108] Co-Mo/TiO_2-ZrO_2-V_2O_5 catalyst[109] has two times higher activity in HDS of heavy diesel and vacuum gas oil and three times higher activity in HDN than commercial Co-Mo catalyst. The surface area and average pore diameter of this catalyst were approximately 120 m^2/g and 90 Å, respectively, and improvement in the physical structure of the support may be required. It was also reported that Co-Mo/TiO_2-ZrO_2, Co-Mo/TiO_2-CeO_2 and Co-Mo/TiO_2-MnO_2 show high initial activities.[110] No-Mo/TiO_2-ZrO_2 exhibited higher activity in HDN of quinoline in the presence of alkylaniline than commercial Ni-Mo/Al_2O_3.[94]

Cui *et al.* prepared Mo catalysts supported on lanthanum-modified alumina (La: 2–20 %, Mo: 6 %).[111] La oxide (LaO) supported on alumina was present as a monomolecular layer strongly attached to the alumina surface. At high La loadings, a second LaO overlayer occurs on top of the first LaO layer and undergoes sulfiding. The Mo oxide reacted with both the Al oxide and the LaO phases and formed essentially monolayer slabs of Mo sulfide after sulfiding. The higher intrinsic activity of the MoS_2 phase in thiophene HDS and hexene hydrogenation was obtained at high La loading; this was attributed to the electronic influence of the underlying phase.

Duchet *et al.* prepared nickel molybdenum sulfide supported on nickel and magnesium aluminates (NiAl$_2$O$_4$ and MgAl$_2$O$_4$).[112] They performed thiophene HDS at atmospheric pressure and HDN of 2,6-diethylaniline (DEA) and 1,2,3,4-tetrahydroquinoline (THQ) at 70 atm in a flow microreactor, and HDN of quinoline or phenanthridine at 140 atm in the presence of DEA in a batch reactor. In some cases for these reactions, catalysts supported on NiAl$_2$O$_4$ were slightly superior to the reference commercial Ni-Mo/Al_2O_3 catalyst whereas those supported on MgAl$_2$O$_4$ were less active. Competition between HDN of THQ and of DEA was observed and was unfavarable to DEA conversion.

Anthony and coworkers prepared silicon-amine intercalated titanates as novel supports for hydrotreating catalysts.[113] The Ni-Mo catalysts with this support were more active than a commercial catalyst for hydrogenation of pyrene, while the activities for HDS and HDN were similar to those of the commercial catalyst. It was reported that Pd and Ni/Mo exchanged hydrous titanium oxides had high activities for hydrogenation and hydroprocessing reactions, respectively, arising principally from the effective dispersion of the active phases and the acidity of the substrate.[114] Kelly et al. also studied the hydrotreating activity of Ni/Mo hydrous titanium oxides.[115] HDS activity increased with Mo loading up to the concentration of ca. 12% investigated. Low Ni loadings of ca. 1% achieved by ion exchange were just as effective as the higher loadings from impregnation. HDS activities of the hydrous titanium oxides increased markedly with calcination temperature up to 600°C, and this could well be associated with the differential scanning calorimetry-observed exothermic structural change of the water-washed substrate at 500°C.

In HDS reactions, silica-supported catalysts generally reveal lower activity than alumina-supported ones. In a recent study of the promoter dual effect (Co + Ni) on HDS activity, a negative effect occurred with silica[116] while a positive effect occurred with alumina.[117,118] However, the Co-Mo catalyst where 5–15% SiO_2 was added to a support Al_2O_3 revealed higher HDS and HDN activities than Co-Mo/ Al_2O_3 without SiO_2.[119] Muralidhar et al. reported that for both Co-Mo and Mo catalysts HDS and HYD activities decreased with an increase in SiO_2 content from 10% to 75% (Fig. 4.6).[71] When 15% SiO_2-Al_2O_3 was used as the support for Ni-Mo catalyst, the HDN activity increased. However, the activity decreased with 25% SiO_2-Al_2O_3. The HDS activity decreased in both cases.[5] These results indicate that the catalytic properties of SiO_2-Al_2O_3-supported molybdenum catalysts depend on the content of SiO_2 and the reactions. On the other hand, SiO_2-Al_2O_3-supported catalysts without molybdenum were reported. Fish et al. performed the gas-phase HDN of quinoline and related compounds with a 50% nickel oxide/aluminate on SiO_2-Al_2O_3 catalyst at 250°C under 1 atm of hydrogen gas.[120] Under these mild conditions, alkylaromatics were the predominant

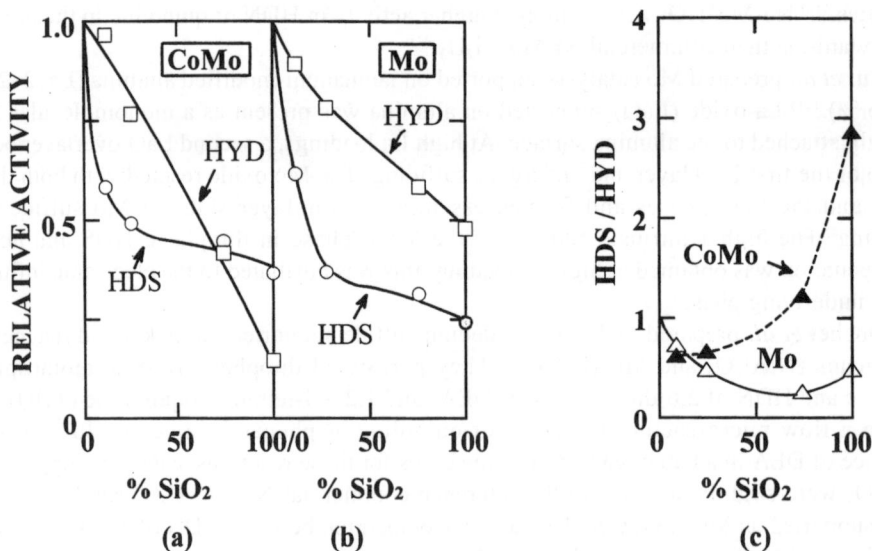

Fig. 4.6   Relative HDS and HYD activities of (a) Co-Mo/SiO_2-Al_2O_3, and (b) Mo/SiO_2-Al_2O_3 as a fumction of SiO_2,
(c) HDS/HYD vs. SiO_2 content.
(From G. Muralidhar et al., J. Catal., 85, 47 (1984))

HDN products. The reaction network of this catalytic process was different from that of commercial HDN processes in which quinoline is hydrogenated fully to decahydroquinoline prior to C-N bond cleavage(see Section 2. 2. 1B2). The activity of nickel oxide/aluminate on SiO_2-Al_2O_3 catalyst is within one order of magnitude of the activity of commercial HDN catalysts. Although this nickel oxide catalyst was irreversibly poisoned by sulfur and slowly deactivated during HDN due to coke formation, the coked catalyst could be regenerated by oxidation in air and subsequent reduction in hydrogen.

Magnesia was used as a support for hydrotreating catalysts. The magnesia-supported sulfide catalysts were less active in hydrotreatment reactions than alumina-supported catalysts,[121–123] probably because the high surface area of magnesia is transformed to low surface area $Mg(OH)_2$ during aqueous impregnation. In recent years, however, Chary et al. reported that Mo was better dispersed on MgO when Mo/MgO was prepared by aqueous impregnation, and that sulfided Mo/MgO catalysts exhibited a slightly higher activity in thiophene HDS than sulfided Mo/Al_2O_3 catalysts.[124] The higher activity was attributed to an increase in dispersion of Mo on MgO. A linear correlation was obtained between oxygen chemisorption and HDS activity. Hillerová et al.[125] also prepared magnesia-supported Ni-Mo sulfide catalysts by non-aqueous impregnation using dimethylsulfoxide and methanol as solvents. The activities of the best Ni-Mo/MgO sample prepared by this method for the single thiophene HDS, parallel thiophene HDS and parallel pyridine HDN were 94%, 290% and 105%, respectively, of the activity of the commercial Ni-Mo/Al_2O_3 catalyst at 320°C. As compared with alumina-supported catalysts, Ni-Mo/MgO catalysts revealed lower inhibition of HDS by parallel HDN and lower selectivity for HDN in parallel HDN and HDS.

C. Zeolites

Zeolites have the potential for application to hydrotreating catalysts because they possess numerous features such as high surface area, shape selectivity, acid-base properties, high thermal stability, increased resistance towards organic sulfur and nitrogen compounds, etc. Further, their crystal structure and catalytic function have been clarified comparatively and it may be possible to design and develop novel hydrotreating catalysts using zeolites on a molecular or atomic basis.

Transition metals containing zeolites have been prepared by various methods, namely impregnation, ion exchange, vapor phase adsorption, solid-solid reaction, direct incorporation into zeolite lattice during the synthesis step. For example, Dai and Lunsford introduced molybdenum ions into Y zeolite by solid-solid exchange method and characterized them by IR, EPR, XPS and XRD analyses.[126] After presulfiding or reduction, the resulting materials are used for HDS. In HDS using zeolite, the initial activity is generally very high.[127] At the initial stage of the reaction, however, the acidic properties of zeolite occurring at the presulfiding or reduction cause rapid deactivation due to coking of the surface. The inhibition of this coking properties may be key to the preparation of zeolite-supported catalysts.

Okamoto et al. prepared zeolite-supported molybdenum catalysts from $Mo(CO)_6$ which were more active in thiophene HDS than those by commercial impregnation.[128–131] They suggested that highly dispersed molybdenum sulfide can be prepared in NaY and KY zeolite and that the basicity of the zeolites controls the strength of interaction between the zeolite oxygens and the subcarbonyl species and thus the final dispersion of molybdenum sulfide. Vrinat et al. reported HDS of dibenzothiophene catalyzed by HY and NaY zeolites-supported cobalt-molybdenum.[132] Cobalt was introduced by ion exchange and molybdenum using molybdenum carbonyl. However, the HDS activities of these catalysts were less than that of commercial Co-Mo/Al$_2$O$_3$ catalyst. Cid et al. prepared molybdena impregnated NaY zeolite catalysts.[133,134] The crystallinity of the samples decreased almost linearly with molybdena loading. A strong interaction was observed between molybdenum species and the zeolite lattice. The same authors reported that, when molybdenum and cobalt were successively supported on NaY zeolite by impregnation, these were situated separately not to interact.[134-137] The relatively low HDS activity of this catalyst may be due to this lack of interaction between cobalt and molybdenum. They also reported that, when Co was supported on NaY zeolite by ion exchange prior to molybdenum deposition, a better catalyst was obtained than by simultaneous impregnation.

Ion-exchanged NiNaY catalysts showed an increase in thiophene HDS and cracking activity with increasing nickel loading, and therefore, with increasing acidity of the zeolites.[138] However, these decreased markedly with pyridine poisoning. Precalcination at 873 K or nonsulfidation pretreatments of the Ni-exchanged zeolites led to significant decrease in HDS activity. It was pointed out that sulfided Ni-species formed on zeolite were predominantly involved in the thiophene HDS. Welters et al. also prepared NaY-supported nickel sulfide catalysts by impregnation and ion-exchange.[139] They found that the catalysts showed large differences in catalytic behavior depending on the preparation method (impregantion vs. ion exchange) and the pretreatment conditions (method of sulfidation). The ion-exchanged catalysts especially showed a high initial activity, but due to the presence of acid sites, deactivation was very strong. The differences in catalytic activity were ascribed not only to variation in overall nickel sulfide dispersion but also to the acidity of the support. Korányi et al. studied adsorption and reaction of thiophene over nickel- and cobalt-containing zeolites.[140,141] When non-sulfided and sulfided NiNaY zeolite catalysts were prepared by ion-exchange and impregantion, the nickel species was located mainly inside the micropores of the zeolites in all samples.[141] The ion-exchanged catalysts exhibited higher HDS activity than the impregnated catalysts; this was attributed to a more homogeneous distribution of nickel species in ion-exchanged samples.

Tatsumi and coworkers[142] investigated HDS of benzothiophene catalyzed by Ni-Mo/NaY zeolite, revealing comparable activity with a conventional Ni-Mo/Al$_2$O$_3$. They prepared this catalyst by introducing NiMoS$_x$ cubane type cluster into NaY zeolite by ion exchange. Ni, Mo, and Ni-Mo containing Y or ZSM-5 zeolite was also described by Davidova et al.[143-145] Welters et al. reported that the increase in initial thiophene HDS activity with increasing support acidity is observed for Co, Ni and Mo sulfide catalysts prepared by impregnation using various zeolite supports (H$_{(x)}$NaY, ZSM-5, Y, HUSY).[146] They proposed that the increased activity is not caused by an increase in the metal sulfide dispersion, but probably by a synergetic effect between the metal sulfide particles and the acidic zeolite support, by an increase in thiophene adsorption in the zeolite pores, or by a direct effect of H$^+$ on the thiophene HDS reaction. Anderson et al. prepared USY and NaY zeolite-supported molybdenum catalysts.[147] Mo/USY catalyst exhibited decreasing conversions with reaction time due to coking while the activity of Mo/NaY remained constant. There was no relationship between dispersion and HDS or hydrogenation selectivity. On the other hand, it has been reported that the thiophene HDS activity of sulfided NiNaY and MoNaY catalysts is strongly determined by the distribution (in- or outside the zeolite pores) and the dispersion of the Ni or Mo species.[148,149] Agudo et al. reported that the activity of Mo/HZSM-5 catalysts depends on the dispersion of Mo species rather than on the acidity of the catalysts.[150] Das *et al.* reported NaX zeolite supported molybdena catalysts.[151]

Cid *et al.* prepared tungsten-modified ultrastable Y (USY) zeolite catalysts by conventional impregnation with ammonium metatungstate solutions at different pHs and calcination temperatures.[152] The incorporation of tungstate ions in the USY followed by calcination at high temperature caused a small loss of crystallinity, particularly for the catalysts impregnated at acidic pH. In this case, most W was inhomogeneously deposited outside the zeolite cavities, but without forming WO_3 species. This location of W and the increased acidity after sulfidation with H_2S generated a substantial increase in HDS activity relative to the parent USY, and the appearance of hydrogenation (HYD) activity. Conversely, in the catalysts prepared at basic pH, a large part of W is inside the zeolite cavities, most probably in the supercages. This W location and the loss of acidity resulted in very low HDS activity and no HYD activity for both calcined and sulfided catalysts. They also prepared nickel-containing ultrastable HY zeolites by ion exchange at pH 2.7, 5.6 and 11 (NH_4OH medium).[153] The zeolite samples having only one type of active site, either acidic (prepared at pH 2.7) or nickel sulfide species (treated with a NaOH solution), showed low thiophene HDS activity. However, those that have both types of sites, such as the samples prepared at pH 5.6 and 11, exhibited high activity for thiophene HDS.

Some studies are using zeolite and other supports simultaneously for the hydrotreating catalysts.[154,155] In these studies, the hydrogenation properties by Co-Mo, Ni-Mo or NiW sulfide was combined with a hydrogenolysis function by the support with acid sites. To improve the HDN activity, Harvey and Matheson combined commercial Ni-Mo/Al_2O_3 to RuY zeolite.[156-158] Lemberton *et al.* investigated HDN of 1,2,3,4-tetrahydroquinoline on nickel and molybdenum sulfides supported on alumina-zeolite mixtures.[159] The number of acid sites and the number of hydrogenating sites changed by varying the weight of zeolite in the support. They found the existence of an optimal ratio (number of acid sites)/(number of hydrogenating sites). This represented the existence of an optimal distance between the two types of sites, for which the decahydroquinoline formed on the hydrogenating sites shifts rapidly on the neighboring acid sites and cracked into the desired reaction product (propylcyclohexane). The alumina-zeolite catalysts exhibited much higher selectivity for HDN than conventional Ni-Mo/Al_2O_3 catalysts.

D. Clays and Natural Minerals

Some clays, pillared clays and natural minerals were used as supports for preparing low cost catalysts.[160-167] Natural bauxite supported Mo, Co-Mo and Ni-Mo revealed lower activities for the HDS and HDN reactions.[160] Some natural clays were used as catalysts and Kaolinite was found to serve as a good catalyst for *n*-butylamine HDN.[161] Further, Schlutz *et al.* reported that MoS_2 supported on acid-treated Brazilian bentonites revealed higher hydrocracking and HDS activities than Al_2O_3 support.[162] Occelli and Rennard used pillared bentonite as a support for Ni-Mo catalysts for the hydrotreatment of vacuum gas oil feedstocks.[163] Warburton used clays pillared by Fe-sulfide for high-pressure demetallization of heavy crude oil.[164] These pillared clays displayed interesting hydrotreating properties. Kloprogge *et al.* prepared nickel sulfide catalysts supported on Al-pillared montmorillonite which exhibited comparable and even higher activities in thiophene HDS than alumina- and carbon-supported nickel sulfide catalysts.[165] The catalytic activity increased by raising the amount of Al pillaring agent while an increase in calcination temperature decreased the activity. Sychev et al. synthesized chromia-pillared and pillared-dealminated clays from different montmorillonites.[166] Chromia-sulfide pillared materials showed a high activity amd selectivity in thiophene HDS and the consecutive hydrogenation of butene. Recently, Hayashi et al. prepared cobalt oxide loaded on high surface area smectites such as montmorillonite, saponite, porous saponite, hectorite and stevensite, and tested them in a pulse flow reactor for thiophene HDS.[167] Co-porous saponite showed the highest activity among Co smectites, Co/Al$_2$O$_3$, and Co-Mo/Al$_2$O$_3$ so far studied. This suggests that the properties of smectite influence catalyst activity.

4.1.2 Effects of Additives on HDS and HDN Reactivities of Conventional Molybdenum-based Catalysts

The addition of phosphorus, boron and fluorine to the molybdenum-based conventional catalysts is a simple method to modify the HDS and HDN reactivities of these catalysts. The modification of supports can change their interaction with the active phase and consequently the catalytic activity. Further, some information about the effect of the supports on the synergy between cobalt or nickel and molybdenum may be provided by studying these changes.

It is well known that not only nickel and molybdenum but phosphorus also is contained in commercial HDN catalysts for the treatment of heavy feedstocks.[168] The positive effects of phosphorus on HDS,[169–200] HDN,[169–177,179,182–184,190–194] hydrogenation,[190,195–198] hydrocracking[190] and hydrodemetallation (HDM)[169] catalyzed by molybdenum-based catalysts have been reported. The roles of phosphorus are reported as follows: The addition of phosphorus enhances the solubility of molybdate in the impregnation solution[170] and improves the mechanical and thermal stability of the support by the formation of AlPO$_4$.[199] Phosphate has a strong interaction with alumina and forms AlPO$_4$.[189,194,200–206] The AlPO$_4$ formed decreases the adsorption of molybdate [189,200–203] and weakens the interaction between molybdenum (nickel) oxide and the support.[189,194,200–203,207,208] These effects may change the number and reactivity of active sites,[209–211] the active phase dispersion,[178,181,189,194,202,203] the surface structure and crystal morphology of the catalysts.[168] For example, it was shown in a high-resolution transmission electron microscopy study that nickel and phosphate contribute to stacking of MoS$_2$ slabs in sulfided Ni-Mo-P/Al$_2$O$_3$ catalysts.[212] Further, it was observed that phosphate changes the acidity of the alumina support,[194,204,213,214] which affects the cracking and isomerization activity of the catalysts[200, 214] or the formation of coke.[213] Eijsbouts et al.[168] reported that the addition of phosphate to Ni-Mo/Al$_2$O$_3$ catalysts increased the HDN conversion significantly although phosphate has almost no effect on the HDS activity. It was suggested that the effect of phosphate is due to a combination of structural and catalytic factors: Phosphate improves the activity by inducing the formation of the type II NiMoS structure although it lowers the activity by inducing a decrease in the dispersion of the NiMoS phases and a segregation of Ni$_3$S$_2$.

The effects of addition of phosphorus to catalysts other than Ni-Mo/Al$_2$O$_3$ catalysts, i.e., W/Al$_2$O$_3$,[215] Ni or Co/Al$_2$O$_3$,[216,217] Ni-W/Al$_2$O$_3$,[218,219] carbon-supported transition metal sulfides,[220,221] etc., have also been reported. Both HDS and HDN were notably enhanced by increasing the phosphorus content in W/Al$_2$O$_3$ catalysts containing various amounts of phosphorus. The effect was somewhat larger for HDS than for HDN.[215] The addition of phosphorus to Ni/Al_2O_3 catalysts promotes simultaneously the gas oil HDS and pyridine HDN reactions.[216] The major promotion effect of phosphorus was observed on HDN rather than on HDS. The result suggests that the HDN improvement is associated with new active sites, most likely acidic $AlPO_4$ groups. Atanasova *et al.* reported that phosphorus had a promoting effect on the HDS activity of $Ni-W/Al_2O_3$ catalysts, which depended on both the phosphorus content and the preparation method.[218] Eijsbouts *et al.*[220] studied the poisoning effect of phosphorus on the thiophene HDS activity of carbon-supported transition metal sulfides, which was different for the thiophene conversion to hydrocarbons and butene hydrogenation, and depended on the transition metal sulfide. Group VIII transition metal sulfides were rather resistant towards poisoning by the phosphorus compound. Only the HDS activity of the carbon-supported nickel catalyst was promoted in the presence of phosphorus. Jhansi Lakshmi *et al.*[221] used amorphous $AlPO_4$ as a carrier to synthesize highly active WO_3 hydrotreating catalysts.^{1}H MAS-NMR and low-temperature oxygen chemisorption have indicated the monolayer loading of WO_3 on $AlPO_4$ as 9 wt%. The presulfided $WO_3/AlPO_4$ catalysts have shown improved HDS activity compared with conventional Al_2O_3 supported catalysts. Phosphate strongly decreased the thiophene HDS activity of sulfided carbon-supported Co and Co-Mo catalysts where metal (II)-phosphate phase responsible for the decrease in HDS activity is formed.[222] Gulková and Vít used a new microporous silica-ceria support in the preparation of NiMo(P) sulfide catalysts.[223] The activities of the catalysts in parallel HDN of pyridine and HDS of thiophene was higher than the activity of commercial $NiMoP/Al_2O_3$ catalyst. The amount of intermediate piperidine produced was significantly lower and the HDN/HDS selectivity was comparable with $NiMoP/Al_2O_3$ catalyst. These properties were ascribed to the specific properties of both support and phosphorus.

It is well known that alumina-boria ($Al_2O_3-B_2O_3$) catalysts are composite oxides which act as solid acids[224–232] and can be used as catalysts for cracking[233,234] and toluene disproportionation and xylene isomerization.[235] It was also shown that $Al_2O_3-B_2O_3$ supported on various combinations of zeolite, nickel oxide, and molybdenum oxide can be used for the hydrocracking of petroleum feedstocks.[236] It has been reported that other catalytic and physical properties of materials with a combination of aluminum and boron are related to their acidities.[237–239] Concerning the hydrotreating reactions, the modification of $Ni-Mo/Al_2O_3$ with boron diminished the interaction of the metals with the support and increased the catalytic activity in the upgrading of a lube oil.[240] The HDS activity of DBT over $Ni-Mo/Al_2O_3-B_2O_3$ reached maximum at a B_2O_3 loading of about 1 mol%, in agreement with the basal dispersion, and thus with the hydrogenation ability of the catalysts.[241] Boron-modified commercial $Ni-Mo/Al_2O_3$ catalysts were also effective for the removal of sulfur and nitrogen in LGO.[242] Further, when nickel and tungsten were supported on $Al_2O_3-B_2O_3$, the catalysts were active for hydrocracking of atmospheric residue and vacuum residue.[243] It was suggested that these catalysts principally converted aromatics in feedstock to saturates and naphthenic compounds in gas oil. The effects of boron on Co-Mo catalysts have also been investigated by several researchers.[244–247] Recently, Tsai *et al.* prepared a series of aluminum borates with various Al/B molar ratios by the precipitation method.[244] When cobalt and molybdenum were supported on aluminum borates, the catalysts were much more active than the conventional $Co-Mo/Al_2O_3$ in HDS and HDM of atmospheric residue. The difference in the desulfurization activity was due to the difference in dispersion and the interaction of Mo species with the support. Li *et al.*[245] reported that, in HDS of gas oil over Co-Mo/alumina-aluminum borate catalysts, there is a correlation between the acidity of the catalysts and their HDS activity. This suggests that the beneficial effect of boron results from an increase in surface acidity and metal dispersion. Ramírez et al.[246] studied the effect of boron addition on the activity and selectivity of hydrotreating Co-Mo/Al$_2$O$_3$ catalysts in detail. Boria was deposited as a monolayer on the surface of the alumina up to a loading of 0.8% B. The incorporation of boron into the catalyst supports led to a change in the distribution of the cobalt and molybdenum oxidic species, resulting in a decrease in the proportion of tetrahedrally coordinated Co^{2+} and Mo^{6+} species in strong interaction with the alumina support. As a result of this change in species of oxidic precursors, the number of cobalt sites probed by nitric oxide in the sulfided catalysts increased with boron content, while the molybdenum sites remained practically unchanged. The net result was a beneficial effect on the catalytic activity of the thiophene HDS which reached a maximum at a boron content of 0.8 wt% B. Studies on the characterization of cobalt oxide supported on boron-modified alumina have also been reported.[248,249]

The effects of addition of fluorine to molybdenum-based catalysts on their physical and catalytic properties have been investigated.[45,73,182,250--266] It is well known that the addition of fluorine to the catalysts enhances its surface acidity and thus increases acid-catalyzed reactions.[254,256,267,268] These studies deal mainly with cracking, hydrocracking,[45,182,250--255] HDS,[45,73,182,250--252,258,259] and hydrogenation.[45,182,251,252,255,258,269] Lycourghiotis and co-workers investigated the effects of the order of deposition of F$^-$ ions on HDS activity of fluorine-promoted Co-Mo/Al$_2$O$_3$ catalysts.[261] They prepared one doped Co-Mo/Al$_2$O$_3$ and three fluorinated catalysts in which the F$^-$ ions were deposited before, after and simultaneously with the Mo^{VI} and Co^{II} phases. It was found that a change in the deposition order provoked a change in the specific surface area, in the dispersity of the supported Mo and Co species and in the intrinsic activity of each HDS active site. The last change was found to be related to the ratio Mo(octahedral)/Mo(tetrahedral). Further, it was inferred that fluorination increased the number of active sites per unit active surface. When fluorine was deposited after Mo and Co, the catalyst exhibited the highest HDS activity and its oxidic precursor state had the lowest (Mo^{VI} octahedral)/(Mo^{VI} tetrahedral) ratio. Jirátová and Kraus[258] prepared Ni-Mo/Al$_2$O$_3$ modified by different ions (F, Cl, SO$_4$, Zn, Mg, Na). They reported that only the catalyst containing fluorine increased the activity of thiophene HDS. Very little has been reported on the effect of F addition to Mo catalysts on HDN reactions. Fierro et al. reported simultaneous high pressure HDS and HDN reactions on F-modified Ni-Mo/Al$_2$O$_3$ catalysts.[270] The addition of F decreased HDS of gas oil and slightly increased HDN of pyridine. Ramírez et al.[77] reported the the effect of fluoride incorporation on the titania support on the surface structure of promoted Co-Mo catalysts and their catalytic activity for thiophene HDS. The promoted catalysts revealed higher activity than the unpromoted catalysts. In high-resolution electron microscopy (HREM), the result was ascribed to the formation of smaller molybdenum crystallites on the promoted supports leading to an increase in the dispersion of the active phase. Jones et al.[182] prepared P and F bi-promoted Ni-Mo catalysts. In the hydroprocessing of a high-nitrogen, quinoline-spiked feed, these catalysts were quite effective in hydrogenation and HDS. The activities of quinoline HDN of these catalysts were superior to those of both P-only and F-only promoted catalysts. This was thought to be because the promotional effect of phosphorus in hydrogenation and the promotional effect of fluoride in C-N bond cleavage are both present in these bi-promoted catalysts. Further, these catalysts have excellent retention of fluoride, better than that for F-only promoted catalysts. They suggested that this occurred because of a modification of the surface fluoride species by the presence of phosphorus.

It has also been reported that the incorporation of Mg^{2+} and Zn^{2+} to alumina prior to Co and Mo affects the physical and catalytic properties.[247,271–275] These ions displace Co^{2+} from the alumina lattice to the surface, increasing the ratio of octahedrally coordinated cobalt.[271] Zn has a higher propensity to form a surface spinel with alumina than Co.[272,273] Further, it was revealed that Zn^{2+} species are preferentially and more strongly adsorbed on an alumina surface than Co^{2+} species, leading to important differences in the distributions of Zn and Co on the alumina surface upon calcination.[274] Fierro et al.[275] reported that, when the ratio of Zn/Co varied in the range 0 to 1 in HDS of gas oil catalyzed by $ZnCo-Mo/Al_2O_3$, the activity enhancement was observed at Zn/Co = 1. It was suggested that, in this case, the incorporation of Zn to alumina also inhibited the formation of inactive $CoAl_2O_4$ and increased the dispersion of Co and the interaction between Co and Mo. Thomas et al. prepared doubly promoted (Zn, Co)Mo/Al_2O_3 catalysts by equilibrium adsorption.[276] The promotional effect in the gas-oil HDS was observed on the catalysts prepared by simultaneous incorporation of Zn and Co. This effect was explained in terms of changes in the distribution and dispersion of promoters. Cambra et al.[277] reported the effect of fluorine on the thiophene HDS activity on a series of doubly promoted $ZnCo-Mo/Al_2O_3$. A maximum of the thiophene conversion was observed at 0.0–4.0 wt% F. Textural and chemical characterization results revealed that alumina reacts with $(NH_4)HF_2$ solution to produce pores of greater size and to plug those of lower size. The activity for quinoline HDN catalyzed by same catalyst systems has also been reported by the same group.[278]

The effect of the addition of various metal oxides to $Co-Mo/Al_2O_3$ on HDS of thiophene in heptane has been reported, and the effect of iron oxide was found to be the highest.[279] Lee et al. studied the promotional effect of tungsten in a $Co-Mo/Al_2O_3$ catalyst.[280] The maximum promotion of HDS and hydrogenation activities occurred at a low content of tungsten corresponding to 0.025 in W/(W + Mo). Oxygen uptake correlated well with catalytic activities. In general, the catalysts prepared by impregnating tungsten onto the $Co-Mo/Al_2O_3$ showed higher activities than the catalysts prepared by impregnating tungsten onto Mo/Al_2O_3 prior to impregnation of cobalt. Weissman et al. reported that gas-oil HDS activity of $Ni-Mo/Al_2O_3$ catalysts was significantly improved when prepared with zirconium alkoxide, but lowered when prepared with Zr nitrate.[281] Investigations by LRS, IR and XPS indicated no change in the structure of the active metals. IR of pyridine and ^{27}Al NMR both showed greatly enhanced Lewis and Brønsted acidities of the Zr alkoxide modified materials as compared with unmodified samples. This indicates an indirect effect of Zr on activity, through modification of the support.

4.1.3 HDS and HDN Reactivities of New Molybdenum Materials

A. Molybdenum Nitride and Carbide

Transition metal nitrides and carbides have long been known as very hard materials with thermal and mechanical stability which have both the physical properties of refractory ceramics and the electronic and magnetic properties of metals.[282,283] They can be used for cutting tools, wear-resistant parts, high-temperature structural components, electronic and magnetic components and superconductors. These materials have had too low surface area to use as catalysts. However, recent progress in the preparation of these materials with high surface area using a temperature-programmed reaction of MoO_3 with NH_3 or CH_4 and H_2[284-290] has enabled their use as catalysts for several hydrogen transfer reactions: ammonia synthesis,[291,292] CO hydrogenation,[293,294] hydrogenation of aromatics and unsaturated hydrocarbons,[295-297] hydrogenation of acetonitrile [298] and hydrogenolysis of saturated hydrocarbons.[294,299,300] The behavior of nitrides and carbides in most of the reactions is similar to that of the group 8 metals, indicating that these materials have the potential of acting as substitutes.[301]

These properties of nitrides and carbides have promoted the development of novel hydrotreating catalysts using these materials.[302-314] Metal nitrides, especially, may be used for HDN because metal sulfides are active for HDS. The predominant phase in molybdenum nitride catalysts is γ-Mo_2N. The molybdenum atoms in γ-Mo_2N are arranged in a face-centered cubic array, with nitrogen atoms randomly distributed in half of the octahedral interstices. These interstices may also be filled with other atoms such as carbon or oxygen. Choi et al. reported that HDN activities of unsupported molybdenum nitrides changed depending on the heating rates and space velocities used, and were superior to those of commercial sulfided Co-Mo/Al_2O_3.[302] Further, the activity decreased with increasing surface area, indicating that HDN would be structure-sensitive over the molybdenum nitrides. The results suggest that supporting active molybdenum nitrides on supports with high surface area may give effective catalysts for HDN. Colling and Thompson reported that the supported molybdenum nitrides were more active for HDN of pyridine than a commercial Ni-Mo/Al_2O_3.[303] Their activities were comparable to those of the unsupported molybdenum nitrides. The catalytic properties of the supported molybdenum nitrides depended on the size and composition of the molybdenum nitride domains. Schlatter et al. reported that supported and unsupported molybdenum nitrides and carbides provide quinoline HDN activity of the same magnitude as commercial sulfided Ni-Mo/Al_2O_3 and show higher propylbenzene/propylcyclohexane ratio in the absence of sulfur in feed than that in the presence of sulfur.[304] Similar results have been reported for HDN catalyzed by molybdenum nitride.[305-309] In these reports, the products were aromatic compounds rather than completely saturated ones, indicating that its hydrogenation ability is rather low. These observations suggest differences in the extent to which simple aromatics and nitrogen heterocycles interact with the Mo_2N surface. In order to understand these interactions, Armstrong et al. investigated temperature-programmed desorption and ^{2}H-NMR spectroscopy of chemisorbed benzene and pyridine on high surface area Mo_2N.[310] The uptake of chemisorbed pyridine on Mo_2N at 295 K was about four times larger than that of benzene and pyridine displaced benzene completely in a competitive adsorption experiment.

Ramanthan and Oyama performed the HDS of DBT, the HDN of quinoline, the HDO of benzofuran, hydrogenation of tetralin in tetradecane simultaneously and showed that the HDN activity of molybdenum carbide was higher than that of sulfided Ni-Mo/Al_2O_3 although its HDS activity was lower.[311] Choi et al. prepared a series of molybdenum carbides by the temperature-programmed carburization of MoO_3 with pure CH_4 or equimolar mixture of CH_4 and H_2.[313] The Mo carbides were very active for pyridine HDN with catalytic properties that were similar to those of Mo nitrides and superior to those of commercial sulfided Co-Mo and Ni-Mo/Al_2O_3 catalysts. Selectivities over the carbides were significantly different from those over the Mo nitride and sulfide catalysts. The Mo carbides produced substantial amounts of cyclopentane while the Mo nitrides and the sulfided catalysts produced mostly pentane. The differences were interpreted in terms of differing bonding geometries for pyridine on the

Mo carbides and nitrides. As for the possibility of using molybdenum carbide, the supported catalyst may also reveal a higher activity than Co-Mo or Ni-Mo catalysts.

Abe et al. reported that bulk Mo_2N, W_2N and VN could be obtained with high surface area and that each of these nitrides exhibited high activity of quinoline HDN, the turnover frequency for this reaction decreasing in the order $Mo_2N > W_2N > VN$.[308] Highly dispersed VN supported on SiO_2 was also found to have specific activity for quinoline HDN identical to that of bulk VN. Yu et al. prepared a new bimetallic V-Mo oxynitride which was found to be a more active catalyst in quinoline HDN than pure VN and Mo_2N and a commercial Ni-Mo/Al_2O_3 catalyst.[314,315]

Supported and unsupported molybdenum nitrides are also active for HDS.[307,308,316–318] Markel and Van Zee prepared the unsupported Mo_2N catalyst with a high surface area for resistance against sulfiding during HDS in a catalytic reactor.[316] When the unsupported Mo_2N catalysts with different surface areas were used in thiophene HDS, the hydrogenation of olefins, HDS products proceeded more rapidly on the catalysts with higher surface area. In contrast, Nagai et al. reported that an alumina-supported molybdenum nitride was more active than alumina-supported molybdenum sulfide in HDS of DBT and produced biphenyl selectively rather than cyclohexylbenzene.[317,318] This indicates that the ability of hydrogenation of supported molybdenum nitride is rather low in HDS as well as in HDN. These results suggest that when the morphology of molybdenum nitride is changed, the balance in the abilities of hydrogenolysis and hydrogenation may be controlled. It has also been shown that both Mo_2N and 27% VN/SiO_2 exhibit high activity for thiophene HDS.[308]

Liaw et al.[319] prepared nanoscale, high surface area bulk molybdenum nitrides using two methods: a laser pyrolysis technique, which was developed by Haggerty[320] to synthesize nanoscale powder, and a temperature programmed reaction mentioned above. The two catalysts were active for simultaneous HDS and HDN reactions of Illinois No. 6 naphtha, and exhibited the same activity and selectivity.

B. Chevrel Phase, Heteropolycompounds and Other Molybdenum-based Materials

Schrader and coworkers reported that superconductor Chevrel phase compounds ($M_xMo_6S_8$: M = Pb, Co, Ni, Sn, In, Ho, La, Dy, Lu, etc.) were active for thiophene HDS.[321–326] Since these compounds showed lower activity in hydrogenation of 1-butene, their catalysts may be effective as less hydrogen-consuming catalysts. Further, they prepared thin film of Chevrel phase compounds on Al_2O_3.[327] The thin film of Chevrel phase compounds $PbMo_6S_8$ revealed a higher activity per unit surface area in the thiophene HDS than MoS_2 sulfided at 1000°C and MoS_2 with Co (Co/Mo = 0.25).

Supported heteropolycompounds including Mo, Ni, Co showed high activity in thiophene HDS.[328,329] HDS activity changed depending on the kinds of supports such as alumina, silica, or titania.[328] In these supports, the structures of heteropolycompounds were different from each other. Further, the activity of silica-supported P-Mo heteropolycompounds in thiophene HDS changed depending on sulfidation conditions. This is because the acid properties of the catalysts changed depending on sulfidation conditions. As the acid properties of the supported heteropolycompounds increase, the activity becomes higher. Therefore, not only do the supported heteropolycompounds have possibilities as novel HDS catalysts, but a detailed investigation of the catalysis of these compounds may also elucidate the mechanism of the generation of the activity using supported Co-Mo HDS catalysts. On the other hand, molybdenum nitride prepared from 12-molybdophosphoric acid was an active catalyst for HDN of indole.[330] It gave comparable activity (per unit mass) to that of Mo-Ni/Al$_2$O$_3$ but consumed less hydrogen.

Ho *et al.* prepared Fe-Mo and Fe-W bulk sulfide catalysts from thermal decomposition of iron bis(diethylenetriamine)thiomolybdate or thiotungstate in H$_2$S/H$_2$ at 325°C.[331–333] This catalyst system gave an unusual combination of high HDN and low HDS activities.[331] The results from a brief accelerated aging experiment showed that this bulk sulfide system is thermally stable. Characterization of the Fe-Mo catalyst indicated that it consists of a single sulfide phase which during activity testing partially transforms into an iron sulfide mixed with an MoS$_2$-like phase. This was also active for hydrogenating aromatics.[332] It was shown that the HDN and HDS volumetric activities of the Fe-Mo catalyst could be significantly increased by promotion with nickel or cobalt.[333] The increased activities were much higher than those observed with existing commercial catalysts. Karroua *et al.* prepared unsupported iron, molybdenum and iron-molybdenum sulfides by the Homogeneous Sulfide Precipitation (HSP) method.[334] This method gave a high proportion of the Fe species attributed to the so called "FeMoS" phase, together with a noncrystalline phase possibly corresponding to FeS$_2$ and some pyrrhotite (Fe$_{1-x}$S). However, the FeMoS phase decomposed extensively during parallel thiophene HDS and hydrogenation of cyclohexene. Catalytic activity did not correlate with the quantity of FeMoS present in the samples, but seemed to vary in parallel with the amount of FeS$_2$ remaining in the working catalyst.

Recently, a single phase molybdenum phosphide, MoP, with a moderate surface area was prepared using temperature-programmed reduction of an amorphous phosphate precursor and the sample thus prepared was found to be a good catalyst for HDN.[335]

4.1.4 Approaches Using Other New Preparation Methods

As mentioned above (Section 4.1.1A), the addition of nitrilotriacetic acid (NTA) to Co-Mo catalysts increased the catalytic activity of thiophene HDS. The effects were reported for Al$_2$O$_3$,[21,22,34,336] SiO$_2$ [21,34] and carbon [21,32,34] supports. Yamada and coworkers prepared Co-Mo/Al$_2$O$_3$ catalysts by an impregnation method using NTA and performed HDS of BT in a pressurized flow reactor (270°C, 5 MPa).[336] The NTA-added catalysts showed *ca.* 20% higher HDS activity than conventional catalysts. XPS measurements for catalyst precursors (before sulfidation) revealed that, when they were not calcined, the Co concentration at the surface of the catalysts prepared with NTA was higher than that of conventional catalysts without NTA. They proposed that the role of NTA was to increase highly active Co concentration at the catalyst surface. Blanchard *et al.* used ethylenediamine (EDA) for the preparation of Co-Mo/Al$_2$O$_3$ catalysts.[337] This method improved the dispersion of cobalt and molybdenum species on the catalyst and increased the activity of thiophene HDS. Yoshimura *et al.* prepared molybdate, nickel-molybdate and cobalt-molybdate/Al$_2$O$_3$ catalysts by an impregnation method using citric acid as well as ammonia as ligands.[338] The Ni-Mo catalyst prepared using citric acid was inferior to the one prepared using ammonia in terms of both hydrogenation and HDN activities. This effect of citric acid was attributed to a decrease in the amount of active NiMoS phase. On the other hand, the Co-Mo catalyst prepared using citric acid was superior to the one prepared using ammonia in terms of HDS activity. The increase in the HDS activity was attributed to a decrease in the lateral size of MoS$_2$-like crystallites.

Zdrazil and coworkers developed a new slurry impregnation method for the preparation of molybdenum-based catalysts.[339,340] Aqueous slurry of powdered MoO_3 was mixed with alumina extrudates and the mixture was refluxed.[339] The low solubility of MoO_3 was sufficient for transport of MoO_3 from powder form via solution to the surface of the support. The activity of thiophene HDS for the catalysts prepared by this method was similar to the activity of the conventional Mo/Al_2O_3 catalyst. The advantages of this method they proposed to be as follows: Calcination, producing nitrogeneous waste gases, is not required; all deposited molybdenum species are adsorbed and not precipitated. Active carbon supported molybdena catalysts prepared by the slurry impregnation revealed four times higher activity than a commercial molybdena alumina catalyst and about the same activity as a carbon-supported catalyst prepared by conventional impregnation.[340] In this case, the advantage of this method is also that molybdena species is deposited by adsorption and that oxidative degradation of the carbon support cannot occur during preparation because calcination is left out.

Lebihan *et al.* prepared molybdenum oxide-alumina catalysts with a wide range of Mo loadings (1–25% Mo) by the sol-gel process.[341] Two methods were used. In one route, alumina was prepared by hydrolysis of aluminium tri-sec-butylate in butanol and butanediol, and molybdenum was deposited by a classical dry imregnation with ammonium heptamolybdate. In the other route, the molybdenum oxospecies was dispersed in butanediol and added to the aluminum alkoxide before hydrolysis. The thiophene HDS activity of the calcined and sulfided catalyst prepared by the latter method increased linearly with the loading of Mo and revealed good linear relationship even at high loadings. This was attributed to the good dispersion of molybdenum which was obtained by the sol-gel process.

Recently, it has been reported that dispersed supported molybdena/alumina catalysts can be prepared from simple physical mixtures by spreading MoO_3 on the surface of an Al_2O_3 support (solid-solid wetting).[342–347] Reddy and Manohar applied this "solid-solid wetting" process as a simple and effective method for the preparation of molybdenum sulfide/alumina catalysts.[348] The spreading of MoS_2 on an Al_2O_3 support under thermal treatment at 723 K in the presence of CS_2-H_2 atmosphere results in a dispersed phase, which exhibits activity for thiophene HDS comparable to that using a conventional catalyst. Moreau *et al.* performed quinoline HDN over a mechanical mixture of sulfided $Co-Mo/Al_2O_3$ and $Ni-Mo/Al_2O_3$ catalysts at 340°C and 70 bar hydrogen pressure.[349] The $Co-Mo/Al_2O_3$-rich mixtures exhibited higher activity for the overall quinoline conversion. Enhancement of the reactivity mainly due to the better hydrogenolysis properties of the catalysts mixtures was observed by a factor of about three. These catalysts were effective for HDN of alkylanilines in the presence of heavier nitrogen compounds. Kim *et al.* prepared new sulfided Mo/Al_2O_3 catalysts by multiple impregnation of previously sulfided catalysts. These displayed a morphology similar to that shown by conventionally prepared catalysts, but exhibited lower NO chemisorption and higher HDS activity.[350] They proposed that the higher activities are due to the creation of more catalytically active sites from the repeated exposure to air during preparation. Rao *et al.* prepared alumina-supported Mo, Co-Mo and Ni-Mo catalysts by the Precipitation From Homogeneous Solution (PFHS) technique using thioacetamide hydrolysis in a single step.[351] These catalysts did not need presulfiding prior to HDS reaction and exhibited higher HDS activity than conventional catalysts.

4.2 Approaches Using Molybdenum Carbonyls

Generally, molybdena-alumina catalysts are prepared using ammonium heptamolybdate. An alternative method is to use supported metal complexes. Metal sulfide complexes have been used for preparing supported and unsupported hydrotreating catalysts.[352-355] Supported metal carbonyl complexes are found to be active in several catalytic reactions,[356-360] e.g., hydrogenation of carbon monoxide,[361-363] metathesis of olefin,[364-365] etc. The reactivities of alumina-supported molybdenum carbonyls have also been reported by several authors.[366,367] However, their reactivities in HDS of thiophenes has been investigated very little, especially in a pressurized flow system. HDN using supported metal carbonyls is also rare.[368] Several groups performed thiophene HDS over catalysts derived from supported-$Mo(CO)_6$ under atmospheric pressure. Okamoto et al. reported that a $Mo(CO)6/Al_2O_3$ with low content of Mo was more active than a conventional MoO_3/Al_2O_3,[130,369] in HDS of thiophene. They also studied HDS of thiophene catalyzed by zeolite-supported $Mo(CO)_6$,[131] but selectivities for products and life of the catalyst were not examined. Anderson et al.[145] and Laniecki and Zmierczak[370] also prepared zeolite-supported molybdenum catalysts for thiophene HDS in a similar way. The enhanced activity was ascribed to a higher dispersion of molybdenum sulfide species. Sugioka et al. also used $Mo(CO)_6$ to prepare zeolite-supported HDS catalysts which showed higher activity in thiophene HDS than alumina-supported ones.[371] Some groups used cluster complexes which may be regarded as models of heterogeneous active phases. Curtis et al. used $(\eta\text{-}C_5H_4Me)_2Mo_2Co_2S_3(CO)_4$ to prepare supported Co-Mo catalysts and found that these catalysts showed activity comparable to that of commercial Co-Mo/Al_2O_3.[372] Brenner et al. prepared catalysts derived from alumina, magnesia or carbon-supported sulfide clusters to better understand the promotional effect in hydrotreatment catalysis.[373] Similarly, Carvill and Thompson prepared alumina-supported catalysts derived from the sulfide clusters $Cp_2Mo_2(\mu\text{-}SH)_2(\mu\text{-}S)_2$, $Cp_2Mo_2Co_2(\mu_3\text{-}S)_2(\mu_4\text{-}S)(CO)_4$, and $Cp_2Mo_2Fe2(\mu3\text{-}S)2(CO)_8$ (Cp = cyclopentadienyl).[374] There were differences between the thiophene HDS product distributions. The sulfide cluster-derived catalysts produced significant amounts of C_2 and C_3 hydrocarbons while the commercial catalyst produced mostly C_4 hydrocarbons. They suggested that C-C bond hydrogenolysis preceded C-S bond cleavage over the sulfide cluster-derived catalysts. Okamoto et al. used $Co_2(CO)_8$ or $Co(NO)(CO)_3$ with $Mo(CO)_6$ to prepare Co-Mo catalysts.[369,375] Mauge et al. also used $Co(CO)_3NO$ to prepare Co-Mo/Al_2O_3 catalysts with high catalytic activity.[376] In these examples, however, the HDS reactions were often performed in a batch system or under atmospheric conditions, and the difference in the reaction conditions could have caused some discrepancy. Further, the HDS reactivity of the catalysts derived from supported metal carbonyls under pressurized flow systems is not yet well known.

On the other hand, the reaction mechanism of HDS of DBT catalyzed by supported molybdenum-based catalysts has been described in Chapter 2 in detail. Since it has often been pointed out that DBT is less reactive than thiophene and benzothiophene, it was presumed that DBT would be one of the key compounds in deep HDS of heavy feedstock. Further, it has been confirmed that DBT is a key compound in deep HDS of light gas oil (LGO). To our knowledge, however, there is very little literature on HDS of DBT catalyzed by supported metal carbonyls. Vrinat et al. reported HDS of DBT catalyzed by $Mo(CO)_6$/Zeolite,[132] and the catalytic activity was found to be much lower than that of the conventional MoO_3/Al_2O_3.

Halbert *et al.* also reported that catalysts derived from an "edge decoration" of MoS_2 with low-valent organometallic complexes such as $Co_2(CO)_8$ showed catalytic activity equal to or greater than that of commercial Ni-Mo/Al_2O_3 in HDS of DBT and gas oil.[377]

In the following section, HDS of DBT catalyzed by alumina-supported metal carbonyl complexes such as $Mo(CO)_6$, $Cr(CO)_6$ and $W(CO)_6$ is described, specifically concerning the catalytic activity and product selectivity. Anionic metal carbonyls with a metal sulfur bond are supported on alumina. The vaporization of the metal species in activation processes can be inhibited with the use of these salts in heterogeneous catalytic reactions,[361-363] because the vapor pressure of these anionic salts is much smaller than that of neutral metal carbonyls. Further, the hydrotreatment of metal carbonyls with metal sulfur bonds will form metal sulfides. For preparation of anionic metal carbonyls with metal sulfur bond, references were made to Beck's work,[378] where $Mo(CO)_5THF$ reacted with sodium ethyl sulfide to form $NaMo(CO)_5SEt$, and to Seyferth's work,[379] where triiron dodecacarbonyl reacted with triethylamine (NEt_3) and ethanethiol (EtSH) to form anionic iron carbonyl species with iron-sulfur bonds. It is important to note here that the catalysts derived from metal carbonyls-NEt_3-EtSH/Al_2O_3 systems, where metal carbonyls react with NEt_3 and EtSH to form anionic metal carbonyls with metal-sulfur bonds, are more effective in HDS of DBT than are conventional impregnated catalysts.[380,381]

4.2.1 Hydrodesulfurization Catalysts Prepared from Alumina-supported Group VI Metal Carbonyl Complexes

A. Alumina-supported Molybdenum Carbonyls

HDS of DBT was carried out into a pressurized fixed-bed flow reactor under the following conditions: temperature, 300°C; 50 kg/cm^2; H_2 18 l/h; WHSV, 16.5 h^{-1}; initial concentration of DBT, 1.0 wt%; solvent xylene, catalyst, 0.5 g. Catalysts derived from alumina-supported molybdenum carbonyl complexes were active in HDS of DBT over 10 h and the major products were BP and CHB. The results from supported $Mo(CO)_6$ systems are shown in Table 4.1. $Mo(CO)_6$/Al_2O_3 was prepared by mixing $Mo(CO)_6$ and Al_2O_3 in THF, which was removed *in vacuo*. When $Mo(CO)_6$/Al_2O_3 was activated by H_2 (run 1), the conversion of DBT and the selectivity for BP were 18% and 89%, respectively. When the same system was activated by H_2S in H_2 (run 2), the catalytic activity increased and the conversion of DBT was 26%. These results show that the formation of molybdenum sulfide at the initial stage is essential in the high conversion of DBT. In $Mo(CO)_6$-Ph_2S/Al_2O_3, Al_2O_3 was impregnated with a pentane solution of $Mo(CO)5Ph_2S$[382] formed from $Mo(CO)_6$ and Ph_2S under UV irradiation. When $Mo(CO)_6$-Ph_2S/Al_2O_3 was activated by H_2 (run 3), the conversion of DBT was 25%, similar to that in run 2. It was suggested that molybdenum sulfide could be formed by hydrogenolysis of $Mo(CO)_5Ph_2S$ on Al_2O_3. In run 4, in which, the $Mo(CO)_6$-Ph_2S/Al_2O_3 system was presulfided by H_2S, the conversion of DBT decreased to 18%. It is likely that excessive sulfidation of molybdenum species decreased the conversion of DBT.

$Mo(CO)_6$-NEt_3-EtSH/Al_2O_3 was prepared as follows. THF solution of $Mo(CO)_6$ was stirred under UV irradiation and reflux for 2 h to give $Mo(CO)_5THF$,[378] which reacted with NEt_3 and EtSH (Mo:N:S = 1:2:2) at room temperature to give a dark brown solution. This solution showed IR absorptions at 1941(vs) and 1871(m) cm^{-1} which were attributed to the

Table 4.1 Hydrodesulfurization of Dibenzothiophene Catalyzed by Alumina-supported Molybdenum Carbonyls [a]

Run	Catalyst	Conv. of DBT (%)	Conv. of DBT to BP (%)	Conv. of DBT to CHB (%)	Selec. for BP (%)	Mo Content (wt%/cat)
1	Mo(CO)$_6$/Al$_2$O$_3$ [b]	18	16	2	89	7.1
2	Mo(CO)$_6$/Al$_2$O$_3$ [c]	26	19	7	72	7.3
3	Mo(CO)$_6$-Ph$_2$S/Al$_2$O$_3$ [b,d]	25	21	4	83	8.1
4	Mo(CO)$_6$-Ph$_2$S/Al$_2$O$_3$ [c,d]	18	14	3	82	7.7
5	Mo(CO)$_6$-NEt$_3$-EtSH/Al$_2$O$_3$ [b,e]	43	32	11	74	7.8
6	Mo(CO)$_6$-NEt$_3$-EtSH/Al$_2$O$_3$ [c,e]	43	27	16	63	7.8
7	Mo(CO)$_6$-NEt$_3$-EtSH/Al$_2$O$_3$ [b,f]	43	34	9	79	7.8
8	Mo(CO)$_6$-NEt$_3$-EtSH/Al$_2$O$_3$ [c,f]	26	14	12	53	7.3
9	Mo(CO)$_6$-NaSEt/Al$_2$O$_3$ [b]	33	30	3	91	—
10	[PPN][Mo(CO)$_5$SH]/Al$_2$O$_3$ [b]	9	8	1	88	—
11	[NEt$_4$][Mo(CO)$_5$(OOCCH$_3$)]/Al$_2$O$_3$ [c]	35	23	12	66	6.2
12	MoO$_3$/Al$_2$O$_3$ [c,g]	38	27	11	71	7.8

[a] 300°C, 50 atm, LHSV 14 h^{-1}, H$_2$ 18 1/h, Cat 0.5 g, Mo 8.0 wt%. [b] Activated by H$_2$. [c] Presulfided by H$_2$S in H$_2$ (H$_2$S 3%). [d] Mo(CO)$_6$:Ph$_2$S = 1:2. [e] Mo(CO)$_6$: NEt$_3$: EtSH = 1:1:1. [f] Mo(CO)$_6$:NEt$_3$:EtSH = 1:2:2. [g] MoO$_3$ 12.5 wt%.
(From A. Ishihara, T. Kabe et al., J. Jpn. Petrol. Inst., **36**, 362 (1993))

formation of [NEt$_3$H][Mo$_2$(CO)$_{10}$SEt] (*vide infra*).[383] Al$_2$O$_3$ was added to the solution. The mixture was stirred for 2 h, after which THF was removed *in vacuo*.

The most active catalysts were derived from the hydrogenolysis and presulfiding of Mo(CO)$_6$-NEt$_3$-EtSH/Al$_2$O$_3$ (Mo:N:S = 1:1:1) (runs 5 and 6) where the conversions of DBT reached 43%. In runs 7 and 8, twofold amounts of NEt$_3$ and EtSH were used in the preparation of the catalyst. The catalyst derived from hydrogenolysis showed 43% conversion of DBT, while the catalyst derived from presulfiding lowered the conversion. This observation was similar to the result in run 4. The conversion of DBT changed depending on the amount of NEt$_3$ and EtSH added at the preparation of the catalyst. In the catalysts derived from hydrogenolysis of the Mo(CO)$_6$-NEt$_3$-EtSH/Al$_2$O$_3$ system, the effects of the amount of NEt$_3$ and EtSH were investigated and the results are shown in Fig. 4.7; the ratio of NEt$_3$ to EtSH was always 1:1. At EtSH:Mo(CO)$_6$ = 1:1, the conversion of DBT reached maximum, but the selectivity for BP was minimum. Further, addition of NEt$_3$ and EtSH slightly decreased conversion. However, catalysts derived from hydrogenolysis of precursors were not affected by the amount of NEt$_3$ and EtSH as much as those from presulfiding. This suggests that catalysts derived from hydrogenolysis always have a certain level of unsaturated sites because the excess of sulfur in a precursor could be flushed by hydrogen sweeping. On the contrary, it seems that the presulfiding procedure could not generate a sufficient level of unsaturated sites when a precursor contained excess sulfur. Since precursors in runs 4 and 8 contained excess sulfur, which decreased active sites on the catalysts by presulfiding, the catalytic activity decreased. Fig. 4.8 shows the effects of the amount of loaded Mo on the conversion of DBT and the selectivity for BP. The conversion of DBT increased with increase in the amount of Mo loaded, reaching the maximum at 12 wt% of Mo loaded, and conversion was kept constant above 12 wt% of Mo. Regarding the use of MoO$_3$/Al$_2$O$_3$ derived from ammonium heptamolybdate, similar results have been reported.[384]

The effect of cations on HDS of DBT was investigated. When sodium ethyl sulfide was

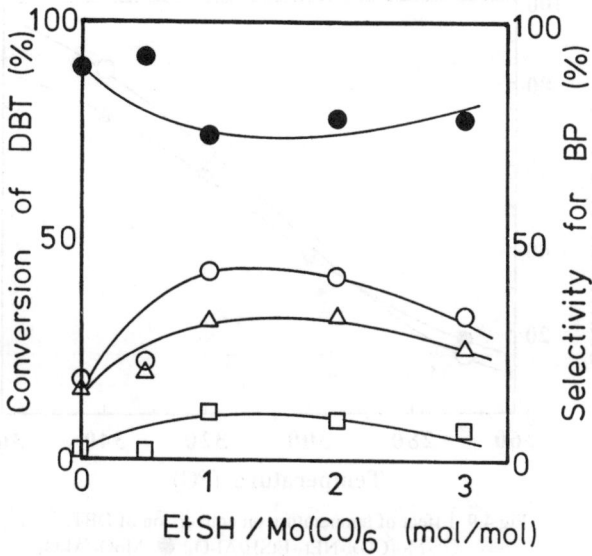

Fig.4.7 Effect of ratio of EtSH/Mo(CO)₆ on conversion of DBT and selectivity for BP.
○: Total Conversion of DBT; △: Conversion of DBT into BP;
□: Conversion of DBT into CHB; ●: Selectivity for BP.
(From A. Ishihara, T. Kabe *et al.*, *J. Jpn. Petrol. Inst.*, **36**, 363 (1993))

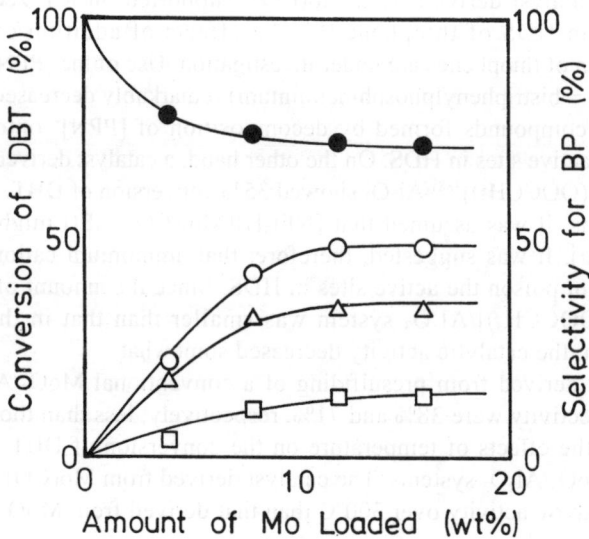

Fig.4.8 Effect of amount of Mo loaded, on conversion of DBT and selectivity for BP.
○: Total Conversion of DBT; △: Conversion of DBT into BP;
□: Conversion of DBT into CHB; ●: Selectivity for BP.
(From A. Ishihara, T. Kabe *et al.*, *J. Jpn. Petrol. Inst.*, **36**, 363 (1993))

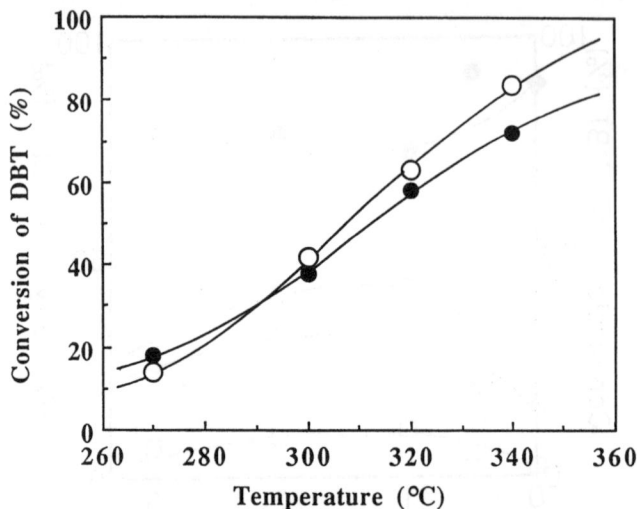

Fig.4.9 Effect of temperature on conversion of DBT.
○: Mo(CO)$_6$-NEt$_3$-EtSH/Al$_2$O$_3$; ●: MoO$_3$/Al$_2$O$_3$.
(From A. Ishihara, T. Kabe *et al.*, *J. Jpn. Petrol. Inst.*, **36**, 364 (1993))

used instead of NEt$_3$ and EtSH (run 9), the catalytic activity slightly decreased although the selectivity for BP increased. There are few examples on the effect of addition of an alkali metal on hydrodesulfurization. Na-doped Co-Mo/Al$_2$O$_3$ did not affect the catalytic activity in HDS of thiophene,[385] while Li-doped Co-Mo/Al$_2$O$_3$ decreased it.[386] Further, Okamoto *et al.* reported that the catalyst derived from Mo(CO)$_6$ supported on KY zeolite showed high catalytic activity in HDS of thiophene.[139] The effects of addition of alkali metals to hydrodesulfurization of thiophenes are under investigation. Use of the [PPN][Mo(CO)$_5$SH][387]/Al$_2$O$_3$ system (PPN = bistriphenylphosphineiminium) remarkably decreased catalytic activity. The deposition of compounds formed by decomposition of [PPN]$^+$ onto the catalyst may cause poisoning of active sites in HDS. On the other hand, a catalyst derived from presulfiding of [NEt$_4$][Mo(CO)$_5$(OOCCH$_3$)][388]/Al$_2$O$_3$ showed 35% conversion of DBT. In Mo(CO)$_6$-NEt$_3$-EtSH/Al$_2$O$_3$ systems, it was assumed that [NEt$_3$H][Mo$_2$(CO)$_{10}$SEt] might be supported on alumina (*vide infra*). It was suggested, therefore, that ammonium cations, such as [NEt$_4$] and [NEt$_3$H], did not poison the active sites in HDS. Since the amount of Mo loaded in the [NEt$_4$][Mo(CO)$_5$(OOCCH$_3$)]/Al$_2$O$_3$ system was smaller than that in the Mo(CO)$_6$-NEt$_3$-EtSH/Al$_2$O$_3$ system, the catalytic activity decreased somewhat.

In the catalyst derived from presulfiding of a conventional MoO$_3$/Al$_2$O$_3$ (run 12), the conversion and selectivity were 38% and 71%, respectively, less than those in runs 5, 6 and 7. Fig. 4.9 shows the effects of temperature on the conversion of DBT in Mo(CO)$_6$-NEt$_3$-EtSH/Al$_2$O$_3$ and MoO$_3$/Al$_2$O$_3$ systems. The catalyst derived from Mo(CO)$_6$-NEt$_3$-EtSH/Al$_2$O$_3$ showed higher catalytic activity over 300°C than that derived from MoO$_3$/Al$_2$O$_3$.

B. Characterization of Supported Molybdenum Catalysts

Figure 4.10 shows the IR spectra of catalyst precursors. Alumina-supported Mo(CO)$_6$ was a pale yellow solid and showed IR absorptions at 1975 and 1952 cm^{-1} (Fig. 4.10(a)). The

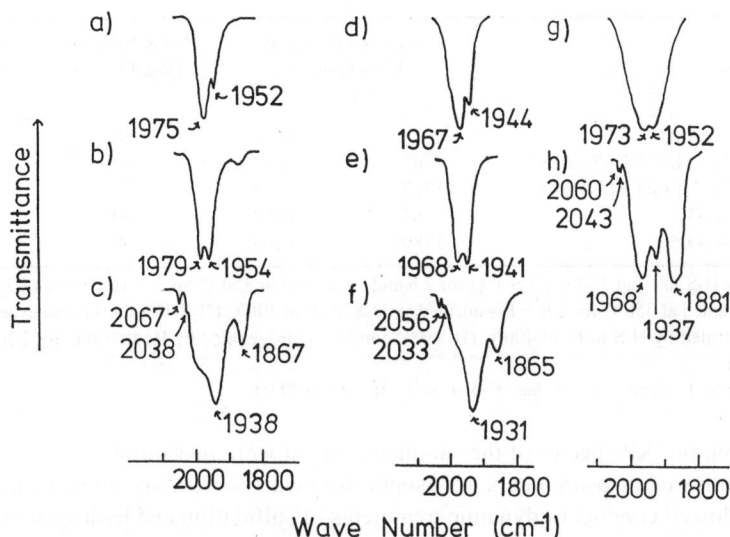

Fig.4.10   FTIR spectra of alumina-supported metal carbonyls.
(a) $Mo(CO)_6/Al_2O_3$; (b) $Mo(CO)_6$-Ph_2S/Al_2O_3; (c) $Mo(CO)_6$-NEt_3-$EtSH/Al_2O_3$; (d) $W(CO)_6/Al_2O_3$;
(e) $W(CO)_6$-Ph_2S/Al_2O_3; (f) $W(CO)_6$-NEt_3-$EtSH/Al_2O_3$; (g) $Cr(CO)_6/Al_2O_3$; (h) $Cr(CO)_6$-NEt_3-$EtSH/Al_2O_3$.
(From A. Ishihara, T. Kabe et al., J. Jpn. Petrol. Inst., 36, 364 (1993))

former is due to $Mo(CO)_6$ species on Al_2O_3 and the latter may be due to $Mo(CO)_5$ species on Al_2O_3. The $Mo(CO)_6$-Ph_2S/Al_2O_3 system showed IR absorption at 1954 cm^{-1} (Fig. 4.10(b)) which is attributed to $Mo(CO)_5Ph_2S$ species on Al_2O_3.[181] The $Mo(CO)_6$-NEt_3-$EtSH/Al_2O_3$ system showed IR absorptions at 2067(w), 2038(w), 1942(vs, br) and 1869(m) cm^{-1} (w = weak; vs = very strong; m = medium; br = broad)(Fig. 4.10(c)). These results indicate that a dinuclear anionic complex $[Mo_2(CO)_{10}SEt]^-$ [383] was formed on alumina. When complexes with a metal-sulfur bond were formed on alumina and activated by hydrogen, higher conversions of DBT were obtained. It was suggested that the activation of these complexes by H_2 form highly dispersed molybdenum sulfide which is active in HDS of DBT. This was confirmed by NO chemisorption experiments and XPS measurements.

Volumetric measurement of NO chemisorption was carried out in a conventional Pyrex glass high vacuum adsorption system. A typical procedure was as follows. Samples of 200 mg were placed in a glass reactor, heated at 5°C/min and activated at 300°C under a flow of H_2 or 3% H_2S/H_2 (30 mL/min) for 2 h. After this treatment, samples were evacuated at 300°C under vacuum (10^{-3} Torr) for 1 h and then cooled to 25°C. When conventional MoO_3/Al_2O_3 was used, a sample was sulfided at 400°C for 3 h, evacuated at 400°C for 1 h then cooled to 25°C. Adsorption of NO was always carried out at 25°C as follows. After a first isotherm, a second one was obtained with an intermediate outgassing under vacuum (10^{-3} Torr) at 25°C for 1 h. The difference in uptake between the first (total adsorption) and the second (reversible adsorption) isotherm was assumed to be the amount of the irreversibly chemisorbed NO. It has been reported that this procedure gives results similar to those obtained by extrapolating to zero NO pressure the linear part (high pressures) of the first isotherm.[389] After NO

Table 4.2 Characterization of Supported Molybdenum Catalysts by Means of NO Chemisorption

Run	Catalyst	NO Chemisorption Wavenumber (cm^{-1})		NO Adsorption (μmol/g-cat)	NO/Mo (mol/mol)
13	Mo(CO)$_6$/Al$_2$O$_3$ [a]	1777	1671	256	0.34
14	Mo(CO)$_6$-2(EtSH-Et$_3$N)/Al$_2$O$_3$ [b]	1802	1682	668	0.82
15	Mo(CO)$_6$-2(EtSH-Et$_3$N)/Al$_2$O$_3$ [a]	1773	1674	321	0.42
16	MoO$_3$/Al$_2$O$_3$ [c]	1786	1688	110	0.14
17	MoO$_3$/Al$_2$O$_3$ [d]	1780	1690	269	0.33

[a] Presulfided by H$_2$S in H$_2$ at 350°C (H$_2$S 3%) for 2 h and evacuated at 350°C for 2. [b] Hydrogenated by H$_2$ at 350°C for 2 h and evacuated at 350°C for 2 h. [c] Presulfided by H$_2$S in H$_2$ at 400°C (H$_2$S 3%) for 3 h and evacuated at 400°C for 2 h. [d] Presulfided by H$_2$S in H$_2$ at 400°C (H$_2$S 3%) for 3 h, hydrogenated by H$_2$ at 400°C for 2 h and evacuated at 400°C for 2 h.

(From A. Ishihara, T. Kabe et al., J. Jpn. Petrol. Inst., **36**, 364 (1993))

chemisorption, the XP spectra of the sampled catalyst were measured.

For the infrared measurements, self-supporting wafers were placed in a special infrared cell which allowed conduct of dynamic treatments of sulfidation and hydrogenation. Samples were sulfided or hydrogenated and evacuated *in situ* under the same conditions as in the volumetric measurements. After cooling to 25°C, samples were then exposed to 10 Torr NO for 15 min. After evacuating for 10 min, FTIR spectra of NO chemisorption were recorded.

Table 4.2 and Fig. 4.11 show the results from NO chemisorption for sulfided and reduced catalysts. Both the wavenumber and the amount of NO chemisorption for sulfided MoO$_3$/Al$_2$O$_3$ (run 16 and Fig. 4.11(d)) were in good agreement with those reported by other workers.[389,390] The wavenumbers of the doublet peak were 1786 and 1688 cm^{-1}, which are assigned to a dinitrosyl complex. The wavenumbers and the shapes of peaks of the other catalysts were very close to those of sulfided MoO$_3$/Al$_2$O$_3$, except for one prepared from hydrogenation of a Mo(CO)$_6$-NEt$_3$-EtSH/Al$_2$O$_3$ system. The catalysts prepared from sulfidation of alumina-supported molybdenum carbonyls (runs 13 and 15), therefore, can be deduced to have a similar structure of active species. On the contrary, the catalyst prepared from hydrogenation of a Mo(CO)$_6$-NEt$_3$-EtSH/Al$_2$O$_3$ system showed different shapes of peaks, while their wavenumbers were similar. The peak at 1682 cm^{-1} was about two times as large as one at 1773 cm^{-1}. It is likely that a potion of the active species for NO chemisorption is different from those of other sulfided catalysts. The amounts of NO adsorption for catalysts derived from supported molybdenum carbonyls were much larger than that for sulfided MoO$_3$/Al$_2$O$_3$. Further, the amount of NO adsorption, for catalyst prepared from hydrogenation of the supported molybdenum carbonyl (run 14), was much larger than those for sulfided ones (runs 13 and 15). The value was higher than that for one prepared by hydrogenation of sulfided MoO$_3$/Al$_2$O$_3$ (run 17). These results indicate that, in the preparation of catalysts, catalysts derived from supported metal carbonyls have a higher dispersion of molybdenum species than those derived from conventional catalysts. The catalysts derived from sulfided supported molybdenum carbonyls other than those from a Mo(CO)$_6$-NEt$_3$-EtSH/Al$_2$O$_3$ system, however, showed lower catalytic activity in HDS of DBT than that of sulfided MoO$_3$/Al$_2$O$_3$. It has been pointed out that there is no linear correlation between the amount of NO adsorption and HDS activity for Ni-Mo or Co-Mo catalysts and that the deviation from linearity is probably attributed to the fact that the total NO uptake involves different contributions of

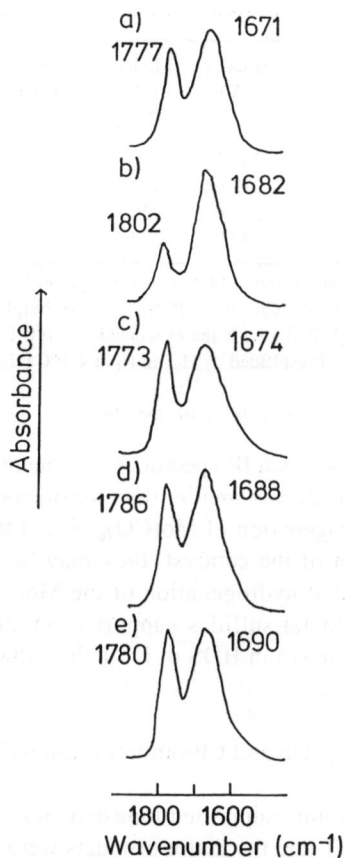

Fig.4.11 FTIR spectra of NO chemisorption on catalysts derived from alumina-supported molybdenum carbonyls.
(a) Derived from sulfidation of $Mo(CO)_6/Al_2O_3$; (b) Derived from hydrogenation of $Mo(CO)_6$-NEt_3-EtSH/Al_2O_3; (c) Derived from sulfidation of $Mo(CO)_6$-NEt_3-EtSH/Al_2O_3; (d) Derived from sulfidation of MoO_3/Al_2O_3; (e) Derived from hydrogenation after sulfidation of MoO_3/Al_2O_3.
(From A. Ishihara, T. Kabe et al., J. Jpn. Petrol. Inst., **36**, 365 (1993))

chemisorbed NO on Mo, Ni (or Co), and Ni-Mo (or Co-Mo) sites.[390,391] In this study, NO chemisorption may reflect the surface coverage or dispersion of Mo, rather than the number of active sites in HDS.

Table 4.3 shows the results from XPS measurements of catalysts used in NO chemisorption experiments. Binding energies for every catalyst were very similar to those of MoS_2. Further, the values of N/Mo and S/Mo ratios were very close to each other, except that of the hydrogenated catalyst (run 19). These findings indicate that the active species of these catalysts for NO chemisorption would be almost the same. In their study of sulfided MoO_3/Al_2O_3 using EXAFS, Clausen et al. reported that Mo was predominantly present as MoS_2-like domains.[391] Topsøe and Topsøe also proposed that the adsorption of NO occurred mainly on Mo atoms present as MoS_2.[390] In our systems, it is likely that MoS_2 on alumina is formed by the sulfidation of supported molybdenum carbonyls. On the contrary, the value of N/Mo ratio for hydrogenated supported molybdenum carbonyls was greater than for others.

Table 4.3 Characterization of Supported Molybdenum Catalysts by XPS [a]

Run	Catalyst	Mo3d$_{3/2}$ (eV)	Mo3d$_{5/2}$ (eV)	N1s (eV)	S2p (eV)	N/Mo [b]	S/Mo [b]
18	Mo(CO)$_6$/Al$_2$O$_3$ [c]	233.2	230.2	395.5	163.5	2.55	3.18
19	Mo(CO)$_6$-2(EtSH-Et$_3$N)/Al$_2$O$_3$ [d]	233.1	230.2	395.7	162.7	4.04	2.09
20	Mo(CO)$_6$-2(EtSH-Et$_3$N)/Al$_2$O$_3$ [c]	232.6	229.6	395.6	162.6	2.50	3.00
21	MoO$_3$/Al$_2$O$_3$ [e]	233.0	229.8	395.9	163.0	2.61	3.06
22	MoS$_2$ [f]	232.8	229.6		162.6		2.67

[a] XPS spectra were measured for the samples used in NO chemisorption. Every binding energy wsa referenced to Carbon C1s 285.0 eV due to adventitious carbon. [b] Ratio of peak heights between Mo3d$_{5/2}$, N1s and S2P. [c] Presulfided by H$_2$S in H$_2$ at 350°C (H$_2$S 3%) for 2 h and evacuated at 350°C for 2 h. [d] Hydrogenated by H$_2$ at 350°C for 2 h and evacuated at 350°C for 2 h. [e] Presulfided by H$_2$S in H$_2$ at 400°C (H$_2$S 3%) for 3 h and evacuated at 400°C for 2 h. [f] neat MoS$_2$.
(From A. Ishihara, T. Kabe et al., J. Jpn. Petrol. Inst., **36**, 365 (1993))

This may reflect the difference between IR spectra of NO chemisorption in Fig. 4.6. The S/Mo ratio for this hydrogenated catalyst was smaller than for others. This indicates that, although this catalyst derived from hydrogenation of Mo(CO)$_6$-NEt$_3$-EtSH/Al$_2$O$_3$ forms molybdenum sulfides during the preparation of the catalyst, they may be different from other sulfided catalysts. The results suggest that hydrogenation of the Mo(CO)$_6$-NEt$_3$-EtSH/Al$_2$O$_3$ system would form highly dispersed metal sulfides supported on alumina such as MoS$_{2-x}$/Al$_2$O$_3$. This species seems to be more active in HDS of DBT than alumina-supported MoS$_2$ derived from sulfidation of MoO$_3$/Al$_2$O$_3$.

C. Alumina-supported Tungsten and Chromium Carbonyls

Catalysts derived from alumina-supported tungsten and chromium carbonyl complexes were also active in HDS of DBT and the major products were BP and CHB. These catalysts were prepared by procedures similar to those for the corresponding molybdenum catalysts. The results from supported W(CO)$_6$ systems are shown in Table 4.4. When W(CO)$_6$/Al$_2$O$_3$ was activated by H$_2$S in H$_2$ (run 23), the conversion of DBT was 8.6%. When W(CO)$_6$-Ph$_2$S/Al$_2$O$_3$ was presulfided by H$_2$S (run 24), the conversion of DBT decreased to 6.4%. The catalyst derived from the hydrogenolysis of W(CO)$_6$-NEt$_3$-EtSH/Al$_2$O$_3$ (run 25) was the most active one in which the conversion of DBT was 9.3%. In the catalyst derived from presulfiding of a conventional WO$_3$/Al$_2$O$_3$ (run 26), the conversion was 10% greater than those derived from supported tungsten carbonyls, while the selectivity for BP was lower.

The results from supported Cr(CO)$_6$ systems are also shown in Table 4.4. When Cr(CO)$_6$/Al$_2$O$_3$ was activated by H$_2$S in H$_2$ (run 27), the conversion of DBT was 4.9%. The catalyst derived from the hydrogenolysis of Cr(CO)$_6$-NEt$_3$-EtSH/Al$_2$O$_3$ (run 28) was more active, the conversion of DBT being 7.3%. In the catalyst derived from presulfiding of a conventional CrO$_3$/Al$_2$O$_3$ (run 29), the conversion was 10.4% greater than those derived from supported chromium carbonyls, while the selectivity for BP was lower.

Figure 4.10 also shows the IR spectra of precursors of tungsten and chromium catalysts. Alumina-supported W(CO)$_6$ was a pale yellow solid and showed IR absorptions at 1967 and 1944 cm^{-1} (Fig. 4.10(d)). The former is attributed to W(CO)$_6$ species on Al$_2$O$_3$ and the latter may be to W(CO)$_5$ species on Al$_2$O$_3$. W(CO)$_6$-Ph$_2$S/Al$_2$O$_3$ system showed IR absorption at 1941 cm^{-1} (Fig. 4.10(e)), which is attributed to W(CO)$_5$Ph$_2$S species on Al$_2$O$_3$.[382] W(CO)$_6$-

Table 4.4 Hydrodesulfurization of Dibenzothiophene Catalyzed by Alumina-Supported Tungsten and Chrominum Carbonyls [a]

Run	Catalyst	Conv. of DBT (%)	Conv. of DBT to BP (%)	Conv. of DBT to CHB (%)	Select. for BP (%)
23	W(CO)₆/Al₂O₃ [c]	8.6	8.0	0.6	93
24	W(CO)₆-Ph₂S/Al₂O₃ [c,d]	6.4	5.5	0.9	86
25	W(CO)₆-NEt₃-EtSH/Al₂O₃ [b,e]	9.3	8.5	0.8	91
26	WO₃/Al₂O₃ [c,f]	10.0	8.2	1.8	82
27	Cr(CO)₆/Al₂O₃ [b]	4.9	4.5	0.4	93
28	Cr(CO)₆-NEt₃-EtSH/Al₂O₃ [c,g]	7.3	6.6	0.7	91
29	CrO₃/Al₂O₃ [b,h]	10.4	9.1	1.3	87

[a] 300°C, 50 atm, LHSV 14 h^{-1}, H₂ 18 l/h, Cat 0.5 g, W 8.0 wt%. [b] Activated by H₂. [c] Presulfided by H₂S in H₂ (H₂S 3%). [d] W(CO)₆: Ph₂S = 1: 2. [e] W(CO)₆: NEt₃: EtSH = 1: 2: 2. [f] WO₃ 10 wt%. [g] Cr(CO)₆: NEt₃: EtSH = 1:2:2. [h] CrO₃ 14 wt%.
(From A. Ishihara, T. Kabe et al., J. Jpn. Petrol. Inst., **36**, 366 (1993))

NEt₃-EtSH/Al₂O₃ system showed IR absorptions at 2056(w), 2033(w), 1931(vs, br) and 1865(m) cm^{-1} (Fig. 4.10(f)). These findings indicate that a dinuclear anionic complex [W₂(CO)₁₀SEt]$^-$ [182] was formed on alumina.

Alumina-supported Cr(CO)₆ was also a pale yellow solid and showed IR absorptions at 1973 and 1952 cm^{-1} (Fig. 4.10(g)). The former is attributed to Cr(CO)₆ species on Al₂O₃ and the latter may be due to Cr(CO)₅ species on Al₂O₃. The Cr(CO)₆-NEt₃-EtSH/Al₂O₃ system showed IR absorptions at 2060(w), 2043(w), 1968(vs, br), 1937(s) and 1881(vs, br) cm^{-1} (Fig. 4.10(h)). This spectrum was very different from those of molybdenum and tungsten systems. Although a dinuclear complex may be formed, it seems that the amount is very small. When complexes with the metal-sulfur bond were formed on alumina and activated by hydrogen, higher conversions of DBT were obtained although they were lower than those of conventional ones in the cases of tungsten and chromium. It was deduced, therefore, that the activation of these complexes by H₂ formed highly dispersed tungsten or chromium sulfide, which are active in HDS of DBT.

4.2.2 Effects of Supports on Catalysts Prepared from Supported Anionic Molybdenum Carbonyls

In the preceding section, it was shown that the use of alumina-supported anionic molybdenum carbonyls inhibited the sintering and sublimation of molybdenum species and maintained the dispersion of molybdenum sulfide on the support. When a support with a surface area larger than that of alumina is used, higher dispersion of metal species can be obtained to provide higher catalytic activity. In this section, the effects of supports on HDS of DBT catalyzed by supported metal carbonyl complexes is described.[381,392,393] SiO₂-Al₂O₃, Al₂O₃, SiO₂, TiO₂, NaY zeolite, HY zeolite, HZSM-5, and active carbon were used as the supports and an anionic carbonyl, [NEt₄][Mo(CO)₅(OOCCH₃)], as the precursor in HDS of DBT catalyzed by supported anionic molybdenum carbonyls. [NEt₄][Mo(CO)₅(OOCCH₃)] (Et = C₂H₅) was synthesized by a method reported elsewhere.[388] SiO₂-Al₂O₃ (JRC-SAL2: 560 m²/g), SiO₂ (JRC-SIO4: 347 m²/g), TiO₂ (JRC-TIO1: 70.8 m²/g), NaY zeolite (JRC-Z-Y4.8: 670 m²/g), HY zeolite (JRC-Z-HY4.8: 663 m²/g) were supplied by the Catalysis Society of

Japan. Al_2O_3 was γ-alumina supplied from Nippon Ketjen (256 m²/g). These supports were dried under vacuum at 350 °C for 4 h prior to use and stored in Ar atmosphere. Preparation methods of catalysts are as follows: $[NEt_4][Mo(CO)_5(OOCCH_3)]$ and a support (SiO_2-Al_2O_3, Al_2O_3, SiO_2, TiO_2, NaY zeolite, HY zeolite, HZSM-5, or active carbon) were stirred in THF at 25°C for 2 h and THF was removed *in vacuo*. $Mo(CO)_6$-NEt_3-$EtSH$/SiO_2-Al_2O_3 and $Mo(CO)_6$/SiO_2-Al_2O_3 systems were also prepared using methods similar to those presented in the preceding section. MoO_3/SiO_2-Al_2O_3 was prepared with ammonium molybdate similar to MoO_3/Al_2O_3.

A. Catalysts Derived from Supported $[NEt_4][Mo(CO)_5(OOCCH_3)]$ Systems

Catalysts derived from supported $[NEt_4][Mo(CO)_5(OOCCH_3)]$ were active for HDS of DBT over 10 h and the major products from HDS were BP and CHB. The results are shown in Table 4.5. When silica-alumina was used as the support, the total conversion of DBT was 85% and the yield of BP and CHB was 58%. The difference between the total conversion (85%) and the yields of BP and CHB (58%) indicates the conversion of DBT by hydrocracking that does not yield BP or CHB. However, products of DBT conversion by hydrocracking were formed in the range C_1–C_{11}. Although a number of peaks of hydrocracking products were detected by GC analysis, each yield was less than 1% except for benzene and toluene, which may be formed by hydrocracking of the solvent, xylene. Benzene and toluene may also be formed by hydrocracking of BP or CHB. However, when the same amount of BP and CHB as DBT was added to a DBT solution, and the same reaction was performed, no detectable increase in the yield of toluene was found. Further, an increase in only a very small amount of benzene, which was within the range of error in GC analysis, was detected, indicating that the cracking products was formed mainly by cracking of DBT directly and not cracking of desulfurization products. Silica-alumina without molybdenum revealed some hydrocracking activity; however, the conversion of DBT was less than 3% at 300 °C. Therefore, it is suggested that the hydrocracking activity of the catalyst prepared from $[NEt_4][Mo(CO)_5(OOCCH_3)]$/$SiO_2$-$Al_2O_3$ was derived from the molybdenum species supported

Table 4.5 Hydrodesulfurization of Dibenzothiophene Catalyzed by Supported $[NEt_4][Mo(CO)_5(CH_3COO)]$ Catalysts [a]

Run	Catalyst	Conv. of [b] DBT (%)	Conv. of DBT to BP (%)	Conv. of DBT to CHB (%)	Rate [c] of HDS (mol/mol-Mo/h)	Mo content (wt%)
1	SiO_2-Al_2O_3	85(58)	26	32	0.71	7.0
2	Al_2O_3	35(35)	23	12	0.48	6.2
3	SiO_2	20(20)	17	3	0.24	7.0
4	TiO_2	31(31)	18	13	0.60	4.4
5	NaY zeolite	14(14)	13	1	0.52	2.3
6	HY zeolite	75(0)	0	0	—	—
7	HZSM-5	14(14)	7	7	—	—
8	Active Carbon	61(31)	19	12	—	—

[a] Reaction temp. 300°C, Pressure 50 kg/cm², WHSV 16.5 h⁻¹, Cat. 0.50 g, H_2 18 l/h, Initial concentration of DBT 1.0 wt%. Presulfided by H_2S in H_2 at 350°C (H_2S 3%). [b] Yields of BP and CHB, i.e., the sum of the convensions of DBT to BP and CHB are given in parentheses. [c] Rate of HDS of DBT into BP and CHB. This value was calculated from the sum of the conversions of DBT toBP and CHB.
(From A. Ishihara, T. Kabe *et al.*, *J. Jpn. Petrol. Inst.*, **37**, 413 (1994))

on silica-alumina. The selectivity for BP i.e., the ratio of BP to the sum of BP and CHB, was less than 50%. Further, the rate of HDS, which was determined using the yields of BP and CHB, was 0.71 mol/mol-Mo/h, larger than that obtained using other supports (*vide infra*).

When Al_2O_3, SiO_2, TiO_2, and NaY zeolite were used instead of SiO_2-Al_2O_3, the yield of BP and CHB decreased; they were 35%, 20%, 31% and 14%, respectively. With these catalysts, however, the hydrocracking activity was not observed, but the selectivities for BP were 66%, 85%, 58% and 93%, respectively, or higher than the selectivity of SiO_2-Al_2O_3 if it is assumed that the selectivity for BP does not change with temperature so much as the conversion of DBT. The rates of HDS decreased in the order TiO_2 > NaY zeolite > Al_2O_3 > SiO_2. Although the amounts of molybdenum species loaded on TiO_2 and NaY zeolite were less than those on other supports, the use of these supports showed higher rates of HDS per molybdenum atom. The results obtained from the use of HY zeolite were different from those of runs 1–5. The total conversion of DBT was 75%, close to that obtained using SiO_2-Al_2O_3. However, a large number of products whose the carbon numbers were smaller than 12 were formed by hydrocracking, each yield being a few percent, and no HDS product such as BP and CHB were detected, indicating that the aromatic structure of DBT was significantly destroyed by the molybdenum species supported on HY zeolite. Using HZSM-5, the conversion of DBT was 14% and hydrocracking activity was not observed. With the use of active carbon, the total yield of BP and CHB was 31%. In this case also, the hydrocracking activity was not observed. The difference between the total conversion and HDS yields is due to the adsorption of DBT and products onto the support because the material balance was not complete within several hours of the HDS run.

B. Catalysts Derived from Silica-alumina supported Systems

HDS catalyzed by SiO_2-Al_2O_3-supported system was performed and the results are shown in Table 4.6. When $Mo(CO)_6$/SiO_2-Al_2O_3 was presulfided with H_2S in H_2 (run 9), the conversion of DBT and the yields of BP and CHB were 31% and 26%. The rate of HDS was 0.42 mol/mol-Mo/h, much smaller than that of run 1. The catalyst prepared by presulfiding of the $Mo(CO)_6$-NEt_3-EtSH/SiO_2-Al_2O_3 system (run 10) revealed catalytic activity as high as that of run 1, indicating that the use of anionic molybdenum carbonyl species was essential for high catalytic activity. The catalyst prepared by hydrogenolysis of the $Mo(CO)_6$-NEt_3-EtSH/SiO_2-Al_2O_3 system (run 11) revealed a total conversion of 46%; however, hydrocracking activity was not observed in this system. This catalyst showed higher selectivity for BP, indicating that activation of the catalyst by hydrogenolysis makes it difficult to form active sites for hydrogenation of the aromatic ring.

For the catalyst derived from presulfiding of conventional MoO_3/SiO_2-Al_2O_3 (run 12), the conversion of DBT reached 91%. However, the yields of BP and CHB (the sum of the conversions of DBT into BP and CHB) and the rate of HDS of DBT into BP and CHB per content of molybdenum were 39% and 0.49 mol/mol-Mo/h, respectively, which were much lower values than those for runs 9 and 10. This indicates that hydrocracking of a significant amount of DBT occurred on the catalyst derived from MoO_3/SiO_2-Al_2O_3. Large amounts of hydrocarbons in the range of C_1–C_{10} were formed but the amount of each was less than 1%. However, the yields of benzene and toluene were more than 1% as determined by GC analysis. Because benzene and toluene can also be formed from xylene used as a solvent, their precise yields by hydrocracking of DBT have not been determined. When the catalysts derived from

Table 4.6 Hydrodesulfurization of Dibenzothiophene Catalyzed by Silica-Alumina-Supported Molybdenum Catalysts.[a]

Run	Catalyst	Conv. of [b] DBT (%)	Conv. of DBT to BP (%)	Conv. of DBT to CHB (%)	Rate [c] of HDS (mol/mol-Mo/h)	Mo content (wt%)
1	$[NEt_4][Mo(CO)_5(CH_3COO)]/SiO_2-Al_2O_3$	85(58)	26	32	0.71	7.0
9	$Mo(CO)_6/SiO_2-Al_2O_3$	31(26)	17	9	0.42	5.3
10	$Mo(CO)_6-EtSH-NEt_3/SiO_2-Al_2O_3$	79(56)	22	34	0.74	6.5
11	$Mo(CO)_6-EtSH-NEt_3/SiO_2-Al_2O_3$[d]	46(46)	40	6	0.58	6.8
12	$MoO_3/SiO_2-Al_2O_3$[e]	91(39)	19	20	0.49	6.8
13	MoO_3/Al_2O_3[e]	38(38)	27	11	0.40	7.8

[a] Reaction temp. 300°C, Pressure 50 kg/cm^2, WHSV 16.5 h^{-1}, Cat. 0.50 g, H$_2$ 18 l/h, Initial concentration of DBT 1.0 wt%. Presulfided by H$_2$S in H$_2$ at 350°C (H$_2$S 3%). [b] Total conversion of DBT. The sum of the conversions of DBT into BP and CHB, which represents the conversion of DBT by HDS, is given in parentheses. The difference between the total conversion and the value in parentheses represents the conversion of DBT by hydrocracking. [c] Rate of HDS of DBT into BP, and CHB. This value was calculated from the sum of the conversion of DBT into BP and CHB. [d] Activated by H$_2$ stream at 350°C. [e] Presulfided by H$_2$S in H$_2$ at 400°C (H$_2$S 3%).
(From A. Ishihara, T. Kabe et al., Chem. Lett., 590 (1993))

silica-alumina supported molybdenum carbonyls were used (runs 1, 9 and 10), the conversion of DBT by hydrocracking was much less than that in the case of MoO$_3$/SiO$_2$-Al$_2$O$_3$. When MoO$_3$/Al$_2$O$_3$ was used (run 13), the conversion of DBT was 38%. However, the hydrocracking of DBT hardly occurred with MoO$_3$/Al$_2$O$_3$ while the HDS rate of DBT into BP and CHB was much less than that of silica-alumina supported anionic molybdenum carbonyls and very close to that of MoO$_3$/SiO$_2$-Al$_2$O$_3$.

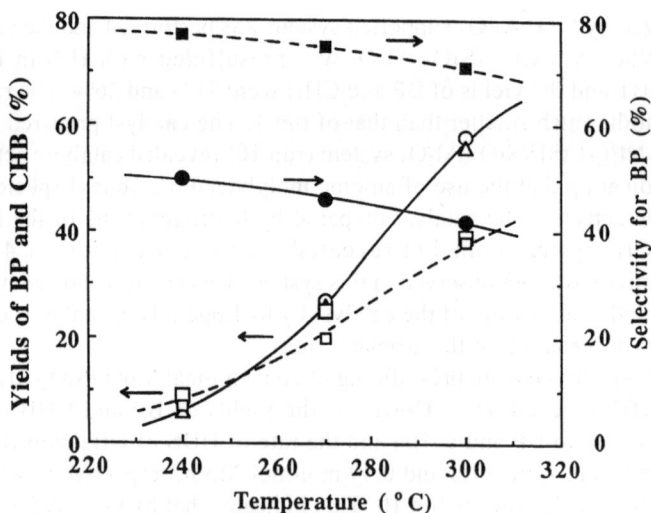

Fig.4.12   Effect of temperature on yields of BP and CHB and selectivity for BP.
Yields of BP and CHB and BCH(%): ◯: $[NEt_4][Mo(CO)_5(CH_3COO)]/SiO_2-Al_2O_3$; ▫: MoO$_3$/SiO$_2$-Al$_2O_3$; △: Mo(CO)$_6$-2(EtSH-NEt$_3$)/SiO$_2$-Al$_2O_3$; ▫: MoO$_3$/Al$_2O_3$; Selectivity for BP (%): ●: $[NEt_4][Mo(CO)_5(CH3COO)]/SiO_2-Al_2O_3$; ■: MoO$_3$/Al$_2O_3$.
(From A. Ishihara, T. Kabe et al., J. Jpn. Petrol. Inst., 37, 415 (1994))

In order to investigate the effect of temperature, the yields of BP and CHB were plotted against temperature in Fig. 4.12. Four types of catalysts derived from [NEt$_4$] [Mo(CO)$_5$(OOCCH$_3$)]/SiO$_2$-Al$_2$O$_3$, Mo(CO)$_6$-NEt$_3$-EtSH/SiO$_2$-Al$_2$O$_3$, MoO$_3$/SiO$_2$-Al$_2$O$_3$, and MoO$_3$/Al$_2$O$_3$ systems were used. Both yields and apparent activation energies of the catalysts derived from [NEt$_4$][Mo(CO)$_5$(OOCCH$_3$)]/SiO$_2$-Al$_2$O$_3$ and Mo(CO)$_6$-NEt$_3$-EtSH/SiO$_2$-Al$_2$O$_3$ systems were very similar to each other. Both yields and apparent activation energies of the catalysts derived from MoO$_3$/SiO$_2$-Al$_2$O$_3$ and MoO$_3$/Al$_2$O$_3$ systems were also very similar to each other. As clearly shown in Fig. 4.7, the apparent activation energies of silica-alumina-supported anionic metal carbonyl systems were much larger than those of the conventional systems. Therefore, the catalytic activities of the former were larger than those of the latter above about 260 °C. When the selectivities for BP of [NEt$_4$][Mo(CO)$_5$(OOCCH$_3$)]/SiO$_2$-Al$_2$O$_3$ and MoO$_3$/Al$_2$O$_3$ were compared, the value of the former was much smaller than that of the latter in the temperature range investigated. It has already been reported that hydrogenation of the aromatic ring, as well as desulfurization, is one of the key reactions for realizing the deep desulfurization of methyl-substituted DBT.[394)] As mentioned above, silica-alumina-supported catalysts have higher selectivities for CHB, that is, higher activities for hydrogenation of the aromatic ring than alumina-supported catalysts, indicating that silica-alumina-supported ones may be applicable in deep desulfurization processes.

C. Characterization of Catalysts Derived from Silica-alumina-supported Systems

The FTIR spectra of silica-alumina-supported molybdenum carbonyls before activation were measured and are shown in Fig. 4.13. When [Mo(CO)$_5$(OOCCH$_3$)]$^-$ was supported on

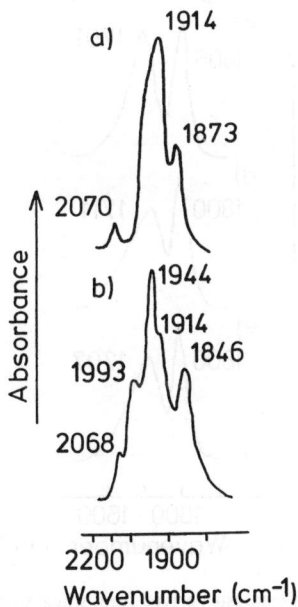

Fig.4.13   FTIR spectra of silica-alumina supported molybdenum carbonyls.
(a) [NEt$_4$][Mo(CO)$_5$(CH$_3$COO)]/SiO$_2$-Al$_2$O$_3$; (b) Mo(CO)$_6$-EtSH-NEt$_3$/SiO$_2$-Al$_2$O$_3$.
(From A. Ishihara, T. Kabe et al., J. Jpn. Petrol. Inst., **37**, 415 (1994))

silica-alumina (Fig. 4.13(a)), three absorptions were observed at 2070 (w), 1914 (vs, b = broad) and 1873 (m) cm^{-1}. It was reported that IR spectrum of [Mo(CO)$_5$(OOCCH$_3$)]$^-$ obtained as a Nujol mull showed four absorptions at 2035 (vw = very week), 1985 (s = strong), 1930 (vs = very strong) and 1857 (m = medium) cm^{-1}.[388] Although the configuration of carbonyl ligands in [Mo(CO)$_5$(OOCCH$_3$)]$^-$ appeared to change with deposition of the complex onto silica-alumina, the IR spectrum confirmed the presence of a mononuclear anionic molybdenum carbonyl on the support. In the Mo(CO)$_6$-NEt$_3$-EtSH/SiO$_2$-Al$_2$O$_3$ system, five absorptions were observed at 2068 (w), 1993 (m), 1944 (vs), 1914 (m, sh = shoulder) and 1846 (m) cm^{-1}. It was also reported that the IR spectrum of [Mo$_2$(CO)$_{10}$SCH$_3$]$^-$ obtained as a tetrahydrofuran solution showed six absorptions at 2059 (w), 2045 (m), 1974 (w, sh), 1942 (vs), 1916 (m) and 1857 (s) cm^{-1}.[182] This also confirmed the presence of a dinuclear molybdenum carbonyl on the support.

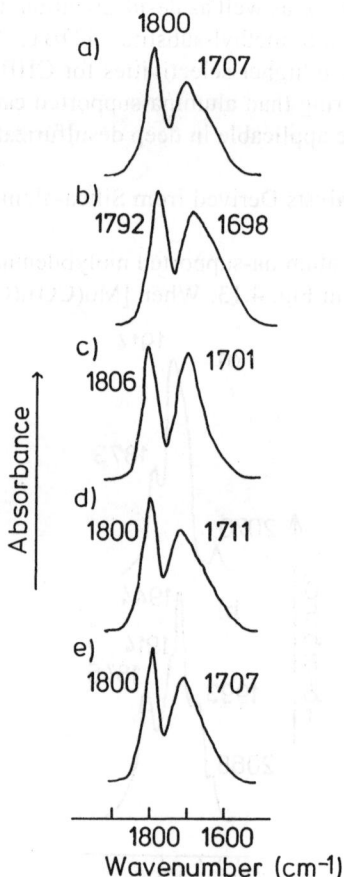

Fig.4.14   FTIR of NO chemisorption of catalysts derived from silica-alumina-supported molybdenum carbonyls.
(a) [NEt$_4$][Mo(CO)$_5$(CH$_3$COO)]/SiO$_2$-Al$_2$O$_3$ presulfided; (b) Mo(CO)$_6$-EtSH-NEt$_3$/SiO$_2$-Al$_2$O$_3$ presulfided;
(c) Mo(CO)$_6$-EtSH-NEt$_3$/SiO$_2$-Al$_2$O$_3$ treated with H$_2$; (d) MoO$_3$/SiO$_2$-Al$_2$O$_3$ presulfided; (e) MoO$_3$/SiO$_2$-Al$_2$O$_3$ treated with H$_2$ after presulfiding.
(From A. Ishihara, T. Kabe et al., J. Jpn. Petrol. Inst., 37, 416 (1994))

Table 4.7 Characterization of Supported Molybdenum Catalysts by NO Chemisorption [a]

Run Catalyst	NO Chemisorption Wavenumber (cm^{-1})		NO Adsorption (μmol/g-cat)	NO/Mo (mol/mol)
14 [NEt$_4$][Mo(CO)$_5$(CH$_3$COO)]/SiO$_2$-Al$_2$O$_3$ [b]	1800	1707	170	0.23
15 Mo(CO)$_6$-EtSH-Et$_3$N/SiO$_2$-Al$_2$O$_3$ [b]	1792	1698	160	0.24
16 Mo(CO)$_6$-EtSH-Et$_3$N/SiO$_2$-Al$_2$O$_3$ [c]	1806	1701	435	0.61
17 MoO$_3$/SiO$_2$-Al$_2$O$_3$ [d]	1800	1711	129	0.18
18 MoO$_3$/SiO$_2$-Al$_2$O$_3$ [e]	1800	1707	347	0.49
19 MoO$_3$/-Al$_2$O$_3$ [d]	1786	1688	110	0.14

[a] All measurements of NO chemisorption were carried out at 25°C. [b] Presulfided by H$_2$S in H$_2$ at 350°C (H$_2$S 3%) for 2 h and evacuated at 350°C for 2h. [c] Activated by H$_2$ at 350°C for 2 h and evacuated at 350°C for 2 h. [d] Presulfided by H$_2$S in H$_2$ at 400°C (H$_2$S 3%) for 3 h and evacuated at 400°C for 2 h. [e] Presulfided by H$_2$S in H$_2$ at 400°C (H$_2$S 3%) for 3 h, treated with H$_2$ at 400°C for 2 h and evacuated at 400°C for 2 h.
(From A. Ishihara, T. Kabe et al., Chem. Lett., 591 (1993))

Further characterization of silica-alumina-supported molybdenum catalysts was performed by NO chemisorption. The FTIR spectra of NO chemisorption on silica-alumina supported molybdenum catalysts are shown in Fig. 4.14 and the wavenumbers of nitrosyl species are listed in Table 4.7. The wavenumbers of the doublet peak for sulfided MoO$_3$/Al$_2$O$_3$ (run 19) were 1786 and 1688 cm^{-1}, which were assigned to the dinitrosyl species adsorbed on the edge of MoS$_2$.[389,390] The wavenumbers of the peaks of silica-alumina-supported catalysts distributed in the ranges 1792–1806 and 1698–1711 cm^{-1} were slightly higher than those of sulfided MoO$_3$/Al$_2$O$_3$ while their shapes were very close to each other. Although molybdenum species of the silica-alumina-supported molybdenum catalysts (runs 14–18) can be deduced collectively to be similar to those of alumina supported ones, they may not be completely the same. Some differences between alumina and silica-alumina-supported catalysts can be found in the amounts of NO chemisorption and in the characteristics of XPS spectra.

The results of NO chemisorption measured are listed in Table 4.7. The amounts of NO adsorption for catalysts derived from silica-alumina-supported molybdenum carbonyls were larger than those for sulfided MoO$_3$/SiO$_2$-Al$_2$O$_3$ (run 17) and MoO$_3$/Al$_2$O$_3$ (run 19). Furthermore, the amount of NO adsorption for the catalyst prepared from hydrogenolysis of the supported molybdenum carbonyl (run 16) was much larger than that for sulfided ones (runs 14 and 15) and that for the one prepared by hydrogen treatment of sulfided MoO$_3$/SiO$_2$-Al$_2$O$_3$ (run 18). These results indicate that the catalysts derived from supported metal carbonyls have a higher dispersion of molybdenum species than conventional ones. It is likely that such a high dispersion of the molybdenum species may be related to the selectivity for BP of the catalyst, which is different from that of sulfided catalysts. The yields of BP and CHB and the total conversion of DBT for the sulfided catalysts (runs 1 and 10) are higher than those for the hydrogenated catalyst (run 11). For run 11, the formation of tetra- or hexahydrodibenzothiophene will be less than that for the sulfided catalysts because hydrogenation of the aromatic ring in DBT is difficult to occur. As a result, the conversion of DBT for the former is higher than that for the latter because HDS of tetra- or hexahydrodibenzothiophene is one order of magnitude faster than that of DBT.[395] This also indicates that hydrogenation of the aromatic ring is effective for achieving higher conversion of DBT. For sulfided catalysts, the relationship between the amount of NO adsorption (NO/Mo) and the HDS rate of DBT into BP and CHB was compared and the results are

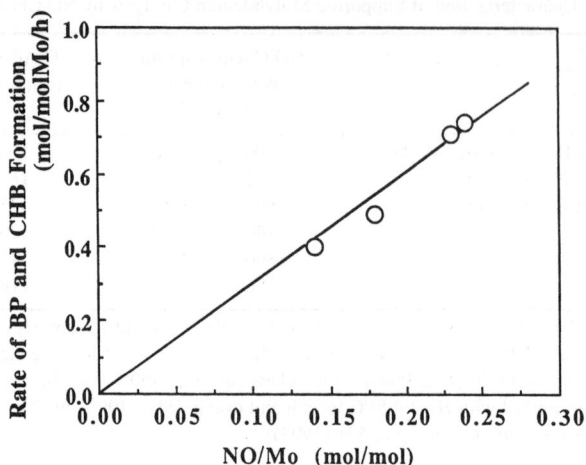

Fig. 4.15 Effect of ratio of NO/Mo on HDS rate.
(From A. Ishihara, T. Kabe et al., Chem. Lette., 592 (1993))

Table 4.8 Characterization of Supported Molybdenum Catalysts by XPS [a]

Run Catalyst	Mo3d$_{3/2}$ (eV)	Mo3d$_{5/2}$ (eV)	N1s (eV)	S2p (eV)	N/Mo [b]	S/Mo [b]
20 [NEt$_4$][Mo(CO)$_5$(CH$_3$COO)]/SiO$_2$-Al$_2$O$_3$ [c]	232.3	229.5	395.4	163.0	2.08	1.92
21 Mo(CO)$_6$-EtSH-Et$_3$N/SiO$_2$-Al$_2$O$_3$ [c]	232.4	229.7	395.9	163.1	2.77	1.84
22 Mo(CO)$_6$-EtSH-Et$_3$N/SiO$_2$-Al$_2$O$_3$ [d]	231.8	229.2	395.4	163.0	3.43	0.71
23 MoO$_3$/SiO$_2$-Al$_2$O$_3$ [e]	232.1	229.3	395.3	162.9	2.50	1.08
24 MoO$_3$/SiO$_2$-Al$_2$O$_3$ [f]	231.8	228.8	395.0	162.6	2.36	0.92
25 MoO$_3$/-Al$_2$O$_3$ [e]	233.0	229.8	395.9	163.0	2.61	3.06

[a] XPS spectra were measured for the samples used in NO chemisorption. Every binding energy was referenced to Carbon C1s 285.0 eV due to adventitious carbon. [b] Ratio of peak heights between Mo3d$_{5/2}$, N1s and S2P. [c] Presulfided by H$_2$S in H$_2$ at 350°C (H$_2$S 3%) for 2 h and evacuated at 350°C for 2. [d] Hydrogenated by H$_2$ at 350°C for 2 h and evacuated at 350°C for 2 h. [e] Presulfided by H$_2$S in H$_2$ at 400°C (H$_2$S 3%) for 3 h and evacuated at 400°C for 2 h. [f] Presulfided by H$_2$S in H$_2$ at 400°C (H$_2$S3%) for 3 h, hydrogenated by H$_2$ at 400°C for 1 h and evacuated at 400°C for 2 h.
(From A. Ishihara, T. Kabe et al., J. Jpn. Petrol. Inst., 37, 417 (1994))

illustrated in Fig. 4.15. When the amount of NO adsorbed per molybdenum loaded increased, the rate of HDS increased linearly. NO chemisorption appeared to reflect the number of active sites for HDS as well as the extent of surface coverage or dispersion of Mo. As described in the preceding section, there was no linear correlation between the amount of NO adsorption and HDS activity for alumina-supported Mo, Ni-Mo or Co-Mo catalyst, and the deviation from linearity was probably attributable to the fact that the total NO uptake involves different contributions of chemisorbed NO on Mo, Ni (or Co), and Ni-Mo (or Co-Mo) sites.[380,390,391] For alumina supports, NO chemisorption appeared to reflect the extent of surface coverage or dispersion of Mo rather than the number of active sites in HDS. In contrast, it was likely that NO chemisorption for the silica-alumina supported catalysts reflected the number of active sites in HDS.

In Table 4.8 are shown the results from XPS measurements of catalysts used in NO chemisorption experiments. Binding energy for every catalyst was very similar to that for sulfided MoO_3/Al_2O_3 (run 25). The values of NO/Mo ratios and N/Mo ratios were close to each other. However, the value of S/Mo of silica-alumina-supported catalysts was smaller than that of MoO_3/Al_2O_3. This indicates that molybdenum species of silica-alumina-supported catalysts was not the same as that of alumina-supported ones. This is due to the formation of MoS_{2-x} on SiO_2-Al_2O_3. For alumina-supported molybdenum catalysts, it has been reported that the adsorption of NO occurs mainly on Mo atoms present as MoS_2,[390] and that Mo is predominantly present as MoS_2-like domains in sulfided MoO_3/Al_2O_3.[391] In the present SiO_2-Al_2O_3-supported systems, however, it was likely that a highly dispersed MoS_{2-x} with many sulfur anion vacancies was formed on silica-alumina by the sulfidation of silica-alumina-supported anionic molybdenum carbonyls.

4.2.3 Hydrodesulfurization Catalysts Prepared from Alumina- and Silica-Alumina-Supported Molybdenum and Cobalt Carbonyls

For a better understanding of HDS reactivity of supported metal carbonyls, HDS of DBT in a pressurized flow system has been discussed systematically in the preceding sections. The catalysts derived from alumina- or silica-alumina-supported anionic molybdenum carbonyls showed higher catalytic activity than the catalysts derived from supported neutral molybdenum carbonyl or from conventional sulfided MoO_3/Al_2O_3. The addition of cobalt compounds to highly dispersed molybdenum sulfide catalysts which are formed by activation of the supported metal carbonyls is expected to increase the HDS activity remarkably. In the following section, the effects of the addition of cobalt carbonyls on catalysts derived from supported anionic molybdenum carbonyls for HDS of DBT are described and the results of these catalysts compared with those of commercial Co-Mo/Al_2O_3. The catalysts derived from silica-alumina-supported molybdenum and cobalt carbonyls revealed much higher catalytic activity than those derived from alumina-supported molybdenum and cobalt carbonyls and those derived from conventional Co-Mo/Al_2O_3. The characterization of the catalysts derived from supported molybdenum and cobalt carbonyls by means of NO chemisorption, FTIR, and XPS showed the formation of highly dispersed cobalt species located on molybdenum sulfide which were active for HDS.[396]

A. Effect of Addition of Cobalt Compounds on Hydrodesulfurization of Dibenzothiophene Catalyzed by Alumina-supported Molybdenum Carbonyls

Effects of the addition of cobalt compounds (cobalt carbonyl ($Co_2(CO)_8$), cobalt nitrate ($Co(NO_3)_2$), cobalt (II) acetylacetonate ($Co(acac)_2$), and cobalt (III) acetylacetonate ($Co(acac)_3$) on HDS of DBT catalyzed by the $Mo(CO)_6$-$EtSH$-Et_3N/Al_2O_3 system were investigated and the results are shown in Table 4.9. The addition of one of these compounds remarkably increased the catalytic activity, which was maintained over 10 h. The major products were BP and CHB. The order of activity was $Co_2(CO)_8 > Co(NO_3)_2 > Co(acac)_2 > Co(acac)_3$. When a molybdenum precursor was changed in run 5, the catalytic activity also increased. These experiments confirmed that the addition of cobalt compounds to catalysts derived from supported molybdenum catalysts increased the activities whatever the precursor used. When supported molybdenum and cobalt carbonyls were tested separately, conversions

Table 4.9 Hydrodesulfurization of Dibenzothiophene Catalyzed by Alumina-Supported Molybdenum and Cobalt Catalysts [a]

Run Catalyst	Conv. of DBT (%)	Conv. of DBT to BP (%)	Conv. of DBT to CHB (%)	Selectivity for BP (%)	Mo Content (wt%)
1 $Mo(CO)_6$-EtSH-Et_3N-$Co_2(CO)_8$/Al_2O_3 [b]	99	87	12	87	8.0
2 $Mo(CO)_6$-EtSH-Et_3N-$Co(NO_3)_2 6H_2O$/Al_2O_3 [b]	98	87	11	88	7.8
3 $Mo(CO)_6$-EtSH-Et_3N-$Co(acac)_2 2H_2O$/Al_2O_3 [b]	86	76	10	88	7.8
4 $Mo(CO)_6$-EtSH-Et_3N-$Co(acac)_3$/Al_2O_3 [b]	79	69	10	87	7.5
5 $[NEt_4][Mo(CO)_5(OOCCH_3)]$-$Co_2(CO)_8$/$Al_2O_3$ [c]	99	88	11	88	8.0
6 $Mo(CO)_6$-EtSH-Et_3N/Al_2O_3 [d]	43	27	16	63	7.8
7 $Co_2(CO)_8$/Al_2O_3	8	7	1	87	—

[a] Reaction Temp. 300°C, Pressure 50 atm, LHSV 14 h^{-1}, Cat 0.5 g, H_2 18 l/h, Amount of $Mo(CO)_6$ 0.11 g. Presulfided by H_2S in H_2 at 350°C (H_2S 3%). [b] Co/Mo = 0.8. [c] Co/Mo = 0.4. [d] Ref. 380, 381.
(From A. Ishihara, T. Kabe et al., J. Jpn. Petrol. Inst., **39**, 28 (1996))

of DBT were 43% and 8% (runs 6 and 7), respectively, as shown in Table 4.9.

The effects of temperature on the catalytic activity was investigated with the catalysts derived from supported carbonyls, and the results compared with those of conventional Co-Mo/Al_2O_3 (KF 124: MoO_3: 12.3 wt%; CoO: 3.8 wt%; 274 m^2/g). Fig. 4.16 shows the results from three catalysts. The catalyst derived from the $Mo(CO)_6$-EtSH-Et_3N-$Co_2(CO)_8$/Al_2O_3 system showed the same catalytic activity and product selectivity as those of the catalyst derived from conventional Co-Mo/Al_2O_3 at all temperatures while the activity from the $Mo(CO)_6$-EtSH-Et_3N-$Co(NO_3)_2$/Al_2O_3 system was lower at low temperatures. In the alumina-supported $[Mo(CO)_6$-EtSH-Et_3N]-$Co_2(CO)_8$ system, the maximum catalytic activity was

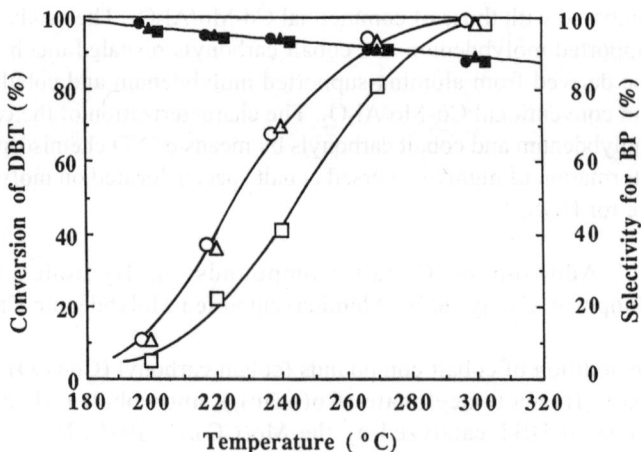

Fig. 4.16 Effect of temperature on the conversion of DBT and selectivity for BP.
○, □, △: Conversion of DBT; ●, ■, ▲: Selectivity for BP;
○, ●: $[Mo(CO)_6$-EtSH-Et_3N]-$Co_2(CO)_8$/Al_2O_3; □, ■: $[Mo(CO)_6$-EtSH-Et_3N]-$Co(NO_3)_2$/Al_2O_3; △, ▲: CoO-MoO_3/Al_2O_3.
Pressure 50 atm, WHSV 16.5 h^{-1}, Cat 0.5 g, H_2 18 l/hr, 1.0 wt% DBT, Solvent: xylene.
(From A. Ishihara, T. Kabe et al., J. Jpn. Petrol. Inst., **39**, 28 (1996))

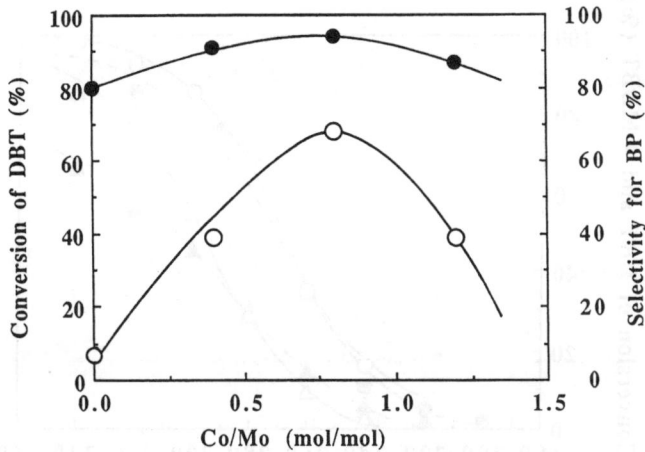

Fig. 4.17 Effect of amount of cobalt on the conversion of DBT and selectivity for BP.
○: Conversion of DBT; ●: Selectivity for BP. [Mo(CO)$_6$-EtSH-Et$_3$N]-Co$_2$(CO)$_8$/Al$_2$O$_3$.
Temperature 240°C, Pressure 50 atm, WHSV 16.5h^{-1}, Cat 0.5g, H$_2$ 18 1/hr, 1.0 wt% DBT,
Solvent: xylene.
(From A. Ishihara, T. Kabe et al., J. Jpn. Petrol. Inst., **39**, 28 (1996))

Table 4.10 Hydrodesulfurization of Dibenzothiophene Catalyzed by Supported Molybdenum and Cobalt Catalysts [a]

Run	Catalyst	Conv. of DBT (%)	Conv. of DBT to BP (%)	Conv. of [b] DBT to CHB (%)	Selectivity [c] for BP (%)	Mo Content (wt%)
8	Mo(CO)$_6$-EtSH-Et$_3$N-Co$_2$(CO)$_8$/SiO$_2$-Al$_2$O$_3$	100	89	11	89	6.7
9	Mo(CO)$_6$-EtSH-Et$_3$N-Co$_2$(CO)$_8$/SiO$_2$	96	82	14	85	7.0
10	Mo(CO)$_6$-EtSH-Et$_3$N/SiO$_2$-Al$_2$O$_3$[d]	56	22	34	39	7.0
11	Mo(CO)$_6$-EtSH-Et$_3$N/SiO$_2$[d]	20	17	3	85	7.0

[a] Reaction Temp. 300°C, Pressure 50 atm, LHSV 14 h^{-1}, Cat 0.5 g, H$_2$ 18 l/h, Amount of Mo(CO)$_6$ 0.11 g. Presulfided by H$_2$S in H$_2$ at 350°C (H$_2$S 3%). [b] Conversion of DBT to bicyclohexylbenzene has been included. [c] The ratio of the conversion of DBT into BP to total conversion of DBT. [d] Ref. 392, 393.
(From A. Ishihara, T. Kabe et al., J. Jpn. Petrol. Inst., **39**, 29 (1996))

shown at Co/Mo = 0.8 as shown in Fig. 4.17, while it was reported that in conventional Co-Mo/Al$_2$O$_3$, the catalytic activity leveled off above Co/Mo ca. 0.6.[377,390] In the present case, it seemed that the catalysts showed maximum activities at higher ratios of Co/Mo because of the higher dispersion of molybdenum species than found in conventional catalysts (see below).

The addition of cobalt carbonyl to silica-alumina- and silica-supported molybdenum carbonyl systems also increased the catalytic activity for HDS of DBT, as shown in Table 4.10. When the effect of temperature on the catalytic activity of the [Mo(CO)$_6$-EtSH-Et$_3$N]-Co$_2$(CO)$_8$/SiO$_2$-Al$_2$O$_3$ system in HDS of DBT was investigated, the conversion of DBT was slightly higher at each temperature than for Co-Mo/Al$_2$O$_3$, as shown in Fig. 4.18. Therefore, the effect of temperature on the conversion of 4-MDBT in the [Mo(CO)$_6$-EtSH-Et$_3$N]-Co$_2$(CO)$_8$/SiO$_2$-Al$_2$O$_3$ system was investigated and compared with that in commercial Co-Mo/Al$_2$O$_3$. In HDS of 4-MDBT, the conversion of 4-MDBT for the [Mo(CO)$_6$-EtSH-Et$_3$N]-Co$_2$(CO)$_8$/SiO$_2$-Al$_2$O$_3$ system was higher than that for Co-Mo/Al$_2$O$_3$ at temperatures above

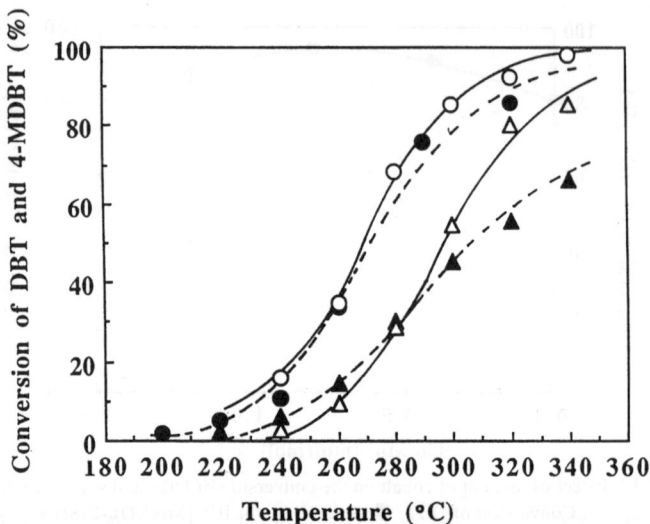

Fig. 4.18 Effect of temperature on the conversion of DBT and 4-MDBT.
Conv. of DBT: ○: [Mo(CO)$_6$-EtSH-Et$_3$N]-Co$_2$(CO)$_8$/SiO$_2$-Al$_2$O$_3$;
●: CoO-MoO$_3$/Al$_2$O$_3$; Conv. of 4-MDBT: △: [Mo(CO)$_6$-EtSH-Et$_3$N]-Co$_2$(CO)$_8$/SiO$_2$-
Al$_2$O$_3$; ▲: CoO-MoO$_3$/Al$_2$O$_3$; Pressure 50 atm, WHSV 70 h^{-1}, Cat 0.2 g, H$_2$ 18 1/hr,
1.0 wt% DBT or 0.1 wt% 4-MDBT, Solvent: decalin.
(From A. Ishihara, T. Kabe et al., J. Jpn. Petrol. Inst., 39, 28 (1996))

280°C. It is evident from Fig. 4.18 that the apparent activation energy of the [Mo(CO)$_6$-EtSH-Et$_3$N]-Co$_2$(CO)$_8$/SiO$_2$-Al$_2$O$_3$ system is larger than that of Co-Mo/Al$_2$O$_3$. Taking into account the Langmuir-Hinshelwood mechanism, it may be assumed that the heat of adsorption of 4-MDBT in the [Mo(CO)$_6$-EtSH-Et$_3$N]-Co$_2$(CO)$_8$/SiO$_2$-Al$_2$O$_3$ system may also be larger than that of Co-Mo/Al$_2$O$_3$ system.

B. Characterization of Catalysts by NO Chemisorption, FTIR and XPS

FTIR spectra of alumina- and silica-alumina-supported molybdenum and cobalt carbonyls are shown in Fig. 4.19. The [Mo(CO)$_6$-EtSH-Et$_3$N]-Co$_2$(CO)$_8$/Al$_2$O$_3$ system showed peaks at 2068 (vw = very weak), 2000 (s = strong), 1962 (vs), 1952 (vs), 1854 (m = medium) and 1815 (w = weak) cm^{-1} (Fig. 4.19(a)). Among them, two peaks at 2000 and 1815 cm^{-1} were due to alumina-supported cobalt species because the intensities of these peaks changed depending on the amount of Co$_2$(CO)$_8$ added. It was confirmed that, in the Co$_2$(CO)$_8$-EtSH-Et$_3$N/Al$_2$O$_3$ system where Co$_2$(CO)$_8$ was supported on alumina after treatment with Et$_3$N and EtSH, two peaks at 2012(vs) and 1813(w) cm^{-1} were observed as shown in Fig. 4.19(d). The pattern of the three peaks at 2068, 1952, and 1854 cm^{-1} was similar to that of [Mo(CO)$_6$-EtSH-Et$_3$N]-Co$_2$(CO)$_8$/SiO$_2$-Al$_2$O$_3$ (Fig. 4.19(b)) and [Mo(CO)$_6$-EtSH-Et$_3$N]/SiO$_2$-Al$_2$O$_3$ (Fig. 4-19(c)) systems. The peaks were assigned to a dinuclear molybdenum carbonyl anion, [Mo$_2$(CO)$_{10}$SEt]$^-$, supported on alumina or silica-alumina as mentioned in the preceding section. In the silica-alumina system, clear peaks of cobalt species were not observed probably because the cobalt carbonyl complex decomposed on silica-alumina to release carbon

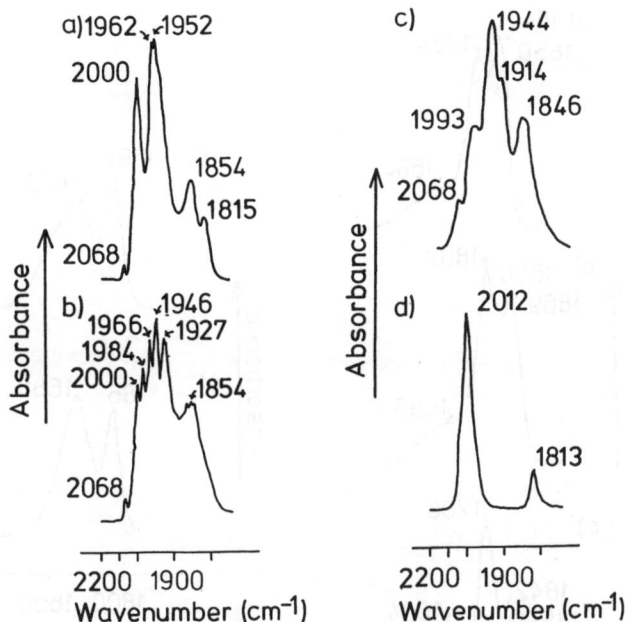

Fig. 4.19 FTIR spectra of alumina and silica-alumina supported molybdenum and cobalt carbonyls.
(a) [Mo(CO)$_6$-EtSH-Et$_3$N]-Co$_2$(CO)$_8$/Al$_2$O$_3$; (b) [Mo(CO)$_6$-EtSH-Et$_3$N]-Co$_2$(CO)$_8$/SiO$_2$-Al$_2$O$_3$;
(c) [Mo(CO)$_6$-EtSH-Et$_3$N]/SiO$_2$-Al$_2$O$_3$; (d) [Co$_2$(CO)$_8$-EtSH-Et$_3$N]/Al$_2$O$_3$.
(From A. Ishihara, T. Kabe et al., J. Jpn. Petrol. Inst., **39**, 30 (1996))

monoxide.

FTIR spectra of NO chemisorption species on the catalysts after sulfidation are shown in Fig. 4.20. Two large and one small absorptions near 1850, 1790, and 1685 cm^{-1} were observed in both [Mo(CO)$_6$-EtSH-Et$_3$N]-Co$_2$(CO)$_8$/Al$_2$O$_3$ (Fig. 4.20(a)) and [Mo(CO)$_6$-EtSH-Et$_3$N]-Co$_2$(CO)$_8$/SiO$_2$-Al$_2$O$_3$ (Fig. 4.20(b)) systems. The two large peaks were due to dinitrosyl species on the surface cobalt species and the small peak was due partly to the dinitrosyl species on molybdenum species. Three absorptions observed in conventional sulfided Co-Mo/Al$_2$O$_3$ (Fig. 4.20(c)) were consistent with those reported in the literature.[390] For a better understanding of these data, it is useful to compare the spectra of NO chemisorption on cobalt promoted catalysts (Figs. 4.20(a)–(c)) with those on unpromoted ones (Figs. 4.20(d)–(e)). When Co$_2$(CO)$_8$ was used, there was only a small amount of coordinatively unsaturated sites on molybdenum sulfide, and most of the adsorption sites for NO chemisorption on Mo species were occupied by cobalt species in [Mo(CO)$_6$-EtSH-Et$_3$N]-Co$_2$(CO)$_8$/Al$_2$O$_3$ and [Mo(CO)$_6$-EtSH-Et$_3$N]-Co$_2$(CO)$_8$/SiO$_2$-Al$_2$O$_3$ systems as shown in Figs. 4.20(a) and (b). This result suggests that the cobalt species located on molybdenum species are directly related to the HDS reaction rather than the surface molybdenum species themselves.

After presulfiding of the precursor, the amount of NO chemisorption was measured, and the results are also shown in Table 4.11. The amount of NO chemisorption in Co-Mo/Al$_2$O$_3$ was in agreement with that reported by Topsøe and Topsøe.[390] The amount of NO chemisorption for the [Mo(CO)$_6$-EtSH-Et$_3$N]-Co$_2$(CO)$_8$/SiO$_2$-Al$_2$O$_3$ system was the largest

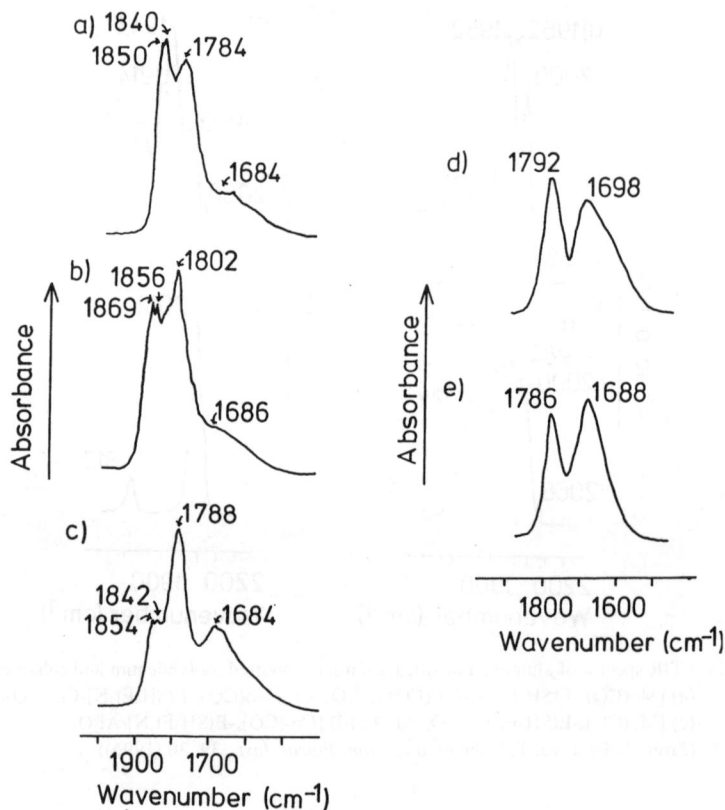

Fig. 4.20 FTIR spectra of NO chemisorption on the catalysts prepared from alumina and silica-alumina-supported molybdenum and cobalt carbonyls.
(a) [Mo(CO)$_6$-EtSH-Et$_3$N]-Co$_2$(CO)$_8$/Al$_2$O$_3$; (b) [Mo(CO)$_6$-EtSH-Et$_3$N]-Co$_2$(CO)$_8$/SiO$_2$-Al$_2$O$_3$; (c) [CoO-MoO$_3$/Al$_2$O$_3$; (d) Mo(CO)$_6$-EtSH-Et$_3$N/SiO$_2$-Al$_2$O$_3$; (e) MoO$_3$/Al$_2$O$_3$.
(From A. Ishihara, T. Kabe et al., J. Jpn. Petrol. Inst., **39**, 30 (1996))

Table 4.11 Characterization of Supported Cobalt and Molybdenum Catalysts by NO Chemisorption [a]

Run	Catalyst	NO Chemisorption Wavenumber (cm^{-1})			NO Adsorption (μmol/g-cat)	NO/Metal [b] (mol/mol)
12	[Mo(CO)$_6$-EtSH-Et$_3$N]-Co$_2$(CO)$_8$/Al$_2$O$_3$	1840	1784		460	0.31
13	[Mo(CO)$_6$-EtSH-Et$_3$N]-Co$_2$(CO)$_8$/SiO$_2$-Al$_2$O$_3$	1856	1802		617	0.41
14	Co-Mo/Al$_2$O$_3$ [c]	1842	1788	1684	260	0.21

[a] Presulfided by H$_2$S in H$_2$ at 350°C (H$_2$S 3%) for 2 h and evacuated at 350°C for 2 h. [b] The sum of molybdenum and cobalt. [c] Presulfided by H$_2$S in H$_2$ at 400°C (H$_2$S 3%) for 3 h and evacuated at 400°C for 2 h.
(From A. Ishihara, T. Kabe et al., J. Jpn. Petrol. Inst., **39**, 30 (1996))

among the catalysts shown in Table 4.11. This showed that molybdenum and cobalt species were highly dispersed on the support at least at the time of presulfiding, because of the larger surface area of silica-alumina and the extent of coordinative unsaturation in molybdenum and cobalt species on silica-alumina supported catalysts (*vide infra*). The amount of NO chemisorption for the [Mo(CO)$_6$-EtSH-Et$_3$N]-Co$_2$(CO)$_8$/Al$_2$O$_3$ system was larger than that of Co-Mo/Al$_2$O$_3$, indicating that molybdenum and cobalt species of the [Mo(CO)$_6$-EtSH-Et$_3$N]-Co$_2$(CO)$_8$/Al$_2$O$_3$ system were also highly dispersed on the support at least at the time of presulfiding. However, the dispersion seemed to decrease to the level of Co-Mo/Al$_2$O$_3$ at the steady state of conversion because the catalytic activity and selectivity for the products in the [Mo(CO)$_6$-EtSH-Et$_3$N]-Co$_2$(CO)$_8$/Al$_2$O$_3$ system were almost the same as those for Co-Mo/Al$_2$O$_3$ at each temperature. On the other hand, although the decrease in dispersion of Co and Mo might also occur in the [Mo(CO)$_6$-EtSH-Et$_3$N]-Co$_2$(CO)$_8$/SiO$_2$-Al$_2$O$_3$ system, it was suggested that higher dispersion in this system could be maintained at the steady state.

XPS was measured for the samples used in NO chemisorption experiments and the results are shown in Table 4.12. Co2p$_{3/2}$, Mo3d$_{3/2}$, and Mo3d$_{5/2}$ spectra of the catalysts used were observed in the ranges, 779.1–780.5, 231.3–232.7 and 228.4–229.8 eV, respectively. These values of binding energy for the catalysts derived from supported metal carbonyls were smaller than those for the conventional catalyst, indicating that the oxidation states of the metals of the catalysts used were lower than those of the conventional catalyst. This was also confirmed by the fact that the ratio of sulfur to the sum of cobalt and molybdenum (S/(Co + Mo)) on the surface of the catalysts was much lower than that for the conventional one. Specifically, the lowest value of S/(Co + Mo) for the silica-alumina-supported species showed a high degree of coordinative unsaturation which would be related to the larger amount of NO chemisorption as mentioned above. It can be assumed that silica-alumina-supported molybdenum and cobalt species may be difficult to sulfurize at the presulfiding step. The authors have already reported a similar result in a previous study in which the S/Mo ratios of silica-alumina-supported molybdenum catalysts were lower than the ratio of conventional molybdena-alumina at the presulfiding step.[393] However, it has not yet been clarified why silica-alumina-supported species are difficult to sulfurize in the presulfiding step.

In the [Mo(CO)$_6$-EtSH-Et$_3$N]-Co$_2$(CO)$_8$/Al$_2$O$_3$ system and Co-Mo/Al$_2$O$_3$, the ratios of peak areas of Mo and Al were equal, indicating that the same amounts of molybdenum

Table 4.12 Characterization of Supported Molybdenum Catalysts by XPS [a]

Run	Catalyst	Al2p (eV)	Co2p$_{3/2}$ (eV)	Mo3d$_{3/2}$ (eV)	Mo3d$_{5/2}$ (eV)	S2p	$\dfrac{Co[b]}{Mo}$	$\dfrac{Mo[b]}{Al}$	$\dfrac{S[b]}{(Co + Mo)}$
15	[Mo(CO)$_6$-EtSH-Et$_3$N]-Co$_2$(CO)$_8$/Al$_2$O$_3$ [e]	75.5	779.6	232.4	229.3	162.6	4.74	0.038	1.09
16	[Mo(CO)$_6$-EtSH-Et$_3$N]-Co$_2$(CO)$_8$/SiO$_2$-Al$_2$O$_3$ [c, d]	75.0	779.1	231.3	228.4	161.7	4.59	0.037	0.31
17	Co-Mo/Al$_2$O$_3$ [e]	75.3	780.5	232.7	229.8	162.7	1.30	0.038	1.91

[a] XPS spectra were measured for the samples used in NO chemisorption. Every binding energy was referenced to Carbon C1s 285.0 eV due to adventitious carbon. [b] Ratio of peak areas between Al2p, Co2p$_{3/2}$, Mo3d and S2p. In Run 16, the both areas of Si2p and Al2p were included in the value of Al, a denominator. [c] Presulfided by H$_2$S in H$_2$ at 350°C (H$_2$S 3%) for 2 h and evacuated at 350°C for 2h. [d] The peak of Si2p was observed at 103.1 eV. [e] Presulfided by H$_2$S in H$_2$ at 400°C (H$_2$S 3%) for 3 h and evacuated at 400°C for 2 h.
(From A. Ishihara, T. Kabe et al., *J. Jpn. Petrol. Inst.*, **39**, 31 (1996))

species were present on the surface. In contrast to this, the ratios of Co/Mo in [Mo(CO)$_6$-EtSH-Et$_3$N]-Co$_2$(CO)$_8$/Al$_2$O$_3$ and [Mo(CO)$_6$-EtSH-Et$_3$N]-Co$_2$(CO)$_8$/SiO$_2$-Al$_2$O$_3$ systems were three times larger than the ratio in Co-Mo/Al$_2$O$_3$. This proved that the amounts of cobalt located on the surface were larger in [Mo(CO)$_6$-EtSH-Et$_3$N]-Co$_2$(CO)$_8$/Al$_2$O$_3$ and [Mo(CO)$_6$-EtSH-Et$_3$N]-Co$_2$(CO)$_8$/SiO$_2$-Al$_2$O$_3$ systems even if the amount of cobalt initially added to the catalysts was considered (Co/Mo in metal carbonyl systems: 0.8; Co/Mo in Co-Mo/Al$_2$O$_3$: 0.6). This also indicates that the use of Co$_2$(CO)$_8$ dramatically increases the surface concentration of cobalt at the time of catalyst preparation although Co concentration on the surface is also higher even in conventional Co-Mo/Al$_2$O$_3$ than in the bulk catalyst. This result from XPS is also supported by the results of NO chemisorption and FTIR. The Co/Mo ratio on the surface of Co-Mo/Al$_2$O$_3$ is approximately consistent with the value cited in the literature.[397] Taking into account the results of FTIR spectra of NO chemisorption in Fig. 4.20, where the amount of NO adsorption on molybdenum species can be deduced to be extremely small compared with that on the cobalt species in supported molybdenum and cobalt carbonyl systems, it is suggested that most of the coordinatively unsaturated sites of molybdenum sulfide present on the catalyst surface can be occupied by cobalt species derived from Co$_2$(CO)$_8$ and that coordinatively unsaturated sites present in such cobalt species may be related to the HDS activity.

It has been reported in Mössbauer emission spectroscopy studies of Co-Mo/Al$_2$O$_3$ that the so-called CoMoS structure, where there is some type of interaction between Co and Mo, is important for HDS activity.[16] It has been also pointed out in an FTIR study of NO chemisorption on Co-Mo/Al$_2$O$_3$ that there is a quite good correlation between the absorbance due to NO on Co species and the amount of Co atoms present as CoMoS, and that most Co atoms chemisorbing NO are likely to be those present as CoMoS.[390] In the preparation of CoMoS, the increase in dispersion of Mo species will increase the dispersion of Co species because the Co species adsorbed on the NO adsorbing sites of molybdenum species to prevent the NO adsorption on molybdenum species. In the present study, although it is not clear whether there are CoMoS phases in supported metal carbonyl derived catalysts or not, a higher dispersion of molybdenum species can be obtained by the use of anionic molybdenum carbonyls and further a large amount of cobalt species adsorbing NO can be formed selectively on the molybdenum species in the catalyst surface by the use of cobalt carbonyl.

4.3 Approaches Using Noble Metals

Although HDS catalysts with higher activity have generally been developed by extending the surface area of alumina and increasing the metal loading in the traditional alumina-supported Co-Mo, Ni-Mo and NiW catalysts, it is difficult to continue increasing the surface area of alumina and there is a limit to this method. Since Pecoraro and Chianelli have reported that ruthenium sulfide is most active for HDS of DBT among the transition metal sulfides in Fig. 4.21,[398] much attention has been focused on HDS, hydrogenation, and HDN catalyzed by noble metal sulfides, especially ruthenium sulfide, to develop a new generation of catalysts with different properties from the present Co-, Ni-, Mo- and W-based ones.[399-418] Lacroix et al.[402] confirmed that unsupported RuS$_2$ was most active, even in a flow system for HDS of DBT, among the transition metal sulfides. Vrinat et al.[403] reported, on HDS of thiophene, that Ni$_x$Ru$_{1-x}$S and Co$_x$Ru$_{1-x}$S among the mixed sulfide solid solutions,[419] prepared from RuS$_2$ and

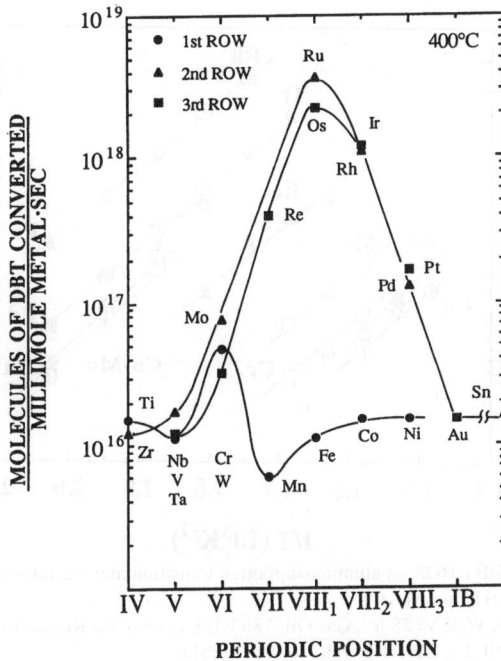

Fig. 4.21 Periodic trends for HDS of DBT/milimole of catalyst at 400°C.[398]
(T. A. Pecoraro, R. R. Chianelli, *J. Catal.*, **67**, 440 (1981))

CoS_2 or NiS_2, exhibited activities higher than that of RuS_2, CoMoS or NiMoS.

The deposition of ruthenium sulfide onto supports with large surface areas, e.g. alumina, silica, silica-alumina, zeolite, etc., can be expected to increase catalytic activity per ruthenium atom. This approach is also economical because ruthenium is one of the noble metals. Some research groups have already performed HDS catalyzed by supported ruthenium sulfide. Harvey and Matheson reported HDS of benzothiophene and HDN of quinoline with catalysts derived from $[Ru(NH_3)_6]^+$, $Ru_3(CO)_{12}$ and $RuCl_3$ supported on alumina and zeolite in a batch system.[156–158,404] Y zeolite-supported ruthenium sulfide showed activity comparable to that of commercial Ni-Mo/Al_2O_3 for quinoline HDN. HDS activity of the catalyst was not so high as its HDN activity. Vissers *et al.*[408] and Ledoux *et al.*[409] prepared carbon-supported ruthenium sulfide catalyst for HDS of thiophene. Mitchell *et al.* performed HDS of thiophene with the catalyst derived from Ru(III) acetate.[406] Kuo *et al.*,[407] De Los Reyes *et al.*,[401] and Geantet *et al.*[410] used catalysts derived from alumina-supported $RuCl_3$ or $Ru_3(CO)_{12}$ in HDS of thiophene. With these catalysts, the HDS activity was rather low probably because sulfidation of ruthenium species was incomplete and RuS_2 on alumina is unstable in hydrogen atmosphere. Further, the catalytic activity of the catalyst derived from the supported $Ru_3(CO)_{12}$ was often lower than that from the supported $RuCl_3$.[75,197,202] The sulfidation condition also affects the hydrogenation and HDS activities for ruthenium catalysts.[410–412] Chary *et al.*[413] compared RuS_2/Al_2O_3 with MoS_2/Al_2O_3 in HDS of thiophene and O_2 chemisorption. Most of these HDS reactions, however, were performed under atmospheric pressure, and so far the activities of the catalysts derived from the supported ruthenium in a pressurized flow system have not been

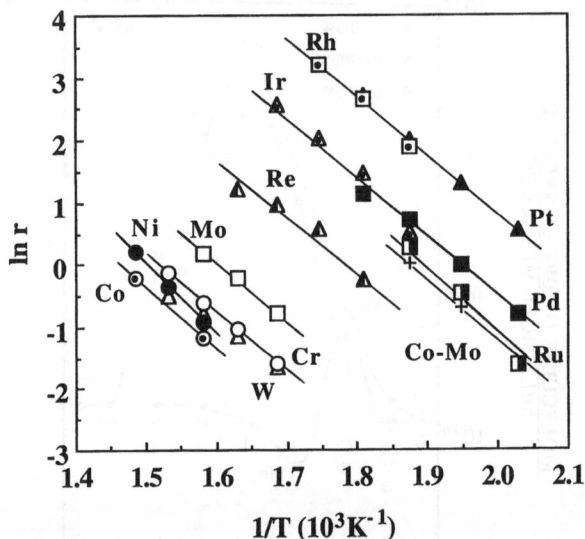

Fig. 4.22   Arrhenius plots in DBT HDS on alumina-supported transition metal catalysts.
r: HDS rate: mol-DBT/mol-Metal·h
Pressure: 50 kg/cm^2, WHSV: 28 h^{-1}, Gas/Oil: 786 NL/L (except for Ru and Co-Mo catalysts: WHSV: 70 h^{-1}, Gas/Oil: 1132 NL/L), DBT 1.0 wt%, solvent decalin.
◯: Cr/Al$_2$O$_3$ (CrO$_3$: 11 wt%, Cr(NO$_3$)$_3$·9H$_2$O); ☐: Mo/Al$_2$O$_3$ (MoO$_3$: 12.4 wt%, (NH$_4$)$_6$Mo$_7$O$_{24}$·4H$_2$O);
△: W/Al$_2$O$_3$ (WO$_3$: 28 wt%, H$_{26}$N$_6$W$_{12}$O$_{41}$); ▲: Re/Al$_2$O$_3$ (Re: 2 wt%, Re$_2$(CO)$_{10}$);
▐: Ru/Al$_2$O$_3$ (Ru: 8 wt%, Ru$_3$(CO)$_{12}$+6CsOH); ⊙: Co/Al$_2$O$_3$ (CoO: 4 wt%, Co(NO$_3$)$_2$·6H$_2$O);
▣: Rh/Al$_2$O$_3$ (Rh: 0.25 wt%, Rh$_6$(CO)$_{16}$+6CsOH); △: Ir/Al$_2$O$_3$ (Ir: 2 wt%, IrCl$_4$·H$_2$O);
●: Ni/Al$_2$O$_3$ (NiO: 3 wt%, Ni(NO$_3$)$_2$·6H$_2$O); ■: Pd/Al$_2$O$_3$ (Pd: 1 wt%, PdCl$_2$);
▲: Pt/Al$_2$O$_3$ (Pt: 1 wt%, H$_2$PtCl$_6$·6H$_2$O);
+: Co-Mo/Al$_2$O$_3$ (MoO$_3$: 12.4 wt%, CoO: 3.8 wt%, (NH$_4$)$_6$Mo$_7$O$_{24}$·4H$_2$O, Co(NO$_3$)$_2$·6H$_2$O).

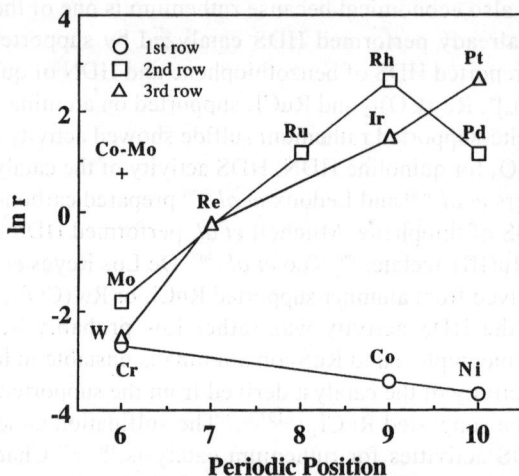

Fig. 4.23   Periodic trends in HDS of DBT catalyzed by various supported transition metals.
r (mol-DBT/mol-Metal·h): HDS rate at 280°C adopted from Fig. 4.22. The value of r for a supported Cr, Mo, W, Ru, Ni or Co-Mo catalyst was estimated on the basis of the data in Fig. 4.22.

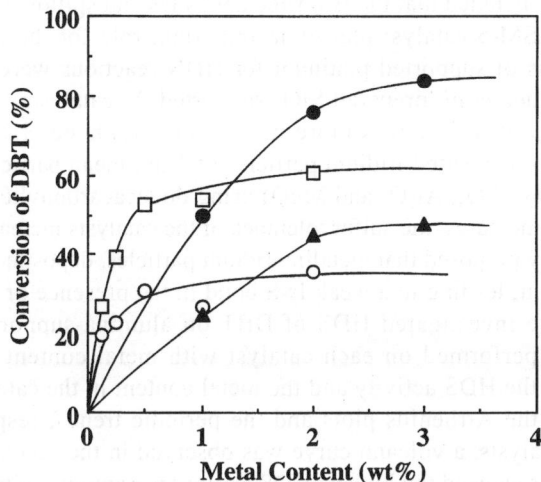

Fig. 4.24   Effect of metal content on conversion of DBT.
Pressure: 50 kg/cm^2, WHSV: 28 h^{-1}, Gas/Oil: 786 NL/L.
Rh/Al$_2$O$_3$(Rh$_6$(CO)$_{16}$+6CsOH): ○: 280°C ; □: 300°C.
Pt/Al$_2$O$_3$(H$_2$PtCl$_6$·6H$_2$O): ▲: 260°C ; ■: 280°C.

clarified as well as those of Co-Mo/Al$_2$O$_3$ and Ni-Mo/Al$_2$O$_3$. Further, an extremely rapid deactivation has been described in the case of Ru/alumina and Ru/Y-zeolite.[414] Recently, Liaw et al.[415] reported HDS and HDN of coal-derived naphtha catalyzed by alumina- and zeolite-supported ruthenium sulfide catalysts in a pressurized flow system. Their HDN activities per gram of catalyst were higher than those of Co-Mo/Al$_2$O$_3$ and Ni-Mo/Al$_2$O$_3$ catalysts, although the HDS activities per gram of catalyst were lower than those of Co-Mo/Al$_2$O$_3$ and Ni-Mo/Al$_2$O$_3$ catalysts. De Los Reyes and Vrinat reported that the alumina-supported Ni-Ru-S catalyst exhibited a very high hydrogenation activity compared with a commercial alumina-supported Ni-Mo catalyst.[417] Markel and Van Zee prepared the HDS catalysts derived from H$_2$SRu$_3$(CO)$_9$, HSRu$_2$Co(CO)$_9$, SRuCo$_2$(CO)$_9$, and [Co$_3$S$_2$(CO)$_7$]$_2$ supported on carbon and alumina.[418] For the catalysts sulfided at 400°C, the mixed metal cluster catalysts were less active for thiophene HDS than catalysts containing only one metal component. The carbon-supported catalysts were more active for thiophene HDS and more selective for butane formation than alumina-supported catalysts.

Noble metals other than ruthenium have also been found to be active for HDS of thiophenes.[398,408,409,420–428] Ledoux et al.[409] also observed a volcano curve between the HDS activity and the periodic position emerging for the second and third row transition metals. These periodic trends on carbon supports were in better agreement with the calculated periodic trends reported by Harris and Chianelli[420] than those in Fig. 4.21. They found that, at very low metal loading, carbon-supported rhodium and iridium sulfides showed the highest activities as opposed to ruthenium in the second row and osmium in the third row, respectively. Vissers et al.[408] reported similar results. Sugioka et al. reported that ultrastable Y (USY) zeolite-supported rhodium and platinum-paradium catalysts, and HZSM-5- and mesoporous silicate-supported platinum catalysts reveal higher activity than commercial Co-Mo/Al$_2$O$_3$.[421–423] In the latter case, they concluded that the Brønsted acid sites and spillover hydrogen formed on Pt particles in Pt/HZSM-5 catalyst played an important role for the thiophene HDS. The catalytic effectiveness of supported platinum for HDN reactions were reported by several authors.[60,425,426] Vázquez et al. prepared SiO_2-supported Pd catalysts. The Pd/SiO_2 catalyst reduced in hydrogen at 673 K was more active for thiophene HDS than the sulfided catalyst.[427] Frety et al. deposited iridium particles of 1 nm mean particle size on carriers of different acidity (SiO_2-Al_2O_3, Al_2O_3 and MgO) using dodecacarbonyltetrairidium.[428] As the acidity of the support increases, the sulfur tolerance of the catalysts increases whereas the HDS ability decreases. They proposed that metallic iridium particles, deposited on an acidic carrier, were electron-deficient, leading to a weak Ir-S bond in the presence of hydrogen.

The authors have investigated HDS of DBT on alumina-supported transition metal catalysts. HDS was performed on each catalyst with metal content in which the linear relationship between the HDS activity and the metal content in the catalyst are found. Figs. 4.22 and 4.23 show the Arrhenius plots and the periodic trends, respectively. Similar to carbon-supported catalysts, a volcano curve was observed in the second row, and alumina-supported Rh catalysts showed the highest HDS activity per amount of metal loaded. Further, group 8–10 transition metals in the second and third rows showed relatively high catalytic activities. As shown in Fig. 4.24, however, the activity of the Rh catalyst increased linearly up to 1 wt% but leveled off over 1 wt%. Pt catalyst showed a similar trend. In contrast, the activities of Ru and Mo catalysts increased linearly with increasing metal content even at more than 8 wt%. Although it is not shown in Fig. 4.24, the activity of the Ru catalyst increases up to 16 wt% (see Fig. 4.26). The results show that not only the intrinsic activity of a metal but also the capacity of a support for metal loading is very important to obtain high activity per catalyst weight.

Bimetallic catalysts using Ru-U, Pd-U, V-U,[429] Co-Ru, Ni-Ru, Fe-Ru, Ru-Mo,[410,412,430] Pd-Ni[431] (Co, Pt, Pd, Ir, or Ru)-Mo,[432,433] (Ru, Rh or Pd)-Fe[434] or Pt-Pd[435] were also studied. Delmon and coworkers[436] prepared biphasic sulfided hydrotreating catalysts. Bulk MoS_2 or WS_2 was mechanically mixed with noble metal-supported phases (PtS/Al_2O_3, Rh_2S_3/Al_2O_3 and PdS/Al_2O_3). These catalysts indicated conspicuous synergy effects in HDS of thiophene and hydrogenation of cyclohexene. A Ru-promoted Co-Mo/Al_2O_3 catalyst exhibited higher activity in thiophene HDS than Co-Mo/Al_2O_3.[437] A sulfided Ru-Mo/Al_2O_3 catalyst also exhibited higher activity in biphenyl HYD and pyridine HDN than commercial Ni-Mo/Al_2O_3 although the effect of the combination of the two phases remains unclear and cannot be assigned to any Ru-Mo-S type structure.[412]

To correlate these results with their remote control theory (see Chapter 3), Delmon and coworkers emphasized that mutual contamination of the sulfides of group VI and group VIII metals was quite unlikely in these experiments because the group VIII metals were located in the pores of the support. They explained the synergy using the volcano curve[59,398,402,405,408,409,438] as shown in Fig. 4.21 for HDS of DBT[398] or Fig. 4.25 for HDN of quinoline[405]: The synergy occurs between metals situated respectively on the left and on the right of the volcano curves. There are two phases in the remote control model: One donor phase on the right-hand part of the volcano curve is able to activate hydrogen to spillover hydrogen which migrate to the other acceptor phase on the left-hand part of the volcano curve. This may increase the number of active sites or induce changes in the structure of the surface. In contrast, Chianelli and coworkers explained the synergy effects by either the formation of mixed sulfides of "pseudobinary sulfides"[439] or an electronic transfer between the sulfides.[440] Although there

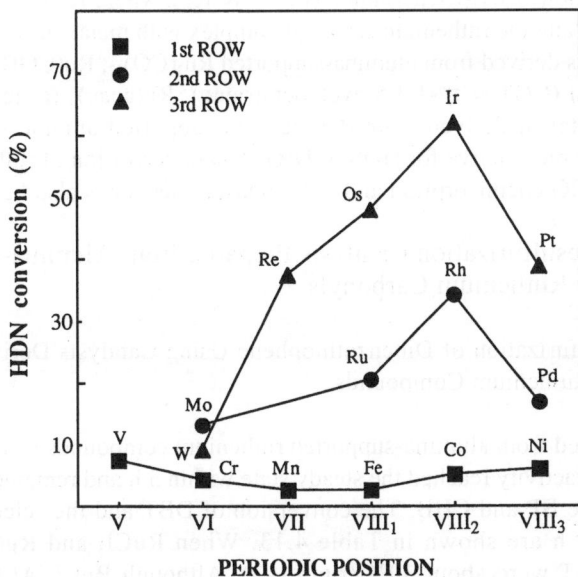

PERIODIC POSITION

Fig. 4.25 Tendencies in the periodical system for the hydrodenitrogenation of qinoline (5.9 mol% quinoline, 93.6 mol% hexadecane, 0.5 mol% carbon disulfide).
Catalyst sulfided with 10% H_2S in H_2, with temperature increasing from 293 to 653 K and then maintained at 653 K for 1 h. Reactions in a stirred autoclave; 653K approximate pressure 5 MPa.[405] Other details, including the preparation of the catalysts, in Refs. 60 and 405.
(From S. Eijsbouts, V. H. J. De Beer and R. Prins, *J. Catal.*, **109**, 218 (1988))

may be lack of agreement in the explanation of the synergy effect, there is no doubt that the strength of the metal-sulfur bond changes when going from one side of the volcano to the other. Thus, ruthenium sulfide may have the most favarable bond strength for HDS reactions, and as shown above, a number of attempts have been made to prepare unsupported and supported ruthenium catalysts.

4.4 Approaches Using Ruthenium Carbonyls

In Section 4.2, it has been mentioned that the catalysts derived from supported anionic molybdenum carbonyls are more active in HDS of DBT than conventional sulfided molybdena alumina and the one derived from alumina-supported neutral molybdenum hexacarbonyl.[380,381,392,393,396] It was shown that the use of alumina-supported anionic molybdenum carbonyls inhibited the sintering and sublimation of molybdenum species to maintain the dispersion of molybdenum sulfide on the support. These techniques were employed in the preparation of supported ruthenium carbonyls for HDS of DBT. Anionic ruthenium carbonyl complexes with a metal-sulfur bond were prepared because these complexes were effective for HDS of DBT catalyzed by supported molybdenum carbonyls.

In this section, HDS of DBT catalyzed by ruthenium catalysts in a pressurized flow reactor is described. The activity of the catalyst derived from the alumina-supported

$Ru_3(CO)_{12}$-ethanethiol(EtSH)-triethylamine(Et_3N) system, where $Ru_3(CO)_{12}$ reacts with EtSH and Et_3N to give an anionic ruthenium carbonyl complex with metal-sulfur bonds, is compared with that of catalysts derived from alumina-supported $Ru_3(CO)_{12}$, Ru(COD)(COT) (COD = η^4-1,5-cyclooctadiene) (COT = η^6-1,3,5-cyclooctatriene), $Ru(acac)_3$ (acac = acetylacetonate) and $RuCl_3$. The catalyst derived from the alumina-supported anionic ruthenium carbonyl was found to be the most active for HDS of DBT. Characterization of ruthenium catalysts by means of NO and CO chemisorption and XPS measurements has also been performed.[441,442]

4.4.1 Hydrodesulfurization Catalysts Prepared from Alumina-supported Anionic Ruthenium Carbonyls

A. Hydrodesulfurization of Dibenzothiophene Using Catalysts Derived from Alumina-supported Ruthenium Compounds

Catalysts derived from alumina-supported ruthenium compounds were active for HDS of DBT. The catalytic activity reached the steady state within 3 h and remained constant for over 10 h. Products were BP and CHB. The conversion of DBT and the selectivity for BP for a reaction time of 3 h are shown in Table 4.13. When $RuCl_3$ and $Ru(acac)_3$ were used, conversions of DBT were about 30% (runs 1–3). Although $RuCl_3/Al_2O_3$ was calcined at 450°C in run 1 but not in run 2, the degree of conversion of DBT between runs 1 and 2 barely differed. The use of organometallic compounds, Ru(COD)(COT) and $Ru_3(CO)_{12}$, increased the conversion of DBT as shown in run 4 (35%) and in run 5 (44%). The effect of the amount of Ru on the conversion of DBT and the effect on selectivity for BP was investigated for $Ru_3(CO)_{12}/Al_2O_3$ and the results are shown in Fig. 4.26. The conversion of DBT increased linearly with increasing amount of Ru up to 8 wt% but only slightly thereafter until 60% (maximum) was attained at 24 wt%; the conversion then decreased. Selectivity for BP decreased with increasing amount of Ru. The results show that high dispersion of ruthenium can be maintained up to 8 wt% of Ru. When an anionic ruthenium carbonyl complex (*vide infra*), formed by the reaction of $Ru_3(CO)_{12}$ with ethanethiol (EtSH) and triethylamine (NEt_3) in THF under reflux, was supported on the alumina ($Ru_3(CO)_{12}$-NEt_3-EtSH/Al_2O_3 system in run 6), the catalytic activity was highest among the catalysts prepared

Table 4.13 Hydrodesulfurization of Dibenzothiophene Catalyzed by Almina-Supported Ruthenium Carbonyls [a]

Run Catalyst	Conv. of DBT (%)	Conv. of DBT to BP (%)	Conv. of DBT to CHB (%)	Selec. for BP (%)
1 $RuCl_3/Al_2O_3$ [b,c]	32	23	9	71
2 $RuCl_3/Al_2O_3$ [b,d]	29	23	6	80
3 $Ru(acac)_3/Al_2O_3$ [e]	32	23	9	71
4 Ru(COD)(COT)/Al_2O_3 [f]	35	31	4	87
5 $Ru_3(CO)_{12}/Al_2O_3$	44	38	6	86
6 $Ru_3(CO)_{12}$-NEt_3-EtSH/Al_2O_3	48	43	5	90

[a] React. Temp. 300°C, 50 atm, LHSV 14 h^{-1}, H_2 18 l/h, Cat 0.5 g, Ru 8.0 wt%; Presulfided by H_2S in H_2 (H_2S 3%) at 300°C. [b] Presulfided by H_2S in H_2 (H_2S 3%) at 400°C. [c] Calcined at 450°C. [d] Not calcined. [e] acac = acetylacetonate. [f] COD = cyclooctadiene; COT = cyclooctatriene.
(From A. Ishihara, T. Kabe et al., *J. Jpn. Petrol. Inst.*, **37**, 302 (1994))

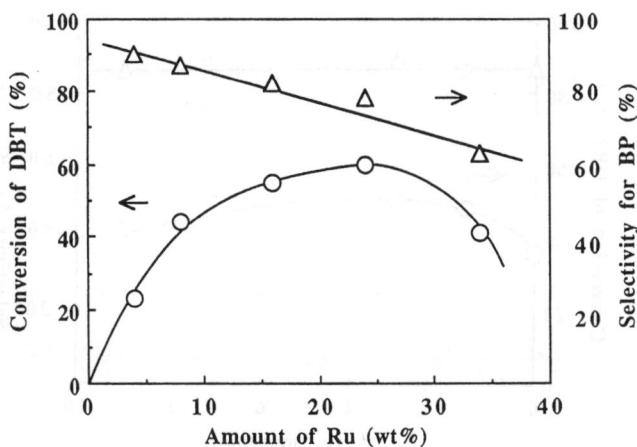

Fig. 4.26 Effect of amount of Ru on conversion of DBT.
(From A. Ishihara, T. Kabe et al., J. Catal., **150**, 213 (1994))

from alumina-supported ruthenium compounds and the conversion of DBT was 48%. Selectivities for BP with organometallic compound-derived catalysts were nearly 90% and higher than those with other catalysts.

Since it can be assumed that the anionic ruthenium carbonyl complex includes a metal-sulfur bond in its structure, activation of the $Ru_3(CO)_{12}$-NEt_3-EtSH/Al_2O_3 system by either H_2 or N_2 stream may be able to form active species in HDS of DBT directly. HDS of DBT was performed after the $Ru_3(CO)_{12}$-NEt_3-EtSH/Al_2O_3 system was activated by either H_2 or N_2 stream. The results are compared with those of $Ru_3(CO)_{12}$/Al_2O_3 system in Table 4.14. When the $Ru_3(CO)_{12}$/Al_2O_3 system was activated by H_2 and N_2, the conversions of DBT decreased to 36% and 33%, respectively. In contrast, when the $Ru_3(CO)_{12}$-NEt_3-EtSH/Al_2O_3 system was activated by H_2 and N_2, conversions of DBT were 47% and 50%, respectively, similar to those in the case of presulfiding. These results indicate that the ruthenium catalysts in a pressurized system require the presence of a small amount of sulfur to obtain and keep the catalytic activity sufficiently. Further, the effect of the amount of EtSH was investigated in the $Ru_3(CO)_{12}$-NEt_3-EtSH/Al_2O_3 system activated by H_2 stream, and the results are shown in Fig. 4.27. In this figure, the amounts of NEt_3 and EtSH were kept equal. The addition of EtSH/Ru_3 1.5 was the minimum needed to convert $Ru_3(CO)_{12}$ to an anionic species completely

Table 4.14 Hydrodesulfurization of Dibenzothiophene Catalyzed by Almina-Supported Ruthenium Carbonyls [a]

Run Catalyst	Conv. of DBT (%)	Conv. of DBT to BP (%)	Conv. of DBT to CHB (%)	Select. for BP (%)
7 $Ru_3(CO)_{12}$/Al_2O_3 [b]	36	31	5	87
8 $Ru_3(CO)_{12}$/Al_2O_3 [c]	33	29	4	88
9 $Ru_3(CO)_{12}$-NEt_3-EtSH/Al_2O_3 [b]	47	41	6	87
10 $Ru_3(CO)_{12}$-NEt_3-EtSH/Al_2O_3 [c]	50	43	7	86

[a] 300°C, 50 atm, LHSV 14 h^{-1}, H_2 18 l/h, Cat 0.5 g, Ru 8.0 wt%; [b] Activated by H_2 at 300°C. [c] Activated by N_2.
(From A. Ishihara, T. Kabe et al., J. Jpn. Petrol. Inst., **37**, 303 (1994))

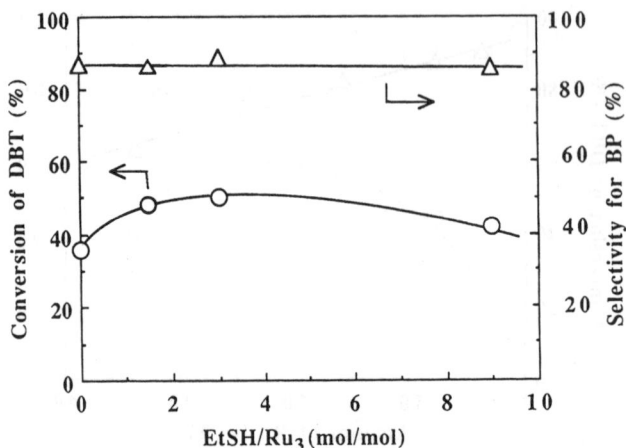

Fig. 4.27 Effect of amount of EtSH on conversion of DBT and selectivity for BP.
(From A. Ishihara, T. Kabe *et al.*, *J. Jpn. Petrol. Inst.*, **37**, 303 (1994))

and enough to increase the conversion of DBT. Further addition of EtSH and NEt$_3$ scarcely affected the conversion of DBT, indicating that excess EtSH or NEt$_3$ does not cause poisoning of the catalyst probably because some portions of EtSH and NEt$_3$, which were not needed to make anionic ruthenium species, were removed during the preparation of the catalyst.

The catalyst derived from activation of the Mo(CO)$_6$-NEt$_3$-EtSH/Al$_2$O$_3$ system and conventional sulfided molybdena-alumina showed 43% and 38% conversions of DBT, respectively. The selectivity for BP in molybdenum systems was lower than 80%.[380,381,392,393] Ruthenium catalysts derived from Ru$_3$(CO)$_{12}$-NEt$_3$-EtSH/Al$_2$O$_3$ and Ru$_3$(CO)$_{12}$/Al$_2$O$_3$ systems showed higher catalytic activity and selectivity for BP than the molybdenum-based catalysts.

B. Characterization of Alumina-supported Ruthenium Catalysts by Means of NO and CO Chemisorption

FTIR spectra of Ru$_3$(CO)$_{12}$-NEt$_3$-EtSH/Al$_2$O$_3$ and Ru$_3$(CO)$_{12}$/Al$_2$O$_3$ systems are shown in Fig. 4.28. Alumina-supported Ru$_3$(CO)$_{12}$ showed IR absorptions at 2058, 2020 and 1998 cm^{-1}, similar to IR absorptions of Ru$_3$(CO)$_{12}$ itself. Ru$_3$(CO)$_{12}$ reacted with NEt$_3$ and EtSH to give a complex which had IR absorptions at 2060 (weak) and 2002 (vs) cm^{-1} in a THF solution. This indicates that an anionic trinuclear ruthenium complex was formed in this reaction. When this complex was supported on Al$_2$O$_3$, IR absorptions at 2056 (medium) and 2000 (vs, broad) cm^{-1} were observed. Although a small amount of the initial complex reacted with alumina, it can be assumed that most of the anionic species with the original structure remained on the alumina. Since hydrogenolysis and thermolysis of these species on Al$_2$O$_3$ revealed high conversion of DBT, it can be assumed that the reaction of Ru$_3$(CO)$_{12}$ with NEt$_3$ and EtSH formed a complex with a ruthenium-sulfur bond and that hydrogenolysis and thermolysis of the complex supported on Al$_2$O$_3$ produced a highly dispersed ruthenium species which is active for HDS of DBT.

The FTIR spectra of NO and CO species adsorbed on ruthenium catalysts were measured and the results are shown in Figs. 4.29 and 4.30. In these spectra, the solid lines represent the

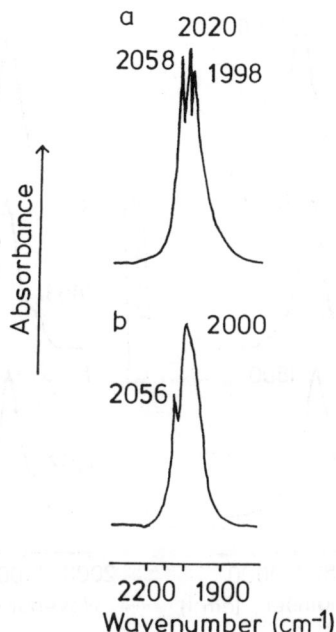

Fig. 4.28 FTIR measurement of supported ruthenium carbonyls.
(a) Ru$_3$ (CO)$_{12}$/Al$_2$O$_3$; (b) Ru$_3$ (CO)$_{12}$-NEt$_3$-EtSH/Al$_2$O$_3$.
(From A. Ishihara, T. Kabe *et al.*, *J. Catal.*, **150**, 214 (1994))

spectra after exposure to NO or CO for 1 h and the dotted ones represent those after subsequent evacuation for 1 h. The difference between the peak areas of the solid and broken lines represents the area of the weakly adsorbed species, while the area of the broken line represents the strongly adsorbed species. The ratio of the area of weakly adsorbed species to the total area is shown in Table 4.15 as reversible chemisorption.

When RuCl$_3$/Al$_2$O$_3$, Ru(acac)$_3$/Al$_2$O$_3$, Ru$_3$(CO)$_{12}$/Al$_2$O$_3$, and Ru$_3$(CO)$_{12}$-NEt$_3$-EtSH/Al$_2$O$_3$ systems were presulfided by H$_2$S/H$_2$, wavenumbers of NO chemisorption were 1788, 1802, 1800, and 1813 cm^{-1}, respectively (Fig. 4.29(a), (b), (c) and (d)). When the Ru$_3$(CO)$_{12}$-NEt$_3$-EtSH/Al$_2$O$_3$ system was treated with H$_2$ and N$_2$ instead of H$_2$S/H$_2$, peaks were observed at 1790 and 1779 cm^{-1} with shoulder peak at 1863 and 1852 cm^{-1}, respectively. The wavenumbers of the catalysts treated with N$_2$ were somewhat lower than those treated with H$_2$, except for the case of RuCl$_3$/Al$_2$O$_3$, indicating that the latter species has a higher electron density. As shown in Table 4.15, the values of reversible chemisorption of presulfided RuCl$_3$/Al$_2$O$_3$, Ru(acac)$_3$/Al$_2$O$_3$, and Ru$_3$(CO)$_{12}$-NEt$_3$-EtSH/Al$_2$O$_3$ systems were nearly 10% higher than those of the others. This may be related to the amount of sulfur on the catalyst, which is shown by XPS measurements of the catalysts below. When catalysts have a higher sulfidation state, the reversible chemisorption of NO appears to increase. With ruthenium catalysts, mononitrosyl species may be formed on the surface. In contrast to this, a dinitrosyl species where two molecular nitrogen monoxides coordinate on an unsaturated site was formed in the case of sulfided molybdenum catalysts, although the shoulder peaks in Fig. 4.29(e) and (f) may indicate one of the pair in the dinitrosyl species.[380,393)]

When RuCl$_3$/Al$_2$O$_3$, Ru(acac)$_3$/Al$_2$O$_3$, Ru$_3$(CO)$_{12}$/Al$_2$O$_3$ and Ru$_3$(CO)$_{12}$-NEt$_3$-EtSH/Al$_2$O$_3$

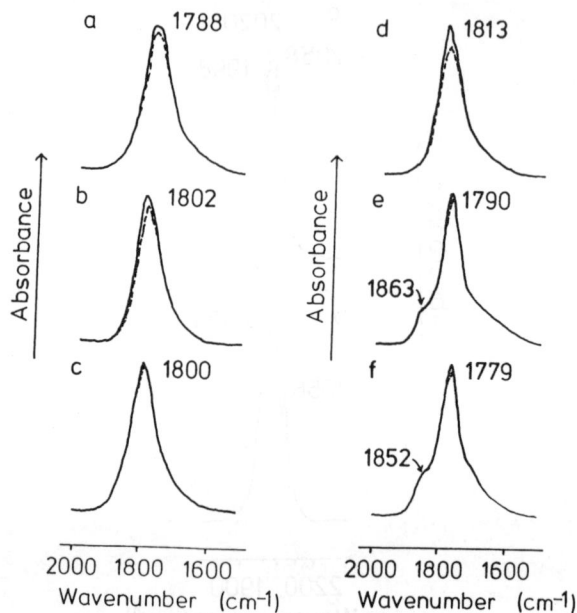

Fig. 4.29   FTIR measurement of NO chemisorbed on catalysts.
(a) RuCl₃/Al₂O₃; (b) Ru(acac)₃/Al₂O₃; (c) Ru₃(CO)₁₂/Al₂O₃; (d) Ru₃(CO)₁₂-NEt₃-EtSH/Al₂O₃ activated by H₂S/H₂; (e) Ru₃(CO)₁₂-NEt₃-EtSH/Al₂O₃ activated by H₂; (f) Ru₃(CO)₁₂-NEt₃-EtSH/Al₂O₃ activated by N₂. (From A. Ishihara, T. Kabe et al., J. Jpn. Petrol. Inst., 37, 304 (1994))

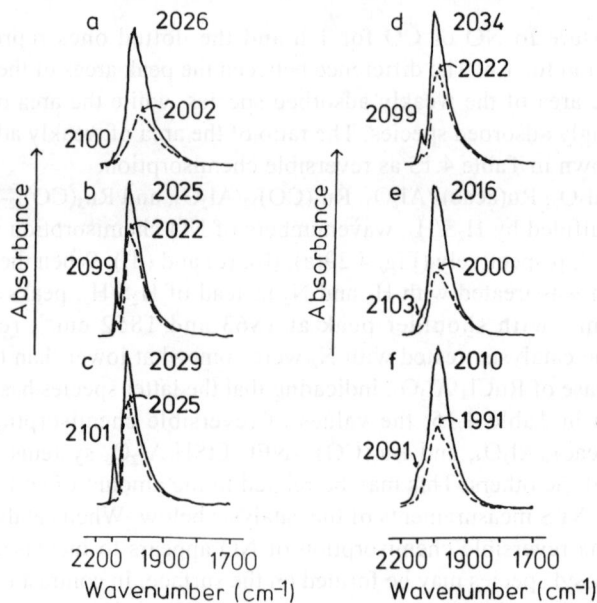

Fig. 4.30   FTIR measurement of CO chemisorbed on catalysts.
(a) RuCl₃/Al₂O₃; (b) Ru(acac)₃/Al₂O₃; (c) Ru₃(CO)₁₂/Al₂O₃; (d) Ru₃(CO)₁₂-NEt₃-EtSH/Al₂O₃ activated by H₂S/H₂; (e) Ru₃(CO)₁₂-NEt₃-EtSH/Al₂O₃ activated by H₂; (f) Ru₃(CO)₁₂-NEt₃-EtSH/Al₂O₃ activated by N₂. (From A. Ishihara, T. Kabe et al., J. Jpn. Petrol. Inst., 37, 305 (1994))

Table 4.15 Characterization of Supported Ruthenium Catalysts by NO Chemisorption

Run Catalyst	Chemisorption [a]		Reversible Chemisorption [b]	
	NO	CO	NO	CO
	($\times 10^2$ μmol/g-cat)	($\times 10^2$ μmol/g-cat)	%	%
11 $RuCl_3/Al_2O_3$ [c]	1.6	1.1	11	61
12 $Ru(acac)_3/Al_2O_3$ [d]	1.8	1.4	8	34
13 $Ru_3(CO)_{12}/Al_2O_3$ [d]	1.8	1.2	1	38
14 $Ru_3(CO)_{12}$-EtSH-Et_3N/Al_2O_3 [d]	2.0	1.6	10	43
15 $Ru_3(CO)_{12}$-EtSH-Et_3N/Al_2O_3 [e]	2.2	2.4	3	58
16 $Ru_3(CO)_{12}$-EtSH-Et_3N/Al_2O_3 [f]	2.6	2.0	1	42

[a] The amounts of chemisorption of NO and CO were measured at 25°C. [b] The value means the ratio of reversible chemisorption to total one and was determined by peak areas measured by FTIR. [c] Presulfided by H_2S in H_2 at 400°C (H_2S 3%) for 3 h and evacuated at 400°C for 1 h. [d] Presulfided by H_2S in H_2 at 300°C (H_2S 3%) for 3 h and evacuated at 300°C for 1 h. [e] Hydrogenated by H_2 at 300°C for 2 h and evacuated at 300°C for 1 h. [f] Treated by N_2 at 300°C for 2 h and evacuated at 300°C for 1 h.
(From A. Ishihara, T. Kabe et al., J. Jpn. Petrol. Inst., **37**, 304 (1994))

systems were presulfided by H_2S/H_2, wavenumbers of CO chemisorption were 2026, 2025, 2029, and 2034 cm^{-1}, respectively (Fig. 4.30(a), (b), (c) and (d)). When the $Ru_3(CO)_{12}$-NEt_3-EtSH/Al_2O_3 system was treated with H_2 and N_2 instead of H_2S/H_2, the peaks were observed at 2016 and 2010 cm^{-1}, respectively. Weak peaks were observed at about 2100 cm^{-1}, indicating that there may be some dicarbonyl species where two molecular carbon monoxides coordinate on an unsaturated site. These weak peaks disappeared with evacuation of the systems and the peaks due to irreversibly adsorbed carbonyl species remained at 2002, 2022, 2025, 2022, 2000, 1991 cm^{-1} in Fig. 4.30(a)–(f), respectively. As shown in NO chemisorption, the wavenumbers of Figs. 4.30(e) and (f) were somewhat lower than those of the former, except for the case of $RuCl_3/Al_2O_3$, indicating that these species have a higher electron density. In contrast to NO chemisorption, reversible chemisorption of carbonyl species was nearly 50% (Table 4.15). It can be assumed that NO is adsorbed on the catalyst surface and poisons the active sites very strongly and selectively while CO is adsorbed weakly.

The amounts of adsorbed NO and CO are listed in Table 4.15. The amounts of NO and CO adsorbed on the catalysts derived from the $Ru_3(CO)_{12}$-NEt_3-EtSH/Al_2O_3 system were larger than the others. The amounts of irreversibly adsorbed NO and CO, which were calculated from the values of the total amount of chemisorption and reversible chemisorption, were plotted against the conversion of DBT in Fig. 4.31. Although there is some scattering of results, the dispersion of active species can be assumed to be related to the catalytic activity. The scattering may be due to the difference in the condition of the active sites before and after HDS reaction. In order to confirm the difference, XPS spectra were measured for the catalysts before and after HDS reaction.

C. Characterization of Alumina-supported Ruthenium Catalysts by XPS

XPS spectra were measured for the samples used in NO chemisorption as catalysts before HDS, and the results are shown in Table 4.16. For supported ruthenium catalysts, S2p spectra were observed at 162.5–162.9 eV, Ru3p$_{3/2}$ at 461.8–462.1 eV, Ru3d$_{3/2}$ at 283.6–284.5 eV, and Ru3d$_{5/2}$ at 279.9–280.6 eV. Ru3p$_{3/2}$, Ru3d$_{3/2}$ and Ru3d$_{5/2}$ spectra of Ru(0) are observed at 461.1, 284.3, and 280.0 eV, respectively, while the Ru3p$_{3/2}$ spectrum of RuS_2 is observed at

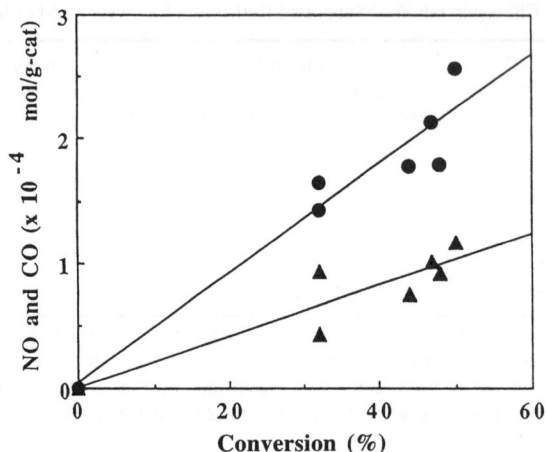

Fig. 4.31 Amount of adsorbed NO and CO vs. conversion.
●: NO; ▲: CO.
(From A. Ishihara, T. Kabe et al., J. Jpn. Petrol. Inst., **37**, 305 (1994))

Table 4.16 Characterization of Supported Ruthenium Catalysts by XPS [a]

Run	Catalyst	Al2p (eV)	S2p (eV)	Ru3p$_{3/2}$ (eV)	Ru3d$_{3/2}$ (eV)	Ru3d$_{5/2}$ (eV)	Ru3p/Al [b]	S/Ru3p [b]	S/Al [b]
17	RuCl$_3$/Al$_2$O$_3$ [c]	74.8	162.8	462.1	284.5	280.6	0.027	2.31	0.062
18	Ru(acac)$_3$/Al$_2$O$_3$ [d]	74.8	162.5	461.9	284.5	279.9	0.035	1.51	0.053
19	Ru$_3$(CO)$_{12}$/Al$_2$O$_3$ [d]	74.6	162.8	462.0	284.3	280.5	0.026	0.83	0.022
20	Ru$_3$(CO)$_{12}$-EtSH-Et$_3$N/Al$_2$O$_3$ [d]	74.8	162.6	462.0	284.5	280.6	0.027	1.64	0.044
21	Ru$_3$(CO)$_{12}$-EtSH-Et$_3$N/Al$_2$O$_3$ [e]	74.8	162.7	461.8	283.6	280.2	0.021	0.49	0.010
22	Ru$_3$(CO)$_{12}$-EtSH-Et$_3$N/Al$_2$O$_3$ [f]	74.8	162.9	461.9	284.4	280.5	0.023	0.67	0.015

[a] XPS spectra were measured for samples used in NO chemisorption. Before the measurement, the samples were sputtered by Ar$^+$ ion for 10 min. Every binding energy was referenced to Oxygen O1s 232.0 eV. [b] Ratio of peak areas between Al2p, S2p and Ru3p$_{3/2}$. [c] Presulfided by H$_2$S in H$_2$ at 400°C (H$_2$S 3%) for 3 h and evacuated at 400°C for 1 h before chemisorption. [d] Presulfided by H$_2$S in H$_2$ at 300°C (H$_2$S 3%) for 3 h and evacuated at 300°C for 1 h before chemisorption. [e] Hydrotreated by H$_2$ at 300°C for 2 h and evacuated at 300°C for 1 h before chemisorption. [f] Treated by N$_2$ at 300°C for 2 h and evacuated at 300°C for 1 h before chemisorption.
(From A. Ishihara, T. Kabe et al., J. Jpn. Petrol. Inst., **37**, 306 (1994))

462.7 eV. Vissers et al. reported that, with the catalyst derived from presulfiding of carbon-supported ruthenium trichloride, Ru3d$_{5/2}$ and S2p spectra were observed at 281.0 and 163.3 eV, respectively.[408] It was assumed that Ru was in a positive valence state coordinated by S ligands and that the major part of the S2p peak was due to the presence of S^{2-} sulfide ions coordinated to the metal ions. Mitchell et al. reported that catalysts derived from presulfiding of alumina-supported ruthenium (III) acetate showed the Ru3d$_{5/2}$, Ru3p$_{3/2}$ and S2p spectra at 280.2–280.4, 461.1–461.2, and 162.2–162.6 eV, respectively, and contained Ru-S species and S^{2-} sulfide ion.[406] They also reported that RuS$_2$ precipitated by H$_2$S from RuCl$_3$ in ethyl acetate showed spectra at 280.1 and 460.8 eV, while RuS$_{0.6}$ from reduced RuCl$_3$ reacted with H$_2$/H$_2$S (10%) at 643 K, 4 h at 280.6 and 461.6 eV. De Los Reyes et al. also reported that catalysts derived from presulfiding of alumina-supported RuCl$_3$ showed Ru3d$_{5/2}$ and S2p at 280.2–280.3 and 162.1–162.5 eV, respectively.[401] The binding energies observed with catalysts prepared in this work are in fairly good agreement with the reported ones. Thus, it is suggested that in the catalysts prepared here, the valence state of Ru is between Ru(0) and RuS$_2$ and that sulfur on the surface is presented as S^{2-} sulfide ions coordinated to the metal ions. However, the values of the binding energies in the catalysts are not so different as to be able to distinguish one catalyst from another. It is pointed out that XPS is not a suitable technique to identify metallic ruthenium because Ru(0) binding energies are very close to those of RuS$_2$.[401]

A clear difference was observed in comparison with the ratio of S/Ru in XP spectra. The sulfidation of RuCl$_3$/Al$_2$O$_3$ and Ru(acac)$_3$/Al$_2$O$_3$ systems proceeded under their presulfiding conditions, and the ratios of S/Ru were 2.31 and 1.51, respectively. However, the Ru$_3$(CO)$_{12}$/Al$_2$O$_3$ system containing Ru(0) species was not sulfided as much as were RuCl$_3$/Al$_2$O$_3$ and Ru(acac)$_3$/Al$_2$O$_3$ systems, and the ratio of S/Ru was 0.83. Sulfidation of the Ru$_3$(CO)$_{12}$-EtSH-Et$_3$N/Al$_2$O$_3$ system proceeded to a considerable extent probably because ethane thiol was added at the time of preparation of the precursor and sulfur atoms were already present on the support prior to presulfiding (S/Ru: 1.64). Ratios of S/Ru for the catalysts derived from H$_2$ and N$_2$ treatments of Ru$_3$(CO)$_{12}$-EtSH-Et$_3$N/Al$_2$O$_3$ system were 0.49 and 0.67, respectively, lower than the ratio for the Ru$_3$(CO)$_{12}$/Al$_2$O$_3$ system. The result also confirms that a complex containing metal-sulfur bonds was formed during the preparation of the precursor. De Los Reyes reported that the S/Ru ratio was 1.8 when RuCl$_3$/Al$_2$O$_3$ was presulfided under the following conditions : 15% H$_2$S/H$_2$, 673 K, 4h.[401] On the other hand, it was reported that the ratio of S/Ru of the catalyst derived from presulfiding (under 10% H$_2$S/H$_2$, 673 K, 1h) of ruthenium sponge was *ca.* 0.3 and that it was difficult to sulfide the ruthenium metal to ruthenium sulfide.[407] Our results were fairly consistent with these reports. The S/Ru ratio in our Ru$_3$(CO)$_{12}$/Al$_2$O$_3$ system was 0.83 and it was higher than that (*ca.* 0.3) in ruthenium sponge reported by Kuo *et al.*[407] because the Ru(0) species derived from the Ru$_3$(CO)$_{12}$/Al$_2$O$_3$ system was highly dispersed on alumina and could be effectively exposed to hydrogen sulfide. It is assumed that sulfidation of Ru$_3$(CO)$_{12}$-EtSH-Et$_3$N/Al$_2$O$_3$ system proceeded similarly because of its high dispersion of ruthenium.

De Los Reyes *et al.* reported that the catalysts prepared by H$_2$S/N$_2$ (15%) and H$_2$S (100%) treatments of RuCl$_3$/Al$_2$O$_3$, whose ratios in XPS were 4.2 and 3.6, respectively, showed higher catalytic activity than that of the catalyst prepared by the H$_2$S/H$_2$ (15%) treatment. It was concluded that sulfidation under H$_2$S/N$_2$ without prereduction gave a well sulfided and highly active RuS$_2$ supported catalyst.[401] However, our results were significantly different. The XPS spectra of ruthenium catalysts after HDS of DBT were recorded and the results are shown in Table 4.17. Although the binding energies of the catalysts were shown at 162.4–162.8 (S2p), 461.8–462.2 (Ru3p$_{3/2}$), 283.6–284.0 (Ru3d$_{3/2}$) and 280.2–280.5 eV (Ru3d$_{5/2}$), it was difficult to distinguish one catalyst from another, as shown by the catalysts before HDS reaction. An interesting result was found in the ratio of S/Ru. It was evident that the ratio of S/Ru after HDS was lower than that before HDS. The value of S/Ru decreased considerably in RuCl$_3$/Al$_2$O$_3$ and Ru(acac)$_3$/Al$_2$O$_3$ systems, slightly in the Ru$_3$(CO)$_{12}$/Al$_2$O$_3$ system, and hardly at all in the Ru$_3$(CO)$_{12}$-EtSH-Et$_3$N/Al$_2$O$_3$ system and appeared to approach about 0.5. This indicates that active ruthenium species on the catalyst are like metallic ruthenium rather than RuS$_2$ and that, irrespective of the precursor, the amount of sulfur remaining on the catalyst surface is very small, at least under the conditions pressurized by H$_2$. Further, it can be assumed that the activation of the Ru$_3$(CO)$_{12}$-EtSH-Et$_3$N/Al$_2$O$_3$ system would give a surface structure like that of the actual active species prior to HDS reaction.

Table 4.17 Characterization of Supported Ruthenium Catalysts by XPS [a]

Run Catalyst	Al2p (eV)	S2p (eV)	Ru3p$_{3/2}$ (eV)	Ru3d$_{3/2}$ (eV)	Ru3d$_{5/2}$ (eV)	Ru3p/Al [b]	S/Ru3p [b]	S/Al [b]
23 RuCl$_3$/Al$_2$O$_3$ [c]	74.9	162.8	461.8	283.8	280.2	0.041	0.52	0.021
24 Ru(acac)$_3$/Al$_2$O$_3$ [d]	75.1	162.8	462.2	283.9	280.5	0.026	0.97	0.025
25 Ru$_3$(CO)$_{12}$/Al$_2$O$_3$ [d]	75.0	162.5	462.1	284.0	280.5	0.025	0.45	0.011
26 Ru$_3$(CO)$_{12}$-EtSH-Et$_3$N/Al$_2$O$_3$ [e]	75.0	162.4	462.0	283.6	280.4	0.022	0.54	0.012

[a] XPS spectra were measured for samples used in HDS reaction of DBT. Prior to measurement, the samples were sputtered by Ar$^+$ ion for 10 min. Every binding energy was referenced to Oxygen O1s 232.0 eV. [b] Ratio of peak areas between Al2p, S2p, and Ru3p$_{3/2}$. [c] Presulfided by H$_2$S in H$_2$ at 400°C(H$_2$S 3%) for 3 h before HDS. [d] Presulfided by H$_2$S in H$_2$ at 300°C (H$_2$S 3%) for 3 h before HDS. [e] Hydrotreated by H$_2$ at 300°C for 2 h before HDS.
(From A. Ishihara, T. Kabe et al., J. Jpn. Petrol. Inst., 37, 307 (1994))

Lacroix et al. reported that, by elemental analysis of bulk RuS$_2$ catalyst[402] the ratios of S/Ru before and after HDS of DBT (flow system, 1 atm) were 2.2 and 1.87, respectively. It was also reported that some Ru(0) species were present in the unsupported RuS$_2$ tested under high hydrogen pressures.[443] These reports may support the results presented here.

D. General Discussion

Concerning the mechanism of HDS of DBT, we can utilize the results of reactions of organometallic complexes (cf. Chapter 2, Section 2.1.5.B), recently reported. A low valence noble metal center has been reported to insert a C-S bond of thiophenes to give metallathiabenzene. Chen et al. synthesized an iridathiabenzene complex in the reaction of an Ir(III) complex, which was coordinated by 2,5-dimethylthiophene with the η^5-bound mode with Na[H$_2$Al(OCH$_2$CH$_2$OMe)$_2$].[444] Jones also reported the formation of metallathiabenzenes in the reaction of (C$_5$Me$_5$)Rh(PMe$_3$)(Ph)H with thiophene, BT, and DBT.[445] In these reports, precursor metal(III) complexes were reduced by a reducing agent or by reductive elimination of the ligands attached to the complex to give metal(I) complexes that reacted with thiophenes to give metallathiabenzene complexes. It is evident that reduction of the metal center induces C-S bond cleavage of thiophenes. These results suggest that C-S bond cleavage of thiophenes through oxidative addition may occur even on the heterogenously reduced ruthenium surface formed on the supported ruthenium catalyst to give both a metal-carbon bond and a metal-sulfur bond (Fig. 4.32).

This mechanism on ruthenium catalysts appears to be different from that on molybdenum-based catalysts. It is believed that HDS is initiated by the adsorption of thiophenes to the anion vacancy located on molybdenum sulfide.[446-449] Schuit et al. suggested that a four-electron reduction process leads to the formation of H$_2$S and butadiene from thiophene after adsorption to the anion vacancy.[446,450,451] From the point of view of anion vacancy, end-on and side-on mechanisms have been proposed. In the end-on mechanism, it is assumed that thiophene adsorbs upright on the surface and that a metal-carbon is not formed after sulfur has been incorporated into an anion vacancy.[447] In the side-on mechanism, thiophene is coordinated at the sulfur anion vacancy through the C$_1$-C$_2$ double bond and a sulfur-sulfur bond and a metal-carbon bond are formed after the C-S bond is cleaved.[448] As shown in these mechanisms of molybdenum-based catalysts, it will be difficult to form both the metal-carbon bond and metal-sulfur bond in the C-S bond cleavage of thiophene. Neither the carbon nor the sulfur

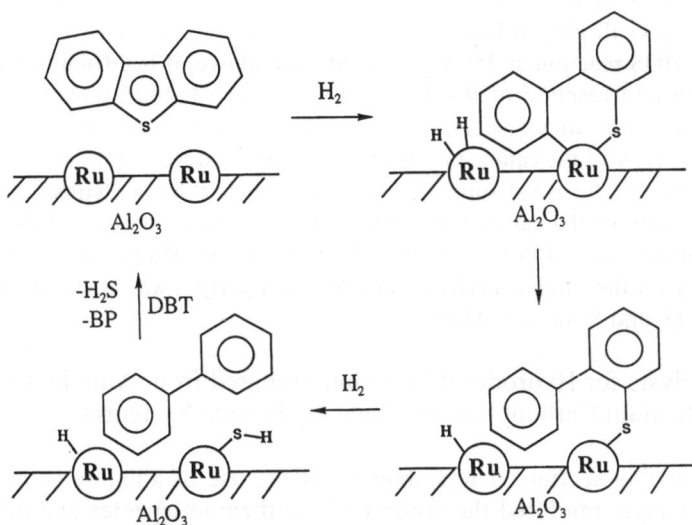

Fig. 4.32 Hydrodesulfurization mechanisms on alumina-supported ruthenium catalysts.
(From A. Ishihara, T. Kabe *et al.*, *J. Jpn. Petrol. Inst.*, **37**, 307 (1994))

atom of thiophene can approach the molybdenum atom because the ionic radius of sulfur is larger than that of molybdenum in MoS_2 and molybdenum atoms are surrounded by sulfur atoms. Further, a reaction like the oxidative addition to form both metal-carbon and metal-sulfur bonds by C-S bond cleavage of thiophene is difficult to produce because the molybdenum in MoS_2 is formally Mo(IV) and in a higher oxidation state.

In organometallic chemistry, some models of HDS on metal sulfides have been reported.[446] For example, the C-S bond of thiophene coordinated with ruthenium was cleaved by the attack of nucleophiles on the carbon attached to sulfur.[452,453] On an HDS catalyst surface, a nucleophilic attack may be carried out by the M-S⁻ group, yielding butadienedithiolate.[453] It has also been reported that a vinylthiolate Mo complex reacts with H_2 to cleave the C-S bond. In these reactions, the metal-carbon bond is not formed with the cleavage of the C-S bond.[454] These reactions may occur under conditions in which the metal is surrounded by sulfur, e.g., metal sulfides. In contrast to these reports, a new mechanism has been proposed by Angelici and coworkers. An iridathiabenzene complex reacted with low-valent iron carbonyls to give metallacyclopentadiene complexes with a new metal-carbon bond.[455] In this reaction, sulfur in one metal center was removed by another low-valent metal center. In heterogenous catalysis, a reaction like this could occur in reduced metal catalysts because their active species can be regarded as an aggregate of several metal centers. In the present HDS catalyzed by ruthenium catalysts under pressure, a reaction like the oxidative addition of DBT might occur on the reduced ruthenium surface to give an intermediate containing both metal-carbon and metal-sulfur bonds, followed by the removal of sulfur to another reduced ruthenium metal center.

De Los Reyes *et al.*[401] reported that the HDS activity of catalysts derived from $RuCl_3/Al_2O_3$ was higher than that from $Ru_3(CO)_{12}/Al_2O_3$. The results were explained by the difference in the dispersion of Ru on Al_2O_3. The difference between the author's results and theirs may be due to the difference in presulfiding and activation conditions. De Los Reyes *et al.* carried out presulfiding of $RuCl_3/Al_2O_3$ and $Ru_3(CO)_{12}/Al_2O_3$ in N_2/H_2S (15%) at 400°C for 4 h and the HDS reaction at 350°C. As mentioned above, it is difficult to sulfide Ru(0), and presulfiding and reaction of $Ru_3(CO)_{12}/Al_2O_3$ at high temperatures may promote the sintering of ruthenium metal species. Harvey and Matheson reported that, in HDN of quinoline, the activity of the catalyst derived from supported $Ru_3(CO)_{12}$ at 350°C was higher than that at 400°C.[157] The use of supported metal carbonyls at higher temperatures appears to decrease the activity of the catalyst. In HDS of DBT using a pressurized flow system, the presulfiding temperature, 300°C, was sufficient to activate $Ru_3(CO)_{12}/Al_2O_3$ because the catalytic activity for the catalyst derived from $Ru_3(CO)_{12}/Al_2O_3$ was higher than the activities from $RuCl_3/Al_2O_3$ and $Ru(acac)_3/Al_2O_3$.

4.4.2 Catalysts for Hydrodesulfurization Prepared from Alumina-supported Ruthenium Carbonyl-alkali Metal Hydroxide Systems

In the preceding section, it was shown that the use of alumina-supported anionic ruthenium carbonyls inhibited the sintering of ruthenium species and maintained the dispersion of ruthenium species on the support. Further, ruthenium species on the support appeared to be ruthenium metal rather than ruthenium sulfide. In this section, HDS of DBT, catalyzed by alkali-promoted ruthenium catalysts, has been investigated in a pressurized flow reactor. The activity of the catalyst derived from the alumina-supported $Ru_3(CO)_{12}$-alkali metal hydroxide system, in which $Ru_3(CO)_{12}$ reacts with alkali metal hydroxide to give an anionic ruthenium hydride complex $M[HRu_3(CO)_{11}]$ (M = alkali metal), is compared with that of catalysts derived from alkali promoted alumina-supported $Ru(acac)_3$ and $RuCl_3$. It was found that the addition of alkali metal remarkably increased the activities of ruthenium catalysts, and that the catalyst derived from the alumina-supported anionic ruthenium carbonyl is the most active for HDS of DBT. Characterization of ruthenium catalysts by means of NO chemisorption and XPS measurement was also performed.[456,457]

A. Effects of Addition of Alkali Metal Salts on the Hydrodesulfurization Reactivities of Alumina-supported Ruthenium Catalysts

Initially, the effects of the addition of NaOH to the alumina-supported ruthenium catalysts were investigated under the following conditions: 300 °C; 50 kg/cm²; H_2 18 l/h; WHSV, 16.5 h⁻¹; initial concentration of DBT, 1.0 wt% in xylene; catalyst, 0.5 g. Alumina-supported ruthenium catalysts examined were active for HDS and major products were BP and CHB. The results are shown in Table 4.18. In the absence of NaOH (runs 1–3), the catalytic activities of catalysts derived from alumina-supported ruthenium decreased in the order $Ru_3(CO)_{12} > Ru(acac)_3 > RuCl_3$. When $Ru_3(CO)_{12}$ was reacted with NaOH then supported on alumina (run 4), the catalytic activity remarkably increased and the conversion of DBT was 71%. At the preparation of the catalys in this system, $Ru_3(CO)_{12}$ reacted with sodium hydroxide to give $Na[HRu_3(CO)_{11}]$ quantitatively; this was confirmed by FTIR (2076 (w), 2022 (vs), 1993 (s), 1964 (m), 1642 (w) cm⁻¹ in methanol.[458] With the use of the $Ru(acac)_3$-$NaOH/Al_2O_3$ system, the activity increased slightly (run 5). When NaOH was added to the calcined $RuCl_3/Al_2O_3$, however, the conversion of DBT decreased (run 6). When $Ru_3(CO)_{12}$ and/or NaOH were supported on alumina in sequence (see footnote in Table 4.18), significant

Table 4.18 Hydrodesulfurization of Dibenzothiophene Catalyzed by Alumina-supported Ruthenium Catalysts. [a]

Run Catalyst	Conv. of DBT (%)	Conv. of DBT to BP (%)	Conv. of DBT to CHB (%)	Selectivity of BP (%)
1 $Ru_3(CO)_{12}/Al_2O_3$	44	38	6	86
2 $Ru(acac)_3/Al_2O_3$ [b]	35	30	5	87
3 $RuCl_3/Al_2O_3$ [c]	29	23	6	80
4 $Ru_3(CO)_{12}$-3NaOH/Al_2O_3	71	67	4	95
5 $Ru(acac)_3$-NaOH/Al_2O_3	43	40	3	93
6 $RuCl_3/Al_2O_3$ + NaOH [d]	15	14	1	95
7 $Ru_3(CO)_{12}/Al_2O_3$ + 3NaOH [e]	43	40	3	94
8 3NaOH/Al_2O_3 + $Ru_3(CO)_{12}$ [f]	32	30	2	94

[a] Reaction Temp 300°C, Pressure 50 kg/cm^2, WHSV 16.5 h^{-1}, Cat. 0.5 g, H_2 18 l/h, Amount of $Ru_3(CO)_{12}$ 0.13 mmol; Presulfided by H_2S in H_2 at 300 °C (H_2S 3%). [b] acac = acetylacetonate. [c] Presulfided by H_2S in H_2 at 400°C (H_2S 3%). [d] Initially $RuCl_3$ was supported on Al_2O_3 then NaOH was added at Ru/Na = 1. [e] Initially $Ru_3(CO)_{12}$ was supported on Al_2O_3 then NaOH was added at Ru/Na = 1. [f] Initially NaOH was supported on Al_2O_3 then $Ru_3(CO)_{12}$ was added at Ru/Na = 1.
(From A. Ishihara, T. Kabe et al., Chem. Lett., 2286 (1992))

increase in the catalytic activity was not observed (runs 7, 8). This suggests that the location of sodium close to ruthenium species by the presulfiding of $Na[HRu_3(CO)_{11}]$ supported on alumina may be intrinsic for high catalytic activity. The addition of NaOH increased the selectivity for BP in every catalyst, independent of the activity. Presence of NaOH appears to poison the active sites for hydrogenation of an aromatic ring, without affecting those for desulfurization. Further, the result indicates that the active sites for hydrogenation of an aromatic ring are different from those for HDS.

The effects of the addition of sodium salts other than NaOH were investigated and listed in Table 4.19. Use of sodium iodide decreased the catalytic activity and did not affect the selectivity for BP. The use of sodium hydrogen carbonate, sodium acetate and sodium hydrogen sulfite did not reveal any remarkable change in the catalytic activity, but slightly increased the selectivity for BP. While both activity and selectivity increased with use of sodium sulfide, the conversion of DBT was less than that with NaOH. It is suggested that the catalytic activity in HDS did not increase since these compounds cannot react with $Ru_3(CO)_{12}$ to give $Na[HRu_3(CO)_{11}]$.

When an alkali metal was changed in the alumina-supported $Ru_3(CO)_{12}$-alkali metal hydroxide system, as shown in Table 4.20, the conversion of DBT increased in the order Li < Na < K < Rb < Cs, which seems to be related to the ion radius of the alkali metal. When the rate of HDS of DBT was plotted against the reciprocal of the ion radius of the alkali metal shown in Fig. 4.33, it markedly increased for large ions such as Cs and Rb, indicating that ion radius affects the catalytic activity. As shown in Fig. 4.34, the conversion of DBT and the selectivity for BP increased with an increase in a M/Ru molar ratio (M = Na or Cs). Both values for Cs were greater than those for Na in every M/Ru value. Further, the conversion showed maximum at M/Ru = 2, which was greater than the amount of alkali metal hydroxide needed to form $M[HRu_3(CO)_{11}]$ (M/Ru = 1), suggesting that excess alkali metals were consumed by alumina. Further addition of alkali metals decreased the conversion, probably because such a large amount of alkali metal would shield active ruthenium species or reduce the reactivity of active sites.

Table 4.19 Hydrodesulfurization of Dibenzothiophene Catalyzed by Alumina-supported Ruthenium Carbonyl-Alkali Metal Compound Systems [a)]

Run Catalyst	Conv. of DBT (%)	Conv. of DBT to BP (%)	Conv. of DBT to CHB (%)	Selectivity for BP (%)
9 $Ru_3(CO)_{12}$-3NaI/Al_2O_3	16	14	2	87
10 $Ru_3(CO)_{12}$-3NaHCO$_3$/Al_2O_3	37	35	2	95
11 $Ru_3(CO)_{12}$-3NaC$_2$H$_3$O$_2$/Al_2O_3	49	47	2	96
12 $Ru_3(CO)_{12}$-3NaHSO$_3$/Al_2O_3	40	37	3	93
13 $Ru_3(CO)_{12}$-3Na$_2$S/Al_2O_3	60	59	1	99

[a)] Reaction temp. 300°C, Pressure 50 kg/cm², WHSV 16.5 h^{-1}, Cat. 0.50 g, H$_2$ 18 l/h; Concentration of DBT 1.0 wt%. Presulfided by H$_2$S in H$_2$ at 300°C (H$_2$S 3%).
(From A. Ishihara, T. Kabe et al., J. Jpn. Petrol. Inst., **39**, 215 (1996))

Table 4.20 Hydrodesulfurization of Dibenzothiophene Catalyzed by Alumina-supported Ruthenium Carbonyl-Alkali Metal Hydroxide Systems [a)]

Run Catalyst	Conv. of DBT (%)	Conv. of DBT to BP (%)	Conv. of DBT to CHB (%)	Selectivity for BP (%)
14 $Ru_3(CO)_{12}$-3LiOH/Al_2O_3	61	57	4	93
15 $Ru_3(CO)_{12}$-3KOOH/Al_2O_3	76	73	3	96
16 $Ru_3(CO)_{12}$-3RbOH/Al_2O_3	82	79	3	96
17 $Ru_3(CO)_{12}$-3CsOH/Al_2O_3	95	93	2	98

[a)] Reaction temp. 300°C, Pressure 50 kg/cm², WHSV 16.5 h^{-1}, Cat. 0.50 g, H$_2$ 18 l/h; Concentration of DBT 1.0 wt%. Presulfided by H$_2$S in H$_2$ at 300°C (H$_2$S 3%) for 3h.
(From A. Ishihara, T. Kabe et al., J. Jpn. Petrol. Inst., **39**, 215 (1996))

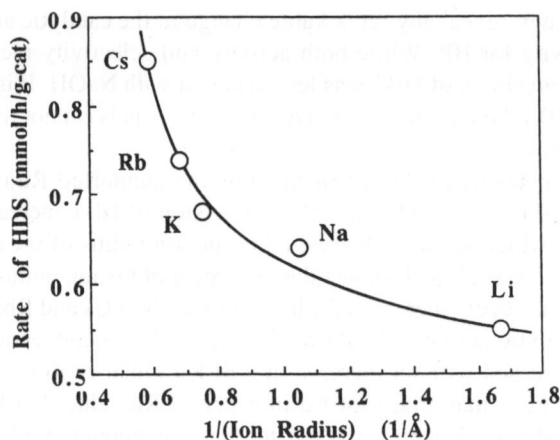

Fig. 4.33   Rate of HDS vs. 1/(ion radius).
Ru$_3$(CO)$_{12}$-3MOH/Al$_2$O$_3$ (M = Li, Na, K, Rb or Cs), 300°C; 50 kg/cm²; H$_2$ 18 l/h; WHSV, 16.5 h^{-1}; initial concentration of DBT in xylene, 1.0 wt%; catalyst, 0.5 g; Ru 8 wt%.
(From A. Ishihara, T. Kabe et al., Chem. Lett., 2286, (1992))

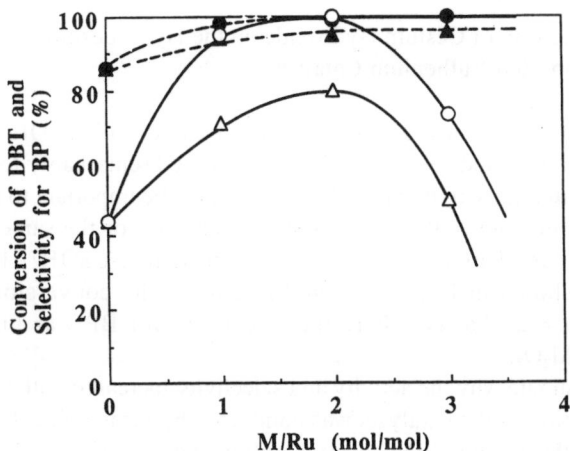

Fig. 4.34   Effect of M/Ru ratio on conversion of DBT and selectivity for BP.
Conversion of DBT: $\bigcirc$: Cs, $\triangle$: Na; Selectivity for BP: $\bullet$: Cs, $\blacktriangle$: Na.
$Ru_3(CO)_{12}$-nMOH/Al_2O_3 (M = Na or Cs), 300°C; 50 kg/cm²; H_2 18 1/h; WHSV, 16.5 h⁻¹; initial
concentration of DBT in xylene, 1.0 wt%; catalyst, 0.5 g; Ru 8 wt%.
(From A. Ishihara, T. Kabe et al., J. Jpn. Petrol. Inst., **39**, 215 (1996))

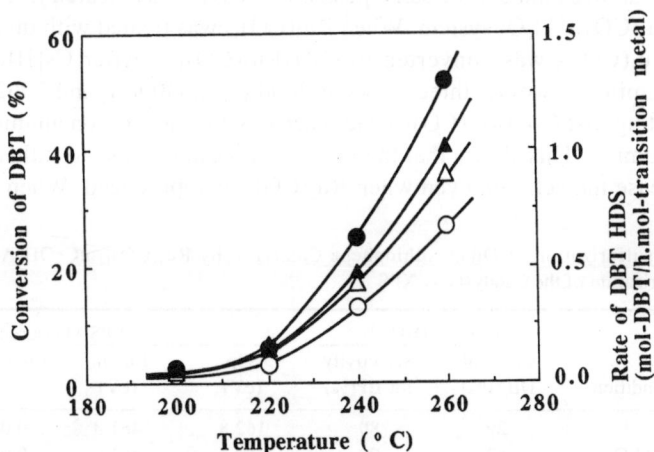

Fig. 4.35   Effect of temperature on conversion of DBT and rate of DBT HDS.
Conversion of DBT: $\bigcirc$: Ru-Cs/Al, $\triangle$: Co-Mo/Al; Rate of DBT HDS: $\bullet$: Ru-Cs/Al, $\blacktriangle$: Co-Mo/Al.
$Ru_3(CO)_{12}$-6CsOH/Al_2O_3 (Ru, 8 wt%); Co-Mo/Al_2O_3 (Mo, 8 wt%, Co/Mo = 0.58); 50 kg/cm²; H_2 18
1/h; WHSV, 70 h⁻¹; initial concentration of DBT in decalin, 1.0 wt%; catalyst, 0.2 g.
(From A. Ishihara, T. Kabe et al., J. Jpn. Petrol. Inst., **39**, 216 (1996))

To compare the activities between catalysts derived from $Ru_3(CO)_{12}$-6CsOH/Al_2O_3 and
a commercial Co-Mo/Al_2O_3, HDS of DBT was performed under the following conditions:
200–260°C; 50 kg/cm²; H_2 18 1/h; WHSV, 70 h⁻¹; initial concentration of DBT, 1.0 wt% in
decalin; catalyst, 0.2 g. The results are shown in Fig. 4.35. Although the conversion of DBT
over the ruthenium catalyst was slightly less than that over Co-Mo/Al_2O_3, HDS rate of DBT
per the supported amount of transition metal of the former was greater than that of the latter.

B. Effects of Addition of Cesium Hydroxide on the Hydrodesulfurization Reactivities of Alumina-supported Ruthenium Catalysts

When $Ru_3(CO)_{12}$, $Ru(acac)_3$ and $RuCl_3$ were treated with CsOH then supported on alumina, catalytic activity and selectivity for BP increased remarkably for all catalysts. This indicates that to obtain high activity and selectivity, it may be important to prepare alkali metal salts of ruthenium compounds at the initial stage. Activity for the same amount of cesium increased in the order $RuCl_3$-$CsOH/Al_2O_3$ < $Ru(acac)_3$-$CsOH/Al_2O_3$ < $Ru_3(CO)_{12}$-$3CsOH/Al_2O_3$. As shown in Fig. 4.34 and Table 4.21, the conversion of DBT reached maximum at Cs/Ru = 2, that is, where the selectivity for BP was higher than 99%, in $Ru_3(CO)_{12}$-$6CsOH/Al_2O_3$.

In order to investigate why the activity and selectivity increased with the addition of alkali metals, characterization of the catalysts was conducted by measurements of FTIR spectra of catalyst precursors, the amount of NO chemisorption and XPS spectra of the catalysts before and after use for HDS.

1. FTIR Spectra of the Catalyst Precursors

The FTIR spectra of the catalyst precursors are shown in Fig. 4.36. $Ru_3(CO)_{12}$ supported on alumina showed three peaks at 1998 (s), 2020 (vs) and 2058 (vs) cm^{-1}. Because $Ru_3(CO)_{12}$ without alumina showed almost the same peaks, $Ru_3(CO)_{12}$ was located just on the alumina surface in the $Ru_3(CO)_{12}/Al_2O_3$ system. When $Ru_3(CO)_{12}$ was treated with threefold CsOH in methanol, all $Ru_3(CO)_{12}$ was converted to $Cs[HRu_3(CO)_{11}]$. After $Cs[HRu_3(CO)_{11}]$ was supported on alumina, however, three peaks at 2000 (s), 2030 (vs) and 2062 (s) cm^{-1} were observed, indicating that $Cs[HRu_3(CO)_{11}]$ had reacted with acid sites on alumina to reproduce $Ru_3(CO)_{12}$ on alumina. This shows that the location of cesium close to ruthenium species is intrinsic to increase the activity, even when $Ru_3(CO)_{12}$ is reproduced. When $Ru_3(CO)_{12}$ was

Table 4.21 Hydrodesulfurization of Dibenzothiophene Catalyzed by $Ru_3(CO)_{12}$-$nCsOH/Al_2O_3$ Systems and Characterization of the Catalysts by XPS

		HDS of DBT [a]		XPS Measurement [b]			
		Conv. of	Selectivity	S2p	Ru3p$_{3/2}$	Ru3p/Al [c]	S/Ru3p [c]
Run	Activation Condition	DBT(%)	for BP(%)	(eV)	(eV)		
18	$RuCl_3/Al_2O_3$	29	80	162.8	461.8	0.041	0.52
19	$RuCl_3$-$CsOH/Al_2O_3$	62	90	162.4	461.6	0.040	2.50
20	$Ru(acac)_3/Al_2O_3$	35	87	162.8	462.2	0.026	0.97
21	$Ru(acac)_3$-$CsOH/Al_2O_3$	74	98	161.0	462.1	0.037	2.49
22	$Ru_3(CO)_{12}/Al_2O_3$	44	86	162.5	462.1	0.025	0.45
23	$Ru_3(CO)_{12}$-$3CsOH/Al_2O_3$	95	98	161.4	461.9	0.027	2.93
24	$Ru_3(CO)_{12}$-$6CsOH/Al_2O_3$	100	99	160.5	462.2	0.016	7.19
25	$Ru_3(CO)_{12}$-$9CsOH/Al_2O_3$	73	100	161.3	461.3	0.027	1.18

[a] Reaction temp. 300°C, Pressure 50 kg/cm², WHSV 16.5 h⁻¹, Cat. 0.50 g, H_2 18 l/h; Concentration of DBT 1.0 wt%. Presulfided by H_2S in H_2 (H_2S 3%). [b] XPS spectra were measured for samples used in HDS of DBT. Before measurement, the samples were sputterd by Ar⁺ ion for 10 min. Every binding energy was referenced to Oxygen O1s 532.0 eV.
[c] Ratio of peak areas between Al2p, S2p, and Ru3p$_{3/2}$.
(From A. Ishihara, T. Kabe et al., J. Jpn. Petrol. Inst., **39**, 216 (1996))

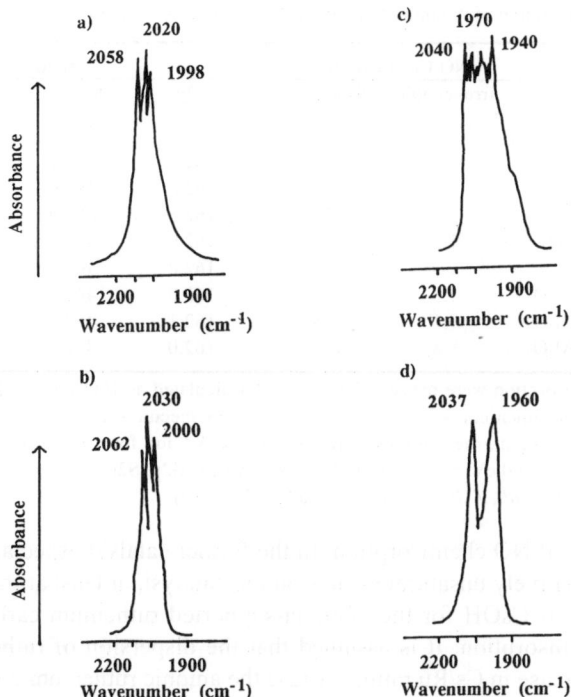

Fig. 4.36   FTIR spectra of $Ru_3(CO)_{12}$-nCsOH/Al_2O_3.
(a) $Ru_3(CO)_{12}$/Al_2O_3; (b) $Ru_3(CO)_{12}$-3CsOH/Al_2O_3; (c) $Ru_3(CO)_{12}$-6CsOH/Al_2O_3; (d) $Ru_3(CO)_{12}$-9CsOH/Al_2O_3.
(From A. Ishihara, T. Kabe et al., J. Jpn. Petrol. Inst., **39**, 216 (1996))

treated with sixfold CsOH in methanol and supported on alumina, $Ru_3(CO)_{12}$ was not reproduced and the anionic species was maintained on alumina, since peaks with wavenumbers lower than those of $Ru_3(CO)_{12}$ were observed. This result shows that maintenance of the anionic species on alumina as well as the location of cesium close to ruthenium species is also essential to increase the activity. When $Ru_3(CO)_{12}$ was treated with ninefold CsOH in methanol and supported on alumina, the anionic species was also maintained as shown in Fig. 4.31(d). It seems, however, that the trinuclear anionic $[HRu_3(CO)_{11}]^-$ cannot be maintained on alumina, probably because the amount of CsOH added was too much to maintain the structure. Although it is not clear whether or not the trinuclear structure is required to obtain high catalytic activity, the addition of an excess amount of cesium would shield the active ruthenium species or reduce the reactivity of active sites and decrease the activity.

2. NO Chemisorption of the Catalysts

The amount of NO chemisorption can be considered to be proportional to that of the coordinatively unsaturated sites on a catalyst. The amount of NO chemisorption was measured because the number of coordinatively unsaturated sites on a catalyst is often related to the catalytic activity. No irreversible chemisorption of NO was observed on CsOH/Al_2O_3. As shown in Table 4.22, the addition of CsOH to $RuCl_3$/Al_2O_3 and $Ru(acac)_3$/Al_2O_3 did not

Table 4.22 Characterization of Alumina-Supported Ruthenium Catalysts by NO Chemisorption and XPS

Run Catalyst	NO Chemisorption [a]		Before HDS [c]			
	Irreversible ($\times 10^2 \mu$mol/g-cat)	Dispersion [b] (%)	S2p (eV)	Ru3p$_{3/2}$ (eV)	Ru3p/Al [d]	S/Ru3p [d]
26 RuCl$_3$/Al$_2$O$_3$	1.4	18	162.8	462.1	0.027	2.31
27 RuCl$_3$-CsOH/Al$_2$O$_3$	0.2	3	162.0	462.0	0.070	1.57
28 Ru(acac)$_3$/Al$_2$O$_3$	1.3	16	162.5	461.9	0.035	1.51
29 Ru(acac)$_3$-CsOH/Al$_2$O$_3$	1.0	13	162.0	462.2	0.041	2.68
30 Ru$_3$(CO)$_{12}$/Al$_2$O$_3$	1.3	16	162.8	462.0	0.026	0.83
31 Ru$_3$(CO)$_{12}$-3CsOH/Al$_2$O$_3$	1.8	23	162.0	462.0	0.027	3.30
32 Ru$_3$(CO)$_{12}$-6CsOH/Al$_2$O$_3$	9.1	115	162.1	462.3	0.015	7.33
33 Ru$_3$(CO)$_{12}$-9CsOH/Al$_2$O$_3$	8.8	111	162.0	462.0	0.025	3.04

[a] Amounts of NO chemisorption were measured at 25 °C. [b] Calculated as 100% at NO/Ru = 1 mol/mol. In this calculation, the amount of ruthenium is 7.9 $\times 10^2 \mu$mol/g-cat. [c] XPS spectra were measured for samples used in NO adsorption. Before measurement, the samples were sputterd by Ar$^+$ ion for 10 min. Every binding energy was referenced to Oxygen O1s 532.0 eV. [d] Ratio of peak areas between Al2p, S2p, and Ru3p$_{3/2}$.
(From A. Ishihara, T. Kabe et al., J. Jpn. Petrol. Inst., 39, 217 (1996))

increase the amount of NO chemisorption. In the former catalyst especially, CsOH seems to obstruct the coordinatively unsaturated sites on the catalyst, at least at the presulfiding step. In contrast, the use of CsOH for the alumina-supported ruthenium carbonyl increased the amount of NO chemisorption. It is assumed that the dispersion of ruthenium species may increase with an increase in Cs/Ru ratio, because the anionic ruthenium complex may interact with acid sites to be anchored to alumina at the presulfiding step. As shown in FTIR measurements, the anionic ruthenium carbonyls were maintained on alumina only in the Ru$_3$(CO)$_{12}$-6CsOH/Al$_2$O$_3$ and Ru$_3$(CO)$_{12}$-9CsOH/Al$_2$O$_3$ systems. In these systems, the dispersion exceeded 100%, indicating that there may be a dinitrosyl species where two molecules of NO are adsorbed on one ruthenium atom. Further, it is also suggested that stable ruthenium sulfide may be present because dinitrosyl species were often observed for metal sulfide.[389,390,442] In order to confirm the presence of ruthenium sulfide, XP spectra of the catalysts were measured before and after HDS.

3. XPS Measurement of the Catalysts Before and After HDS Reaction
    XP spectra were measured for the catalysts before and after HDS reaction and the results are shown in Tables 4.21 and 4.22. The catalysts before HDS were the samples used for NO chemisorption. The S2p and Ru3p$^{3/2}$ XP spectra from the alumina-supported ruthenium used for HDS showed peaks in the ranges 160.5–162.8 eV and 461.3–462.2 eV, respectively. These results suggest that the ruthenium species for every catalyst are those between ruthenium metal (461.0 eV) and RuS$_2$ (462.7 eV). The binding energies of Ru3p$_{3/2}$ for the Ru$_3$(CO)$_{12}$-6CsOH/Al$_2$O$_3$ system before and after HDS were the highest among the catalysts examined, while those for some catalysts were not affected by the addition of alkali metal. This shows that the ruthenium species for the Ru$_3$(CO)$_{12}$-6CsOH/Al$_2$O$_3$ system may be close to RuS$_2$, rather than ruthenium metal. Further, the binding energies of S2p decreased with alkali metal for both catalysts before and after HDS, indicating that the sulfur atom in the catalysts with alkali metal would have higher electron density than that without alkali metal. The values of Ru3p/Al and S/Ru3p were obtained from the peak areas of XP spectra. Although the values of Ru3p/Al for alumina-supported RuCl$_3$ and Ru(acac)$_3$ catalysts were higher than

those for alumina-supported $Ru_3(CO)_{12}$, the catalytic activities for the former catalysts were lower than those for the latter, indicating that the dispersion of the former ruthenium species was not as high as that of the latter while the bulk of the ruthenium species for the former catalysts existed in higher concentrations. The most important phenomenon was observed for the S/Ru3p ratio values. When cesium existed in the catalysts, the values of S/Ru3p remarkably increased, as shown in Table 4.21, for the catalysts after HDS, indicating that the presence of cesium was able to stabilize ruthenium sulfide to accommodate sulfur on the catalyst surface. These values of S/Ru3p were plotted against the conversions of DBT shown in Fig. 4.37. Fig. 4.37 shows the conversion of DBT increasing with increase in the S/Ru3p ratio. This suggests that the catalytic activity increases when ruthenium sulfide is stabilized in the presence of cesium. Further, it seems that the presence of alkali metal strengthens the bond of ruthenium and sulfur to stabilize ruthenium sulfide, and the carbon-sulfur bond scission may proceed more easily than ruthenium species that are sulfided less and close to low valent ruthenium metal.

It has been reported that when alumina-supported ruthenium chloride was presulfided by H_2S/H_2, H_2S and H_2S/N_2, the treated catalysts revealed 1.8, 3.6 and 4.2 of S/Ru in XPS measurement, respectively.[401] The two latter catalysts were sulfurized completely and showed catalytic activity in HDS of thiophene higher than the former, which was not sulfurized completely. In this case, the value of S/Ru in XPS was also much greater than the stoichiometric value of ruthenium disulfide, 2. In the present study, when alumina-supported ruthenium chloride was presulfided by H_2S/H_2, the treated catalyst revealed 2.31 of S/Ru in XPS measurement, as shown in Table 4.22. Further, in cesium-promoted $Ru_3(CO)_{12}/Al_2O_3$ systems, both catalysts, before and after HDS, showed a maximum value of S/Ru3p at Cs/Ru = 2, which was also much larger than the stoichiometric value of ruthenium disulfide, 2. This suggests that in this catalyst system, not only are ruthenim disulfide, 2. This suggests that in this catalyst system, not only are ruthenium species sulfurized completely, but also a significant amount of sulfur formed by the HDS reaction may be accommodated on the catalyst and coat ruthenium atoms, because HDS of DBT is much more rapid than the

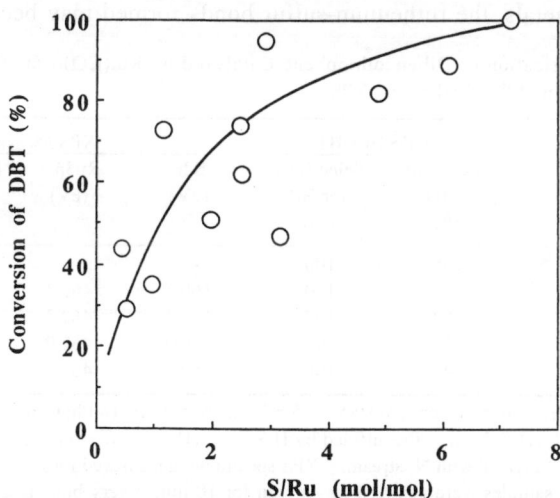

Fig. 4.37 Plots of S/Ru ratio in XPS vs. conversion of DBT.
(From A. Ishihara, T. Kabe et al., J. Jpn. Petrol. Inst., **39**, 218 (1996))

formation of hydrogen sulfide by the reaction of surface sulfur with hydrogen. The coating of the ruthenium atoms with large amounts of sulfur may result in higher values of S/Ru3p and lower values of Ru3p/Al. The further addition of cesium decreased S/Ru3p ratio, however, indicating that excess of CsOH may prevent the formation of ruthenium sulfide by shielding the coordinatively unsaturated sites on ruthenium species or reduce the reactivity of active sites to decrease the sulfur accommodation. Although the formation of cesium sulfide can also be considered, the amount of sulfur accommodated on a catalyst as cesium sulfide may not be significant, since the value of S/Ru3p at Cs/Ru = 3 was lower than that at Cs/Ru = 2.

C. Effects of the Conditions of Activation on Hydrodesulfurization Using Catalyst Derived from $Ru_3(CO)_{12}$-6CsOH/Al_2O_3

It has been reported that the method used to activate the catalyst precursor largely affects the HDS activity.[401] The authors investigated the effects of the conditions of activation on the activity using $Ru_3(CO)_{12}$-6CsOH/Al_2O_3 and measured XPS spectra of the catalysts used for HDS. The results are shown in Table 4.23. When the $Ru_3(CO)_{12}$-6CsOH/Al_2O_3 system was activated with H_2S in H_2 (H_2S 50%), the conversion of DBT was 98%, similar to that in run 24, in which H_2S in H_2 (H_2S 3%) was used. The concentration of H_2S does not seem to affect the activity. When H_2S in N_2 (H_2S 50%) and H_2S 100% were used instead of H_2S in H_2, the conversions of DBT decreased slightly to 89% and 82%, respectively, compared with that in run 24. When N_2 and H_2 were used, the conversions decreased markedly. In XPS measurement of these catalysts, S/Ru3p ratios also decreased in comparison with that in run 24. It is likely that the catalyst precursor cannot be sulfurized in the absence of H_2S to release carbon monoxide ligands. In contrast, in the presence of H_2S, it is suggested that anionic ruthenium carbonyls such as $[HRu_3(CO)_{11}]^-$, etc. in $Ru_3(CO)_{12}$-6CsOH/Al_2O_3 react with H_2S to release carbon monoxide ligands easily and form ruthenium-sulfur bonds during the preparation of the catalyst. In this process, H_2S adsorbed on ruthenium would oxidize Ru(0) species to an oxidation state close to Ru(IV) for RuS_2 with the release of a hydrogen molecule. In the presence of alkali metals, the ruthenium-sulfur bonds formed may become stable to keep

Table 4.23 Hydrodesulfurization of Dibenzothiophene Catalyzed by $Ru_3(CO)_{12}$-6CsOH/Al_2O_3 Systems and Characterization of the Catalysts by XPS

		HDS of DBT [a]		XPS Measurement [g]			
Run	Activation Condition	Conv. of DBT (%)	Selectivity for BP (%)	S2p (eV)	Ru3p$_{3/2}$ (eV)	Ru3p/Al [h]	S/Ru3p [h]
34	H_2S in H_2 (H_2S 50%) [b]	98	100	—	—	—	—
35	H_2S in N_2 (H_2S 50%) [c]	89	100	160.7	462.3	0.014	6.11
36	H_2S (100%) [d]	82	100	161.1	462.3	0.015	4.87
37	H_2 [e]	51	99	160.0	462.0	0.023	1.98
38	N_2 [f]	47	100	159.8	461.8	0.026	3.18

[a] Reaction temp. 300°C, Pressure 50 kg/cm², WHSV 16.5 h⁻¹, Cat. 0.50 g, H_2 18 l/h; Concentration of DBT 1.0 wt%.
[b] Presulfided by H_2S in H_2 (H_2S 50%). [c] Presulfided by H_2S in N_2 (H_2S 50%). [d] Treated with H_2S (100%) stream. [e] Treated with H_2 stream. [f] Treated with N_2 stream. [g] XPS spectra were measured for samples used in HDS of DBT. Before measurement, the samples were sputtered by Ar⁺ ion for 10 min. Every binding energy was referenced to Oxygen O1s 532.0 eV. [h] Ratio of peak areas between Al2p, S2p, and Ru3p$_{3/2}$.
(From A. Ishihara, T. Kabe et al., J. Jpn. Petrol. Inst., **39**, 219 (1996))

active ruthenium sulfide even under hydrogen pressure.

D. Role of Alkali Metal

Although HDS of thiophenes using ruthenium sulfide catalysts has been reported, the effects of addition of alkali metals in those catalysts on the HDS reactions are not well known. Göbölös *et al.* compared the activities in HDS of thiophene catalyzed by ruthenium catalysts supported on NaY, KY, HY and Na$_2$S-doped KY zeolites.[416] The activity decreased in the order Na$_2$S-doped KY > KY > NaY > HY, and the Na$_2$S-doped KY showed activity three times higher than that of KY. The order of the activity was explained by the strength of Brønsted acids. The authors concluded that the addition of Na$_2$S neutralized the Brønsted acidity of zeolite to increase the activity as a result. The effects of the kind and amount of alkali metal, however, were not examined, and the activity was rather low in comparison with a commercial catalyst[401] because the amount of ruthenium loaded on the support was too low, 2 wt%. Further, there was no reference to the effect of alkali metal on the stabilization of ruthenium sulfide.

In alumina-supported ruthenium catalysts, the HDS activity is often rather low[157,401,406,407] compared with commercial catalysts, probably because sulfidation of ruthenium species is incomplete and RuS$_2$ on alumina is unstable in hydrogen atmosphere. It has also been reported that catalysts derived from supported Ru$_3$(CO)$_{12}$ showed lower activity for HDS of thiophene in atmospheric pressure of hydrogen than catalysts derived from supported RuCl$_3$.[157,401,407] In contrast, the authors reported that in a pressurized flow system, the catalysts derived from alumina-supported ruthenium carbonyls showed activity in HDS of DBT higher than those from alumina-supported RuCl$_3$ and Ru(acac)$_3$.[441,442] Alumina-supported ruthenium carbonyls gave the active species, which shows the high activity for HDS under high pressure, more easily at the presulfiding step than alumina-supported RuCl$_3$.[441,442] In the present study,

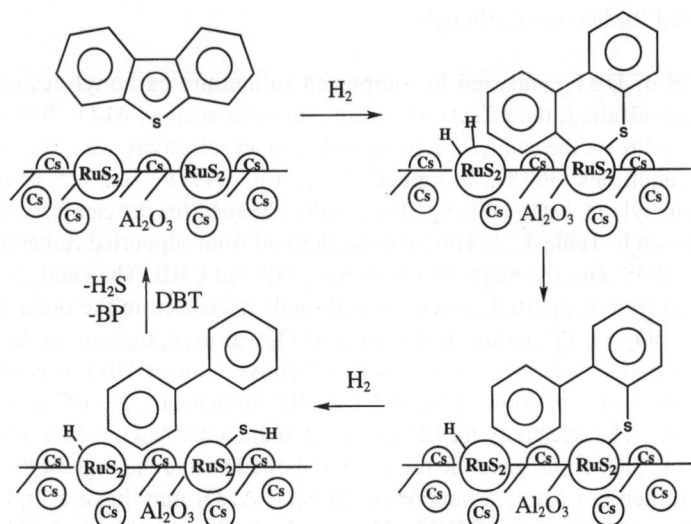

Fig. 4.38 Mechanism of hydrodesulfurization catalyzed by alumina-supported ruthenium-alkali metal systems. (From A. Ishihara, T. Kabe *et al.*, *J. Jpn. Petrol. Inst.*, **39**, 219 (1996))

however, it has been clarified that modification of $Ru_3(CO)_{12}$ with alkali metal hydroxides improved the activity and selectivity of the catalyst obtained. Ruthenium species on the catalyst, in the presence of alkali metal, existed in the oxidation state close to RuS_2 rather than ruthenium metal. The mechanism of reaction is illustrated in Fig. 4.38. It is suggested that RuS_2, which could be stabilized on alumina with alkali metal even in pressurized hydrogen, may reveal activity comparable to that of $Co-Mo/Al_2O_3$.

4.4.3 Effects of Supports on Activities of HDS Catalysts Prepared from Supported Ruthenium Carbonyl-Cesium Hydroxide Systems

Some research groups have attempted HDS catalyzed by supported ruthenium sulfide using carbon,[408,409] alumina,[157,401,406,407,413,459] and zeolite.[157,416,460] In the preceding section, when the HDS of DBT catalyzed by alkali-promoted ruthenium catalysts was investigated in a pressurized flow reactor, it was noted that the addition of alkali metal increased remarkably the activities of the ruthenium catalysts and that cesium-promoted catalysts revealed activities comparable to the activity of $Co-Mo/Al_2O_3$.[456,457] Further, it seemed that, in the presence of alkali metal, ruthenium sulfide was formed on the support even under high hydrogen pressure. In this section, the effects of supports on the activity and product selectivity in the HDS of DBT were investigated for supported ruthenium carbonyl-cesium hydroxide systems.[461] It was found that, in the presence of cesium, the alumina-supported catalyst revealed the highest activity while the silica-alumina supported one revealed the highest activity in the absence of cesium. XPS measurements showed that the addition of an appropriate amount of cesium stabilized ruthenium sulfide on the support even under hydrogen pressure. Supports used were γ-alumina (260 m^2/g), $SiO_2-Al_2O_3$ (JRC-SAL2: 560 m^2/g; Si/Al = 5.3), SiO_2 (JRC-SIO4: 347 m^2/g), TiO_2 (JRC-TIO1: 70.8 m^2/g), and NaY zeolite (JRC-Z-Y4.8: 670 m^2/g).

A. Effects of Supports on the Hydrodesulfurization Activities of Catalysts Derived from Supported Ruthenium Carbonyls

In the HDS of DBT catalyzed by supported ruthenium carbonyl-cesium hydroxide systems after presulfiding, the effects of various supports such as Al_2O_3, $SiO_2-Al_2O_3$, SiO_2, TiO_2 and NaY zeolite on the catalytic activity and product selectivity were investigated under the following conditions: 300°C; 50 kg/cm^2; H_2 18 l/h; WHSV, 16.5 h^{-1}; concentration of DBT, 1.0 wt% in xylene; catalyst, 0.5 g. The results obtained for the catalysts in the absence of CsOH are shown in Table 4.24. The catalysts derived from supported ruthenium carbonyls were active for HDS, and the major products were BP and CHB. The catalytic activities of catalysts derived from supported ruthenium carbonyls decreased in the order $SiO_2-Al_2O_3$ > Al_2O_3 > TiO_2 > SiO_2 > NaY zeolite. When $SiO_2-Al_2O_3$ was used, the sum of the yields of BP and CHB due to HDS was 58% while the total conversion of DBT was 62%. The 4% difference was due to the hydrocracking (HC) of DBT to benzene. No HC was observed for the other supports. The selectivity for BP decreased in the order Al_2O_3 > NaY zeolite > TiO_2 > SiO_2 > $SiO_2-Al_2O_3$. $SiO_2-Al_2O_3$ gave the highest yield of CHB and the highest total conversion of DBT. The ruthenium catalyst supported on $SiO_2-Al_2O_3$ may have a higher ability to hydrogenate the aromatic ring of DBT. The tetrahydrodibenzothiophene (4-HDBT) and hexahydrodibenzothiophene (6-HDBT) formed would be easier to desulfurize than DBT because, as has been reported, the rate of HDS of 4-HDBT or 6-HDBT is four times higher

Table 4.24 Hydrodesulfurization of Dibenzothiophene Catalyzed by Supported Ruthenium Carbonyl Systems [a]

Run Catalyst	Conv. of DBT (%)	Conv. of DBT to BP (%)	Conv. of DBT to CHB (%)	Selectivity for BP (%)
1 $Ru_3(CO)_{12}/Al_2O_3$	44	38	6	86
2 $Ru_3(CO)_{12}/SiO_2\text{-}Al_2O_3$	62	36	22	57
3 $Ru_3(CO)_{12}/SiO_2$	26	17	9	64
4 $Ru_3(CO)_{12}/NaY$ zeolite	17	12	5	70
5 $Ru_3(CO)_{12}/TiO_2$	33	22	11	65

[a] Reaction temp. 300°C; Pressure 50 kg/cm^2; WHSV 16.5 h^{-1}; Cat. 0.50 g; H$_2$ 18 l/h; initial concentration of DBT 1.0 wt%; presulfided by H$_2$S in H$_2$ at 300°C (H$_2$S 3%) for 3h.
(From A. Ishihara, T. Kabe et al., J. Jpn. Petrol. Inst., **39**, 405 (1996))

Table 4.25 Hydrodesulfurization of Dibenzothiophene Catalyzed by $Ru_3(CO)_{12}\text{-}nCsOH/Al_2O_3$, $SiO_2\text{-}Al_2O_3$ and SiO_2 Systems and Characterization of the Catalysts by XPS

Run Activation Condition	HDS of DBT [a]		XPS Measurement [b]			
	Conv. of DBT (%)	Selectivity for BP (%)	S2p (eV)	Ru3p$_{3/2}$ (eV)	Ru/Al [c] or Ru/Si	S/Ru [c]
6 $Ru_3(CO)_{12}/Al_2O_3$	44(7)	86(100)	162.5	462.1	0.025	0.45
7 $Ru_3(CO)_{12}\text{-}3CsOH/Al_2O_3$	95(32)	98(100)	161.4	461.9	0.027	2.93
8 $Ru_3(CO)_{12}\text{-}6CsOH/Al_2O_3$	100(41)	99(100)	160.5	462.2	0.016	7.19
9 $Ru_3(CO)_{12}\text{-}9CsOH/Al_2O_3$	73(25)	100(100)	161.3	461.3	0.027	1.18
10 $Ru_3(CO)_{12}/SiO_2\text{-}Al_2O_3$	62(7)	57(72)	160.2	462.6	0.045	0.21
11 $Ru_3(CO)_{12}\text{-}6CsOH/SiO_2\text{-}Al_2O_3$	86(17)	96(100)	160.9	462.4	0.082	0.66
12 $Ru_3(CO)_{12}\text{-}9CsOH/SiO_2\text{-}Al_2O_3$	100(28)	98(100)	160.9	462.4	0.052	1.34
13 $Ru_3(CO)_{12}\text{-}12CsOH/SiO_2\text{-}Al_2O_3$	90(24)	98(100)	160.7	462.7	0.038	2.75
14 $Ru_3(CO)_{12}/SiO_2$	26(7)	64(100)	161.5	463.2	0.026	0.26
15 $Ru_3(CO)_{12}\text{-}3CsOH/SiO_2$	97(23)	99(100)	161.3	462.6	0.025	0.96
16 $Ru_3(CO)_{12}\text{-}6CsOH/SiO_2$	98(36)	100(100)	160.3	462.9	0.033	1.27
17 $Ru_3(CO)_{12}\text{-}9CsOH/SiO_2$	100(30)	100(100)	160.4	462.5	0.049	0.79

[a] Reaction temp. 300°C; pressure 50 kg/cm^2; WHSV 16.5 h^{-1}; Cat. 0.50 g; H$_2$ 18 l/h; initial concentration of DBT 1.0 wt%; data at reaction temp. 240°C are given in parentheses; presulfided by H$_2$S in H$_2$ (H$_2$S 3%). [b] XPS spectra were measured for samples used in HDS of DBT; before measurement, the samples were sputterd by Ar$^+$ ion for 10 min; every binding energy was referenced to Carbon C1s 285.0 eV. [c] The ratio of peak areas between Al2p, Si2p, and Ru3p$_{3/2}$ are given. Ru/Al was calculated for Al$_2$O$_3$ and Ru/Si was calculated for SiO$_2$-Al$_2$O$_3$ and SiO$_2$.
(From A. Ishihara, T. Kabe et al., J. Jpn. Petrol. Inst., **39**, 405 (1996))

than that of DBT.[395] The authors have described in the previous Section 4.2.2 that when anionic molybdenum carbonyl complexes instead of $Ru_3(CO)_{12}$ are supported on SiO$_2$-Al$_2$O$_3$, a higher yield of CHB is also obtained.[392,393]

Subsequently, the effects of the addition of cesium hydroxide on HDS using the catalysts derived from Al$_2$O$_3$-, SiO$_2$-Al$_2$O$_3$-, and SiO$_2$-supported $Ru_3(CO)_{12}$ systems were investigated and the results are shown in Table 4.25. When $Ru_3(CO)_{12}$ reacted with a threefold amount of CsOH and then supported on alumina (run 7), the catalytic activity increased remarkably and the conversion of DBT at 300 °C was 95%. With the use of SiO$_2$-Al$_2$O$_3$ and SiO$_2$, the conversion of DBT increased to 86% at Cs/Ru = 2 and 97% at Cs/Ru = 1, respectively. During the preparation of the catalyst, $Ru_3(CO)_{12}$ reacted with a threefold amount of cesium hydroxide to give $Cs[HRu_3(CO)_{11}]$ quantitatively; this was confirmed by FTIR (2076 (w), 2022 (vs), 1993 (s), 1962(m), 1647(m, br) cm^{-1} in methanol).[456-458] The authors have described in the previous Section 4.4.2 that, when sodium hydroxide was added to alumina-supported $Ru_3(CO)_{12}$, $Ru(acac)_3$ (acac = acetylacetonate), and $RuCl_3$, the catalyst derived from $Ru_3(CO)_{12}$-NaOH/Al_2O_3 was the most active among the three.[456,457] Further, when $Ru_3(CO)_{12}$ and NaOH were separately supported on alumina, no significant increase in catalytic activity was observed. It was indicated that presulfiding of $Na[HRu_3(CO)_{11}]$ supported on alumina was essential for high catalytic activity. In the present study, the presulfiding of $Cs[HRu_3(CO)_{11}]$ on a support is essential for high catalytic activity probably because the cesium is located close to the ruthenium on the support.

As shown in Fig.4.39 where the conversions of DBT on the catalysts using various supports are plotted against Cs/Ru ratio, the maximum activity was obtained at Cs/Ru = 2 in the alumina-supported $Ru_3(CO)_{12}$-nCsOH system. Further addition of CsOH decreased the conversion of DBT (Table 4.25 and Fig. 4.39). The selectivity for BP was 100% at 240°C in all cesium-promoted systems and more than 98% even at 300°C. The results from SiO_2-Al_2O_3- and SiO_2-supported systems are compared with those from Al_2O_3-supported ones in Fig. 4.39. With the use of SiO_2-Al_2O_3 and SiO_2, the conversion of DBT at 240°C increased with increase in the amount of cesium added and reached maxima 28% at Cs/Ru = 3 and 36% at Cs/Ru = 2, respectively. The maximum value of the conversion decreased in the order Al_2O_3 > SiO_2 > SiO_2-Al_2O_3. As opposed to this, the conversion of DBT did not increase with the addition of CsOH when NaY zeolite was used. With the use of NaY zeolite, neither sodium nor cesium worked as a promoter in this catalyst system even if $Cs[HRu_3(CO)_{11}]$ was used as the precursor. Ruthenium sulfide may have scattered from NaY zeolite because the color of the used catalyst was light gray to white. When supports other than NaY zeolite were used, the catalysts used were always black in color, indicating the presence of ruthenium compounds. It was reported by Göbölös et al. that the activity in HDS of thiophene increased

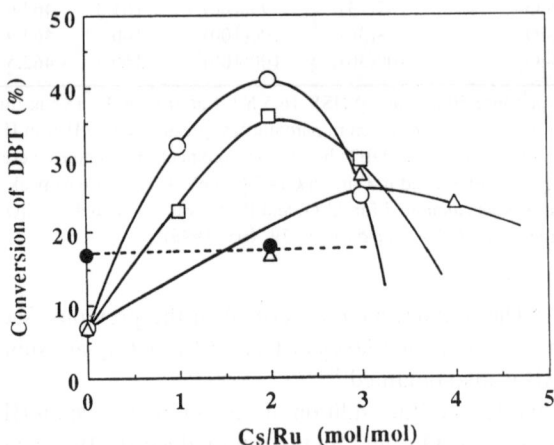

Fig. 4.39   Effect of supports on conversion of DBT.
240°C: ○: Al_2O_3, □: SiO_2, △: SiO_2-Al_2O_3; 300°C: ●: NaY zeolite.
$Ru_3(CO)_{12}$-nCsOH/supports; 50 kg/cm^2; H_2 18 1/h; WHSV 16.5 h^{-1}; concentration of DBT in xylene, 1.0 wt%; catalyst, 0.5 g; Ru 8 wt%.
(From A. Ishihara, T. Kabe et al., J. Jpn. Petrol. Inst., **39**, 406 (1996))

with addition of Na_2S to ruthenium catalysts supported on KY zeolite.[416] In their study, only 2 wt% of Ru was used and the Na/Ru ratio was 1 to 2. The amount of sodium added to the catalyst was much smaller than that used in our study. Their results, therefore, could not readily be compared with our results because such factors as the amount of ruthenium loaded, the reaction conditions, the substrate were not the same.

B. XPS Measurement of Catalysts after HDS Reaction

XP spectra were measured for the catalysts after HDS reaction and the results are shown in Table 4.25. The S2p and Ru3p$_{3/2}$ XP spectra from the supported ruthenium used for HDS showed peaks in the ranges 160.2–162.5 eV and 461.3–463.2 eV, respectively. These results suggest that ruthenium species for most catalysts are in the oxidation state between ruthenium metal and RuS_2. Although the binding energies of S2p and Ru3p$_{3/2}$ were affected by addition of cesium, it was difficult to observe a general tendency in the changes of these binding energies with changes in support and amount of cesium. On the other hand, clear peaks of cesium were not observed for the catalysts used in the present work, indicating that most of the cesium was highly dispersed in the support and that its surface concentration was very low. For example, Fig. 4.40 shows Si 2s1/2 and S 2p peaks in the XPS chart of Ru$_3$(CO)$_{12}$-mCsOH/SiO$_2$-Al$_2$O$_3$ (m = 0, 6, 9 or 12) where the peak of Si 2p1/2 can be used as a reference peak. Although the peaks of Cs 4p1/2 (172 eV) and Cs 4p$_{3/2}$ (162 eV) may be observed in the XPS chart, increase in the peaks around 172 eV could not be observed with increase in the amount of cesium added. In contrast, the peak around 161 eV increased with increase in the amount of cesium added. As a result, it was assumed that the peak around 161 eV would be the peak of S 2p rather than that of Cs 4p$_{3/2}$. For Al$_2$O$_3$ and SiO$_2$, similar results were observed.

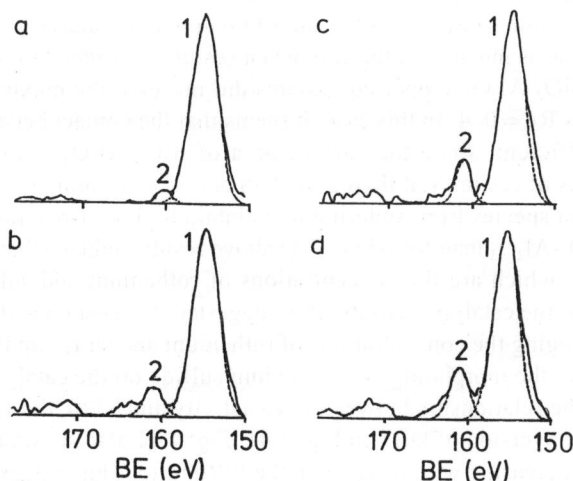

Fig. 4.40   Si2s$_{1/2}$ and S2P peaks in the XPS chart.
1: Si2s$_{1/2}$; 2: S2p, (a) m = 0; (b) m = 6; (c) m = 9; (d) m = 12 in Ru$_3$(CO)$_{12}$-mCsOH/SiO$_2$-Al$_2$O$_3$.
In each figure, the peak height of 1 was adjusted to a constant value. The peak areas of S2p given by the dotted line of 2 were determined by the curve fitting treatment.
(From A. Ishihara, T. Kabe et al., J. Jpn. Petrol. Inst., **39**, 407 (1996))

Further, when these supports were used, the peak areas around 161 eV at Cs/Ru = 3 were much smaller than those at Cs/Ru = 2, indicating that the contribution of Cs $4p_{3/2}$ to the peak around 161 eV would be very slight.

The Ru/Al or Ru/Si and S/Ru ratios were obtained from the peak areas of the XP spectra (Table 4.25). Although the values of Ru/Si for SiO_2-Al_2O_3- and SiO_2-supported catalysts were higher than the Ru/Al values for Al_2O_3-supported catalysts, some of the former catalysts showed lower catalytic activities than the latter. The results indicated that the dispersion of ruthenium species for some SiO_2-Al_2O_3- and SiO_2-supported catalysts was not so high as that of cesium-promoted alumina catalysts while the bulk of the ruthenium species for SiO_2-Al_2O_3- and SiO_2-supported catalysts existed in higher concentrations. One of the most important phenomena was observed with the S/Ru values, which increased with the addition of cesium. When cesium was added to the alumina-supported ruthenium catalysts, the S/Ru values increased remarkably, as shown in Table 4.25, indicating that the presence of cesium stabilized the ruthenium sulfide on the catalyst surface. The alumina-supported catalysts showed the maximum S/Ru value at Cs/Ru = 2. It is likely that in this catalyst system, not only are the ruthenium species sulfurized close to RuS_2, but also a significant amount of the sulfur formed by the HDS reaction may accumulate on the catalyst and may coat the ruthenium atoms, because HDS of DBT was probably much more rapid than the formation of hydrogen sulfide by the reaction of surface sulfur with hydrogen. The coating of the ruthenium atoms by large amounts of sulfur results in a higher value of S/Ru3p and a lower value of Ru3p/Al. This lower value of Ru3p/Al would also reflect the higher dispersion of ruthenium species. It is suggested that the presence of cesium may prevent ruthenium species from sintering by inhibiting the reduction of ruthenium sulfide species to ruthenium metal to maintain the higher dispersion and activity. Further addition of cesium decreased the value of S/Ru3p probably because HDS activity was not high and accumulation of sulfur did not occur. It appears that the S/Ru value may be related to the HDS activity. Similar results were observed for the addition of cesium to SiO_2-Al_2O_3- and SiO_2-supported catalysts. However, the S/Ru values did not increase as much as in the case when cesium was added to the Al_2O_3-supported catalysts. Further, SiO_2-Al_2O_3-supported systems did not give the maximum value of S/Ru value in the range Cs/Ru = 0–4. In this case, it seems that the contact between ruthenium and cesium was not sufficient, since the surface area of SiO_2-Al_2O_3 is twice that of Al_2O_3. Therefore, the effects of cesium stabilizing the R-S bonds, promoting C-S bond scission and preventing ruthenium species from sintering to maintain higher dispersion and HDS activity were smaller for SiO_2-Al_2O_3 than for Al_2O_3. The above results indicate that the values of S/Ru and Ru/Al or Ru/Si, which are the concentrations of ruthenium and sulfur on the catalyst surface, are related to the catalytic activity. It is suggested that cesium and supports affect the HDS activity by changing the concentrations of ruthenium and sulfur on the catalyst surface; this may be related to the morphology of ruthenium sulfide on the catalyst.

To investigate the relationship between HDS activity and S/Ru, the values of S/Ru were plotted against the conversion of DBT in Fig. 4.41. Fig. 4.41 also shows the tendency of the conversion of DBT increasing with increase in the S/Ru ratio. This suggests that the catalytic activity increases when ruthenium sulfide is stabilized in the presence of cesium. Further, it seems that the presence of this promoter strengthens the bond between ruthenium and sulfur to stabilize ruthenium sulfide in which the carbon-sulfur bond scission may proceed more easily than that in less sulfided ruthenium species close to the low valent ruthenium metal. Among the catalysts examined in this study cesium-promoted $Ru_3(CO)_{12}/Al_2O_3$ catalysts

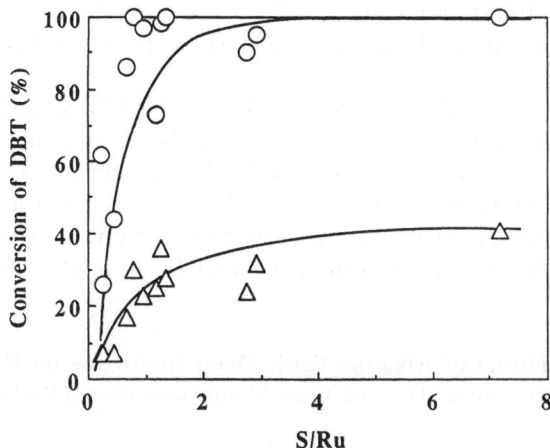

Fig. 4.41   Plots of S/Ru ratio in XPS vs. conversion of DBT.
Ru$_3$(CO)$_{12}$-mCsOH/supports: ◯: 300°C; △: 240°C, m = 0, 3, 6, 9, or 12;
supports = Al$_2$O$_3$, SiO$_2$ Al$_2$O$_3$, or SiO$_2$.
(From A. Ishihara, T. Kabe et al., J. Jpn. Petrol. Inst., 39, 407 (1996))

showed the maximum value of S/Ru at Cs/Ru = 2. This may be explained by the additive effects resulting from the combined use of ruthenium, cesium, and alumina.

C. The Role of Cesium

As described in the previous section, Göbölös et al. reported that the addition of Na$_2$S neutralized the Brønsted acidity of the zeolite and thus increased the activity. In the present study, the neutralization of Brønsted acidity of the supports also seems to be related to the increase in the HDS activity. For example, it can be assumed that SiO$_2$-Al$_2$O$_3$-supported catalysts need larger amounts of added cesium than Al$_2$O$_3$- or SiO$_2$-supported ones to neutralize Brønsted acids to obtain maximum activity. However, all the results obtained here cannot be explained merely by neutralization of Brønsted acidity of the supports. For example, the effect of the addition of cesium on SiO$_2$-supported catalysts cannot be explained only by the neutralization of Brønsted acidity of the supports because such neutralization is not important for SiO$_2$. There must be something more besides neutralization of Brønsted acidity of the supports to the role of cesium.

In this section, it was noted that modification of Ru$_3$(CO)$_{12}$ with cesium hydroxide improved the properties of the catalyst obtained. Ruthenium species on the catalyst which showed higher catalytic activity in the presence of cesium existed in the oxidation state close to RuS$_2$ rather than to ruthenium metal. It was suggested that the addition of an appropriate amount of cesium stabilized RuS$_2$ on the support even under hydrogen pressure, promoted the C-S bond scission, and prevented the ruthenium species from sintering to maintain higher dispersion and HDS activity.

4.4.4 Hydrodesulfurization Catalysts Prepared from Alumina-supported Ruthenium Carbonyl-alkaline Earth Metal Hydroxide Systems

As described in the preceding two sections (4.4.2 and 4.4.3), the addition of alkali metal to supported ruthenium catalysts remarkably increased the activities in the HDS of DBT in a pressurized flow reactor. Alkali metal may stabilize ruthenium sulfide on supports. In this section, the addition of alkaline earth metal hydroxides such as $Ca(OH)_2$, $Sr(OH)_2$ and $Ba(OH)_2$ to the alumina-supported ruthenium carbonyl is described. Alkaline earth metals as well as alkali metals effectively promote the HDS of DBT. The result with barium is compared with that with cesium.[462]

A. Effects of Addition of Alkaline Earth Metal Hydroxide on Hydrodesulfurization Reactivities of Catalysts Derived from Alumina-supported Ruthenium Carbonyls

In HDS of DBT catalyzed by alumina-supported ruthenium carbonyl-alkaline earth metal hydroxide systems, the effects of the addition of alkaline earth metal hydroxides on the HDS reactivities were investigated and the results are listed in Table 4.26 and Fig. 4.42. Products were BP and CHB. As shown in the previous sections, in the absence of alkaline earth metal hydroxide, the activity was rather low and the conversion of DBT was 44%. When $Ru_3(CO)_{12}$ was made to react with a 1.5-fold amount of alkaline-earth metal hydroxide ($Ca(OH)_2$, $Sr(OH)_2$ or $Ba(OH)_2$) and was then supported on alumina (runs 2, 6, 8), the catalytic activity increased remarkably and the conversions of DBT were 62%, 67% and 75% at 300 °C, respectively. During the preparation of these catalysts, $Ru_3(CO)_{12}$ reacted with more than 1.5-fold of alkaline earth metal hydroxide to give $M[HRu_3(CO)_{11}]_2$ (M = Ca, Sr or Ba)

Table 4.26 Hydrodesulfurization of Dibenzothiophene Catalyzed by $Ru_3(CO)_{12}$-$M(OH)_2$/Al_2O_3 systems and Characterization of the Catalysts by XPS [a]

		HDS of DBT [b]		XPS Measurement [c]			
Run	Catalyst Precursor	Conv. of DBT(%)	Selectivity for BP(%)	S2p (eV)	Ru3p$_{3/2}$ (eV)	Ru/Al [d]	S/Ru [d]
1	$Ru_3(CO)_{12}$/Al_2O_3	44(7)	86(100)	162.5	462.1	0.025	0.45
2	$Ru_3(CO)_{12}$-1.5Ca(OH)$_2$/Al_2O_3	62(13)	93(100)	162.3	462.0	0.046	1.02
3	$Ru_3(CO)_{12}$-3.0Ca(OH)$_2$/Al_2O_3	72(18)	90(94)	162.8	462.5	0.040	2.77
4	$Ru_3(CO)_{12}$-4.5Ca(OH)$_2$/Al_2O_3	69(11)	90(100)	162.4	462.1	0.040	1.24
5	$Ru_3(CO)_{12}$-6.0Ca(OH)$_2$/Al_2O_3	77(12)	91(100)	161.6	461.4	0.066	1.02
6	$Ru_3(CO)_{12}$-1.5Sr(OH)$_2$/Al_2O_3	67(13)	92(100)	162.5	462.0	0.038	1.57
7	$Ru_3(CO)_{12}$-3.0Sr(OH)$_2$/Al_2O_3	73(24)	95(100)	162.1	461.9	0.052	1.73
8	$Ru_3(CO)_{12}$-1.5Ba(OH)$_2$/Al_2O_3	75(10)	93(97)	162.2	461.7	0.051	0.92
9	$Ru_3(CO)_{12}$-3.0Ba(OH)$_2$/Al_2O_3	91(12)	93(96)	162.0	461.6	0.049	1.83
10	$Ru_3(CO)_{12}$-4.5Ba(OH)$_2$/Al_2O_3	65(12)	95(100)	162.2	461.8	0.056	3.02
11	$Ru_3(CO)_{12}$-6.0Ba(OH)$_2$/Al_2O_3	51(9)	96(100)	162.2	462.2	0.044	1.95

[a] M = Ca, Sr or Ba. [b] Reaction temp. 300°C, Pressure 50 kg/cm^2, WHSV 16.5 h^{-1}, Cat. 0.50 g, H$_2$ 18 l/h; Initial concentration of DBT 1.0 wt%. Presulfided by H$_2$S in H$_2$ (H$_2$S 3%). The result for the conversion of DBT and selectivity for BP at 240°C are given in parentheses. [c] XPS spectra were measured for samples used in HDS of DBT. Before measurement, the samples were sputterd by Ar$^+$ ion for 10 min. Every binding energy was referenced to Carbon C1s 285.0 eV. [d] Ratio of peak areas between Al2p, S2p, and Ru3p$_{3/2}$.
(From A. Ishihara, T. Kabe et al., J. Jpn. Petrol. Inst., **40**, 518 (1997))

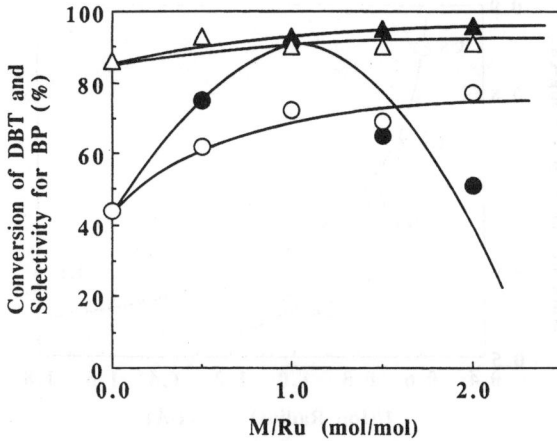

Fig. 4.42 Effect of M/Ru ratio on conversion of DBT and selectivity for BP.
Conversion of DBT: $\bigcirc$: Ca; $\bullet$: Ba; Selectivity for BP: $\triangle$: Ca; $\blacktriangle$: Ba.
$Ru_3(CO)_{12}$-nM(OH)$_2$/Al$_2$O$_3$ (M = Ca or Ba); 300°C, 50 kg/cm^2; H$_2$ 18 1/h; WHSV, 16.5 h^{-1}; concentration of DBT in xylene, 1.0 wt%; catalyst, 0.5 g; Ru 8 wt%.
(From A. Ishihara, T. Kabe et al., J. Jpn. Petrol. Inst., **40**, 518 (1997))

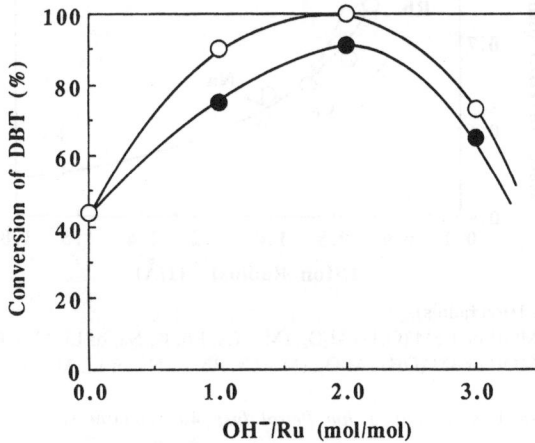

FFig. 4.43 Effect oh OH$^-$/Ru ratio on conversion of DBT.
$\bigcirc$: Cs; $\bullet$: Ba.
$Ru_3(CO)_{12}$-nCsOH or nBa(OH)$_2$/Al$_2$O$_3$; 300°C, 50 kg/cm^2; H$_2$ 18 1/h; WHSV, 16.5 h^{-1}; concentration of DBT in xylene, 1.0 wt%; catalyst, 0.5 g; Ru 8 wt%.
(From A. Ishihara, T. Kabe et al., J. Jpn. Petrol. Inst., **40**, 519 (1997))

quantitatively, as confirmed by FTIR (2074 (w), 2018 (vs), 1993 (vs), 1962(m), 1667(m, br) cm^{-1} in methanol).[456–458,461]

The authors described in the previous sections that the formation and presulfiding of Cs[HRu$_3$(CO)$_{11}$] on alumina was essential to obtain high catalytic activity in HDS of DBT catalyzed by alumina-supported ruthenium carbonyl-cesium hydroxide systems.[456,457,461] In these catalysts, the catalytic activity was assumed to be high probably because cesium was located close to ruthenium on alumina. The formation and presulfiding of M[HRu$_3$(CO)$_{11}$]$_2$ as

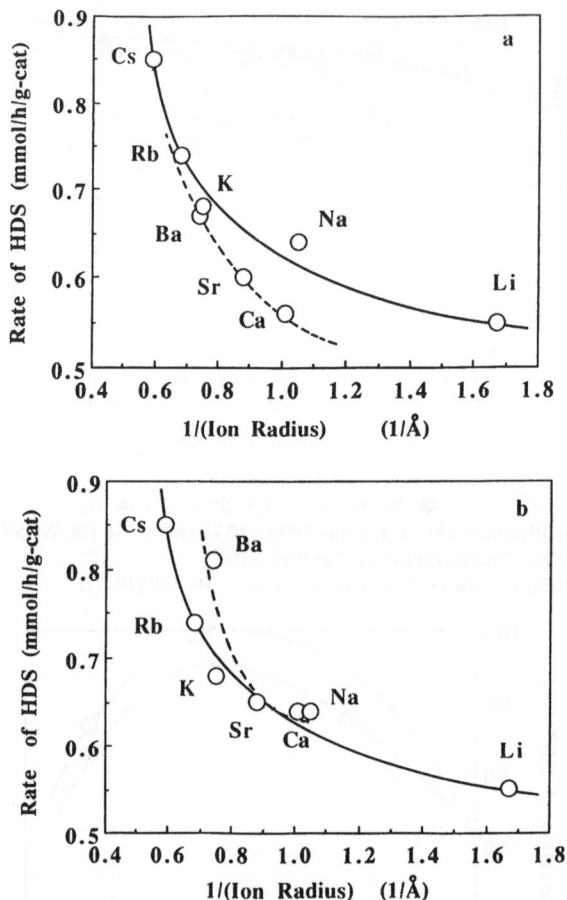

Fig. 4.44   Rate of HDS vs. 1(ion radius).
(a) $Ru_3(CO)_{12}$-3MOH or 1.5M'$(OH)_2$/Al_2O_3; (M = Cs, Rb, K, Na, or Li; M' = Ba, Sr or Ca), OH/Ru = 1.
(b) $Ru_3(CO)_{12}$-3MOH or 3M'$(OH)_2$/Al_2O_3; (M = Cs, Rb, K, Na, or Li; M' = Ba, Sr or Ca), M/Ru = M'/Ru = 1.
(From A. Ishihara, T. Kabe *et al.*, *J. Jpn. Petrol. Inst.*, **40**, 520 (1997))

well as $Cs[HRu_3(CO)_{11}]$ supported on alumina may also be essential to obtain high catalytic activity probably because alkaline earth metal is located close to ruthenium on alumina. As shown in Fig.4.42 where the conversions of DBT are plotted against the ratios of M/Ru (M = Ca or Ba), the maximum activity was obtained at Ba/Ru = 1 in alumina-supported $Ru_3(CO)_{12}$-nBa$(OH)_2$ systems. Further addition of Ba$(OH)_2$ decreased the conversion of DBT. In contrast, the conversion of DBT gradually increased with the use of Ca$(OH)_2$ and no clear maximum was observed in the range Ca/Ru = 0–2.0. The difference in the catalytic activity between Ba and Ca may occur from the difference in the ion radius. This is discussed below. The selectivity for BP, i.e., the ratio of the conversion of DBT to BP to the total conversion of DBT, was 100% at 240°C in most alkaline earth metal-promoted systems while they increased slightly with increase in the M/Ru ratio at 300°C compared with the unpromoted catalyst (86% in $Ru_3(CO)_{12}$/Al_2O_3 (run 1)).

To compare the results from alkaline earth metal with those from cesium, the conversion of DBT was plotted in Fig. 4.43 against OH^-/Ru, which means the molar ratio of hydroxide ion added to ruthenium on a catalyst. The maximum of the activity was observed at $OH^-/Ru = 2$ in both cases, indicating that the amount of added hydroxide ion is related to the catalytic activity. At $OH^-/Ru = 2$, the molar amount of added cesium is two times larger than that of added barium. The difference in the conversion at maximum may reflect this ratio.

To add further insight to this problem, the HDS rate was plotted against the reciprocal value of the ion radius of alkali metal or alkaline earth metal in Fig. 4.44. The data with alkali metals are the results given in Section 4.4.2. The catalytic activity for alkaline earth metal-promoted catalysts as well as alkali metal-promoted ones increased with increase in ion radius of alkaline earth metal. However, the plots for the former seem to form a somewhat different curve from those for the latter. When the amount of hydroxide ion added is constant in Fig. 4.44(a), the dotted line of alkaline-earth metal deviated from the solid line of alkali metal and only barium could be put near the solid line. On the contrary, when the amount of added alkali or alkaline earth metal is constant in Fig. 4.44(b), all plots except barium are approximately located on one solid line. These results indicate that not only the amount of hydroxide ion added but also the amount of alkali metal or alkaline earth metal added may be related to the catalytic activity. The presence of two factors may generate the two kinds of curves for alkali metals and alkaline earth metals, respectively, in Fig. 4.44(a) or 4.44(b). However, it is suggested that the catalytic activity is related to the ion radius since these two curves exist very near each other.

B. FTIR Measurement of Ruthenium Carbonyl Species on the Catalyst Precursors

FTIR spectra of ruthenium carbonyl species on alkaline earth metal-promoted alumina-supported ruthenium carbonyls were measured and the results are listed in Table 4.27. As described previously, $Ru_3(CO)_{12}$ supported on alumina showed three bands at 1998 (s), 2020 (vs) and 2058 (vs) cm^{-1} (Fig. 4.45(a)). Because $Ru_3(CO)_{12}$ showed almost the same bands in the absence of alumina, $Ru_3(CO)_{12}$ was located only on the alumina surface in the $Ru_3(CO)_{12}/Al_2O_3$ system. $Ru_3(CO)_{12}$ reacted with 1.5-fold $M(OH)_2$ (M = Ca, Sr or Ba) in methanol to form $M[HRu_3(CO)_{11}]_2$. After $M[HRu_3(CO)_{11}]_2$ was supported on alumina, peaks similar to those for the $Ru_3(CO)_{12}/Al_2O_3$ system were observed (runs 13, 17 and 19), indicating that $Ru_3(CO)_{12}$ was again regenerated on alumina. These catalyst systems increased the conversions of DBT in comparison with the non-promoted system even if $Ru_3(CO)_{12}$ was regenerated on alumina during the preparation of the catalysts. This result suggests that the location of alkaline earth metal close to ruthenium species is essential for increasing the activity. Similar results were observed in the case of cesium promoted systems.[457] When $Ru_3(CO)_{12}$ reacted with more than threefold $M(OH)_2$ in methanol and was supported on alumina (Fig. 4.45(b)), $Ru_3(CO)_{12}$ was not regenerated and anionic species were maintained on alumina, since peaks with lower wavenumbers than those of $Ru_3(CO)_{12}$ were observed. This shows that the addition of the amount of base can neutralize the acid sites on alumina and maintain the anionic ruthenium species on alumina. It is suggested that the maintenance of the anionic species on alumina as well as the location of alkaline earth metals close to ruthenium species is essential to increase the activity. When $Ru_3(CO)_{12}$ was reacted with excess $M(OH)_2$ in methanol and supported on alumina, two major peaks like those in Fig. 4.45(b) were observed finally in every system while some shoulder peaks were included. Although anionic

Table 4.27 Characterization of the $Ru_3(CO)_{12}$-$Ca(OH)_2$, $Sr(OH)_2$, $Ba(OH)_2$ or $CsOH/Al_2O_3$ Systems by FTIR [a]

Run	Catalyst Precursor	FTIR Wavenumber (cm^{-1})			
12	$Ru_3(CO)_{12}/Al_2O_3$	2058(vs)	2020(vs)	1998(vs)	
13	$Ru_3(CO)_{12}$-1.5$Ca(OH)_2/Al_2O_3$	2058(vs)	2022(s,sh)	1998(s)	1989(s,sh)
14	$Ru_3(CO)_{12}$-3.0$Ca(OH)_2/Al_2O_3$	2051(vs)	1981(s)		
15	$Ru_3(CO)_{12}$-4.5$Ca(OH)_2/Al_2O_3$	2051(vs)	2022(s,sh)	1993(s)	1983(s,sh)
16	$Ru_3(CO)_{12}$-6.0$Ca(OH)_2/Al_2O_3$	2051(vs)	2022(s,sh)	1981(s)	
17	$Ru_3(CO)_{12}$-1.5$Sr(OH)_2/Al_2O_3$	2056(vs)	2020(vs)	2000(s)	
18	$Ru_3(CO)_{12}$-3.0$Sr(OH)_2/Al_2O_3$	2054(vs)	1979(s)		
19	$Ru_3(CO)_{12}$-1.5$Ba(OH)_2/Al_2O_3$	2062(vs)	2031(s,sh)	1983(s,br)	
20	$Ru_3(CO)_{12}$-3.0$Ba(OH)_2/Al_2O_3$	2049(vs)	1973(s)		
21	$Ru_3(CO)_{12}$-4.5$Ba(OH)_2/Al_2O_3$	2049(vs)	1971(s)		
22	$Ru_3(CO)_{12}$-6.0$Ba(OH)_2/Al_2O_3$	2045(vs)	1970(s)		

[a] KBr. vs = very strong, s = strong, br = broad, sh = shoulder.
(From A. Ishihara, T. Kabe et al., J. Jpn. Petrol. Inst., **40**, 519 (1997))

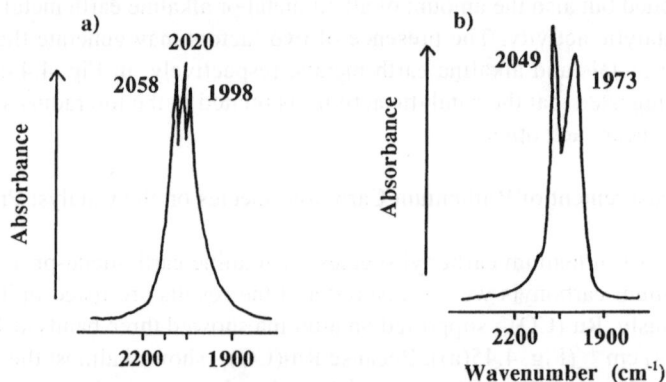

Fig. 4.45 FTIR spectra of alumina-supported ruthenium catalysts.
(a) $Ru_3(CO)_{12}/Al_2O_3$, (b) $Ru_3(CO)_{12}$-3$Ba(OH)_2/Al_2O_3$.
(From A. Ishihara, T. Kabe et al., J. Jpn. Petrol. Inst., **40**, 520 (1997))

species existed on alumina, it seems that the trinuclear anionic $[HRu_3(CO)_{11}]^-$ cannot be maintained on alumina and that dicarbonyl species are formed. When the amount of added $Ba(OH)_2$ increased in the range above Ba/Ru = 1 (Table 4.27), wavenumbers of two peaks shifted to lower values, indicating that the anionic ruthenium species with the higher content of $Ba(OH)_2$ have the higher electron density. In contrast to the cases of $Ba(OH)_2$, the increase in the amount of added $Ca(OH)_2$ did not make wavenumbers shift to lower values (Table 4.27). These results may be related to those from the HDS activity. The catalytic activity decreased dramatically with excess $Ba(OH)_2$ when the wavenumbers of the carbonyl peaks shifted to lower values, that is, when the electron density of the ruthenium species became higher. However, the activity did not change largely with excess $Ca(OH)_2$, and the wavenumbers of the carbonyl peaks did not change either. The electron density may be related to the bond strength of the ruthenium-sulfur (Ru-S) bond. Addition of an appropriate amount of $Ba(OH)_2$ strengthens the Ru-S bond to promote the C-S bond scission of DBT while the addition of an excess amount of Ba(OH)$_2$ makes the Ru-S bond too strong to form hydrogen sulfide and active sites (anion vacancies). Besides these electronic effects, the addition of excess barium may cover the active ruthenium species to decrease the activity. It is not clear whether the trinuclear structure is needed to obtain high catalytic activity.

C. XPS Measurement of the Catalysts after HDS Reaction

XP spectra were measured for the catalysts after HDS reaction and the results are shown in Table 4.26. The S2p and Ru3p$_{3/2}$ XP spectra from the supported ruthenium used for HDS showed peaks in the ranges 161.6–162.8 eV and 461.4–462.5 eV, respectively. These results suggest that the ruthenium species for most catalysts are between ruthenium metal (461.0 eV) and RuS$_2$ (462.7 eV). Although the binding energies of S2p and Ru3p$_{3/2}$ were affected by the addition of alkaline earth metals, no clear relationship between the changes of these binding energies with changes in the amount of alkaline earth metal could be found.

The Ru/Al and S/Ru ratios were obtained from the peak areas of the XP spectra. Although the Ru/Al values obtained for alkaline earth metal-promoted catalysts were higher than the Ru/Al values for cesium-promoted ones (less than 0.027),[457] some of the former showed lower catalytic activities than the latter. This indicates that the dispersion of ruthenium species for alkaline earth metal-promoted catalysts was not so high as that of cesium-promoted alumina catalysts while the bulk of the ruthenium species for the former catalysts existed in higher concentration. An important phenomenon was observed with the S/Ru ratio values, which increased with the addition of alkaline earth metals as well as cesium.[457] When alkaline earth metal was added to the alumina-supported ruthenium catalysts, the S/Ru values increased remarkably, indicating that the presence of an alkaline earth metal stabilizes ruthenium sulfide on the catalyst surface. For example, the barium-promoted catalysts showed the maximum S/Ru value at Ba/Ru = 1.5 although the maximum activity was observed at Ba/Ru = 1.0. It seems that, since the release of sulfur as hydrogen sulfide was very rapid at Ba/Ru = 1.0, the S/Ru value was lower than that at Ba/Ru = 1.5. At Ba/Ru = 1.5, the Ru-S bond was too strong to release sulfur as hydrogen sulfide rapidly and maintain the maximum activity. Further addition of Ba decreased the S/Ru value probably because the accumulation rate of sulfur on the catalysts decreased with the low activity and excess of barium-covered ruthenium species.

The S/Ru value appears to be related to the HDS activity. To investigate the relationship between the HDS activity and S/Ru, conversion of DBT was plotted against S/Ru values in Fig. 4.46. Since alkaline earth metal may cover the ruthenium sulfide at higher M/Ru ratios, data for 0, 0.5 and 1.0 of M/Ru were used. Fig. 4.46 shows that the conversion of DBT increases with increase in the S/Ru ratio, although there are scatters especially in data at 300°C. The results suggest that the catalytic activity increases when ruthenium sulfide is stabilized in the presence of alkaline earth metals. Further, it seems that the presence of these promoters strengthens the bond of ruthenium and sulfur to stabilize ruthenium sulfide. The carbon-sulfur bond scission on the stabilized ruthenium sulfide may proceed more easily than that on less sulfided ruthenium species close to low valent ruthenium metal.

Although HDS of thiophenes using ruthenium sulfide catalysts has been reported, the effects of addition of alkaline earth metal as well as alkali metal salts on the HDS reactions are not well known in those catalysts. In the present study, however, it has been clarified that the modification of Ru$_3$(CO)$_{12}$ with alkaline earth metal hydroxides improved the properties

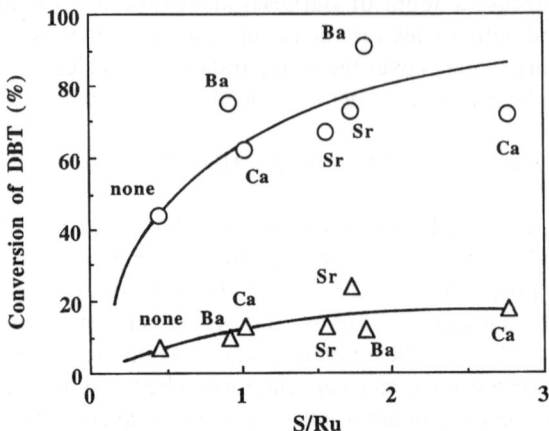

FIg. 4.46 Plots of S/Ru ratio in XPS vs. conversion of DBT.
$Ru_3(CO)_{12}$-m$M(OH)_2/Al_2O_3$; $\bigcirc$: 300°C; 240°C, m = 0, 1.5, or 3.0, M = Ca, Sr or Ba.
(From A. Ishihara, T. Kabe et al., J. Jpn. Petrol. Inst., **40**, 521 (1997))

of the catalyst obtained. Ruthenium species on the catalysts in the presence of alkaline earth metals existed in an oxidation state close to RuS_2 than to ruthenium metal. It was suggested that the addition of an appropriate amount of alkaline earth metals stabilizes ruthenium sulfide on a support even under hydrogen pressure to increase the HDS activity.

4.4.5 Hydrodesulfurization of ^{35}S-labeled Dibenzothiophene on Alumina-supported Ruthenium Sulfide-Cesium Catalysts

In the preceding sections, it was suggested that the addition of an appropriate amount of alkali metals or alkaline earth metals stabilizes ruthenium sulfide on a support even under hydrogen pressure to increase the HDS activity.[456,457,461,462] For the elucidation of the HDS mechanism it is essential to know the behavior of sulfur on HDS catalysts. In order to explore the behavior of sulfur species on the catalyst, radioisotope tracer methods using radioactive ^{35}S have been developed by several researchers and described in Chapter 3 in detail.[463-469] In these reports, the use of ^{35}S tracer such as ^{35}S-labeled H_2S, CS_2, or thiophene was believed to enable to clarify the behavior of sulfur in catalysts. However, it is not well known how sulfur in catalysts acts in the practical performance of HDS. Kabe and coworkers performed the HDS of ^{35}S-labeled dibenzothiophene ([^{35}S]DBT) on the sulfided Co-Mo/Al_2O_3, Ni-Mo/Al_2O_3, Mo/Al_2O_3, Co/Al_2O_3 and Ni/Al_2O_3, and obtained new insight into the mechanism of HDS by monitoring the radioactivities of the unreacted [^{35}S]DBT and the formed [^{35}S]H_2S.[470-477] It was found that the sulfur of DBT was not directly released as H_2S but initially accommodated on the catalyst and that ^{35}S accommodated on the catalyst could not be removed without the incorporation of sulfur from HDS of sulfur compounds. Based on quantitative analysis on the rate of [^{35}S]H_2S formation from the catalyst, it was postulated that the sulfur on the sulfided catalyst was labile and the amount of labile sulfur on the catalysts varied with the reaction conditions. This method made it possible to understand more exactly how sulfur in DBT is translated to H_2S and how sulfur in the sulfided catalyst participates in the actual HDS reaction.

In this section, HDS of [35S]DBT catalyzed by Al_2O_3-supported ruthenium sulfide-cesium catalysts is described to elucidate the behavior of sulfur on the ruthenium catalysts and the role of cesium in HDS. Initially the amount of labile sulfur and the release rate constant of H_2S on presulfided ruthenium catalysts were estimated. Then the total amount of sulfur incorporated to the catalyst was estimated by direct HDS reaction of [35S]DBT without presulfiding of ruthenium catalysts.[478,479]

[35S]DBT was synthesized by the following method. In order to obtain 35S-labeled sulfur, the commercial toluene solution of 35S (total radioactivity: 1 mCi) was mixed sufficiently with 8.7 g sulfur (32S), then the toluene in the mixture was evaporated at room temperature, and the sulfur mixture was dried for 24 hours *in vacuo* until the toluene was entirely removed. Using this 35S-labeled sulfur, DBT was synthesized according to the method developed by Gilman *et al.*[480] After the crude DBT was crystallized from ethanol, colorless needles (purity more than 99.9%) were obtained. The principal preparation methods of catalysts, apparatus, reaction procedures and analytical methods have been described in the preceding sections.

Two typical operation procedures were applied. Operation procedure 1: (a) A decalin solution of 1 wt% [32S]DBT was pumped into the reactor until the conversion of DBT became constant (about 3.5 h). (b) After that, the decalin solution of 1 wt% [35S]DBT (6000 dpm/g) was substituted for that of [32S]DBT. The reaction with [35S]DBT was performed until the amount of [35S]H_2S released into exit of the reactor became constant (about 7 h). (c) Then the reactant solution was returned again to the decalin solution of 1 wt% [32S]DBT. This reaction of [32S]DBT was continued for about 8 h. Operation Procedure 2: After the catalyst precursor was placed in a pressurized fixed-bed flow reactor, the reactor was pressurized with hydrogen and heated at 5°C/min to the expected temperature, and then the solution containing [35S]DBT was supplied with the feed pump without the presulfiding. The reaction with [35S]DBT was performed at the expected temperature until the amount of [35S]H_2S released into exit of the reactor became constant. Then the reactant solution of [35S]DBT was replaced by [32S]DBT solution. The reaction was continued for about 8 h. A liquid scintillation counter[481-483] was employed for measuring radioactivities of unreacted [35S]DBT in liquid product and the [35S]H_2S formed in scintillator solution.

A. HDS of [35S]DBT on Ruthenium Catalysts with Presulfiding

Initially HDS of [35S]DBT was performed on ruthenium catalysts with presulfiding of $Ru_3(CO)_{12}$-nCsOH/Al_2O_3 systems (n = 0, 3, 6 or 9) according to operation procedure (OP) 1. Fig. 4.47 shows the changes in the radioactivities of unreacted [35S]DBT and [35S]H_2S produced, and the conversion of DBT with the reaction time at 280°C using $Ru_3(CO)_{12}$-6CsOH/Al_2O_3 system. After the conversion in the HDS of [32S]DBT reached a constant value and the solution of [32S]DBT was replaced by that of [35S]DBT, the radioactivities of unreacted [35S]DBT in liquid products increased with the reaction time and reached a steady state immediately. In contrast, the time delay in the case of [35S]H_2S was about 6 hours. Then, the decalin solution of [32S]DBT was again substituted for [35S]DBT in a similar way. The time delay for decrease of radioactivities of unreacted [35S]DBT from the steady state to normal state was also 30 minutes while that for the [35S]H_2S was about 6 hours. This result indicates that the sulfur in DBT is not directly released as H_2S, but accommodated on the catalyst, and this is consistent with the results obtained using Al_2O_3-supported molybdenum catalysts.[470-474]

Data for the catalyst derived from the $Ru_3(CO)_{12}$-6CsOH/Al_2O_3 system were treated as

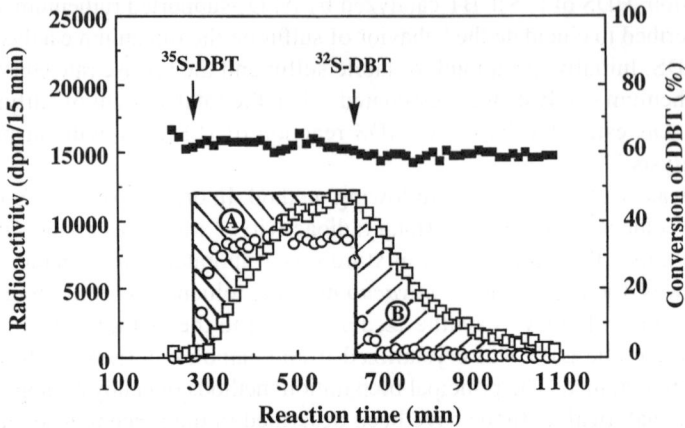

Fig. 4.47 Variation in radioactivities with reaction time.
○: [³⁵S]DBT; □: [³⁵S]H₂S; ■: Conversion.
Ru₃(CO)₁₂-6CsOH/Al₂O₃; 280°C, 50 kg/cm², WHSV = 14 h⁻¹, Gas Flow 18 1/h.
(From A. Ishihara, W. Qian, T. Kabe et al., J. Jpn. Petrol. Inst., 41, 53 (1998))

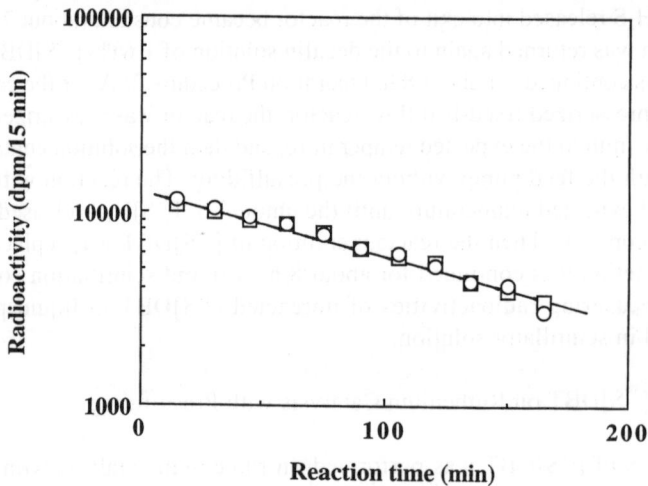

Fig. 4.48 First-order plots of release rate of [³⁵S]H₂S.
□: Decreasing period; ○: Increasing period.
Ru₃(CO)₁₂-6CsOH/Al₂O₃; 280°C, 50 kg/cm², WHSV = 14 h⁻¹, Gas Flow 18 1/h.
(From A. Ishihara,W. Qian, T. Kabe et al., J. Jpn. Petrol. Inst., 41, 53 (1998))

follows. The first-order plot (Fig. 4.48) of the radioactivity of the decreasing period of [³⁵S]H₂S in Fig. 4.47 indicates a linear relationship revealed as

$$\ln y = \ln z - kt \tag{4.1}$$

where y represents the radioactivity of [³⁵S]H₂S (dpm/min), z the radioactivity of [³⁵S]H₂S at steady state (dpm/min), k the rate constant of the release of [³⁵S]H₂S (min⁻¹), t reaction time (min) (see Chapter 3). The first-order plot of the values, where each radioactivity of the increasing period of $[^{35}S]H_2S$ in Fig. 4.47 is subtracted from that at steady state, also shows the linear relationship (o) in Fig. 4.48 and two slopes at this temperature overlap each other. This indicates that the rate of release of $[^{35}S]H_2S$ is equal to that of $[^{32}S]H_2S$. The value of k represents the relative ease of migration of sulfur on the catalyst.[470–474] The increasing period of $[^{35}S]H_2S$ is the process where ^{32}S on the catalyst is replaced by ^{35}S. In contrast, the decreasing period of $[^{35}S]H_2S$ is the process where ^{35}S on the catalyst is replaced by ^{32}S. Therefore, area A, which represents the amount of ^{35}S incorporated into the catalyst, is equivalent to area B, which represents the amount of ^{35}S released from the catalyst. Further, when $[^{32}S]DBT$ was substituted for $[^{35}S]DBT$ after the radioactivities of $[^{35}S]H_2S$ reached the steady state, all ^{35}S accommodated on the catalyst was released as $[^{35}S]H_2S$. From the total amount of radioactivity of this released $[^{35}S]H_2S$ (Area B, x dpm), the amount of labile sulfur on the catalyst (S_0), which may represent the amount of the active sites, could be calculated.[470–474] Since all ^{35}S on the catalyst originated from the desulfurization of $[^{35}S]DBT$, the concentration of ^{35}S in sulfur introduced to the catalyst by the HDS of DBT at the steady state should be equal to the concentration of ^{35}S in sulfur of $[^{35}S]DBT$ because the isotope effect between ^{35}S and ^{35}S can be assumed to be very small. The concentration of ^{35}S in sulfur of $[^{35}S]DBT$ could be defined as $^{35}S_{DBT} / S_{DBT}$ (dpm/g), where $^{35}S_{DBT}$ is radioactivities in 1 mol of DBT (dpm/mol) and S_{DBT} is the amount of sulfur in 1 mol DBT (g/mol). According to this, the amount of labile sulfur on the catalyst (S_0) can be presented by $(x)/(^{35}S_{DBT} / S_{DBT})$.

Figure 4.49 shows the changes in the radioactivities of produced $[^{35}S]H_2S$ with the reaction time at 320°C using $Ru_3(CO)_{12}$-nCsOH/Al_2O_3 systems (n = 0, 3, 6 or 9). The first order plots of these catalysts in the decreasing period of $[^{35}S]H_2S$ are also shown in Fig. 4.50. The values of k and S_0 for these catalysts were estimated in similar methods described above by using these data and are shown in Table 4.28. As described in the previous sections,[456,457,461,462] the conversion of DBT increased with increasing amount of cesium added and reached maximum at Ru:Cs = 1:2 although further addition of cesium decreased the

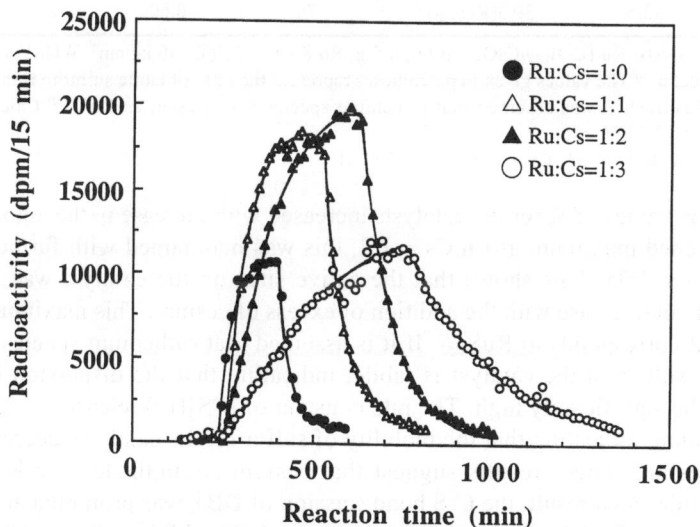

Fig. 4.49 Variation in radioactivities of $[^{35}S]H_2S$ with reaction time.
$Ru_3(CO)_{12}$-nCsOH/Al_2O_3(n = 0, 1, 2 or 3); 320°C, 50 kg/cm², WHSV = 14 h^{-1}, Gas Flow 18 1/h.
(From A. Ishihara, W. Qian, T. Kabe et al., J. Jpn. Petrol. Inst., **41**, 54 (1998))

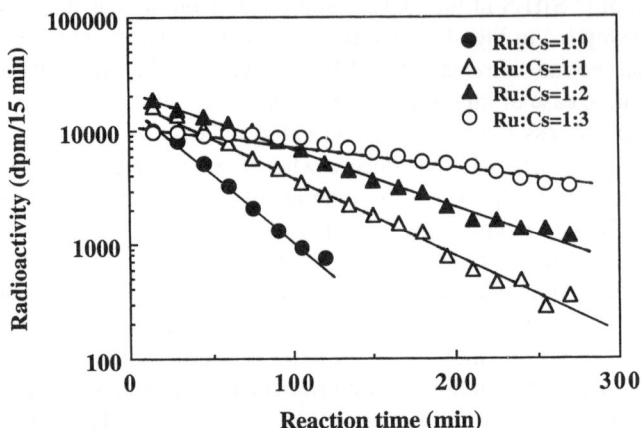

Fig. 4.50   First-order plots of release rate of $[^{35}S]H_2S$.
Ru$_3$(CO)$_{12}$-nCsOH/Al$_2$O$_3$(n = 0, 1, 2 or 3) with preseulfiding; 320°C, 50 kg/cm^2, WHSV = 14 h^{-1}, Gas
Flow 18 1/h.
(From A. Ishihara, W. Qian, T. Kabe $et\ al.$, $J.\ Jpn.\ Petrol.\ Inst.$, **41**, 54 (1998))

Table 4.28 Hydrodesulfurization of ^{35}S-labeled dibenzothiophene over Ru$_3$(CO)$_{12}$-nCsOH/Al$_2$O$_3$ catalysts [a]

Run	Catalyst (Ru:Cs)	Conversion (%)	Labile Sulfur [b] S_0(mg/g-cat)	Ratio of [c] Labile Sulfur S_0/S_t(%)	Rate Constant of H$_2$S Release k($\times 10^{-2}$/min)	HDS Rate [d] ($\times 10^{-1}$mg /g-cat/min)	$S_0 \times k$ ($\times 10^{-1}$mg) /g-cat/min)
1	1:0	50.1	9.2(RuS$_{0.36}$)	18	2.34	2.03	2.15
2	1:1	87.7	17.2(RuS$_{0.68}$)	34	1.92	3.56	3.30
3 [e]	1:2	60.0	30.4(RuS$_{1.20}$)	60	0.79	2.43	2.40
4 [f]	1:2	70.0	34.3(RuS$_{1.35}$)	68	0.86	2.84	2.95
5	1:2	94.0	38.5(RuS$_{1.52}$)	76	1.01	3.81	3.89
6	1:3	52.5	39.3(RuS$_{1.55}$)	78	0.60	2.13	2.36

[a] Procedure 1. Catalysts: Ru$_3$(CO)$_{12}$-nCsOH/Al$_2$O$_3$, 0.5 g, Ru 8 wt%, 320°C, 50 kg/cm^2, WHSV 14 h^{-1}, H$_2$ 18 1/h,
DBT 1 wt% in decalin. [b] The values given in parentheses represent the ratio of labile sulfur to ruthenium. [c] For the
value of S_t (total sulfur), it was assumed that ruthenium species was present as RuS$_2$. [d] Calculated from the
conversion. [e] 280°C. [f] 300°C.
(From A. Ishihara, T. Kabe $et\ al.$, $Chem.\ Lett.$, 744 (1996))

conversion. The values of S_0 for the catalysts increased with increase in the amount of cesium
added and reached maximum at Ru:Cs = 1:2; this was maintained with further addition of
cesium (Ru:Cs = 1:3). This shows that the active sites on the catalyst was not poisoned
because S_0 did not decrease with the addition of excess of cesium. This maximum value of S_0
at Ru:Cs = 1:2 corresponds to RuS$_{1.52}$. If it is assumed that ruthenium species is present as
RuS$_2$, 76% of sulfur on the catalyst is labile, indicating that the dispersion of ruthenium
species could be significantly high. The rate constant of $[^{35}S]H_2S$ release (k) decreased with
increasing cesium, indicating that the mobility of sulfur on the catalysts decreased with the
addition of cesium. These results suggest that cesium strengthened the Ru-S bonds of
ruthenium sulfide. As a result, the C-S bond scission of DBT was promoted and the activity
increased. However, Ru-S bonds become too strong with the addition of excess of cesium. The
formation of H$_2$S and the regeneration of the coordinatively unsaturated sites were prevented,
which results in decreasing the activity. It is also suggested that there is an optimum Ru-S

bond strength for obtaining high catalytic activity.[484]

The amount of labile sulfur represents the amount of active sites which can give the coordinatively unsaturated sites while the rate constant of the release of $[^{35}S]H_2S$ represents the mobility of such active sites. As shown in Table 4.28, the HDS rate calculated from the conversion is in good agreement with the product ($S_0 \times k$) of the amount of labile sulfur on the catalyst and the rate constant of the release of $[^{35}S]H_2S$, indicating that the amount of labile sulfur and the rate constant of $[^{35}S]H_2S$ release which control the HDS rate, can be determined separately by using the ^{35}S tracer method shown here.

B. HDS of $[^{35}S]DBT$ on Ruthenium Catalysts without Presulfiding

The amount of labile sulfur could be determined in HDS of $[^{35}S]DBT$ on ruthenium catalysts with presulfiding. However, total amounts of sulfur accumulated on the catalysts could not be estimated. To do this, HDS of $[^{35}S]DBT$ on ruthenium catalysts was performed directly without presulfiding. Fig. 4.51 shows the changes in the radioactivities of unreacted $[^{35}S]DBT$ and produced $[^{35}S]H_2S$ with the reaction time at 320°C using $Ru_3(CO)_{12}$-$6CsOH/Al_2O_3$. The radioactivity of $[^{35}S]DBT$ gradually increased and reached the steady state at about 200 min. In contrast, radioactivity of $[^{35}S]H_2S$ was not detected for about 100 initial minutes, but then increased and reached the steady state. The amount of total sulfur incorporated into the catalysts (S_t) corresponds to the sum of the radioactivities of areas A and B and can be calculated. Further, after the radioactivity of $[^{35}S]H_2S$ reached the steady state, the solution of $[^{35}S]DBT$ was replaced by that of $[^{32}S]DBT$. Similar to OP 1, the radioactivity of $[^{35}S]DBT$ decreased immediately and a time delay for decrease of radioactivities of $[^{35}S]H_2S$ from the steady state to normal state was observed. From the data for this decreasing period of $[^{35}S]H_2S$, the release rate constant of $[^{35}S]H_2S$ and the amount of labile sulfur could be calculated and the results for cesium-promoted ruthenium catalysts are listed in Table 4.29. Compared with the catalysts with presulfiding, the catalysts without presulfiding showed lower activities and the maximum of the conversion was observed at Ru:Cs = 1:1. Further, all the values for the labile sulfur and the rate constant of H_2S release decreased in comparison

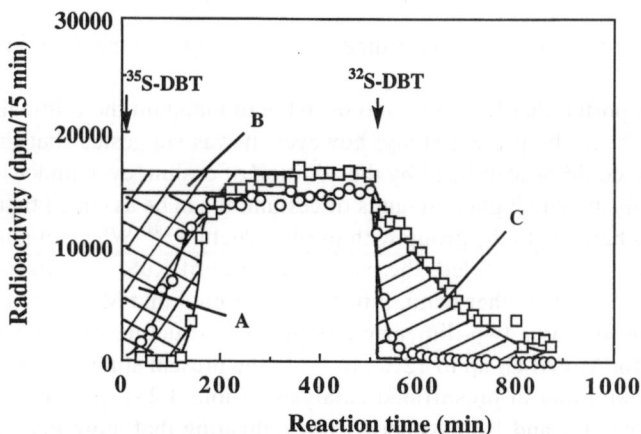

Fig. 4.51   Variation in radioactivities with reaction time.
○: $[^{35}S]DBT$; □: $[^{32}S]H_2S$;
$Rus(CO)_{12}$-$6CsOH/Al_2O_3$; 320°C, 50 kg/cm², WHSV = 14 h⁻¹, Gas Flow 18 1/h.
(From A. Ishihara, W. Qian, T. Kabe et al., J. Jpn. Petrol. Inst., **41**, 55 (1998))

Table 4.29 Hydrodesulfurization of ^{35}S-labeled Dibenzothiophene Catalyzed by Alumina-supported Ru$_3$(CO)$_{12}$-nCsOH Systems [a]

Run	Catalyst (Ru:Cs)	Conversion (%)	Total Sulfur [b] S_t(mg/g-cat)	Labile Sulfur [c] S_0(mg/g-cat)	Ratio of [d] Labile Sulfur S_0/S_t(%)	Rate Constant of H$_2$S Release k ($\times 10^{-2}$/min)
7	1:0	38.5	21.5(Ru$s_{0.85}$)	8.33(Ru$s_{0.33}$)	39	1.87
8	1:1	63.5	28.1(Ru$s_{1.11}$)	13.7(Ru$s_{0.54}$)	49	1.86
9	1:2	50.8	46.1(Ru$s_{1.82}$)	26.7(Ru$s_{1.05}$)	58	0.86
10	1:3	18.9	67.5(Ru$s_{2.66}$)	22.6(Ru$s_{0.89}$)	33	0.41

[a] Procedure 2. Catalysts: Ru$_3$(CO)$_{12}$-nCsOH/Al$_2$O$_3$. 320°C, 50 kg/cm^2, WHSV 14 h^{-1}, H$_2$ 18 1/h, DBT 1 wt% in decalin. [b] The values given in parentheses represent the ratio of total sulfur to ruthenium. [c] The values given in parentheses represent the ratio of labile sulfur to rutheniumin. [d] The total sulfur (S_t) in this calculation was S_t estimated in procedure 2.
(From A. Ishihara, T. Kabe et al., J. Jpn. Petrol. Inst., **41**, 55 (1998))

with those of sulfided catalysts. When the catalysts were not presulfided, activation of the catalysts was not enough to achieve maximum activity. In presulfiding of the catalysts, smaller H$_2$S may be adsorbed more easily on the catalysts and able to expel carbon monoxide ligands. In contrast, it may be more difficult for larger DBT to gain access to the surface and expel carbon monoxide ligands. It seems that the phenomenon occurs especially for the catalysts at Ru:Cs = 1:2 or 1:3, which have larger amounts of alkali metals. Therefore, the activity may decrease at over Ru:Cs = 1:2. The amount of S_t increased with increasing cesium added. S_t at Ru:Cs = 1:3 corresponds to RuS$_{2.66}$, which exceeds the usual RuS$_2$, indicating that some cesium may be sulfided. In these catalysts, the trend shown by the changes in the amount of labile sulfur and the rate constant of H$_2$S release was very similar to that of presulfided catalysts. That is, with the addition of cesium, the amount of labile sulfur increased, giving the maximum value and the rate constant of H$_2$S release decreased monotonically. These results indicate that the addition of cesium also strengthened the Ru-S bond and promoted the C-S bond scission of DBT to increase the activity and that the excess of cesium prevented the desorption of hydrogen sulfide to decrease the activity.

C. Structure of Cesium-promoted Ruthenium Catalysts and the Reaction Mechanisms

It has been reported that RuS$_2$ was too unstable to maintain the original structure under HDS conditions.[398] In the present study, however, it was suggested that the Ru-S bond in ruthenium sulfide could be stabilized by the addition of cesium even under HDS conditions. Especially in catalysts with higher amounts of cesium, it can be assumed that RuS$_2$ is present on alumina. RuS$_2$ belongs to the group with pyrite structure.[398] When an alumina-supported ruthenium catalyst was presulfided, the pyrite structure of RuS$_2$ was also observed on the alumina support.[459] On the other hand, it has been reported that MoS$_2$ phase appeared to be predominantly present as a two-dimension single slab structure oriented flat-wise on the alumina support for Mo/Al$_2$O$_3$ up to 12% Mo.[485] In the present study, when RuS$_2$ is assumed to be present on alumina in presulfided catalysts (Table 4.28), the ratios of S_0/S_t for the catalysts at Ru:Cs = 1:2 and 1:3 exceeded 75%, indicating that more than 75% of sulfur in RuS$_2$ is labile and that RuS$_2$ in these catalysts is successfully dispersed on alumina. For a better understanding of the structure of the catalyst and the mechanism, it was assumed that the

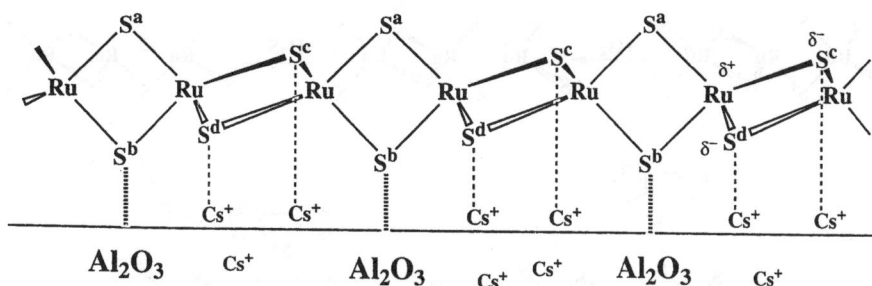

Fig. 4.52 Structure of RuS$_2$ on Cs-promoted RuS$_2$/Al$_2$O$_3$ catalyst.
(From A. Ishihara, W. Qian, T. Kabe *et al.*, *J. Jpn. Petrol. Inst.*, **41**, 56 (1998))

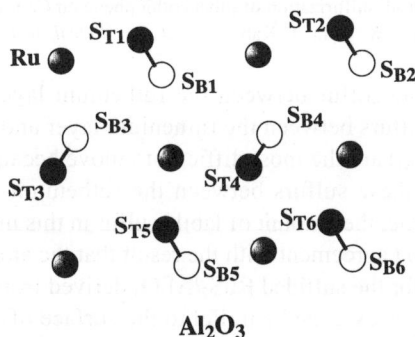

Fig. 4.53 Structure of RuS$_2$ extended parallel to the surface of Al$_2$O$_3$.
⬤ : Ruthenium, ○,⬤: Sulfur. Basal plain consists of Al$_2$O$_3$.

monolayer dispersion would be maintained and that RuS$_2$ phase would be present as a two-dimensional structure oriented flat-wise on the alumina support in Fig. 4.52. Since the locations of sulfur on the surface of alumina were different from each other, the labile capacity of sulfur would be different from each other. The sulfur between the ruthenium layer and alumina surface (S^b) may be the most difficult to move and the sulfur over the ruthenium layer (S^a) may be the most labile. Two sulfurs in other sites (S^c and S^d), which form a triangle with Ru parallel to the alumina surface, may have middle labile capacity. This explains why the amount of labile sulfur changes depending on the reaction temperature (see Table 4.28). If the sulfur between ruthenium and alumina surface, S^b, is not labile, the amount of labile sulfur in the sulfided Ru-Cs/Al$_2$O$_3$ would be 75% of total sulfur. This is in good agreement with the result that the amount of labile sulfur is 76% of the total sulfur assumed for the sulfided RuS$_2$/Al$_2$O$_3$ derived from Ru$_3$(CO)$_{12}$-6CsOH/Al$_2$O$_3$.

Taking into account the pyrite structure of RuS$_2$, another structure may be proposed as in Fig. 4.53 where RuS$_2$ phase is also present as a two-dimensional structure oriented flat-wise on the alumina support. In Fig. 4.53, alumina is in the bottom layer. The second layer is parallel to that of alumina and includes six sulfur atoms from S_{B1} to S_{B6}. The third layer and the top layer included ruthenium atoms and six sulfur atoms from S_{T1} to S_{T6}, respectively. Since the location of sulfur on the surface of alumina was different from each other, the labile capacity of sulfur would also be different. The sulfur over the ruthenium layer (S_{T1}–S_{T6})

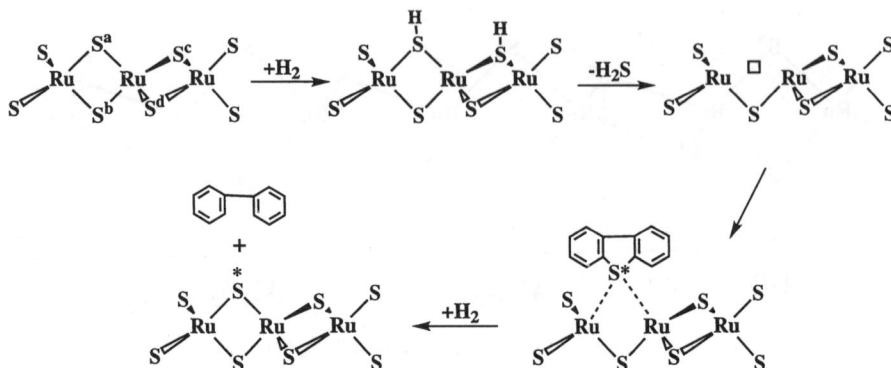

Fig. 4.54 Mechanism of hydrodesulfurization of dibenzothiophene on Cs-promoted RuS$_2$/Al$_2$O$_3$ catalyst. (From A. Ishihara, W. Qian, T. Kabe et al., J. Jpn. Petrol. Inst., **41**, 56 (1998))

may be more labile than the sulfur between the ruthenium layer and the alumina surface (S_{B1}–S_{B6}). Among the six sulfurs between the ruthenium layer and alumina surface (S_{B1}–S_{B6}), three sulfurs (S_{B1}, S_{B3} and S_{B4}) are the most difficult to move because each is surrounded with three ruthenium atoms. If these sulfurs between the ruthenium layer and alumina surface (S_{B1}, S_{B3} and S_{B4}) are not labile, the amount of labile sulfur in this model surface would be 75% of total sulfur. This is in good agreement with the result that the amount of labile sulfur is 76% of the total sulfur assumed for the sulfided RuS$_2$/Al$_2$O$_3$ derived from Ru$_3$(CO)$_{12}$-6CsOH/Al$_2$O$_3$. When the structure of RuS$_2$ is extended parallel to the surface of Al$_2$O$_3$ beyond the structure in Fig. 4.53, the amount of labile sulfur according to this concept would be much less than 75%. Therefore, highly dispersed RuS$_2$ species may not be larger than the bulk of the structure in Fig. 4.53.

The situation of cesium is ambiguous in the catalysts derived from Ru$_3$(CO)$_{12}$-nCsOH/Al$_2$O$_3$. XPS analysis for these catalysts did not show a clear peak of cesium.[457] Therefore, it seems that cesium is located between RuS$_2$ and alumina to stabilize Ru-S bonds or in the inside of alumina to neutralize the acid sites of alumina, as shown in Fig. 4.52. In RuS$_2$ sulfur is anionic and ruthenium is cationic. Such anionic sulfur may interact with cesium cation to stabilize Ru-S bonds. Fig. 4.54 shows the mechanism of HDS of DBT on Cs-promoted RuS$_2$/Al$_2$O$_3$ using the structure of RuS$_2$ in Fig. 4.52. Initially RuS$_2$ reacts with hydrogen to produce an anion vacancy and hydrogen sulfide. [^{35}S]DBT adsorbs on the anion vacancy and reacts with hydrogen to produce BP and ^{35}S-labeled RuS$_2$. It can be assumed that S^c and S^d react similarly. In the absence of cesium, labile sulfurs such as S^a, S^c and S^d would be removed by the reaction with hydrogen before one vacancy is occupied by a sulfur of DBT.

4.5 Other Approaches

Unsupported niobium sulfides were used first for the isomerization and hydrogenation of 1-betene [486] and later for the characteristic hydrotreating reactions.[487] Compared with MoS$_2$ and WS$_2$, unsupported niobium sulfides exhibited higher activities. They were effective especially for hydrogenation in the presence of H$_2$S in contrast to the conventional catalysts.

Further, they also showed particular ability to cleave C-C and C-N bonds.[488,489] Concerning supported niobium sulfide systems used for hydrotreatment, Ledoux et al. reported that niobium sulfide supported on carbon at low concentrations exhibited lower HDS activity.[20] It was reported that sulfided Ni-Nb/SiO$_2$ catalysts exhibited a synergistic effect in HDS,[490] although Hillerova did not observe this effect in the same system.[491] Breysse et al. reported that carbon-supported niobium sulfide catalysts were effective for C-N bond scission reactions.[492] There are discrepancies in the reported results, indicating that the catalytic properties for these supported niobium sulfide systems may be closely related to the preparation parameters. Allali et al. optimized the preparation of carbon-supported niobium sulfide catalysts concerning, i) the nature of the soluble precursor, ii) the drying process, and iii) the sulfiding treatment.[493] The best result was obtained and the activity was superior to that of a supported MoS$_2$ catalyst when the catalyst was prepared with niobium oxalate as the impregnation salt, drying at room temperature, and presulfiding with N$_2$/H$_2$S at 400°C. Alumina-supported niobium sulfide catalyst could be sulfided only under more severe conditions (CS$_2$ under pressure). This difficulty may be attributed to the strong support-cation interactions. The above results indicate the importance of the support and the sulfidation method on the genesis of a niobium sulfide active phase.

Unsupported vanadium sulfide was also employed in hydrotreating reactions.[398,488,494-496] In a study by Pecoraro and Chianelli, the activity of vanadium sulfide VS$_x$ in HDS of DBT was similar to those of Co$_9$S$_8$ and Ni$_3$S$_2$ and five times lower than that of MoS$_2$.[398] Alumina-supported VS$_x$ catalyst exhibited substantial activity in HDS.[494] However, the activity was very susceptible to poisoning. Ledoux et al. prepared an alumina-supported vanadium sulfide by impregnation with an aqueous solution of NH$_4$VO$_3$.[495] The thiophene HDS activity of the catalyst was 30 times lower than that of a Ni-Mo/Al$_2$O$_3$ catalyst. Guillard et al. prepared pure V$_2$S$_3$ by decomposition of ammonium thiovanadate at relatively low temperature.[496] Although HDS properties of V$_2$S$_3$ were of the same order, the hydrogenation performance of various cyclic molecules was superior to those of molybdenum or tungsten sulfides. The catalytic properties of vanadium sulfide system are very complex and appear to change depending on the type of this sulfide.

There are very few examples dealing with rare earth metals in hydrotreating catalysts in academic studies. Lanthanum oxide-modified alumina was described above.[111] Moon and Ihm reported thiophene HDS over unsupported nickel-neodymium bimetallic catalysts.[497,498] The bimetallic catalysts of NdNi prepared by coprecipitation underwent a drastic change to oxides and sulfides during calcination and presulfiding. The activity was increased with addition of neodymium to nickel catalysts and was in good correlation with the surface area of catalysts. Nd worked as a structural promoter and appeared to prevent the sintering of Ni and to keep the high surface area of NdNi bimetal catalyst. Some patents reported the use of rare earth metals for hydrotreating reactions.[497] Rare earth metals reduced the hydrogen pressure required for HDS and enhanced the resistance to coke deposition and hydrogen sulfide. Rare earth metals also improved the activity of HDN and hydrodemetallation (HDM). Another group also evaluated the rare earth-nickel-based compounds and alloys as sulfur-resistant HDN catalysts.[499] A mixture of these solids and molybdenum sulfide further increased both sulfur resistance and activity of pyridine HDN. In a recent approach, HDS using hydrogen storage alloys as stoichiometric reagents, typically LaNi$_5$, was investigated.[500] The alloys were useful reagents for selective elimination of sulfur from sulfur-containing aromatic compounds.

Andreev et al. reported that alumina-supported NiPS$_3$ exhibited high catalytic activity in thiophene HDS comparable with that of Co-Mo/Al$_2$O$_3$ and Ni-Mo/Al$_2$O$_3$.[501] Polycrystalline NiPS$_3$ was prepared by annealing a mixture of powdered nickel, sulfur, and red phosphorus in an evacuated quartz tube at 700°C for 120 h. The catalyst was prepared by mechanical mixing of 10 wt% NiPS$_3$ with 90 wt% Al$_2$O$_3$. The promoting effect of phosphorus on Ni-Mo HDS catalysts was explained by the formation of highly dispersed NiPS$_3$ under the reaction conditions.

References

1. J. K. Minderhoud and J. A. R. van Veen, *Fuel Process. Technol.*, **35**, 87 (1993).
2. B. Delmon, *Catal. Lett.*, **22**, 1 (1993).
3. R. R. Chianelli, M. Daage and M. J. Ledoux, *Adv. Catal.*, **40**, 177 (1994).
4. B. C. Gates, J. R. Katzer and G. C. A. Schuit, *Chemistry of Catalytic Processes*, p.390, McGraw-Hill: New York (1979).
5. M. Breysse, J. L. Portefaix and M. Vrinat, *Catal. Today*, **10**, 489 (1991).
6. H. Topsøe, B. S. Clausen, N. Burrusci, R. Candia and S. Morup, in: *Preparation of Catalysts II*, (B. Delmon, P. Grange, P. Jacobs and G. Poncelet eds.), p.429, Elsevier, Amsterdam (1979).
7. G. C. Stevens and T. Edmonds, in: *Preparation of Catalysts II*, (B. Delmon, P. Grange, P. Jacobs, G. Poncelet eds.), p.507, Elsevier, Amsterdam (1979).
8. J. C. Duchet, E. M. Van Oers, V. H. J. de Beer and R. Prins, *J. Catal.*, **80**, 386 (1983).
9. V. H. J. de Beer, J. C. Duchet and R. Prins, *J. Catal.*, **72**, 369 (1981).
10. A. J. Bridgewater, R. Burch and P. C. H. Mitchell, *Appl. Catal.*, **4**, 267 (1982).
11. G. Muralidhar, B. E. Concha, G. L. Bartholomew and C. H. Bartholomew, *J. Catal.*, **89**, 274 (1984).
12. M. Breysse, B. A. Bennett, D. Chadwick and M. Vrinat, *Bull. Chem. Chim. Belg.*, **90**, 1271 (1981).
13. H. Topsøe, B. S. Clausen, N. Topsøe and E. Pedersen, *Ind. Eng. Chem. Fundam.*, **25**, 25 (1986).
14. H. Topsøe and B. S. Clausen, *Appl. Catal.*, **25**, 273 (1986).
15. H. Topsøe, R. Candia, N. Topsøe and B. S. Clausen, *Bull. Chem. Chim. Belg.*, **93**, 783 (1984).
16. C. Wivel, R. Candia, B. S. Clausen, S. Mørup and H. Topsøe, *J. Catal.*, **68**, 453 (1981).
17. R. Candia, B. S. Clausen and H. Topsøe, *J. Catal.*, **77**, 564 (1982).
18. R. Candia, O. Sorensen, J. Villadsen, N. Topsøe, B. S. Clausen and H. Topsøe, *Bull. Chem. Chim. Belg.*, **93**, 763 (1984).
19. W. Niemann, B. S. Clausen and H. Topsøe, *Catal. Lett.*, **4**, 355 (1990).
20. M. J. Ledoux, O. Michaux, G. Agostini and P. Panissod, *J. Catal.*, **102**, 275 (1986).
21. J. A. van Veen, E. Gerkema, A. M. van der Kraan and A. Knoester, *J. Chem. Soc., Chem. Commun*, 1684 (1987).
22. J. A. van Veen, E. Gerkema, A. M. van der Kraan, P. A. J. M. Hendriks and H. Beens, *J. Catal.*, **133**, 112 (1992).
23. C. K. Groot, V. H. J. de Beer, R. Prins, M. Stolarski and W. S. Niedzwiedz, *Ind. Eng. Chem. Prod. Res. Dev.*, **25**, 522 (1986).
24. V. H. J. de Beer, F. J. Derbyshire, C. K. Groot, R. Prins, A. W. Scaroni and J. M. Solar, *Fuel*, **63**, 1095 (1984).
25. E. J. M. Hensen, M. J. Vissenberg, V. H. J. de Beer, J. A. R. van Veen and R. A. van Santen, *J. Catal.*, **163** (2), 429 (1996).
26. J. P. R. Vissers, C. K. Groot, E. M. van Oers, V. H. J. de Beer and R. Prins, *Bull. Soc. Chim. Belg.*, **93**, 813 (1984).
27. J. P. R. Vissers, V. H. J. de Beer and R. Prins, *J. Chem. Soc., Faraday Trans.1*, **83**, 2145 (1987).
28. J. P. R. Vissers, B. Scheffer, V. H. J. de Beer, J. A. Moulijn and R. Prins, *J. Catal.*, **105**, 277 (1987).
29. S. M. A. M. Bouwens, D. C. Koningsberger, V. H. J. de Beer and R. Prins, *Bull. Soc. Chim. Belg.*, **96**, 951 (1987).
30. S. M. A. M. Bouwens, D. C. Koningsberger, V. H. J. de Beer and R. Prins, *Catal. Lett.*, **1**, 55 (1988).
31. S. M. A. M. Bouwens, R. Prins, V. H. J. de Beer and D. C. Koningsberger, *J. Phys. Chem.*, **94**, 3711 (1990).
32. S. M. A. M. Bouwens, D. C. Koningsberger and V. H. J. de Beer, *J. Phys. Chem.*, **95**, 123 (1991).
33. S. M. A. M. Bouwens, F. B. M. van Zon, M. B. van Dijk, A. M. van der Kraan, V. H. J. de Beer, J. A. R. van Veen and D. C. Koningsberger, *J. Catal.*, **146**, 375 (1994).
34. A. Calafat, J. Laine, A. Lopez-Agudo and J. M. Palacios, *J. Catal.*, **162** (1), 20 (1996).
35. S. P. A. Louwers and R. Prins, *J. Catal.*, **133** (1), 94 (1992).
36. A. W. Scaroni, R. G. Jenkins and P. L. Walker Jr., *Appl. Catal.*, **14**, 173 (1985).
37. F. J. Derbyshire, V. H. J. de Beer, G. M. K. Abosti, A. W. Scaroni, J. M. Solar and D. J. Skrovanek, *Appl. Catal.*, **27**, 117 (1986).
38. A. N. Startsev, S. A. Shkuropat, V. I. Zaikovskii, E. M. Moroz, Yu. I. Yermakov, G. V. Plaksin, M. S.

Tsekhanovich and V. F. Surovikin, *Kinet. Katal.*, **29**, 398 (1988).

39. Yu. I. Yermakov, A. N. Startsev, S. A. Shkuropat, G. V. Plaksin, M. S. Tsekhanovich and V. F. Surovikin, *React. Kinet. Catal. Lett.*, **36** (1), 65 (1988).
40. B. Scheffer, P. Arnoldy and J. A. Moulijn, *J. Catal.*, **112**, 516 (1988).
41. P. Arnoldy, E. M. van Oers, V. H. J. de Beer, J. A. Moulijn and R. Prins, *Appl. Catal.*, **48**, 241 (1989).
42. W. L. T. M. Ramselaar, R. H. Hadders, E. Gerkema, V. H. J. de Beer, E. M. van Oers and A. M. van der Kraan, *Appl. Catal.*, **51**, 263 (1989).
43. W. L. T. M. Ramselaar, M. W. J. Craje, E. Gerkema, V. H. J. de Beer and A. M. van der Kraan, *Appl. Catal.*, **54**, 217 (1989).
44. G. M. K. Abotsi and A. W. Scaroni, *Fuel Proc. Tech.*, **22**, 107 (1989).
45. P. M. Boorman, K. Chong, R. A. Kydd and J. M. Lewis, *J. Catal.*, **128**, 537 (1991).
46. J. M. Solar, F. J. Derbyshire, V. H. J. de Beer and L. R. Radovic, *J. Catal.*, **129**, 330 (1991).
47. B. M. Reddy and V. S. Subramanyam, *Appl. Catal.*, **27**, 1 (1986).
48. F. P. Daly, J. S. Brinen and J. L. Schmitt, *Appl. Catal.*, **11**, 161 (1984).
49. T. I. Korányi, V. Rozanov, R. Kremo and Z. Paal, *J. Mol. Catal.*, **63**, 31 (1990).
50. J. Frimmel and M. Zdrazil, *J. Catal.*, **167** (1), 286 (1997).
51. E. Hillerová and M. Zdrazil, *Catal. Lett.*, **8**, 215 (1991).
52. K. V. R. Chary, H. Ramakrishna and G. Murali Dhar, *J. Mol. Catal.*, **68**, L25 (1991).
53. J. Laine, F. Severino, M. Labady and J. Gallardo, *J. Catal.*, **138**, (1), 145 (1992).
54. M. W. J. Craje, V. H. J. de Beer, J. A. R. van Veen and A. M. van der Kraan, *Appl. Catal.*, A, **100** (1), 97 (1993).
55. S. K. Ihm, Y. H. Moon and C. D. Ihm, *Stud. Surf. Sci. Catal.*, **75**, New Frontiers in Catalysis, Pt.C, 1923 (1993).
56. A. Martin-Gullon, C. Prado-Burguete and F. Rodriguez-Reinoso, *Carbon*, **31** (7), 1099 (1993).
57. J. Laine, F. Severino and M. Labady, *J. Catal.*, **147**, (1), 355 (1994).
58. J. Laine, M. Labady, F. Severino and S. Yunes, *J. Catal.*, **166** (2), 384 (1997).
59. S. Eijsbouts, C. Dudhakar, V. H. J. de Beer and R. Prins, *J. Catal.*, **127** (2), 605 (1991).
60. S. Eijsbouts, V. H. J. de Beer and R. Prins, *J. Catal.*, **127** (2), 619 (1991).
61. M. J. Ledoux, O. Michaux and G. Agostini, *J. Catal.*, **102**, 430 (1986).
62. E. Hillerová, Z. Vít, M. Zdrazil, S. A. Shkuropat, E. N. Bogdanets and A. N. Startsev, *Appl. Catal.*, **67** (2), 231 (1991).
63. A. Drahoradova, Z. Vít and M. Zdrazil, *Fuel*, **71** (4), 455 (1992).
64. P. M. Boorman and K. Chong, *Energy & Fuels*, **6** (3), 300 (1992).
65. A. Calafat, J. Laine and A. Lopez-Agudo, *Catal. Lett.*, **40** (3,4), 229 (1996).
66. Z. Vít, *Fuel*, **72**, 105 (1993).
67. M. J. Ledoux and B. Djellouli, *Appl. Catal.*, **67**, 81 (1990).
68. R. M. M. de Agudelo and A. Morales, *Proc. 9th Int. Cong. Catal.*, **1**, 42 (1988).
69. J. Sobczak, Z. Vít and M. Zdrazil, *Appl. Catal.*, 45, L23 (1988).
70. K. Segawa, T. Soeya and D. S. Kim, *J. Jpn. Petrol. Inst.*, **33**, 6, 347 (1990).
71. M. Vrinat, M. Breysse, S. Fuentes, M. Lacroix and J. Ramírez, *IX Simposio, Iberoamericano de Catalisis* (O. Bermudo, G. Del Angel and R. Gomez eds.), p.1029, F. Cossio (1988).
72. J. Ramírez, S. Fuentes, G. Diaz, M. Vrinat, M. Breysse and M. Lacroix, *Appl. Catal.*, **52**, 211 (1989).
73. G. Muralidhar, F. E. Massoth and J. Shabtai, *J. Catal.*, **85**, 44 (1984).
74. A. Fernandez, J. Leurer, A. R. Gonzalez-Elipe, G. Munuera and H. Knozinger, *J. Catal.*, **112**, 489 (1988).
75. A. Spojakina, S. Damyanova, D. Shopov, T. Shokhireva and T. Yurieva, *React. Kinet. Catal. Lett.*, **27**, 333 (1985).
76. S. Matsudo and A. Kato, *Appl. Catal.*, **8**, 149 (1983).
77. J. Ramírez, R. Cuevas, L. Gasque, M. Vrinat and M. Breysse, *Appl. Catal.*, **71**, 351 (1991).
78. G. C. Bond and S. F. Tahir, *Appl. Catal.*, A, **105**, 289 (1993).
79. D. S. Kim, Y. Kurusu, I. E. Wachs, F. D. Hardcastle and K. Segawa, *J. Catal.*, **120**, 325 (1989).
80. N. Spanos, H. K. Matralis, C. H. Kordulis and A. Lycourghiotis, *J. Catal.*, **136**, 432 (1989).
81. R. B. Quincy, M. Houalla, A. Proctor and D. M. Hercules, *J. Catal.*, **125**, 214 (1990).
82. A. N. Desikan, L. Huang and S. T. Oyama, *J. Chem. Soc., Faraday Trans.*, **88**, 3357 (1992).
83. S. Damyanova and A. Spojakina, *React. Kinet. Catal. Lett.*, **51**, 465 (1993).
84. K. Y. S. Ng and E. Gulari, *J. Catal.*, **92**, 340 (1985).
85. Y. Okamoto, A. Maezawa and T. Imanaka, *J. Catal.*, **120**, 29 (1989).
86. K. C. Pratt, J. V. Sanders and V. Christov, *J. Catal.*, **124**, 416 (1990).
87. D. Hamon, M. Vrinat, M. Breysse, B. Durand, M. Jebrouni, M. Roubin, P. Magnoux and T. Des Courieres, *Catal. Today*, **10**, 613 (1991).
88. D. Hamon, M. Vrinat, M. Breysse, B. Durand, F. Beauchesne and T. Des Courieres, *Bull. Soc. Chim. Belg.*, **100**, 933 (1991).
89. M. Breysse, D. Hamon, M. Lacroix and M. Vrinat, *Proc. JECAT'91*, p. 36 (1991).
90. J. C. Duchet, M. J. Tilliette, D. Cornet, L. Vivier, G. Perot, L. Bekakra, C. Moreau and G. Szabo, *Catal. Today*,

10, 579 (1991).

91. J. L. Portefaix, M. Cattenot, J. A. Dalmon and C. Mauchausse, in: *Advances in Hydrotreating Catalysis*, (M. L. Occelli, and R. G. Anthony eds.), p.67, Elsevier, Amsterdam (1989).
92. M. Vrinat, D. Hamon, M. Breysse, B. Durand and T. Des Courieres, *Catal. Today*, **20**, 273 (1994).
93. P. Afanasiev, C. Geantet and M. Breysse, *J. Catal.*, **153**, 17 (1995).
94. C. Moreau, L. Bekakra, P. Geneste, J. L. Olive, J. C. Duchet, M. J. Tilliette and J. Grimblot, *Bull. Soc. Chim. Belg.*, **100**, 11–12, 841 (1991).
95. K. Saiprasad Rao, H. Ramakrishna and G. Murali Dhar, *J. Catal.*, **133**, 146 (1992).
96. A. Nishijima, H. Shimada, T. Sato, Y. Yoshimura and J. Hiraishi, *Polyhedron*, **5**, 243 (1986).
97. J. B. McVicker and J. J. Ziemiak, *J. Catal.*, **95**, 473 (1985).
98. A. Stranick, M. Houalla and D. M. Hercules, *J. Catal.*, **125**, 214 (1990).
99. H. Tanaka, M. Boulinguiez and M. Vrinat, *Catal. Today*, **29** (1-4), 209 (1996).
100. Z. B. Wei, W. Yan, H. Zhang, T. Ren, Q. Xin and Z. Li, *Appl. Catal.*, A, **167** (1), 39 (1998).
101. J. Ramírez and A. Gutierrez-Alejandre, *J. Catal.*, 170 (1), 108 (1997).
102. E. Olguin, M. Vrinat, L. Cedeno, J. Ramírez, M. Borque and A. Lopez-Agudo, *Appl. Catal.*, **165** (1-2), 1 (1997).
107) F. Kameda, K. Hoshino, S. Yoshinaka and K. Segawa, *J. Jpn. Petrol. Inst.*, **40** (3), 205 (1997).
103. J. Ramírez, L. Ruiz-Ramírez, L. Cedeno, V. Harle, M. Vrinat and M. Breysse, *Appl. Catal. A*, **93**, 163 (1993).
104. A. Spojakina, S. Damyanova and K. Jiratova, *Proc. XII Simposio Iberoamericano de Catalisis* (L. Nogueria, Y. L. Zam eds.), p.571, Brasil (1990).
105. S. Damyanova, A. Spojakina and K. Jiratova, *Appl. Catal. A: Gen.*, **125**, 257 (1995).
106. C. Martin, I. Martin, V. Rives, S. Damyanova and A. Spojakina, *React. Kinet. Catal. Lett.*, **54**, (1), 203 (1995).
107. F. Kameda, K. Hoshino, S. Yoshinaka and K. Segawa, *J. Jpn. Petrol. Inst.*, **40** (3), 205 (1997).
108. C. Pophal, F. Kameda, K. Hoshino, S. Yoshinaka and K. Segawa, *Catal. Today* 39 (1-2), 21 (1997).
109. I. Wang and R. C. Chang, *J. Catal.*, **117**, 266 (1989).
110. F. P. Daly, *J. Catal.*, **116**, 600 (1989).
111. J.-W. Cui, F. E. Massoth and N-Y. Topsøe, *J. Catal.*, **136**, 361 (1992).
112. J. C. Duchet, N. Gnofam, J. L. Lemberton, G. Perot, L. Bekakra, C. Moreau, J. Joffre, S. Kasztelan and J. Grimblot, *Catal. Today*, **10**, 593 (1991).
113. S. Udomsak, R. G. Anthony and S. E. Lott, *Appl. Catal. A: Gen.*, **122**, 111 (1995).
114. L. E. Hayes, C. E. Snape and S. Affrossman, *Prepr.-Am. Chem. Soc., Div. Fuel. Chem.*, **36**, (4), 1817 (1991).
115. B. Kelly, S. Martin, S. Affrossman and C. E. Snape, *Prepr.-Am. Chem. Soc., Div. Pet. Chem.*, **39**, (4), 593 (1994).
116. J. Laine, J. L. Brito and F. Severino, *J. Catal.*, **131**, 385 (1991).
117. C. Caceres, J. L. G. Fierro, A. Lopez-Agudo, F. Severino and J. Laine, *J. Catal.*, **97**, 219 (1986).
118. A. Lopez-Agudo, J. L. G. Fierro, C. Caceres, J. Laine and F. Severino, *Appl. Catal.*, **30**, 185 (1987).
119. I. Akitsuki, T. Miura, H. Kaya and T. Itoh, *Ketzen Catalyst Symp.*, H-8, 1988.
120. R. H. Fish, J. N. Michaels, R. S. Moore and H. Heinemann, *J. Catal.*, **123**, 74 (1990).
121. H. Shimada, T. Sato, Y. Yoshimura, J. Hiraishi and A. Nishijima, *J. Catal.*, **110**, 275 (1988).
122. H. Hattori, K. Yamashita, K. Kobayashi, T. Tanabe and K. Tanabe, *Proc. 1987 Int. Conf. Coal Sci.* (J. A. Moulijn, K. N. Nater and H. A. G. Chermin eds.), p.285, Elsevier, Amsterdam (1987).
123. V. Kolousek, P. Palka, E. Hillerová and M. Zdrazil, *Collect. Czech. Chem. Commun.*, **56**, 580 (1991).
124. K. V. R. Chary, H. Ramakrishna, K. S. Rama Rao, G. Murali Dhar and P. Kanta Rao, *Catal. Lett.*, **10**, 27 (1991).
125. E. Hillerová, Z. Vít and M. Zdrazil, *Appl. Catal. A: Gen.*, **118**, 111 (1994).
126. P. S. E. Dai and J. H. Lunsford, *J. Catal.*, **64**, 173 (1980).
127. M. Sugioka, *J. Jpn. Petrol. Inst.*, **33**, 280 (1990).
128. Y. Okamoto and T. Imanaka, *J. Jpn. Petrol. Inst.*, **36** (3), 182 (1993).
129. Y. Okamoto, A. Maezawa, H. Kane and T. Imanaka, *in Proc. 9th Int. Congr. Catal.*, Calgary, 1988, (M. J. Phillips and M. Ternan, eds.), **Vol. 1**, Chem. Inst. Can., Ottawa (1988).
130. A. Maezawa, M. Kitamura, K. Wakamoto, Y. Okamoto and T. Imanaka, *Chem. Express*, **3**, 1 (1988).
131. Y. Okamoto, A. Maezawa, H. Kane and T. Imanaka, *J. Mol. Catal.*, **52**, 337 (1989).
132. M. L. Vrinat, C. G. Gachet and L. de Mourgues, *Catalysis by Zeolite*, (B. Imelik ed.), p.219, Elsevier, Amsterdam, (1980).
133. R. Cid, F. J. Gil Llambias, J. L. G. Fierro, A. Lopez-Agudo and J. Villasenor, *J. Catal.*, **89**, 478 (1984).
134. R. Cid, F. J. Gil Llambias, M. Gonzalez and A. Lopez-Agudo, *Catal. Lett.*, **24**, 147 (1994).
135. R. Cid, J. Villasenor, F. Orellana, J. L. G. Fierro and A. Lopez-Agudo, *Appl. Catal.*, **18**, 357 (1985).
136. R. Cid, F. Orellana and A. Lopez-Agudo, *Appl. Catal.*, **32**, 327 (1987).
137. A. López-Agudo, R. Cid, F. Orellana and J. L. G. Fierro, *Polyhedron*, **5**, 187 (1986).
138. R. Cid, J. L. G. Fierro and A. Lopez-Agudo, *Zeolites*, **10**, 95 (1990).
139. W. J. J. Welters, G. Vorbeck, H. W. Zandbergen, J. W. de Haan, V. H. J. de Beer and R. A. van Santen, *J. Catal.*, 150, 155 (1994).
140. T. I. Korányi, N. H. Pham, A. Jentys and H. Vinek, *Stud. Surf. Sci. Catal.*, **106**, 509 (1997).

141. T. I. Korányi, L. J. M. van de Ven, W. J. J. Welters, J. W. de Haan, V. H. J. de Beer and R. A. van Santen, *Catal. Lett.*, **17**, 105 (1993).
142. T. Tatsumi, M. Taniguchi, S. Yasuda, Y. Ishii, T. Murata and M. Hidai, *Appl. Catal.*, *A*, **139** (1-2), L5 (1996).
143. N. Davidova, P. Kovacheva and D. Shopov, *Stud. Surf. Sci. Catal.*, **24**, 659 (1985).
144. N. Davidova, P. Kovacheva and D. Shopov, *Stud. Surf. Sci. Catal.*, **28**, 811 (1986).
145. N. Davidova, P. Kovacheva and D. Shopov, *Zeolites*, **6**, 304 (1985).
146. W. J. J. Welters, V. H. J. de Beer and R. A. van Santen, *Appl. Catal. A: Gen.*, **119**, 253 (1994).
147. J. A. Anderson, B. Pawelec, J. L. G. Fierro, P. L. Arias, F. Duque and J. F. Cambra, *Appl. Catal. A: Gen.*, **99**, 55 (1993).
148. W. J. J. Welters, G. Vorbeck, H. W. Zandbergen, L. J. M. van de Ven, E. M. van Oers, J. W. de Haan, V. H. J. de Beer and R. A. van Santen, *J. Catal.*, **161** (2), 819 (1996).
149. G. Vorbeck, W. J. J. Welters, L. J. M. van de Ven, H. W. Zandbergen, J. W. de Haan, V. H. J. de Beer and R. A. van Santen, *Zeolites and Related Microporous Materials: State of the Art 1994*, (J. Weitkamp, H. G. Karge, H. Pfeifer and W. Holderich eds.), *Stud. Surf. Sci. Catal.*, **84**, 1617 (1994).
150. A. L. Agudo, A. Benitez, J. L. G. Fierro, J. M. Palacios, J. Neira and R. Cid, *J. Chem. Soc., Faraday Trans.*, **88**, (3), 385 (1992).
151. D. Das, M. Duttagupta and S. K. Palit, *Bull. Chem. Soc. Jpn.*, **67**, 2906 (1994).
152. R. Cid, J. Neira, J. Godoy, J. M. Palacios, S. Mendioroz and A. Lopez-Agudo, *J. Catal.*, **141**, 206 (1993).
153. R. Cid, J. Neira, J. Godoy, J. M. Palacios and A. Lopez-Agudo, *Appl. Catal. A: Gen.*, **125**, 169 (1995).
154. R. S. Mann, I. S. Sambi and K. C. Khulbr, *Ind. Eng. Chem. Res.*, **27**, 1788 (1988).
155. M. L. Occelli and T. P. Debies, *J. Catal.*, **97**, 357 (1986).
156. T. G. Harvey and T. W. Matheson, *J. Chem. Soc., Chem. Commun.*, 188 (1985).
157. T. G. Harvey and T. W. Matheson, *J. Catal.*, **101**, 253 (1986).
158. T. W. Matheson, K. C. Pratt and T. G. Harvey, Int. patent, BO1J29/12, CO1G 65/14, 27 March, 1986.
159. J. L. Lemberton, N. Gnofam and G. Perot, *Appl. Catal. A: Gen.*, **90**, 175 (1992).
160. A. Iannibello, S. Marengo and A.Girelli, *Appl. Catal.*, **3**, 261 (1982).
161. Y. Sakata and C. E. Hamrin Jr., *Ind. Eng. Chem., Prod. Res. Dev.*, **22**, 250 (1983).
162. G. W. Schultz, H. Shimada, Y. Yoshimura, T. Sato and A. Nishijima, *Bull. Chem. Soc. Jpn.*, **58**, 1077 (1985).
163. M. L. Occelli and R. J. Rennard, *Catal. Today*, **2**, 309 (1988).
164. C. I. Warburton, *Catal. Today*, **2**, 271 (1988).
165. J. T. Kloprogge, W. J. J. Welters, E. Booy, V. H. J. de Beer, R. V. van Santen, J. W. Geus and J. B. H. Jansen, *Appl. Catal.*, A., **97**, 77 (1993).
166. M. Sychev, V. H. J. de Beer, A. Kodentsov, E. M. van Oers and R. A. van Santen, *J. Catal.*, **168** (2), 245 (1997).
167. E. Hayashi, E. Iwamatsu, M. E. Biswas, S. A. Ali, Y. Yamamoto, Y. Sanada, A. K. K. Lee, H. Hamid and T. Yoneda, *Chem. Lett.*, 433 (1997).
168. S. Eijsbouts, J. N. M. van Gestel, J. A. R. van Veen, V. H. J. de Beer and R. Prins, *J. Catal.*, **131**, 412 (1991) and literature cited therein.
169. A. Morales,R. E. Galiasso, M. M. Ramírez de Agudelo, J. A. Salazar and A. R. Carrasquel, US Pat. 4,520,128 (1985).
170. L. A. Pine, US Pat. 4,003,828 (1975).
171. G. A. Mickelson, US Pat. 3,749,663 (1973).
172. G. A. Mickelson, US Pat. 3,749,664 (1973).
173. G. A. Mickelson, US Pat. 3,755,148 (1973).
174. G. A. Mickelson, US Pat. 3,755,150 (1973).
175. G. A. Mickelson, US Pat. 3,755,196 (1973).
176. B. A. Kerns and O. A. Larson, US Pat. 3,446,730 (1966).
177. J. N. Haresnape and J. E. Morris, Brit. Pat. 701,217 (1953).
178. D. Chadwick, D. W. Aitchison, R. Badilla-Ohlbaum and L. Josefsson, *Stud. Surf. Sci. Catal.*, **16**, 323 (1982).
179. R. E. Tischer, N. K. Narain, G. J. Stiegel and D. L. Cillo, *Ind. Eng. Chem. Res.*, **26**, 422 (1987).
180. Y. Okamoto, I. Gomi, Y. Mori, T. Imanaka and S. Teranishi, *React. Kinet. Catal. Lett.*, **22**, 417 (1983).
181. P. Antanasova, T. Halachev, J. Uchytil and M. Kraus, *Appl. Catal.*, **38**, 235 (1986).
182. J. M. Jones, R. A. Kydd, P. M. Boorman and P. H. van Rhyn, *Fuel*, **74** (14), 1875 (1995).
183. P. Atanasova and T. Halachev, *Appl. Catal.*, **48**, 295 (1989).
184. A. Morales and M. M. Ramírez de Agudelo, *Appl. Catal.*, **23**, 23 (1986).
185. J. L. G. Fierro, A. Lopez Agudo, N. Esquivel and R. Lopez Cordero, *Appl. Catal.*, **48**, 353 (1989).
186. P. Atanasova, T. Halachev, J. Vchytil and M. Kraus, *Appl. Catal.*, **38**, 235 (1988).
187. J. M. Lewis and R. A. Kydd, *J. Catal.*, **136**, 478 (1992).
188. P. Atanasova, J. Vchytil, M. Kraus and T. Halachev, *Appl. Catal.*, **65**, 53 (1990).
189. S. I. Kim and S. I. Woo, *J. Catal.*, **133**, 124 (1992).
190. J. M. Lewis, R. A. Kydd, P. M. Boorman and P. H. van Rhyn, *Appl. Catal.*, **84**,103 (1992).
191. J. A. R. van Veen, H. A. Colijn, P. A. J. M. Hendriks and A. J. van Welsenes, *Fuel Proc. Techn.*, **35**, 137 (1993).

192. M. Jian and R. Prins, *Catal. Today*, **30** (1-3), 127 (1996).
193. P. G. Vazquez, M. G. Gonzalez, M. N. Blanco and C. V. Caceres, *Stud. Surf. Sci. Catal.*, **91**, 1121 (1995).
194. M. M. Ramírez de Agudelo, A. Morales, in: *Proc. 9th Int. Congr. Catal., Calgary, 1988*, (M. J. Phillips and M. Ternan, eds.), **Vol.1**, p.42, Chem. Inst. Can., Ottawa 1988.
195. E. C. Housm and R. Lester, Brit. Pat. 807,583 (1959).
196. R. Lopez Cordero, N. Esquivel, J. Lazaro, J. L. G. Fierro and A. Lopez Agudo, *Appl. Catal.*, **48**, 341 (1989).
197. O. Poulet, R. Hubaut, S. Kasztelan and J. Grimblot, *Bull. Soc. Chim. Belg.*, **100**, 857 (1991).
198. R. Hubaut, O. Poulet, S. Kasztelan, E. Payen and J. Grimblot, *Prepr.-Am. Chem. Soc., Div. Pet. Chem.*, **39** (4), 548 (1994).
199. P. D. Hopkins and B. L. Meyers, *Ind. Eng. Chem. Prod. Res. Dev.*, **22**, 421 (1983).
200. K. Gishti, A. Iannibello, S. Marengo, G. Morelli and P. Titarelli, *Appl. Catal.*, **12**, 381 (1984).
201. W. C. Cheng and N. P. Luthra, *J. Catal.*, **109**, 163 (1988).
202. R. Lopez Cordero, S. Lopez Guerra, J. L. G. Fierro and A. Lopez Agudo, *J. Catal.*, **126**, 8 (1990).
203. P. J. Mangnus, J. A. R. van Veen, S. Eijsbouts, V. H. J. de Beer and J. A. Moulijn, *Appl. Catal.*, **61**, 99 (1990).
204. A. Morales, M. M. Ramírez de Agudelo and F. Hernandez, *Appl. Catal.*, **41**, 261 (1988).
205. G. Haller, B. McMillan and J. Brinen, *J. Catal.*, **97**, 243 (1986).
206. R. Lopez Cordero, F. J. Gil Llambias, J. M. Palacios, J. L. G. Fierro and A. Lopez Agudo, *Appl. Catal.*, **56**, 197 (1989).
207. J. A. R. van Veen, P. A. J. M. Hendriks, R. R. Andrea, E. J. G. M. Romers and A. E. Wilson, *J. Phys. Chem.*, **94**, 5275 (1990).
208. J. A. R. van Veen, P. A. J. M. Hendriks, R. R. Andrea, E. J. G. M. Romers and A. E. Wilson, *J. Phys. Chem.*, **94**, 5282 (1990).
209. H. Topsøe, B. S. Clausen, N. Topsøe and P. Zeuthen, *Stud. Surf. Sci. Catal.*, **53**, 77 (1990).
210. S. Eijsbouts, L. van Gruijthuijsen, J. Volmer, V. H. J. de Beer and R. Prins, *Stud. Surf. Sci. Catal.*, **50**, 79 (1989).
211. R. C. Ryan, R. A. Kemp. J. A. Smegal, D. R. Denley and G. E. Spinnler, *Stud. Surf. Sci Catal.*, **50**, 21 (1989).
212. R. A. Kemp, R. C. Ryan and J. A. Smegal, in: *Proc. 9th Int. Congr. Catal., Calgary, 1988* (M. J. Phillips and M. Ternan eds.), **Vol.1**, p.128, *Chem. Inst. Can.*, Ottawa (1988).
213. C. W. Fitz and H. F. Rase, *Ind. Eng. Chem. Prod. Res. Dev.*, **22**, 40 (1983).
214. T. H. Chao, US Pat., 4, 629, 717 (1986).
215. J. Cruz Reyes, M. Avalos-Borja, R. López Cordero and A. Lopez Agudo, *Appl. Catal. A: Gen.*, **120**, 147 (1994).
216. A. Lopez Agudo, R. Lopez Cordero, J. M. Palacios and J. L. Fierro, *Bull. Soc. Chim. Belg.*, **104** (4-5), 237 (1995).
217. W. R. A. M. Robinson, J. N. M. van Gestel, T. I. Korányi, S. Eijsbouts, A. M. van der Kraan, J. A. R. van Veen and V. H. J. de Beer, *J. Catal.*, **161** (2), 539 (1996).
218. P. Atanasova, T. Tabakova, Ch. Vladov, T. Halachev and A. Lopez Agudo, *Appl. Catal.*, **161** (1-2), 105 (1997).
219. P. Atanasova and T. Halachev, *Appl. Catal. A*, **108**, 123 (1994).
220. S. Eijsbouts, J. N. M. van Gestel, E. M. van Oers, R. Prins, J. A. R. van Veen and V. H. J. de Beer, *Appl. Catal. A: Gen.*, **119**, 293 (1994).
221. L. Jhansi Lakshmi, P. Kanta Rao, V, M. Mastikhin and A. V. Nosov, *J. Phys. Chem.*, **97**, 11373 (1993).
222. S. M. A. M. Bouwens, A. M. van der Kraan, V. H. J. de Beer and R. Prins, *J. Catal.*, **128**, 559 (1991).
223. D. Gulková and Z. Vít, *Appl. Catal. A: Gen.*, **125**, 61 (1995).
224. K. Tanabe, *Solid Acids and Bases*, Academic Press: New York, 1970.
225. C. L. Thomas, *Ind. Eng. Chem.*, **41**, 2564 (1949).
226. K. Ikebe, *J. Chem. Soc. Jpn., Ind. Chem. Sec.* (*Kogyo Kagaku Zasshi*), **61,** 575 (1958) [in Japanese].
227. K. Ikebe, N. Hara, K. Mita and K. Shimizu, *J. Fuel Soc. Jpn.*, (*Nenryo Kyokaishi*), **37**, 257 (1958) [in Japanese].
228. T. Curtin, J. B. McMonagle and B. K. Hodnett, *Appl. Catal.*, A, **93**, 91 (1992).
229. S. Engels, E. Herold, H. lausch, H. Mayr, H.-W. Meiners and M. Wilde, *Stud. Surf. Sci. Catal.*, **75**, 2581 (1993).
230. T. Okuhara, H. Tamura and M. Misono, *J. Catal.*, **95**, 41 (1985).
231. V. Vorover, A. Agzamkhodzhaeva, V. Mikita and M. Abidova, *Kinet. Katal.*, **25**, 154 (1984).
232. Y. W. Chen and C. Li, *Catal. Lett.*, **13**, 359 (1992).
233. K. Tanabe, M. Misono, Y. Ono and H. Hattori, *New Solid Acids and Bases*, Kodansha, Tokyo (1989).
234. K. Ikebe, *J. Chem. Soc. Jpn., Ind. Chem. Sec.* (*Kogyo Kagaku Zasshi*), **61**, 437 (1958) [in Japanese].
235. Y. Izumi and T. Shiba, *J. Chem. Soc. Jpn.*, **37**, 1797 (1964).
236. L. A. Pine, US Pat. 3,993,557, 1976.
237. S. Sato, S. Hasebe, H. Sakurai, K. Urabe and Y. Izumi, *Appl. Catal.*, **29**, 107 (1987).
238. W. J. Wang and Y. W. Chen, *Catal. Lett.*, **10**, 297 (1991).
239. K. P. Peil, L. G. Galya and G. Marcelin, *J. Catal.*, **115**, 441 (1989).
240. H. Lafitau, E. Neel and J. C. Clement, *Stud. Surf. Sci. Catal.*, **1**, 393 (1976).
241. D. Li, T. Sato, M. Imamura, H. Shimada and A. Nishijima, *J. Catal.*, **170** (2), 357 (1997).

242. E. C. DeCanio and J. G. Weissman, *Colloid Surfaces, A: Physicochem. Eng. Aspects*, **105**, 123 (1995).
243. K. Seimiya, M. Hashimoto, S. Suzuki, M. Kameyama, Y. Noguchi and K. Nita, *J. Jpn. Petrol. Inst.*, **33** (1), 52 (1990) [in Japanese].
244. M.-C. Tsai, Y.-W. Chen, B. C. Kang, J.-C. Wu and L.-J. Leu, *Ind. Eng. Chem. Res.*, **30**, 1801 (1991).
245. C. Li, Y. W. Chen, S. J. Yang and J. C. Wu, *Ind. Eng. Chem. Res.*, **32**, 1573 (1993).
246. J. Ramírez, P. Castillo, L. Cedeno, R. Cuevas, M. Castillo, J. M. Palacios and A. Lopez Agudo, *Appl. Catal. A: Gen.*, **132**, 317 (1995).
247. J. Trawczynski and J. Walendziewski, *Appl. Catal. A: Gen.*, **119**, 59 (1994).
248. M. Houalla and B. Delmon, *Appl. Catal.*, **1**, 285 (1981).
249. M. A. Stranick, M. Houalla and D. M. Hercules, *J. Catal.*, **104**, 396 (1987).
250. P. M. Boorman, J. F. Kriz, J. R. Brown and M. Ternan, *Proceedings of the Fourth International Conference on the Chemistry and Uses of Molybdenum*; p.192, Climax Molybdenum Co.: Ann Arbor, MI; (1982).
251. P. M. Boorman, J. F. Kriz, J. R. Brown and M. Ternan, *Proceedings of the 8th International Congress on Catalysis*, **Vol.2**, p.281, Berlin (1984).
252. P. M. Boorman, R. A. Kydd, T. S. Sorensen, K. Chong, J. M. Lewis and W. S. Bell, *Fuel*, **71**, 87 (1992).
253. P. M. Boorman, R. A. Kydd, Z. Sarbak and A. Somogybari, *J. Catal.*, **96**, 115 (1985).
254. P. M. Boorman, R. A. Kydd, Z. Sarbak and A. Somogybari, *J. Catal.*, **100**, 287 (1986).
255. P. M. Boorman, R. A. Kydd, Z. Sarbak and A. Somogybari, *J. Catal.*, **106**, 544 (1987).
256. G. L. Tejuca, C. H. Rochester, A. Lopez Agudo and J. L. G. Fierro, *J. Chem. Soc., Faraday Trans. 1*, **79**, 2543 (1983).
257. J. L. G. Fierro, A. López Agudo, G. L. Tejuca and C. H. Rochester, *J. Chem. Soc., Faraday Trans. 1*, **81**, 1203 (1985).
258. K. Jiratova and M. Kraus, *Appl. Catal.*, **27**, 21 (1986).
259. Ch. Papadopoulou, A. Lycourghiotis, P. Grange and B. Delmon, *Appl. Catal.*, **38**, 255 (1988).
260. H. K. Matralis, A. Lycourghiotis, P. Grange and B. Delmon, *Appl. Catal.*, **38**, 273 (1988).
261. H. Matralis, Ch. Papadopoulou and A. Lycourghiotis, *Appl. Catal. A: Gen.*, **116**, 221 (1994).
262. J. M. Lewis, R. A. Kydd and P. M. Boorman, *J. Catal.*, **120**, 413 (1989).
263. Z. Sarbak, *Appl. Catal., A*, **159** (1-2), 147 (1997).
264. Z. Sarbak, *Appl. Catal.*, **164** (1-2), 13 (1997).
265. Ch. Papadopoulou, H. Matralis, A. Lycourghiotis, P. Grange and B. Delmon, *J. Chem. Soc., Faraday Trans.*, **89**, 3157 (1993).
266. Ch. Kordulis, A. Gouromihou, A. Lycourghiotis, Ch. Papadopoulou and H. Matralis, *Appl. Catal.*, **67**, 39 (1990).
267. T. R. Hughes, H. M. White and R. J. White, *J. Catal.*, **13**, 58 (1969).
268. P. D. Scokart, S. A. Selim, J. P. Damon and P. G. Rouxhet, *J. Colloid Interface Sci.*, **70**, 209 (1979).
269. J. Ramírez, R. Cuevas, A. Lopez Agudo, S. Mendioroz and J. L. G. Fierro, *Appl.Catal.*, **57**, 223 (1990).
270. J. L. G. Fierro, R. Cuevas, J. Ramírez and A. Lopez Agudo, *Bull. Soc. Chim. Belg.*, **100**, (11-12), 945 (1991).
271. C. L. Kibby and H. E. Swift, *J. Catal.*, **45**, 231 (1976).
272. M. Lo Jacono and M. Schiavello, in; *Preparation of Catalysts*, (B. Delmon, P. A. Jacobs and G. Poncelet eds.), p.473, Elesevier, Amsterdam (1976).
273. B. R. Strohmeier and D. M. Hercules, *J. Catal.*, **86**, 266 (1984).
274. H. J. Thomas, M. N. Blanco, C. V. Cáceres, N. Firpo, F. J. Gil, Llambias, J. L. G. Fierro and A. Lopez Agudo, *J. Chem. Soc., Faraday Trans.*, **86**, 2765 (1990).
275. J. L. G. Fierro, A. Lopez Agudo, P. Grange and B. Delmon, in *Proc. 8th Int. Congr. Catal.*, *1984*, **Vol.2**, p.363, Verlag Chemie, Weinheim (1988).
276. H. J. Thomas, C. V. Caceres, M. N. Blanco, J. L. G. Fierro and A. Lopez Agudo, *J. Chem. Soc., Faraday Trans.*, **90** (14), 2125 (1994).
277. J. F. Cambra, P. L. Arias, M. B. Guemez, J. A. Legarreta and J. L. G. Fierro, *Ind. Eng. Chem. Res.*, **30**, 2365 (1991).
278. M. B. Guemez, J. F. Cambra, P. L. Arias, J. A. Legarreta and J. L. G. Fierro, *Fuel*, **74**, (2), 285 (1995).
279. C. R. Lahiri and P. N. Nandi, *Indian J. Technol.*, **24**, 252 (1986).
280. D. K. Lee, I. C. Lee, S. K. Park, S. Y. Bae and S. I. Woo, *J. Catal.*, **159** (1), 212 (1996).
281. J. G. Weissman, E. C. DeCanio and J. C. Edwards, *Catal. Lett.*, **24**, 113 (1994).
282. L. E. Toth, *Transition Metal Carbides and Nitrides*; Academic Press, New York (1971).
283. E. K. Storms, T*he Refractory Carbides*, Academic Press, New York, 1967.
284. S. T. Oyama, J. C. Schlatter, J. E. Metcalfe III and J. M. Lambert Jr., *Ind. Eng. Chem. Res.*, **27**, 1639 (1988).
285. R. S. Wise and E. J. Markel, *J. Catal.*, **145**, 344 (1994).
286. J. G. Choi, R. L. Curl and L. T. Thompson, *J. Catal.*, **146**, 218 (1994).
287. M. J. Ledoux, J. Guille, S. Hanzter, S. Marin and C. Pham-Huu, *Extended Abstracts, Proceedings MRS Symposium S*; p.135, Boston, Nov 26-Dec 1, 1990; MRS: (1990).
288. J. S. Lee, S. T. Oyama and M. Boudart, *J. Catal.*, **106**, 125 (1987).

289. L. Volpe and M. Boudart, *J. Solid State Chem.*, **59**, 332 (1985).
290. L. Volpe, S. T. Oyama and M. Boudart, *Preparation of Catalysts III*; p.147, Elsevier, New York (1983).
291. M. Boudart, S. T. Oyama and L. LecLercq, *Proc. 7th Int. Cong. Catal.*; Tokyo, 1980 (T. Seiyama and K. Tanabe eds.), **Vol.1**, p.578, Elsevier, Amsterdam (1981).
292. L. Volpe and M. Boudart, *J. Phys. Chem.*, **90**, 4874 (1986).
293. G. S. Ranhotra, A. T. Bell and J. A. Reimer, *J. Catal.*, **108**, 40 (1987).
294. M. Saito and R. B. Anderson, *J. Catal.*, **63**, 438 (1980).
295. I. Kojima, E. Miyazaki, Y. Inoue and I. Yasumori, *J. Catal.*, **73**, 128 (1982).
296. J. S. Lee, M. H. Yeom, K. Y. Park, I.-S. Nam, J. S. Chung, T. G. Kim and S. H. Moon, *J. Catal.*, **128**, 126 (1991).
297. G. S. Ranhotra, G. W. Haddix, A. T. Bell and J. A. Reimer, *J. Catal.*, **108**, 24 (1987).
298. G. W. Haddix, A. T. Bell and J. A. Reimer, *J. Phys. Chem.*, **93**, 5859 (1987).
299. J. S. Lee, S. Locatelli, S. T. Oyama and M. Boudart, *J. Catal.*, **125**, 157 (1990).
300. F. H. Ribeiro, R. A. Dalla Betta, M. Boudart, J. Baumgartner and E. Iglesia, *J. Catal.*, **130**, 86 (1991).
301. S. T. Oyama and G. L. Haller, *Catalysis* (London), **5**, 333 (1981).
302. J. G. Choi, J. R. Brenner, C. W. Colling, B. G. Demczyk, J. L. Dunning and L. T. Thompson, *Catal. Today*, **15**, 201 (1992).
303. C. W. Colling and L. T. Thompson, *J. Catal.*, **146**, 193 (1994).
304. J. C. Schlatter, S. T. Oyama, J. E. Metcalfe III and J. M. Lambert Jr., *Ind. Eng. Chem. Res.*, **27**, 1648 (1988).
305. M. Nagai and T. Miyao, *Catal. Lett.*, **15**, 105 (1992).
306. K. S. Lee, H. Abe, J. A. Reimer and A. T. Bell, *J. Catal.*, **139**, 34 (1993).
307. H. Abe and A. T. Bell, *Catal. Lett.*, **18**, 1 (1993).
308. H. Abe, T. Cheung and A. T. Bell, *Catal. Lett.*, **21**, 11 (1993).
309. D. Sajkowski and S. T. Oyama, *Prepr. Pap.-Am. Chem. Soc., Div. Pet. Chem.*, **35**,156 (1990).
310. P. A. Armstrong, A. T. Bell and J. A. Reimer, *J. Phys. Chem.*, **97**, 1952 (1993).
311. S. Ramanathan and S. T. Oyama, *J. Phys. Chem.*, **99**, 16365 (1995).
312. C. W. Colling, J.-G. Choi and L. T. Thompson, *J. Catal.*, **160** (1), 35 (1996).
313. J. G. Choi, J. R. Brenner and L. T. Thompson, *J. Catal.*, **154**, 33 (1995).
314. C. C. Yu, S. Ramanathan, F. Sherif and S. T. Oyama, *J. Phys. Chem.*, **98**, 13038 (1994).
315. S. Ramanathan, C. C. Yu and S. T. Oyama, *J. Catal.*, **173** (1), 10 (1998).
316. E. J. Markel and J. W. Van Zee, *J. Catal.*, **126**, 643 (1990).
317. M. Nagai, T. Miyao and T. Tuboi, *Catal. Lett.*, **18**, 9 (1993).
318. M. Nagai and S. Omi, *J. Jpn. Petrol. Inst.*, **38** (6), 363 (1995).
319. S.-J. Liaw, A. Raje, X. X. Bi, P. C. Eklund, U. M. Graham and B. H. Davis, *Energy & Fuels*, **9**, 921 (1995).
320. J. S. Haggerty, *Laser Induced Chemical Process*, (J. I. Steinfeld ed.), Plenum Press, New York (1981).
321. K. F. McCarty and G. L. Schrader, in *Proc. 8th Int. Congr. Catal.*, (E. Ertl ed.), Dechema, Berlin (1984).
322. K. F. McCarty and G. L. Schrader, *Ind. Eng. Chem. Prod. Res. Dev.*, **23**, 519 (1984).
323. K. F. McCarty, J. W. Anderegg and G. L. Schrader, *J. Catal.*, **93**, 375 (1985).
324. M. E. Ekman, J. W. Anderegg and G. L. Schrader, *J. Catal.*, **117**, 246 (1989).
325. S. J. Hilsenbeck, R. E. McCarley, R. K. Thompson, L. C. Flanagan and G. L. Schrader, *J. Mol. Catal. A: Chem.*, **122** (1), 13 (1997).
326. S. J. Hilsenback, R. E. McCarley, A. I. Goldman and G. L. Schrader, *Chem. Mater.*, **10** (1), 125 (1998).
327. I. M. Schewe-Miller, K. F. Koo, M. Columbia, F. Li and G. L. Schrader, *Chem. Mater.*, **6**, 2327 (1994).
328. A. Spojakina and S. Damyanova, *React. Kinet. Catal. Lett.*, **53**, (2), 405 (1994).
329. A. Griboval, P. Blanchard, E. Payen, M. Fournier and J.-L. Dubois, *Chem. Lett.*, 1259 (1997).
330. S. Li and J. S. Lee, *J. Catal.*, **173** (1), 134 (1998).
331. T. C. Ho, A. J. Jacobson, R. R. Chianelli and C. R. F. Lund, *J. Catal.*, **138**, 351 (1992).
332. T. C. Ho, *Ind. Eng. Chem. Res.*, **32**, 1568 (1993).
333. T. C. Ho, R. R. Chianelli and A. J. Jacobson, *Appl. Catal. A: Gen.*, **114**, 127 (1994).
334. M. Karroua, J. Ladriere, H. Matralis, P. Grange and B. Delmon, *J. Catal.*, **138**, 640 (1992).
335. W. Li, B. Dhandapani and S. T. Oyama, *Chem. Lett.*, 207 (1998).
336. T. Shimizu, S. Kasahara, T. Kiyohara, K. Kawahara and M. Yamada, *J. Jpn. Petrol. Inst.*, **38** (6), 384 (1995) [in Japanese].
337. P. Blanchard, E. Payen, J. Grimblot, O. Poulet and R. Loutaty, *Stud. Surf. Sci. Catal.*, **106**, 211 (1997).
338. Y. Yoshimura, N. Matsubayashi, T. Sato, H. Shimada and A. Nishijima, *Appl. Catal. A: Gen.*, **79**, 145 (1991).
339. E. Hillerová, H. Morishige, K. Inamura and M. Zdrazil, *Appl. Catal. A*, **156** (1), 1 (1997).
340. M. Zdrazil, *Appl. Catal. A: Gen.*, **115**, 285 (1994).
341. L. Lebihan, C. Mauchausse, L. Duhamel, J. Grimblot and E. Payen, *J. Sol-Gel Sci. Technol.*, **2**, 837 (1994).
342. Y. Xie and Y. Tang, *Adv. Catal.*, 37, 1 (1990) and references cited therein.
343. Y. Xie, L. Gui, Y. Liu, B. Zhao, N. Yang, Y. Zhang, Q. Guo, L. Duan, H. Huang, X. Gai and Y. Tang, *Proc. 8th Int. Congr. Catal.*, **vol. 5**, p.147, Berlin, Verlag Chemie, Weinheim (1984).

344. J. Leryer, M. I. Zaki and H. Knozinger, *J. Phys. Chem.*, **90**, 4775 (1986).
345. J. Leryer, R. Margrat, E. Taglauer and H. Knozinger, *Surf. Sci.*, **201**, 603 (1988).
346. T. I. Korányi, Z. Paal, J. Leryer and H. Knozinger, *Appl. Catal.*, **64**, L5 (1990).
347. J. Leryer, D. Mey and H. Knozinger, *J. Catal.*, **124**, 349 (1990).
348. B. M. Reddy and B. Manohar, *J. Chem. Soc., Chem. Commun.*, 1435 (1991).
349. C. Moreau, L. Bekakra, R. Durand and P. Geneste, *Catal. Today*, **10**, 681 (1991).
350. C.-S. Kim, F. E. Massoth, C. Geantet and M. Breysse, *New Frontiers in Catalysis*, (L. Guczi ed.), p.1935, Elsevier Science Publishers (1993).
351. K. Somasekhara Rao, V. V. D. N. Prasad, K. V. R. Chary and P. Kanta, *Preparation of Catalysts V* (G. Poncelet, P. A. Jacobbs, P. Grange and B. Delmon eds.), p.611, Elsevier, Amsterdam (1991).
352. E. I. Stiefel, W.-H. Pan, R. R. Chianelli and T. C. Ho, U.S. Patent 4,581,125, 1986.
353. E. I. Stiefel, T. R. Halbert, C. L. Coyle, L. Wei, W.-H. Pan, T. C. Ho, R. R. Chianelli and M. Daage, *Polyhedron*, **8**, 1625 (1989).
354. W. Eltzner, M. Breysse, M. Lacroix and M. Vrinat, *Polyhedron*, **5**, 203 (1986).
355. A. Muller, E. Diemann, A. Branding, F. W. Baumann, M. Breysse and M. Vrinat, *Appl. Catal.*, **62**, L13 (1990).
356. Y. I. Yermakov, *Catal. Rev.-Sci. Eng.*, **13**, 77 (1976).
357. D. C. Bailey and S. H. Langer, *Chem. Rev.*, **81**, 109 (1981).
358. J. Phillips and J. A. Dumesic, *Appl. Catal.*, **9**, 1 (1984).
359. R. F. Howe, in: *Tailored Metal Catalysts* (Y. Iwasawa ed.), p.141, Reidel, Dordrecht (1986).
360. M. Ichikawa, in: *Tailored Metal Catalysts*, (Y. Iwasawa ed.), p.183, Reidel, Dordrecht (1986).
361. A. Ishihara, T. Mitsudo and Y. Watanabe, *J. Jpn. Petrol Inst.*, **33**, 28 (1990) and literature cited therein.
362. A. Ishihara, T. Mitsudo, N. Morita and Y. Watanabe, *J. Jpn. Petrol. Inst.*, **33**, 327 (1990).
363. T. Mitsudo, A. Ishihara and Y. Watanabe, *Ind. Eng. Chem. Res.*, **29**, 163 (1990).
364. R. L. Banks and G. C. Bailey, *Ind. Eng. Chem. Prod. Res. Dev.*, **3**, 170 (1964).
365. R. L. Banks, *Chemtech*, 112 (1986).
366. A. Brenner and R. L. Burwell Jr., *J. Am. Chem. Soc.*, **97**, 2565 (1975).
367. J. Goldwasser, S. M. Fang, M. Houalla and W. K. Hall, *J. Catal.*, **115**, 34 (1989).
368. M. M. Luchsinger, B. Bozkurt, A. Akgerman, C. P. Janzen, W. P. Addiego and M. Y. Darensbourg, *Appl. Catal.*, **68**, 229 (1991).
369. Y. Okamoto, M. Odawara, H. Onimatsu and T. Imanaka, *Ind. Eng. Chem. Res.*, **34**, 3703 (1995).
370. M. Laniecki and W. Zmierczak, *Zeolite*, **11**, 18 (1991).
371. M. Sugioka, Y. Takase and K. Takahashi, *Proc. of JECAT'91*, 224 (1991).
372. M. D. Curtis, J. E. Penner-Hahn, J. Schwank, O. Baralt, D. J. McCabe, L. Thompson and G. Waldo, *Polyhedron*, **7**, 2411 (1988).
373. J. R. Brenner, B. T. Carvill and L. T. Thompson Jr., *Appl. Organomet. Chem.*, **6**, 463 (1992).
374. B. T. Carvill and L. T. Thompson, *Appl. Catal.*, **75**, 249 (1991).
375. Y. Okamoto and H. Katsuyama, *AIChE J.*, **43**, (11A), 2809 (1997).
376. F. Mauge, A. Vallet, J. Bachelier, J. C. Duchet and J. C. Lavalley, *J. Catal.*, **162**, 88 (1996).
377. T. R. Halbert, T. C. Ho, E. I. Stiefel, R. R. Chianelli and M. Daage, *J. Catal.*, **130**, 116 (1991).
378. B. Beck and S. Tadros, *Z. Anorg. Allg. Chem.*, **375**, 231 (1970).
379. D. Seyferth, G. B. Womack, and J. C. Dewan, *Organometallics*, **4**, 398, (1985).
380. A. Ishihara, M. Azuma, M. Matsushita and T. Kabe, *J. Jpn. Petrol. Inst.*, **36** (5), 360 (1993).
381. A. Ishihara, N. Nomura, M. Matsushita, K. Shirouchi and T. Kabe, *New Aspects of Spillover Effect in Catalysis*, (T. Inui *et al.* eds.) p.357, Elsevier Science B.V., 1993.
382. M. Herberhold and G. Suess, *J. Chem. Res. (S)*, 246 (1977).
383. J. K. Ruff and R. B. King, *Inorg. Chem.*, **8**, 180 (1969).
384. A. Maezawa, M. Kitamura, Y. Okamoto and T. Imanaka, *Bull. Chem. Soc. Jpn.*, **61**, 2295 (1988).
385. A. Lycourghiotis and D. Vattis, *React. Kinet. Catal. Lett.*, **21**, No. 1/2, 23 (1982).
386. A. Lycourghiotis, D. Vattis, G. Karaiskakis and N. Katsanos, *J. Less-Common Met.*, **86**, 137 (1982).
387. R. G. W. Gingerich and R. J. Angelici, *J. Am. Chem. Soc.*, **101**, 5604 (1979).
388. G. Doyle, *J. Organonomet. Chem.*, **84**, 323 (1975).
389. A. Lopez Agudo, F. J. Gil Llambias, J. M. D. Tascon and J. L. G. Fierro, *Bull. Soc. Chim. Belg.*, **93**, 719 (1984).
390. N. Topsøe and H. Topsøe, *J. Catal.*, **84**, 386 (1983).
391. B. S. Clausen, H. Topsøe, R. Candia, J. Villadsen, B. Lengeler, J. Als-Nielsen and F. Christensen, *J. Phys. Chem.*, **85**, 3868 (1981).
392. A. Ishihara, K. Shirouchi and T. Kabe, *Chem. Lett.* 589 (1993).
393. A. Ishihara, K. Shirouchi and T. Kabe, *J. Jpn. Petrol. Inst.*, **37** (4), 411 (1994).
394. T. Kabe, A. Ishihara and Q. Zhang, *Appl. Catal., A Gen* **97**, L1 (1993).
395. M. Houalla, N. K. Nag, A. V. Sapre, D. H. Broderick and B. C. Gates, *AICHE, J.*, **24** (6), 1015 (1978).
396. A. Ishihara, M. Matsushita, K. Shirouchi, Qing Zhang and T. Kabe, *J. Jpn. Petrol. Inst.*, **39** (1), 26 (1996).

397. Y. Okamoto, T., Imanaka and S. Teranishi, *J. Catal.*, **65**, 448 (1980).
398. T. A. Pecoraro and R. R. Chianelli, *J. Catal.*, **67**, 430 (1981).
399. J. Passaretti, R. R. Chianelli, A. Wold, K. Dwight and J. Covino, *J. Solid State Chem.*, **64**, 365 (1986).
400. A. Bellaloui, L. Mosoni, M. Roubin, M. Vrinat, M. Lacroix and M. Breysse, *C. R. Acad. Sci. Paris*, **304**, 1163 (1987).
401. J. A. De Los Reyes, S. Göbölös, M. Vrinat and M. Breysse, *Catal. Lett.*, **5**, 17 (1990).
402. M. Lacroix, N. Boutarfa, C. Guillard, M. Vrinat and M. Breysse, *J. Catal.*, **120**, 473 (1989).
403. M. Vrinat, M. Lacroix, M. Breysse, L. Mosoni and M. Roubin, *Catal. Lett.*, **3**, 405 (1989).
404. S. W. Oliver, T. D. Smith, J. Pilbrow, T. G. Harvey, T. W. Matheson and K. C. Pratt, *Inorg. Chim. Acta*, **117**, L9 (1986).
405. S. Eijsbouts, V. H. J. de Beer and R. Prins, *J. Catal.*, **109**, 217 (1988).
406. P. C. H. Mitchell, C. E. Scott, J. P. Bonnelle and J. G. Grimblot, *J. Catal.*, **107**, 482 (1987).
407. Y. Kuo, R. A. Cocco and B. J. Tatarchuk, *J. Catal.*, **112**, 250 (1988).
408. J. P. R. Vissers, C. K. Groot, E. M. van Oers, V. H. J. de Beer and R. Prins, *Bull. Soc. Chim. Belg.*, **93**, 813 (1984).
409. M. J. Ledoux, O. Michaux, G. Agostini and P. Panissod, *J. Catal.*, **102**, 275 (1986).
410. C. Geantet, J. A. De Los Reyes, M. Vrinat and M. Breysse, *Proc. JECAT'91*, 228 (1991).
411. J. A. De Los Reyes, M. Vrinat, C. Geantet and M. Breysse, *Catal. Today*, **10** (4), 645 (1991).
412. C. Geantet, S. Göbölös, J. A. De Los Reyes, M. Cattenot, M. Vrinat and M. Breysse, *Catal. Today*, **10** (4), 665 (1991).
413. K. V. R. Chary, S. Khajamasthan and V. Vijayakumar, *J. Chem. Soc., Chem. Commun.*, 1339 (1989).
414. S. Göbölös, M. Lacroix, T. Decamp, M. Vrinat and M. Breysse, *Bull. Soc. Chim. Belg.*, **100**, 907 (1991).
415. S-J. Liaw, R. Lin, A. Raje and B. H. Davis, *Appl. Catal., A*, **151** (2), 423 (1997).
416. S. Göbölös, M. Breysse, T. Cattenot, M. Decamp, M. Lacroix, J. L. Portefaix and M. Vrinat, in: *Advances in Hydrotreating Catalysts* (M. L. Occelli and R. G. Anthony eds.), p.243, Elsevier Science Publishers B. V., Amsterdam (1989).
417. J. A. De Los Reyes and M. Vrinat, *Appl. Catal. A: Gen.*, **103**, 79 (1993).
418. E. J. Markel and J. W. Van Zee, *J. Mol. Catal.*, **73**, 335 (1992).
419. A. Bellaloui, L. Mosoni, M. Roubin, M. Vrinat, M. Lacroix and M. Breysse, *C. R. Acad. Sci. Paris*, **304**, 1163 (1987).
420. S. Harris and R. R. Chianelli, *J. Catal.*, **86**, 400 (1984).
421. M. Sugioka, F. Sado, Y. Matsumoto and N. Maesaki, *Catal. Today*, **29**, 255 (1996).
422. M. Sugioka, C. Tochiyama, Y. Matsumoto and F. Sado, *Stud. Surf. Sci. Catal.*, **94**, 544 (1995).
423. M. Sugioka, L. Andalaluna, S. Morishita and T. Kurosaka, *Catal. Today*, **39** (1-2), 61 (1997).
424. R. Navarro, B. Pawelec, J. L. G. Fierro, P. T. Vasudevan, J. F. Cambra and P. L. Arias, *Appl. Catal., A*, **137** (2), 269 (1996).
425. M. J. Dees, A. J. den Hartog and V. Ponec, *Appl. Catal.*, **72**, 343 (1991).
426. Triyono and R. Kramer, *Appl. Catal. A: Gen.*, **100**, 145 (1993).
427. A. Vazquez, F. Pedraza and S. Fuentes, *J. Mol. Catal.*, **75**, 63 (1992).
428. R. Frety, P. N. Da Silva and M. Guenin, *Appl. Catal.*, **57**, 99 (1990).
429. M. J. Ledoux, C. Kippelen, G. Maire, G. Szabo, J. Goupyl and O. Krause, *Bull. Soc. Chim. Belg.*, **100**, 873 (1991).
430. P. C. H. Mitchell, C. E. Scott, J. P. Bonnelle and J. Grimblot, *J. Catal.*, **107**, 482 (1987).
431. T. Wada, K. Kaneda, S. Murata and M. Nomura, *Catal. Today*, **31** (1-2), 113 (1996).
432. R. Navarro, B. Pawelec, J. L. G. Fierro and P. T. Vasudevan, *Appl. Catal., A*, **148** (1), 23 (1996).
433. M. Dobrovolszky, K. Matusek, Z. Paal and P. Tetenyi, *J. Chem. Soc., Faraday Trans.*, **89**, (16), 3137 (1993).
434. A. Guerrero-Ruiz and I. Rodríguez-Ramos, *React. Kinet. Catal. Lett.*, **41**, (1), 167 (1990).
435. C.-A. Jan, T.-B. Lin and J.-R. Chang, *Ind. Eng. Chem. Res.*, **35** (11), 3893 (1996).
436. S. Giraldo de León, P. Grange and B. Delmon, *Catal. Lett.*, **47** (1), 51 (1997).
437. F.-S. Xiao, Q. Xin and X.-X. Guo, *React. Kinet. Catal. Lett.*, **46**, 351 (1992).
438. M. J. Ledoux and B. Djellouli, *J. Catal.*, **115**, 580 (1989).
439. R. R. Chianelli, T. A. Pecoraro, T. R. Halbert, W.-H. Pan and E. I. Steifel, *J. Catal.*, **86**, 226 (1984).
440. S. Harris and R. R. Chianelli, *J. Catal.*, **98**, 17 (1986).
441. A. Ishihara, M. Nomura and T. Kabe, *J. Catal.*, **150**, 212 (1994).
442. A. Ishihara, M. Nomura and T. Kabe, *J. Jpn. Petrol. Inst.*, **37** (3), 300 (1994).
443. S. Kasztelan, H. Toulhoat, J. Grimblot and J. P. Bonnelle, *Appl. Catal.*, **13**, 127 (1984).
444. J. Chen, L. M. Daniels and R. J. Angelici, *J. Am. Chem. Soc.*, **112**, 199, (1990).
445. W. D. Jones and L. Dong, *J. Am. Chem. Soc.*, **113**, 559, (1991).
446. R. Prins, V. H. J. De Beer and G. A. Somorjai, *Catal. Rev.-Sci. Eng.*, **31**(1&2), 1-41 (1989).
447. J. M. J. G. Lipsch and G. C. A. Schuit, *J. Catal.*, **15**, 179 (1969).
448. H. Kwart, G. C. A. Schuit and B. C. Gates, *J. Catal.*, **61**, 128 (1980).

449. N. C. Sauer, E. J. Markel, G. L. Schrader and R. J. Angelici, *J. Catal.,* **117**, 295 (1978).
450. V. H. J. De Beer and G. C. A. Schuit, in: *Preparation of Catalysts* (B. Delmon, P. A. Jacobs and G. Poncelet eds.), p.343. Elsevier, Amsterdam (1976).
451. B. C. Gates, J. R. Katzer and G. C. A. Schuit, in: *Chemistry of Catalytic Processes*, p.423. McGraw-Hill, New York (1979).
452. J. W. Hachgenei and R. J. Angelici, *Angew. Chem.*, **99**, 947 (1987).
453. G. H. Spies and R. J. Angelici, *Organometallics,* **6**, 1897 (1987).
454. R. T. Weberg, R. C. Haltiwanger, J. C. V. Laurie and M. Rakowski DuBois, *J. Am. Chem. Soc.*, **108**, 6242 (1986).
455. J. Chen, L. M. Daniels and R. J. Angelici, *J. Am. Chem. Soc.*, **113**, 2544 (1991).
456. A. Ishihara, M. Nomura and T. Kabe, *Chem. Lett.*, 2285 (1992).
457. A. Ishihara, M. Nomura, N. Takahama, K. Hamaguchi and T. Kabe, *J. Jpn. Petrol. Inst.*, **39** (3), 211 (1996).
458. B. F. G. Johnson, J. Lewis, P. R. Raithby and G. Suess, *J. Chem. Soc., Dalton Trans.*, 1356, (1979).
459. J. A. De Los Reyes, M. Vrinat, C. Geantet, M. Breysse and J. G. Grimblot, *J. Catal.,* **142, 455** (1993).
460. M. Breysse, C. Geantet, M. Lacroix, J. L. Portefaix and M. Vrinat, *ACS Prep. Div. Petrol. Chem.*, 587 (1994).
461. A. Ishihara, M. Nomura, K. Shirouchi and T. Kabe, *J. Jpn. Petrol. Inst.*, **39** (6), 403 (1996).
462. A. Ishihara, K. Hamaguchi, K. Shirouchi and T. Kabe, *J. Jpn. Petrol. Inst.*, submitted
463. H. R. Lukens, J. R. G. Meisenheimer and J. N. Wilson, *J. Phys. Chem.*, **66**, 469 (1962).
464. K.A. Pavlova, B. D. Panteleea, E. N. Deryagina and I. V. Kalechits, *Kinet. Katal.*, **6**, 3, 493 (1965).
465. I. V. Kalechits and E. N. Deryagina, *Kinet. Katal.*, **8**, 3, 604 (1969).
466. C. G. Gachet, E. Dhainaut, L. de Mourgues, J. P. Candy and P. Fouilloux, *Bull. Soc. Chim. Belg.*, **90** (12), 1279 (1981).
467. G. V. Isagulyants, A. A. Greish and V. M. Kogan, *Symposium of International Catalyst Annual Conference in Canada*, p. 35, 1988.
468. D.M. Tetenyi and P.P. Zolton, *Chem. Eng. Commun.*, **83**, 1 (1989).
469. A. J. Gellman, M. E. Bussell and G. A. Somorjai, *J. Catal.,* **107**, 103 (1987).
470. T. Kabe, W. Qian, S. Ogawa and A. Ishihara, *J. Catal.,* **143**, 239 (1993).
471. W. Qian, A. Ishihara, S. Ogawa and T. Kabe, *J. Phys. Chem.*, **98**, 3, 907 (1994).
472. T. Kabe, W. Qian and A. Ishihara, *J. Phys. Chem.*, **98**, 3, 912 (1994).
473. T. Kabe, W. Qian and A. Ishihara, *J. Catal.,* **149**, 171 (1994).
474. T. Kabe, W. Qian, W. Wang and A. Ishihara, *Catal. Today*, **29**, 197 (1996).
475. W. Qian, Q. Zhang, Y. Okoshi, A. Ishihara and T. Kabe, *J. Chem. Soc., Faraday Trtans.*, **93** (9), 1821 (1997).
476. W. Qian, A. Ishihara, G. Wang, T. Tsuzuki, M. Godo and T. Kabe, *J. Catal.,* 1997 in press.
477. T. Kabe, W. Qian, K. Tanihata, A. Ishihara and M. Godo, *J. Chem. Soc., Faraday Trans.*, 1997 in press.
478. A. Ishihara, M., Yamaguchi, H. Godo, W. Qian, M. Godo and T. Kabe, *Chem. Lett.*, 743 (1996).
479. A. Ishihara, M., Yamaguchi, H. Godo, W. Qian, M. Godo and T. Kabe, *J. Jpn. Petrol. Inst.*, submitted
480. H. Gilman and A. L. Jacoby, *J. Org. Chem.*, **4**, 108 (1939).
481. Y. Kobayashi and D. V. Maudsley, *Biological Applications of Liquid Scintillation Counting*, Academic Press, New York (1974).
482. D. L. Horrocks, *Applications of Liquid Scintillation Counting*, Academic Press, New York (1974).
483. *Liquid Scintillation Counting* (M. Crook and P. Johnson eds.) **Vol. 4**, Heyden and Son, London (1977).
484. R. R. Chianelli, *Catal. Rev.-Sci. Eng.*, **26**, 361 (1984).
485. N.-Y. Topsøe and H. Topsøe, *J. Catal.,* **139**, 631 (1993).
486. D. A. Lewis and C. N. Kenney, *Trans. Inst. Chem. Eng.*, **59**, 186 (1981).
487. M. Danot, J. Afonso, J. L. Portefaix, M. Breysse and T. des Courieres, *Catal. Today*, **10**, 629 (1991).
488. M. Breysse, J. Afonso, M. Lacroix, J. L. Portefaix and M. Vrinat, *Bull. Soc. Chim. Belg.*, **100**, 923 (1991).
489. M. Cattenot, J. L. Portefaix, J. Afonso, M. Breysse, M. Lacroix and G. Perot, *J. Catal.,* **173** (2), 366 (1998).
490. V. N. Rodin, A. N. Starsev, V. I. Zaikovskii, and Yu. I. Yermakov, *React. Kinet. Catal. Lett.*, **32** (2), 419 (1986).
491. E. Hillerová, J. Sedlacek and M. Zdrazil, *Collect. Czech. Chem. Commun.*, **52**, 1748 (1987).
492. M. Breysse, T. des Courieres, M. Danot, C. Geantet and J. L. Portefaix, U. S. Patent 5,157,009.
493. N. Allali, A.-M. Marie, M. Danot, C. Geantet and M. Breysse, *J. Catal.,* **156**, 279 (1995).
494. L. A. Rankel and L. D. Rollman, *Fuel*, **62**, 44 (1983).
495. M. J. Ledoux, O. Michaux and S. Hantzer, *J. Catal.,* **106**, 525 (1987).
496. C. Guillard, M. Lacroix, M. Vrinat, M. Breysse, B. Mocaer, J. Grimblot, T. des Couriers and D. Faure, *Catal. Today*, **7**, 587 (1990).
497. Y.-H. Moon and S.-K. Ihm, *Catal. Lett.*, **22**, 205 (1993) and literature cited therein.
498. Y.-H. Moon and S.-K. Ihm, *Catal. Lett.*, **42** (1,2), 73 (1996).
499. A. Kherbeche, A. Benharref and R. Hubaut, *React. Kinet. Catal. Lett.*, **57** (1), 13 (1996).
500. S. Nakagawa, T. Ono, S. Murata, M. Nomura and T. Sakai, *J. Jpn. Petrol. Inst.*, **41** (1), 45 (1998).
501. A. Andreev, Ch. Vladov, L. Prahov and P. Atanasova, *Appl. Catal. A: Gen.*, **108**, L97 (1994).

5

Process Engineering

5.1. Hydrotreatment of Petroleum

5.1.1 The Role of Hydroprocessing/Hydrotreating

Hydroprocessing includes both hydrotreating and hydrocracking. Hydrotreating mainly consists of hydrodesulfurization (HDS), hydrodenitrogenation (HDN) and hydrogenation, and essentially does not change molecular size distribution. Hydrocracking changes the molecular size distribution and makes smaller size molecules. Depending on the extent of cracking, mild hydrocracking (MHC) using less severe conditions may be distinguished from hydrocracking. This review focuses mainly on hydrotreating. Other reviews are available for hydrotreating[1-3] and hydrocracking.[4,5] All the reactions included in hydroprocessing are exothermic, as shown in Table 5.1,[6] so control of temperature in the reactor, especially catalyst bed, is very important in practical operation. Although equilibrium constants decrease at higher temperatures, the heteroatom removal reactions are favorable under practical operating conditions. Hydrogenation of aromatics, however, is limited by thermodynamics at high temperatures and lower hydrogen pressures.

Typical properties of petroleum crudes are shown in Table 5.2. Generally, petroleum crudes are complex mixtures of various organic compounds, and the major components are hydrocarbons. Large amounts of heteroatoms are also included and their concentration changes depending on the origin. While sulfur is generally the most abundant heteroatom, the concentration of sulfur is specifically higher in crudes from the Middle East where 50% of the petroleum deposits of the world exist. The compounds with these heteroatoms are distributed over the entire boiling range, and the concentration of heteroatoms increase with increasing boiling point. For example, the concentrations of sulfur in straight-run distillate fractions of Arabian Light and Arabian Heavy increase in the order kerosene < gas oil < atmospheric resid, as shown in Table 5.2. Although the nature of such components are not yet well known in detail, recent developments in analytical methods have improved the situation, as noted in Chapter 2. In lighter fractions, sulfur is present in the form of thiols, sulfides, disulfides and thiophenes while various alkylbenzothiophenes and alkyldibenzothiophenes are included in the heavier gas oil fraction. The components in atmospheric and vacuum resids, which include large amounts of not only sulfur and nitrogen but also nickel and vanadium, have not yet been well characterized.

Thus, petroleum crudes, which contain various impurities shown above, are separated along with the outline of process flow in refining shown in Fig. 5.1. They are refined by

Table 5.1 Hydrotreating and Hydrocracking of Organic Compounds

Classification and Equation	ΔH^0@700°K kcal/mol	$\log_{10}K$ @500°K	@700°K
a. Desulfurization			
$C_2H_5SH + H_2 \rightarrow C_2H_6 + H_2S$	−16.77	+7.06	+5.01
$C_2H_5SC_2H_5 + 2H_2 \rightarrow 2C_2H_6 + H_2S$	−27.99	+18.52	+9.11
(thiophene-S ring) $+ 2H_2 \rightarrow C_4H_{10} + H_2S$	−28.73	+8.79	+5.26
(benzothiophene-S ring) $+ 4H_2 \rightarrow C_4H_{10} + H_2S$	−66.98	+12.07	+3.85
b. Denitrogenation			
(pyridine ring) $+ 5H_2 \rightarrow C_5H_{12} + NH_3$	−93.90	+16.97	+6.30
c. Deoxygenation			
(phenol, OH) $+ H_2 \rightarrow$ (benzene) $+ H_2O$	−30.51	+21.12	+14.38
d. Hydrogenation and Hydrocracking of Hydrocarbons			
$C_6H_{12} + H_2 \rightarrow C_6H_{14}$ (Hexene-1)	−33.84	+7.42	+3.39
$C_4H_6 + 2H_2 \rightarrow C_4H_{10}$ (1,3-Butadiene)	−65.21	+14.54	+6.9
(benzene) $+ H_2 \rightarrow$ (cyclohexane)	−45.35	+2.47	−1.57
(naphthalene) $+ H_2 \rightarrow$ (butylbenzene, C_4H_9)	−43.85	+2.13	−2.66

(From *Sekiyu Seisei Process* (Sekiyu Gakkai *ed.*), 41 (1998))

hydrotreating with various purposes shown in Table 5.3, and give various products. The requirements for product distribution and the composition in products vary in different localities or by environmental legislation. The recent increase in the use of oil products for transportation has brought about not only decrease in the use of fuel oils but also increasing demand for the lighter fractions. Further, the conversion of heavier fractions to lighter fractions is also on the rise, and the role of hydrotreatment in the refinery is becoming more important.

In Fig. 5.1, crude oil is initially separated into naphtha (boiling range: 30–180°C), kerosene (170–250°C), gas oil (240–350 °C) and atmospheric residue (more than 350°C) by atmospheric distillation. These products are so-called straight-run fractions. Atmospheric distillation ends around 350°C and the atmospheric residue is further separated by vacuum distillation into a vacuum gas oil (VGO) and a vacuum residue (VR). The atmospheric residue occupies about 50% of Arabian crude oils as shown in Table 5.2. Each fraction includes a different type and concentration of sulfur and nitrogen compounds and subsequent hydrotreating has different purposes, as shown in Table 5.3.

Table 5.2 Typical Properties of Petroleum Crude

	Arabian Light	Arabian Heavy
Specific Gravity (15/4°C)	0.85	0.89
API degree	34.4	27.9
Sulfur (wt%)	1.7	2.7
Nitrogen (wt%)	0.1	0.15
Conradson Carbon (wt%)	3	6
V + Ni (ppm)	22	66
Boiling Properties		
Naphtha (vol%)	25	20
(IBP–180)		
Kerosene (vol%)	14	10
(180–250)		
Gas Oil (vol%)	14	11
(250–320)		
Atm. Residue (vol%)	48	57
Sulfur content in various fractions		
Kerosene	0.11	0.17
Gas Oil	1.0	1.1
Atm. Residue	3.0	4.0

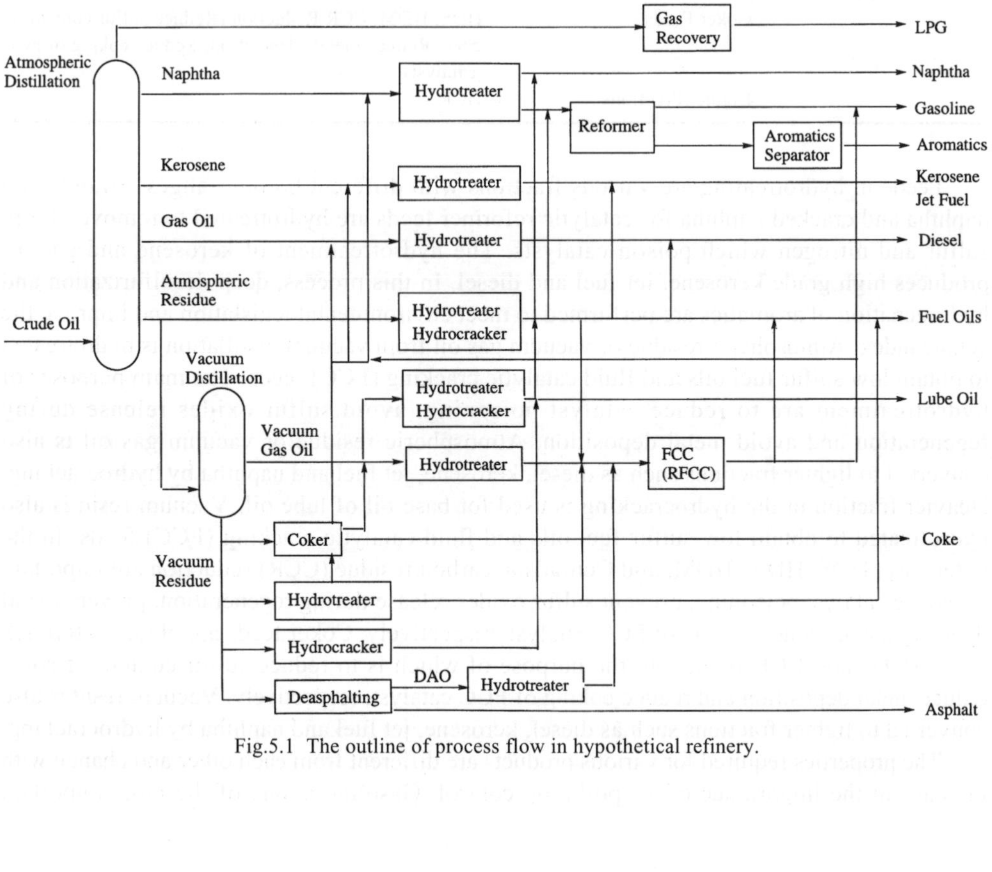

Fig.5.1 The outline of process flow in hypothetical refinery.

Table 5.3 Various Purposes for Hydrotreating in Refinery

Feeds	Products	Purposes
Naphtha	Catalystic Reformer Feeds	HDS, HDN, Reduce olefin content (Avoid catalyst poisoning)
Kerosene Gas Oil	Diesel	HDS, Hydrogenation of aromatics (Meet environmental legislation, Improve cetane index)
	Kerosene, Jet Fuel	Reduce aromatics (Improve smoke point)
Atmospheric Residue Vacuum Gas Oil	Fuel Oils	HDS
	FCC Feeds	HDS, HDN, HDM (Reduce catalyst poisoning, Avoid sulfur oxides release during regeneration, Avoid metal deposition)
	HCR Feeds	HDS, HDN, HDM (Reduce catalyst poisoning, Avoid metal deposition)
	Diesel	HDS, Reduce aromatics, HCR
	Kerosene, Jet Fuel	HDS, Reduce aromatics, HCR
	Naphtha	HCR
	Lube Oils	Reduce aromatics, HCR, HDN (improve stability)
Vacuum Residue	Fuel Oils	HDS
	FCC Feeds	HDS, HDN, HDM, CCR Reduction (Reduce catalyst poisoning, Avoid sulfur oxides release during regeneration, Avoid metal deposition, Reduce coking of FCC catalyst)
	Coker Feeds	HDS, HDM, CCR Reduction (Reduce sulfur content of coke, Reduce metals deposition, Reduce coking of FCC catalyst)
	Lighter Fractions	HCR

Feeds in hydrotreating are various fractions with different boiling ranges. Straight-run naphtha and cracked naphtha for catalytic reformer feeds are hydrotreated to remove olefin, sulfur and nitrogen which poison catalysts. The hydrotreatment of kerosene and gas oil produces high grade kerosene, jet fuel and diesel. In this process, deep desulfurization and hydrogenation of aromatics are performed to meet environmental legislation and improve the cetane index. Atmospheric residue or vacuum gas oil from vacuum distillation is hydrotreated to obtain low sulfur fuel oils and fluid catalytic cracking (FCC) feeds. The main purposes of hydrotreatment are to reduce catalyst poisoning, avoid sulfur oxides release during regeneration and avoid metal deposition. Atmospheric residue or vacuum gas oil is also converted to lighter fractions such as diesel, kerosene, jet fuel and naphtha by hydrocracking. Heavier fraction in the hydrocracking is used for base oil of lube oil. Vacuum resid is also hydrotreated to obtain low-sulfur fuel oils and fluid catalytic cracking (FCC) feeds. In the latter case, HDS, HDN, HDM, and Conradson carbon residue (CCR) reduction are important to reduce catalyst poisoning, prevent sulfur oxides release during regeneration, prevent metal deposition and reduce coking of FCC catalyst, respectively. Coker feeds are obtained through HDS, HDM and CCR reduction, the purpose of which is to reduce sulfur content of coke, reduce metal deposition and reduce coking of FCC catalyst, respectively. Vacuum resid is also converted to lighter fractions such as diesel, kerosene, jet fuel and naphtha by hydrocracking.

The properties required for various products are different from each other and change with increase in the importance of air pollution control. Gasoline is one of the most important products and requires higher octane numbers. However, the use of alkyl lead to increase octane number is decreasing and the allowable concentration of aromatics, which also increase the octane number, has been the subject of environmental legislation. The decrease in aromatic content affects the utilization of catalytic reformer. The addition of high octane number compounds such as methyl-tert-butyl ether (MTBE) to gasoline will increase. To produce reformulated gasoline, significant modification in the refinery is required so the requirements for hydrotreating will also change.

Diesel may be a more attractive fuel than gasoline since it has higher fuel efficiencies. The demand for diesel is increasing and in the near future a shortage of diesel is anticipated.[7] To make up for the shortage of diesel, an increase in the diesel fraction in the hydrocracking of VGO may be needed. A recent approach has focused on catalyst development for this process. For example, the catalytic performance of Ni-Mo supported on the mesoporous crystalline MCM-41 aluminosilicate for mild hydrotreatment of VGO was compared with that of an amorphous silica-alumina and a USY zeolite.[8] The MCM-41 based catalyst showed superior HDS, HDN and HC activities compared with the latter two catalysts and gave better selectivity to middle distillates. The better performance of the former catalysts was explained by their higher surface area, the presence of uniform pores in the mesophase range, mild acidity, and stability. In contrast with gasoline, a higher cetane number is required for diesel fuels. Although the straight-run fractions have been used for diesel fuels, light cycle oil (LCO) from FCC process and coker gas oil (LCGO) must be used to compensate for the future demand for this fraction. LCO as well as LCGO contain higher concentrations of aromatics and cannot be used until aromatics are hydrogenated. In a recent study on the hydrotreatment of LCO, the cetane index was improved in the range from 7.3 to 10.0 for a Ni-W/Al$_2$O$_3$ catalyst and from 6.1 to 10.1 for a Ni-Mo/Al$_2$O$_3$ catalyst, indicating that hydrotreatment of LCO can increase the extent of its blending ratio into the diesel pool.[9] Color degradation during deep desulfurization of diesel fuel is also a matter of the utmost concern in the Japanese market, and several investigators have examined this problem.[10–12] From deep desulfurization tests of various feedstocks, Takatsuka et al. suspected that the color bodies were derived from a newly formed structure of polyaromatics from desulfurized aromatic compounds.[10] They also showed that higher hydrogen pressure suppressed color degradation of diesel fuel to a great extent during deep desulfurization.

Lowering the concentration of sulfur in fuel oil is achieved by either direct desulfurization or indirect desulfurization. In direct desulfurization, atmospheric residue and vacuum residue are desulfurized without pretreatment such as deasphalting. Although the operation condition is severer than that in the indirect desulfurization, fuel oil with lower concentrations of sulfur can be produced by direct desulfurization. In the indirect desulfurization, VGO obtained from vacuum distillation is hydrodesulfurized, and the VGO produced is mixed with untreated vacuum residue to give fuel oil with low sulfur concentration. Though the operation condition is milder than that in the direct desulfurization, the sulfur concentration in fuel oil obtained from the indirect desulfurization is higher than that from the direct desulfurization since vacuum residue is not hydrotreated. The use of fuel oil is decreasing to avoid various environmental problems such as air pollution.

From the viewpoint of environmental conservation, deep desulfurization and aromatics saturation of diesel fuels are now key reactions to reduce NO$_x$ and particulate matter. These reactions are required for reducing the concentration of sulfur compounds to prevent engine corrosion when an exhaust gas recirculation (EGR) system is introduced to the diesel engine in order to reduce NO_x concentration in diesel exhaust. Further, sulfur components present in the exhaust gas poison NO_x treating catalysts, and deep desulfurization is necessary. Moreover, aromatic saturation enables the reduction of particulate matter in the exhaust gas. The reduction of particulate matter is also expected to promote the reduction of NO_x concentration in exhaust gas.

5.1.2 Upgrading of Heavy Crudes

Heavy crudes such as atmospheric and vacuum residues comprise almost half to more than half of raw petroleum crude, and the upgrading of these materials is very important because of limited fossil fuel resources. Representative properties of atmospheric and vacuum residues are shown in Table 5.4.[13,14] Atmospheric and vacuum residues are commonly higher in viscosity and pour point and include concentrated matters such as sulfur, nitrogen, heavy metals, and Conradson carbon residue (CCR) which become environmentally harmful. Among such matter, sulfur and nitrogen are distributed extensively from the lighter fraction to the heavier fraction, while heavy metals such as vanadium and nickel and CCR are concentrated only in the heavier fraction.

Vacuum residue can be divided to saturates, aromatics, resins and asphaltenes.[15] The oil fraction of saturates and aromatics are propane soluble. Resins are propane insoluble and pentane soluble, and asphaltenes are pentane insoluble. The combination of the oil fraction and resins is called maltene. Nickel and vanadium, which cause the deactivation of the catalyst by pore mouth plugging and covering active sites, form porphyrin complexes in resins and asphaltenes.[16] In general, when asphaltene content is higher, the content of heavy metals and CCR is higher. The structure of asphaltenes has been investigated for a long time. Dickies and Yen have proposed an original model of an average structure in which condensed polynuclear aromatics with porphyrin structure associate to form a unit micelle.[17] The unit micelle further aggregates to form a larger micelle. Asphaltenes have higher aromaticity and are difficult to dissolve by themselves in the oil fraction with relatively high aliphaticity. Since resins are the fraction between the oil fraction and asphaltenes, so have affinity to both. Thus resins form the external shell of asphaltenes which can be dispersed in the oil fraction in the colloidal state. The average structure of asphaltenes recently proposed by Speight[18] is different from the original model. This model molecule is somewhat smaller and not so condensed as the original model. In the upgrading of heavy crudes, asphaltenes are believed to cause coking and subsequent deactivation of the catalyst. Therefore, the treatment of asphaltenes is a key reaction for the upgrading of heavy crudes.

Atmospheric and vacuum residues are approximately 1.5–1.9 in H/C ratio and most of these are converted by two different routes, carbon rejection and hydrogen addition to increase the H/C ratio of products. The residues are also converted by partial oxidation to give synthesis gas (H_2 and CO gases). Table 5.5 shows the upgrading processes of residues.[13,14] Fig. 5.2 shows the plots of CCR contents vs. metal contents for atmospheric residues and vacuum residues in 119 kinds of petroleum crudes except those in the USA.[13] These residues are classified into five types in Fig. 5.2. Table 5.6 shows the definition of the five types, the ratio of the number of residue belonging to each type, and the typical residue upgrading process suitable for each type of residue.[13] This table provides a standard for choosing the upgrading process by properties of residue. The carbon rejection route includes coking processes, visbreaking, deasphalting and fluidized catalytic cracking (FCC). The hydrogen addition

Table 5.4 Properties of Typical Residues

	China		South		Middle East		Middle South America	
	Sheng Li	Da Qing	Minas	Duri	Arabian Light	Arabian Heavy	Maya	Ithmus
Atmospheric Residue (343°C+)								
Yields (vol.%)	73.29	70.49	56.8	77.4	44.71	53.76	60.63	43.23
Specific Gravity (15/4°C)	0.942	0.897	0.896	0.946	0.9538	0.9855	1.0035	0.9594
Viscosity (cst60°C)		9.60E+01	5.28E+01	5.58E+02	1.03E+02	1.02E+03	1.40E+04	2.47E+02
Sulfur (wt%)	1.226	0.1148	0.1325	0.2128	3.059	4.274	4.5192	2.6314
Nitrogen (wt%)	3.29	0.2187	0.1839	0.4023	0.1749	0.2512	0.53109	0.3385
Metal V (wtppm)		0	1	2.4	27.17	96.67	462.67	102.27
Ni (wtppm)	20.63	5.45	15	40.85	6.72	27.75	77	19.36
Conradson Carbon (wt%)	8.6	4.2	4.6	6.74	7.8	13.3	15.8	9
Asphaltene (wt%)	1.16	0.13	0.62	2	2	9.7	8.5	1.92
Vacuum Residue (565°C+)								
Yields (vol.%)	36.51	31.1	18.06	43.58	14.97	25	36.43	14.94
Specific Gravity (15/4°C)	0.9881	0.9378	0.9374	0.9641	1.0217	1.0505	1.0512	1.0312
Viscosity (cst60°C)		1.81E+03	1.96E+03	7.61E+04	3.33E+04	3.90E+06	2.57E+09	2.82E+0
Sulfur (wt%)	1.66	0.16	0.19	0.23	4.165	5.63	5.4	3.828
Nitrogen (wt%)		0.4	0.43	0.6	0.34	0.425	0.761	0.6683
Metal V (wtppm)	41.98	0	2.62	4.19	75.62	194.2	735	274
Ni (wtppm)	6.33	11.82	44.48	71.19	18.64	55.72	122	52
Conradson Carbon (wt%)	16.4	9	10.8	11.75	16.5	23.7	25.2	23.6
Asphaltene (wt%)	2.62	0.21	2.76	3.5	5.2	19.6	15.77	5.23

(From Y. Hori and T. Takatsuka, *Petrotech*, **18**, 293 (1995))

Table 5.5 Classification of Residue Upgrading Processes

| | Hydrogen Addition | | | | Carbon Rejection | | | | | | Gasification |
	Ebullated Bed	Moving Bed	Fixed Bed	Slurry Phase	Visbreaking Tubular	Visbreaking Soaker	Coking Batch	Coking Continuous	Fluid Catalytic Cracking	Solvent Deasphalting	Partial Oxidation
Single Process	H-OIL (HRI/Texaco) LC-FINING (Lummus/Oxy/Amoco)	BUNKER (Shell) OCR (Chevron) HYVAHL-M (IFP)	ABC (Chiyoda) HYVAHL-F (IFP) LRHDS (Shell) RCD UNIBON (UOP) RDS/VRDS (Chevron) R-HYC (Idemitsu) RESID UNIONFINING (Unocal) RHC (Cosmo)	CANMET (Petro Canada/Laval In) HDH (Intevep) MRH (Idemitsu/Kellogg) SOC (Asahi/JE) MICROCAT-RC (Exxon) VCC (Veba Oel) (HC)$_3$ (Aostra/ARC)	OPEN ART	TERVAHL (IFP) HSC (TEC) VISBREAKER (Shell/Lummus)	DELAYED COKER (Conoco/Foster Wheeler/Kellogg/Lummus/UOP/Koa Oil Co./ETC.) EUREKA (Kureha/Chiyoda)	ET-II (Fuji/Chiyoda) FLUID COKER (Exxon)	ART (Engel/Kellogg) RFCC (Exxon/Kellogg/Shell/UOP/S&W/Texaco/IFP/Total)	DEMEX (UOP) ROSE (Kellogg) SOLVAHL (IFP)	SGP (Shell) TGP (Texaco)

Combination Process

- HYCON (BUNKER+Fixed Bed) (Shell)
- OCR+RDS/VRDS (Chevron)
- MICROCAT-RC (Slurry+Fixed Bed) (Exxon)
- U-CAN (Unocal/Petro Canada)
- VCC (Slurry+Fixed Bed) (Veba)
- VisABC (ABC+Visbreaking) (Chiyoda)
- Visbreaking+Delayed Coker
- ASCOT (SDA+Delayed Coker) (Foster Wheeler)
- FLEXICOKER (Fluid Coker+Gasification) (Exxon)
- SDA+Gasification
- Delayed Coker+Gasification
- Slurry+Gasification

(From Y. Hori and T. Takatsuka, *Petrotech*, **18**, 290 (1995))

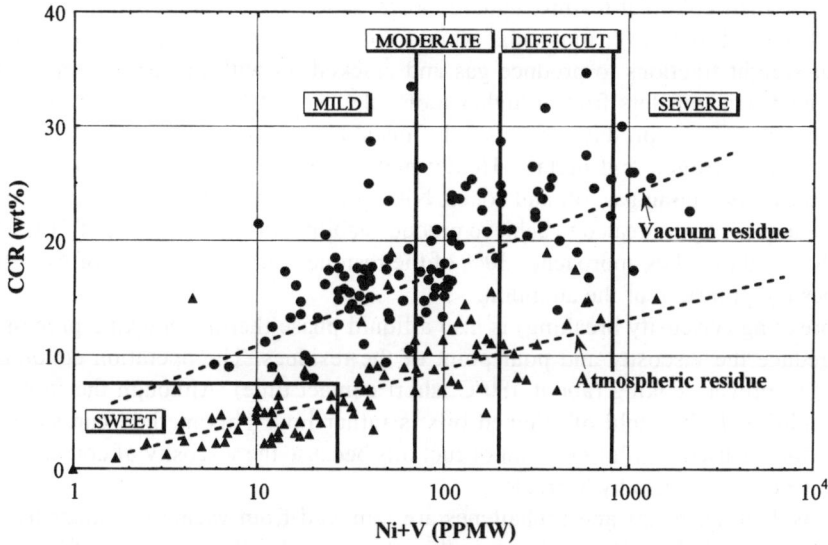

Fig.5.2 Relationship between CCR and metal contents of residue.
(From Y. Hori and T. Takatsuka, *Petrotech*, **18**, 293 (1995))

Table 5.6 Residue Type and Typical Upgrading Processes

Residue Type	Properties of Each Residue Type			Ratio of the Number of Each Residue Type		Typical Residue Type in Upgrading Processes				
						RFCC	Hydrogen Addition			Coker
	Ni+V	CCR	SULFUR	AR	VR	Without Direct	Fixed Bed	Moving Bed	Slurry	SDA
	(ppmw)	(wt%)	(wt%)	(%)	(%)	Desulfurization		Ebullated Bed		
SWEET	< 25	< 7	< 0.5	25	0	○				
MILD	< 70	—	—	48	49		○			
MODERATE	70–200	—	—	16	25		○	○	○	○
DIFFICULT	200–800	—	—	10	21			○	○	○
SEVERE	> 80	—	—	1	5				○	○

(From Y. Hori and T. Takatsuka, *Petrotech*, **18**, 293 (1995))

route includes HDS and HDM using a fixed-bed, an ebullating-bed, a moving-bed or a slurry-bed reactor. The catalytic process is limited by the type of residue. The residue FCC and the fixed-bed reactor can be used only for the sweet type of residue and for residue with amount of metal of less than 200 wtppm, respectively. The ebullating-bed, moving-bed and slurry-bed reactors can be used for residues with higher contents of metals. Coker and deasphalting processes are not limited by the metal content.

A. Carbon Rejection Routes

The coking processes are thermal cracking processes (delayed coking, fluid coking, and flexi-coking) at higher temperatures (around 500°C) and without catalyst to produce lighter hydrocarbons (gas, naphtha to gas oil) and cokes (Fig. 5.1). When asphaltene micelles are thermally treated, resins forming the external shell of the micelle undergo decomposition, dealkylation, and dehydrocyclization. The aliphatic properties of resins decrease so asphaltene micelles lose their colloidal feature, undergo aggregation and separation, and form sludges. On the other hand, the lower molecular weight radicals formed extract hydrogen in higher molecular weight fractions to produce gas and cracked oil with higher H/C ratios. In this process, hydrogen transfers from a higher molecular weight fraction to a lower molecular weight fraction. Coker products are usually unstable and include significant amounts of olefins, sulfur, nitrogen and metals. Heavy metals are concentrated to cokes. Sulfur is distributed evenly among gas, oil and cokes. Nitrogen is found more in the heavier fraction. Coker products are hydrotreated to achieve product stability, to reduce coking of FCC catalyst, etc. As the yield of cokes approaches 30% of the residue feed, the presence of its utilization determines the propriety of the adoption.

Visbreaking (viscosity breaking) is also a liquid phase thermal cracking process and is used to reduce the viscosity and pour point of the residues. The operation conditions are kept mild to prevent coking (about 480°C, short contact time). Although the formation of cokes is inhibited, the yield of cracked oils is rather low. This method can save gas oil fraction which is used as diluting agent of fuel oils because the viscosity of cracked products is lower. Products are also hydrotreated.

In deasphalting, resins and asphaltenes are removed from vacuum residues by solvent extraction. Although there is a weakness in this process in that the separation of large amount of recycled solvent consumes a vast amount of energy, the application of a low solvent ratio and the separation of solvent in critical state reduced the problem to an acceptable level. Deasphalting using propane is representative in petroleum processing and removal of resins and asphaltenes. When pentane is used to increase the yield of the paraffinic product with low metal content, maltenes are extracted as a deasphalted oil (DAO), and asphaltenes are removed. Although the selectivity for sulfur and nitrogen removal is not so high, heavy metals and CCR can be removed to a relatively higher extent in the deasphalting. This is because deasphalting does not accompany chemical changes that occur in cracking, and metals and CCR are concentrated to asphaltenes. However, DAO still includes 20–30 wtppm of nickel and vanadium and 5–8 wt% of CCR as well as 2–4% of sulfur. Therefore, DAO is further hydrotreated for use as FCC feeds, that is, to reduce the content of sulfur, nitrogen, metals and CCR.

Another carbon rejection route is residue fluid catalytic cracking (RFCC). This process was developed originally from the FCC process which cracks gas oil and vacuum gas oil to give gasoline. In this process, residues undergo not only cracking but also hydrogen transfer, isomerization, dehydrogenation, polymerization and cyclization on solid acid catalysts to give LPG, cracked gasoline, cracked gas oil and cokes. The properties of solid acid catalysts such as acid strength and acid density and the diffusion of substrates to the catalytic sites affect the catalytic activity and selectivity for products. Ultrastable Y (USY) zeolites, which are stable for hydrothermal reactions and metal deposition, are used for catalysts. In the coker process, the cracking of paraffinic and naphthenic parts of the substrates does not occur easily. In contrast, reactions in the RFCC process occur on the catalyst very rapidly and parts other than the aromatic structure are cracked to a lighter fraction than C_4 olefins without appropriate control of the reaction conditions. In general, the reaction condition is controlled to obtain the maximum yield of gasoline. Under these conditions, paraffinic parts with alkyl substituents and naphthenic structure cannot remain, and products such as light cycle oil (LCO) and heavy cycle oil (HCO) are rich in the aromatic fraction while gasoline obtained from the cracking of paraffinic and naphthenic structure is rich not in aromatics but olefins.

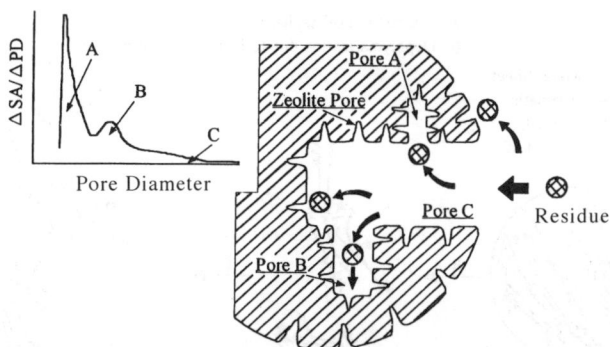

Fig.5.3 Pore structure model of FCC catalyst and effective pore for residue cracking.
Pore B: Preferable pore for residue precracking.
(From S. Sato, T. Takatsuka et al., Sekiyu Gakkaishi, **29**, 64 (1986))

Figure. 5.3 shows the pore structure model of the FCC catalyst and effective pore for residue cracking.[19] Residues are aggregates of heavy hydrocarbons the molecular size of which is larger than 10 Å. Although FCC reactions occur through the diffusion of feeds to zeolite pore (Pore A), most of the molecules in residues are too big to diffuse into Pore A. Therefore, the precracking of residual feeds in matrix pores or mesopores (Pore B) with relatively larger pore size is necessary, while Pore C is too large to obtain sufficient surface area. Since asphaltenes are too large to diffuse to Pore B, the cracking of asphaltenes and hydrogenation of polyaromatic hydrocarbons in asphaltenes are needed to obtain smaller paraffinic structure.

B. Hydrogen Addition Routes

An alternative route to increase the H/C ratio of residues is hydrogen addition. The hydrogen addition for residues upgrading includes hydrotreating and hydrocracking. Hydrogen is consumed for not only HDS, HDN, and HDM but also aromatic hydrogenation and hydrocracking of heavier fractions. As described above, hydrotreatment gives fuel oils with low sulfur concentration, and FCC and deasphalting feeds with lower concentration of sulfur, metal and CCR. Hydrocracking gives middle distillates with low sulfur concentration. Crudes are becoming rich in heavier fractions while lighter fractions are favorable for products. Therefore, higher ability for desulfurization, demetallation and cracking is required for hydrotreating and hydrocracking processes.

In this process, the front part of the reactor is filled by demetallation catalysts with larger pore size and the rear part is filled by HDS catalysts with high surface area. The front part not only works as a guard reactor for demetallation but also decomposes asphaltenes for assisting HDS and hydrocracking in the rear part. To obtain highly desulfurized and demetallated products from residues, desulfurization and demetallation of asphaltenes are needed because asphaltenes contain two to 10 times higher concentrations of sulfur, nitrogen and metals than other fractions in residues. Fig. 5.4 shows the mechanism of hydrocracking of asphaltenes. Initially asphaltenes micelles, aggregates of asphaltenes with layered sheet structure, are relatively easily decomposed by the elimination of vanadium. Eliminated vanadium is accumulated on the catalyst. Subsequent desulfurization causes depolymerization

a : Destruction of Asphaltene Micelle
b : Depolymarization due to Heteroatom Removal

—M— Metal Sheet
——— Aromatic
∿∿∿ Aliphatic
∿∿ Weak Link

a →

b →

Asphaltene Micelle

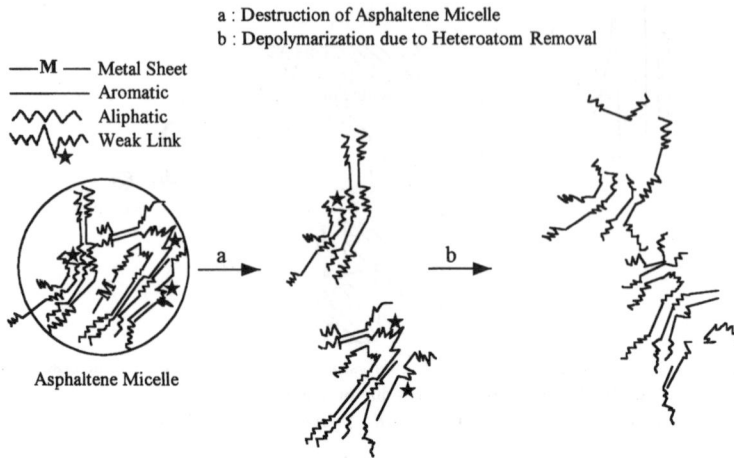

Fig.5.4 Proposed mechanism of asphaltene cracking.
(From C. Takeuchi, M. Nakamura et al., Ind. Eng. Chem., Process Des. Dev., **22**, 240 (1983))

of asphaltenes to give a unit structure of polyaromatic hydrocarbon which is similar to the structure of maltenes in residues.[20,21]

The products from the hydrogen addition routes have higher quality than those from the carbon rejection routes. However, the latter routes are still very important from an economical point of view. The choice of processes changes with various factors such as hydrogen availability, the requirement for quality and kind of product etc. Consequently, some combination of both carbon rejection processes and hydrogen addition processes may be employed in the same refinery at the same time.

Many recent studies have focused on catalyst development for the upgrading as well as the hydrotreatment and hydrocracking of residues.[22–31] Peureux et al. reported that, when optimized, dispersed catalysts such as phosphomolybdic acid and molybdenum naphthenate allowed deep hydroconversion of heavy oil residues with efficient inhibition of coke production.[22] Lee et al. also investigated the residual oil HDS using dispersed catalysts.[23] The combination of Co and Mo was the most promising dispersed catalyst system for the HDS of heavy oil. Further, active carbon packed in the reactor was very effective in recovering the dispersed catalysts from reactant oils. Ying et al. used fibrillar alumina, which has both large surface area (138 m^2/g) and large average pore diameter (252Å), as a carrier material to prepare large-pore catalysts for hydroprocessing of residual oils.[24] The Ni-Mo catalysts prepared using fibrillar alumina was more active than a commercial catalyst with similar loading of nickel and molybdenum for HDM, equally active for HDS, and less active for HDN. Guohe et al. reported HDN of basic nitrogen in Gudao vacuum residue using MMo/Al_2O_3 where M is Ti, Cr, Fe, Co, Ni, Cu, or Zn.[25] The observed order of basic nitrogen removal activity was Ni-Mo > Co-Mo > FeMo > CrMo > ZnMo > MnMo > TiMo > CuMo. Chen et al. investigated the effect of boria content on HDS of residue oils over Ni-Mo/Al_2O_3-aluminum borate catalysts.[26] These catalysts were much more active than the conventional Ni-Mo/Al_2O_3 catalysts. The dispersion of active sulfide phase as well as the hydrogenation ability of Ni-Mo/Al_2O_3 catalyst was increased by the incorporation of adequate boria content. Chen et al. also reported HDS of residual oils over Co-Mo/Al_2O_3-aluminum phosphate catalysts in a trickle bed reactor.[27] They proposed that larger surface area, smaller acid amount, and weaker interaction of Al_2O_3-aluminum phosphate supports made the metal disperse more highly, produced more active sites, and resulted in a high initial HDS activity. Yang *et al.* prepared macropore (MAP) catalysts which had much larger macropores than a commercial reference catalyst.[28] The improved HDN conversion of vacuum residue was achieved with the MAP catalyst. Papayannakos *et al.* studied the kinetics of viscosity reduction and carbon residue reduction of a heavy residue during HDS.[29] First-order kinetics could be applied for these reductions. The activation energies for viscosity reduction, Conradson carbon reduction and Ramsbottom carbon reduction were 97.5, 79.1 and 77.0 kJ/mol, respectively. In parallel with sulfur and nitrogen removal, reductions of even 95% could be achieved in the viscosity and carbon residue of the residual oil. Kushiyama *et al.* investigated the effect of phosphorus addition on HDS activity of *in situ* formed dispersed Mo-Co-S catalysts in heavy oil hydrotreatment.[30,31] The addition of phosphorus improved the HDS activity. The increment of sulfur removal attainable by phosphorous addition was larger for heavy oils of higher vanadium content. These phenomena were explained by assuming the formation of oil-insoluble P-V compounds and their separation from the active Mo-Co-S catalyst phases.

C. Gasification with Partial Oxidation

Another method for upgrading heavy crudes besides carbon rejection and hydrogen addition is gasification with partial oxidation which gives synthesis gas, hydrogen and carbon monoxide. Synthesis gas can be used for the production of hydrogen, methanol and ammonia or combined cycle power generation. The H_2/CO ratio of the synthesis gas generated does not depend on pressure but on the H/C ratio of feed and the ratio of steam. For example, in the case of vacuum residue, the ratios of CO, H_2 and CO_2 are 48%, 44% and 5%, respectively.[32] The composition of synthesis gas can be controlled by the shift reaction with the addition of steam and carbon dioxide. Since catalysts are not used in this process and raw materials are not limited, it is noticeable that this process can be used for treatment and utilization of residues from upgrading processes and waste matter in the refinery which are difficult to treat or utilize.

5.1.3 Hydroprocessing of Other Hydrocarbon Sources

Oil shale,[33–35] tar sand[33–35] and coal[33] can be regarded as alternative energies to petroleum. Shale oils, tar sand bitumens and coal liquids are obtained from oil shale, tar sand and coal, respectively. Hydrotreatment of these hydrocarbon sources other than petroleum is important for the utilization and processing of these resources, both as fuel and nonfuel materials. Some approaches to this subject are presented below.

A. Oil Shale

Oil shale has been distributed to the relatively extensive area, different from petroleum. The reserves of oil shale in the USA are twice the proven reserves of petroleum in the world. Commercial production of shale oil, which is obtained through oil shale retorting, has been ongoing in China for a long time and has recently begun in Brazil. There are also deposits of oil shale in Australia and Russia. Since the shale oil produced is used as fuel oil, however, there are few actual results for the upgrading of shale oil.

There are different types of retorts for oil shale. These retorts can be classified into mainly two general types. One is the retorting of mined oil shale on the ground which includes the external-combustion type retort and the internal-combustion type retort. The other is the retorting of oil shale by partial oxidation under the ground.

As shown in Table 5.7, oil shales consist of a major component, minerals, and polymerized organic compounds, kerogens.[36) Oil shale is estimated by weight% of oil produced by Fischer assay. Kerogens are thermally decomposed to give shale oil at near 500°C. The properties of shale oil ((Table 5.8) are as follows: 1) The boiling range is extensive from propane to 540°C, 2) the amounts of kerosene, gas oil and VGO fractions are relatively rich and a slight amount of vacuum residue is included, and 3) the aromaticity is low, linear paraffin is abundant, and pour point is high. Since not only unsaturated hydrocarbons but also a large amount of arsenic-, nitrogen- and oxygen-containing compounds are included, however, hydrotreatment of shale oil is very difficult. Arsenic remarkably deactivates the hydrotreating catalysts, and further unsaturated hydrocarbons and nitrogen and oxygen compounds reduce the stability of the hydrotreated products and produce precursors of coke to accelerate contamination of the catalyst and the reactor. Nitrogen compounds poison hydrocracking catalysts downstream. Therefore, it is reasonable to assume that unsaturated hydrocarbons, nitrogen and oxygen compounds and arsenic are hydrotreated simultaneously.

It is relatively easy to make fuels for transportation by hydrotreatment of distillates such as naphtha, kerosene and gas oil. However, it is important to use the major component of oil shale, atmospheric residue, and some combinations of processes proposed[37) are as follows: 1) coker-hydrotreating; 2) hydrotreating-hydrocracking; 3) hydrotreating-fluid catalytic cracking. It is difficult to obtain a higher conversion of residue with the combination of coker and hydrotreating. Further, coke remains in the coker process and insufficient denitrogenation gives a colored product. Therefore, in order to obtain transportation fuel two-process combinations including hydrotreating and hydrocracking or fluid catalytic cracking are chosen. The role of hydrotreating is to reduce the amounts of arsenic, unstable unsaturated hydrocarbons, nitrogen, oxygen, etc. and to facilitate hydrocracking or FCC downstream.

Although these cracking processes are same as those in the ordinary hydroprocessing of petroleum, hydrotreating is somewhat different since a large amount of arsenic is contained in shale oil, as shown in Table 5.8. The concentration of arsenic is in the range of a few ppm to more than 100 ppm, which is much larger than that in petroleum. Arsenic is present in the form of inorganic compounds such as As_2O_5 or organic compounds such as methylarsonic acid

Table 5.7 Properties of Oil Shales

	Colorado	Condor
Ash (wt%)	66	81
Total Carbon Content (wt%)	17.7	10.4
Inorganic Carbon	3.6	1.7
Organic Carbon	13.8	8.7
Aromaticity of Organic Carbon (fa)	0.25	0.33
Fischer Assay Oil Yields (wt%)	10.3	6.2

(From H. Ozaki, *Kagaku Kogyo*, **24**, 536 (1997))

Table 5.8 Properties of Shale Oils

	Colorado	Condor	Mao Ming
API Degree	21.9	25.7	28.2
Elemental Analysis (%)			
C	84.3	84.7	83.4
H	11.0	11.5	11.8
S	0.76	0.78	0.42
N	1.92	1.30	0.17
O	2.0	2.0	3.2
Metal (ppm)			
V	1 >	1 >	1 >
Ni	2.8	6.6	1.1
Fe	21	6.6	7.8
As	35	2.6	4.6
Viscosity @50°C	8.35	5.60	5.60
Distillation (%)			
Naphtha IBP–160°C	10	10	10
Kerosene 160–240°C	15	16	16
Gas Oil 240–360°C	26	32.5	32.5
Residue 360°C+	49	41.5	41.5
90% Distillate °C	505	487	487

(From H. Ozaki, *Kagaku Kogyo*, **24**, 537 (1997))

and phenylarsonic acid.[38] Hisamitsu *et al.* reported the effects of arsenic compounds on hydrotreating of vacuum distillates derived from Colorado shale oil and Middle East crude oil on Ni-Mo/Al$_2$O$_3$.[39] The deactivation rate of the catalyst for HDN was found to be higher with the shale oil derivative than with that of petroleum. In addition, the poisoning effect of triphenyl arsenic on catalyst deactivation was significantly accelerated by adding 30 ppm of the arsenic compound to the feed oil. It was suggested that arsenic deactivated the catalyst by bonding to metal sulfide catalyst and by expelling sulfur related to the formation of the active sites.

Some portion of the arsenic compounds in shale oil can be precipitated by noncatalytic thermal treatment at temperatures above 316°C.[40] However, it may be more practical to deposit the arsenic on the adsorbent or catalyst in a pretreatment reactor. More than 90% arsenic in Colorado shale oil can be removed by Mo-Co catalyst,[41] Mo-Ni/Al$_2$O$_3$[42] or Ni-Mo/Al$_2$O$_3$[43] catalyst while more than 50% arsenic can be removed by only alumina.[37] On the other hand, thermally unstable olefinic compounds present in shale oil deteriorate the storage stability of the oil and also tend to give rise to coke deposition, resulting in such difficulties as fouling of equipment and plugging of the catalyst bed. It has been reported that most of the unstable olefinic compounds could be removed by hydrotreating over a conventional Ni-Mo/Al$_2$O$_3$ catalyst in the range 250°C to 300°C.[44] It has also been suggested that reaction temperatures near 300 °C are suitable for pretreatment of shale oil to selectively hydrogenate the unstable olefins and to remove the arsenic concurrently without causing significant coking problems. Further, in the HDN reaction catalyzed by Ni-Mo-P/Al$_2$O$_3$, it was confirmed that increase in the reaction pressure not only accelerates the HDN reaction, but also suppresses catalyst deactivation. Reaction pressures above 100 kg/cm^2 have been found desirable to reduce the nitrogen content of the shale oil to around the 500 ppm level, which is considered to be the economically optimum level from the industrial point of view.[45] Many recent studies deal with the HDN of shale oil.[46,47]

B. Tar Sand

Tar sand similar to petroleum but different from shale oil is unevenly distributed in Canada and the reserves are at least one third as much as the proven reserves of petroleum in the world.[34,35,48] Therefore, tar sand as well as shale oil have been the focus of much attention as petroleum alternatives since the two oil crises of the 1970's. Tar sand is a black tarlike mixture which consists of sand, clay, water and bitumens. Most tar sand oils such as Athabasca bitumens produced in Canada are consumed as synthetic crude oil after upgrading. Tar sand oil produced in Venezuela is mixed with about 30% water to give an emulsion, oil-water mixture (Orimulsion),[49] which is exported and used for boiler fuels.

There are two methods to mine tar sand: opencut mining and the *in-situ* method. In opencut mining, bitumens are recovered from mined tar sand by adding heat and water of steam on the ground or by extraction with solvents. The Syncrude and Suncor projects have adopted this method. The *in-situ* method is further divided to two methods: steam drive method and the *in-situ* combustion method. In the steam drive method, pressurized high temperature steam is injected into the tar sand deposit to decrease the viscosity of bitumens and recover them. In the *in-situ* combustion method, air or oxygen is injected into the tar sand deposit and some of the bitumens are burnt to generate heat, decreasing the viscosity of the bitumens. The Cold Lake, Wolf Lake and Peace River Complex projects have selected the *in-situ* method.

The properties of bitumens (tar sand oils) are shown in Table 5.9. The major fractions of bitumens are residues with much smaller amounts of distillates. They contain not only large amounts of sulfur, nitrogen, and heavy metals, but fine clay is also included. Except for the presence of clay, bitumens are similar to atmospheric residue and vacuum residue fractions derived from petroleum crudes. Therefore, both carbon rejection and hydrogen addition processes, which are also used for upgrading of petroleum heavy crude described in Section

Table 5.9 Properties of Bitumens

	Athabasca Bitumen	Cold Lake Bitumen	Kindersley Bitumen
API Degree	10.2	10.6	13.1
Sufur (wt%)	4.3	4.3	3.2
Nitrogen (ppm)	3650	3600	2700
Kinematic Viscosity (cSt 50°C)	1310	3000	411
Metal V+Ni (ppm)	267	245	135
Asphaltene (wt%)	8.1	8.6	5.9
Conradson Carbon Residue (wt%)	13.3	12.4	11.9
Yield (wt%)			
Naphtha	2.2	0.2	1.7
Kerosene	5.3	2.2	3.5
Gas Oil	12.1	8.5	17.1
Vacuum Gas Oil	31.8	37.1	30.3
Vacuum Residue	48.6	52.0	47.4

(From H. Ozaki, *Kagaku Kogyo*, **24**, 536 (1997))

5.1.2, are used for the upgrading of bitumens. Many recent studies have focused on catalyst development for the upgrading of bitumens[50] and the hydrotreatment and hydrocracking of bitumens and bitumen-derived oils.[51–61]

C. Coal

Coal is expected to be an alternative energy to petroleum because coal deposits are one order of magnitude larger than those of petroleum. Further, unlike petroleum coal is evenly distributed and exists in extensive amounts in North America, Asia and the Pacific area, Russia, Europe, Africa, etc. As shown in Fig. 5.5, it has been predicted that most petroleum and natural gas will be exhausted by the end of the 21st century and that atomic and solar energy will not become major energy sources replacing petroleum, even though the use of these energies are gradually increasing.[62] Thus the use of coal may increase from the beginning of the 21st century and comprise 60% of energy consumption in the world by the end of 21st century. However, it is more difficult to treat solid coal than liquid petroleum and problems related to environmental conservation must be solved in the use of coal. To counter these problems, coal liquefaction technologies have been developed to treat coal like petroleum in the USA (H-coal, SRC-2, EDS-2, and CC-ITSL), UK (LSE), Germany (new IG) and Japan (NEDOL).[63]

The main purpose of coal liquefaction is to obtain liquid fuels such as gasoline and diesel oil by converting coal to lower weight molecules and upgrading these molecular weight products. The H/C ratio of liquid fuels is close to 2 while that of coal is near 1 because coal has a structure in which polycondensed aromatic hydrocarbons are linked by ether and methylene bonds. Therefore, to produce liquid fuels from coal, one atom of hydrogen must be added to one atom of carbon. There are two-stages for this hydrogen addition: first, hydrocracking of ether and methylene bonds in coal structure give lower weight molecules and convert solid coal into liquid state (direct liquefaction), so that not only HDS and HDN but

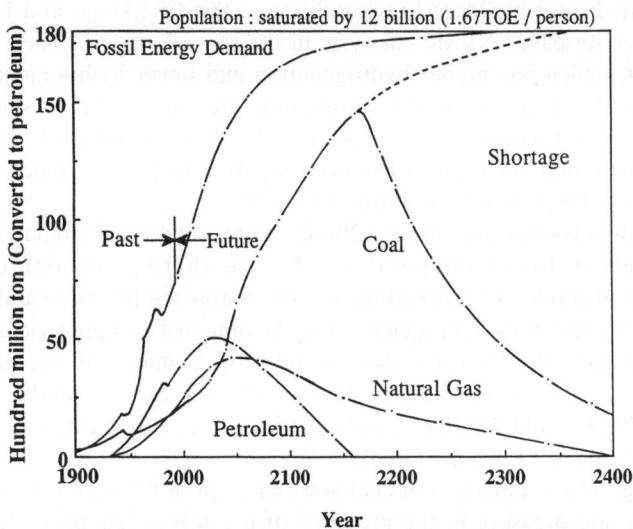

Fig.5.5 Evaluation of exhaustion of fossil energy.
(From H. Tominaga, *Chemistry and Education*, **46**, 147 (1998))

hydrogenation of aromatics is also achieved to give hydrotreated oil in the second stage (upgrading).[64–67]

Direct coal liquefaction was initially developed by Bergius in Germany.[68,69] In direct coal liquefaction, coal is thermally decomposed and the radicals of coal molecule produced are stabilized by hydrogen addition to give lower weight molecules. As coal thermally decomposes near 400°C, catalysts with hydrogenation activity at such high temperatures are fewer in number and their activity is relatively low. The separation of catalysts is also difficult due to the presence of ash in the residue of distillation of coal liquids. Therefore, cheaper expendable iron catalysts are generally used. As sulfur has been proved to be a promoter, iron sulfides or combinations of iron compounds and sulfur are used as catalysts. A number of attempts have been made to develop iron catalysts for coal liquefaction.[70] Among them, it has been reported that a highly dispersed catalyst derived from iron pentacarbonyl is one of the most active catalysts for the hydroliquefaction of a variety of coal samples.[71] It is assumed that pyrrhotite ($Fe_{1-x}S$) is the active species in coal liquefaction. Kabe and coworkers have developed the radioisotope tritium and[14]C tracer methods to elucidate the mechanism of coal liquefaction.[72–82] In one of their recent works, they proposed that the catalyst derived from iron pentacarbonyl was successfully dispersed in coal, and directly acted on the coal to increase both the rate of coal conversion and the tritium transfer from the gas phase to the coal.[79]

Since the coal liquid produced still includes a large amount of polycondensed aromatic and heterocyclic compounds, these are hydrotreated over molybdenum-based hydrotreating catalysts such as $Ni-Mo/Al_2O_3$.[83–85] Recently, much attention has been focused on the development of new catalysts for the upgrading of coal and coal liquids.[86–89] However, it is very difficult to increase the cetane number of the gas oil fraction of coal liquid.[63] When sufficient hydrogenation is achieved, removal of heteroatoms has already been completed in many cases, indicating that hydrogenation of polycondensed aromatic compounds may be very important in coal liquefaction and upgrading of coal liquid. Shimada et al. proposed a two-stage catalytic process for the upgrading of coal-derived liquids, where hydrogenation and HDN are promoted over $Ni-Mo/Al_2O_3$ catalysts in the first stage and hydrocracking is catalyzed over zeolite-based Ni-Mo catalysts in the second stage.[90] They developed a Ni-W/Al_2O_3 catalyst, which has higher hydrogenation and lower hydrocracking activities, to upgrade coal liquids. Because of these properties, the Ni-W catalyst showed high HDN activity and slow deactivation for feedstocks with a large amount of catalyst poisons.[91] Besides the conventional catalysts, it has been reported that impregnated nickel sulfate is active for the HDS of a high organic sulfur lignite.[92]

Coal can be dissolved in solvent near 200°C. If polycondensed aromatic compounds can be hydrogenated at lower temperatures before thermal decomposition occurs, polycondensation of products and cracking of side chains are inhibited and coal liquids the aromatics of which are deeply hydrogenated may be obtained in higher yield. In this method, the catalyst must have the hydrogenation activity of aromatics in the presence of large amounts of sulfur, nitrogen and oxygen compounds, and ash. Recently, Kotanigawa has reported that Ru/Mn_2O_3-NiO, Ru/Mn_2O_3-ZnO and $Ru/Mn_2O_3-La_2O_3$ catalysts are active for the hydrogenation of aromatics in a coal liquid.[63]

Coprocessing, where a mixture of coal and heavy oil or bitumen is hydrogenated under high temperature and pressure in the presence of a catalyst, has been studied by several investigators.[93] CANMET process is known to be one of most effective coprocessing processes.[94,95] In the coprocessing, the yield of distillates is expected to be higher than that predicted from the linear addition of the distillates given by the independent processing of coal and petroleum-derived liquid or bitumen. It is assumed that a significant part of the coprocessed liquid products is derived from coal in the coprocessing. Therefore, the upgrading of coprocessed oils is necessary in order to use liquid products from coprocessing as a liquid transportation fuel. Recently it was reported that the liquid products from hydrotreating of coprocessed oil met specifications for synthetic crude oil.[96] It was also reported that there were significant mass-transfer limitations in the catalytic upgrading of coprocessed liquids.[97] A two-stage catalytic process for the upgrading of coprocessed oil as well as coal-derived liquids was reported. When the optimized Ni-Mo/Al$_2$O$_3$ and Ni-W/zeolite catalysts were used in the first stage and second stage upgrading, respectively, a gasoline fraction was efficiently produced from middle distillates of coprocessing oil.[98]

5.2 Hydrotreating Processes

5.2.1 Hydrotreating and Hydrocracking

The flow-sheet of a typical hydrotreating process is illustrated in Fig. 5.6. The mixture of feedstock, recycle gas and make-up gas pressurized to the expected pressure is heated in a furnace to the reaction temperature and introduced into a fixed-bed reactor containing the catalyst. In the processing of distillates, gas or liquid-phase oil is flowed downward through the solid catalyst particles in the reactor. The effluent from the reactor passes through the heat exchanger to be cooled by the reactor feed. After cooling, the reaction products are separated into liquid and gas via a high-pressure separator and low-pressure separator. Liquid products are separated by fractional distillation. Gas is rich in hydrogen concentration and this is used as recycle hydrogen after washing to remove H$_2$S.

To produce the deeply hydrotreated LGO, there are several methods, as shown in Fig. 5.7. The number of reactors can be increased or LHSV may be decreased to achieve deep desulfurization of LGO with the existing technology. As the demand for middle distillate is increasing, light cycle oil (LCO) and coker gas oil (CGO) as well as LGO must be used. Table 5.10 shows the properties of various gas oils. LCO derived from FCC or RFCC have lower sulfur content and higher olefin and aromatic content, and is badly colored. The cetane number of LCO is less than 40. CGO has much sulfur and nitrogen and is the color even worse. Both LCO and CGO do not have storage stability. Therefore, these materials cannot be used without hydrotreatment for upgrading. To improve the situation, new processes to prepare deeply hydrotreated gas oil have been developed. The process flow of a two-stage deep desulfurization process, which is one of the new processes, is shown in Fig. 5.8.[99] In this process, deep desulfurization of less than 0.05 wt% sulfur is achieved at higher temperatures, higher SV and lower pressures in the first reactor, and the color is removed from highly colored gas oil at lower temperatures (200–300°C) in the second reactor. H$_2$S is not removed between the first reactor and the second reactor. Since noble metal catalysts do not work well with higher than ppm levels of concentration of sulfur, they cannot be used.[100] It is reported that Ni-Mo and Co-Mo catalysts can work for the upgrading of colored oil even under higher concentrations of H$_2$S,[101] and these are used in the second reactor. In another two-stage deep desulfurization process, H$_2$S is removed between the first reactor and the second reactor. In this case, noble metal catalysts can be used in the second reactor.[102]

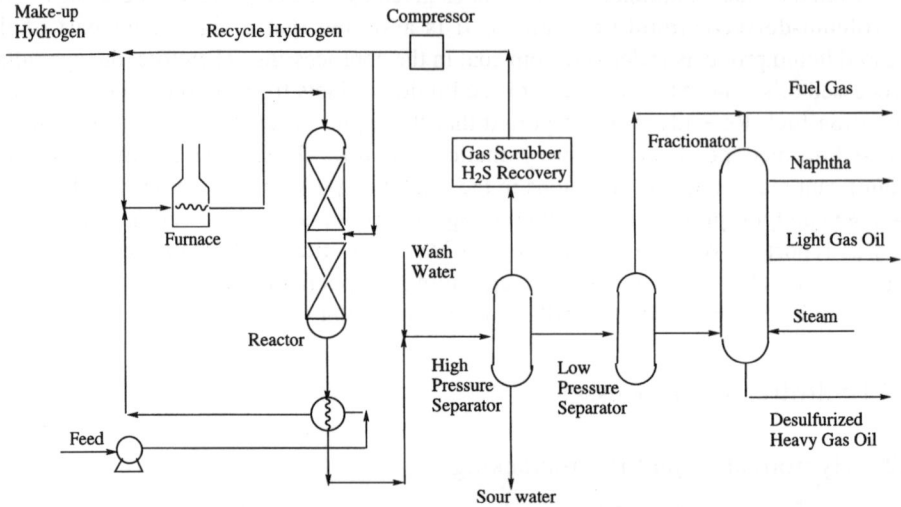

Fig.5.6 Typical vacuum gas oil hydrodesulfurization process.

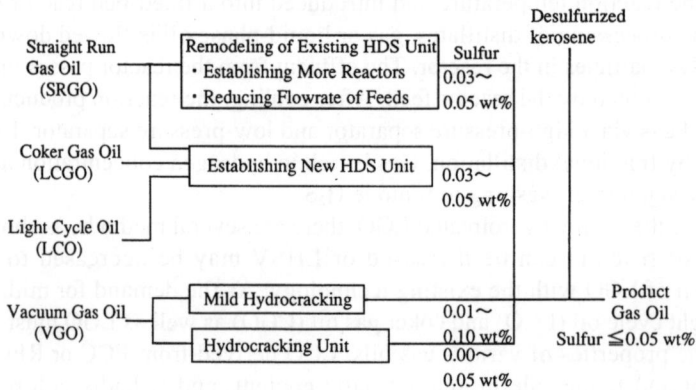

Fig.5.7 Production of deeply desulfurized gas oil.
(From M. Ushio and M. Hatayama, *Petrotech*, **17**, 702 (1994))

In the hydrocracking process shown in Fig. 5.9,[103] high grade middle distillates such as kerosene, jet fuel, gas oil are effectively produced from HGO, VGO, LCO, DAO, etc. Heavy naphtha produced simultaneously is rich in naphthene fraction and is very suitable for feeds of catalytic reforming. Unconverted fraction is hydrotreated significantly and is suitable for basic materials of lubrication oils. The optimal process (one-step process or two-step process) and appropriate reaction conditions are selected taking into consideration the difficulty of conversion of feeds or the desired products. In the one-step process, two reactors of hydrotreating and hydrocracking are combined and the conversion reaches 90%. In the two-step process (Fig. 5.9), products treated in the reactor of the first step are separated by distillation in which the bottom oil is further converted in the reactor of the second step. This process gives the maximum yields of middle distillates and enables the treatment of feeds which contain higher concentrations of nitrogen compounds or which are more difficult to

Table 5.10 Properties of Gas Oils

Origin	Straight Run Gas Oil (SRGO) Atom.Distillation	Light Cycle Oil (LCO) Distillate Fraction from FCC (Collected in 1991)	Light Cycle Oil (LCO) Distillate Fraction from FCC (Collected in 1992)	Light Cycle Oil (LCO) Distillate Fraction from RFCC	Coker Gas Oil (LCGO) Distillate Fraction from Residue Coker
Density (g/cm³)	0.8553	0.8970	0.8812	0.8793	0.8422
Sulfur (wt%)	1.22	0.15	0.07	0.09	1.38
Basic Nitrogen (ppm)	130	190	73	290	310
Composition (vol%)					
Aromatics	36	67	59	55	44
Olefin	1	7	4	6	0
Saturate	63	26	37	39	56
Distillation (°C)					
10%	286.0	194.5	196.0	214.5	219.0
50%	312.5	237.5	242.5	264.5	271.0
90%	348.5	308.0	307.5	328.0	365.5
Cetane Number	58.0	24.0	30.5	37.0	53.0
ASTM Color	L 0.5	3.5	L 1.0	L 1.5	L 5.5

(From M. Ushio and M. Hatayama, *Petrotech*, **17**, 702 (1994))

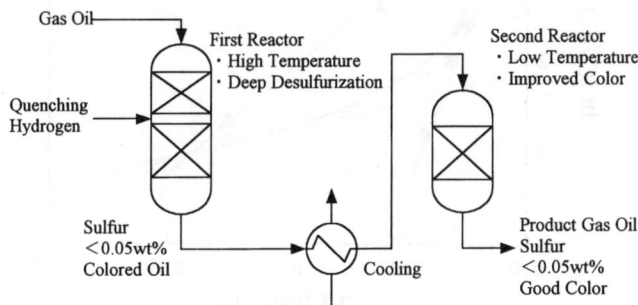

Fig.5.8 Two-stage desulfurization.
(From M. Ushio and M. Hatayama, *Petrotech*, **17**, 702 (1994))

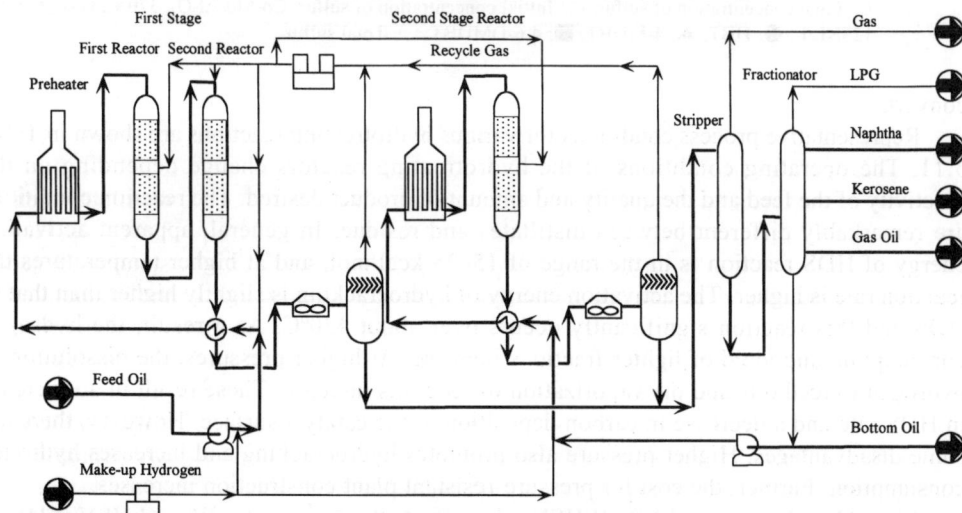

Fig.5.9 IFP hydrocracking process flow (two-stage).
(From M. Ishii, *Petrotech*, **17**, 175 (1994))

Table 5.11 Representative Process Conditions for Various Hydrotreating Reactions

Hydrotreating Process	Temperature (°C)	Pressure (atm)	LHSV (h^{-1})	Hydrogen Consumption (Nm^3 m^{-3})
Naphtha	260–350	10–35	2–10	2–10
Lighter fraction [a]	290–400	20–55	1.5–6	5–50
Heavier fraction [b]	340–425	50–140	1–4	50–180
Residue	340–425	55–170	0.2–1	100–220
Distillate HCR	260–480	35–200	0.5–10	150–400
Residue HCR	400–440	100–200	0.2–1	150–300

[a] kerosene, atmospheric gas oil. [b] vacuum gas oil

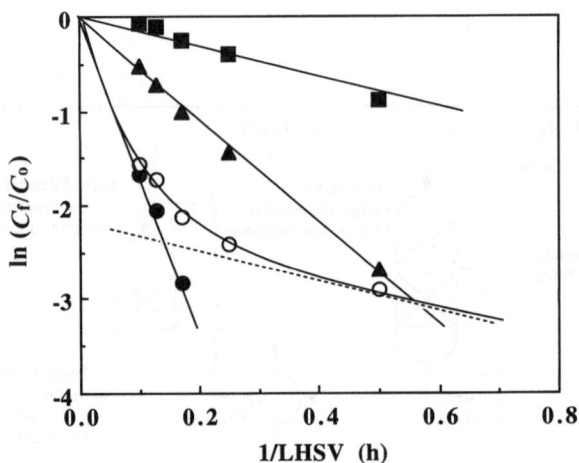

Fig.5.10   The effect of contact time on the remaining sulfur concentration.
C_f: Final concentration of sulfur, C_o: Initial concentration of sulfur, Co-Mo/Al$_2$O$_3$, 370°C, Gas/Oil = 125NL/L, ●: DBT, ▲: 4-MDBT, ■: 4,6-DMDBT, ○: Total sulfur

convert.

Representative process conditions for various hydrotreating reactions are shown in Table 5.11. The operating conditions of the hydrotreating reactors change depending on the reactivity of the feed and the quality and amount of product desired. The reaction conditions are remarkably different between distillates and residue. In general, apparent activation energy of HDS reaction is in the range of 15–25 kcal/mol, and at higher temperatures the reaction rate is higher. The activation energy of hydrocracking is slightly higher than that of HDS and this reaction significantly occurs over about 430°C. As a result, the hydrogen consumption and yield of lighter fractions increase. At higher pressures, the dissolution of hydrogen to feed oils and the vaporization of feed oils increase. These result in an increase in HDS rate and a decrease in carbon deposition to the catalyst surface. However, there are some disadvantages. Higher pressure also promotes hydrocracking and increases hydrogen consumption. Further, the cost for pressure-resistant plant construction increases.

Liquid hourly space velocity (LHSV) also affects the conversion. When LHSV is lower, that is, the contact time (1/LHSV) is higher, the conversion increases. Fig. 5.10 shows the effect of the contact time on the concentration of remaining sulfur in HDS of LGO. In general, HDS rate of total sulfur compounds behaves like a second-order reaction with respect to the total sulfur concentration, as shown in Fig. 5.10. On the other hand, the first-order plot of the sulfur concentration of each compound, DBT, 4-MDBT or 4,6-DMDBT, shows the linear relationship. The result indicates that HDS rate of each component follows the first-order reaction with respect to the sulfur concentration of each compound. The rate constant of each compound, which is estimated from the slope of the straight line, is very different from each other. HDS rate of total sulfur compounds at higher LHSV (lower contact time, 1/LHSV) is very close to that of DBT, which is relatively easy to desulfurize in HDS of LGO. However, HDS rate of total sulfur compounds at lowered LHSV (higher contact time, 1/LHSV) is very close to that of 4,6-DMDBT, which is most difficult to desulfurize in HDS of LGO. These results indicate that since the HDS rate constant of each compound is remarkably different, HDS rate of total sulfur compounds apparently behaves like a second-order reaction with respect to the total sulfur concentration. In contrast to HDS rate, HDN rate of total nitrogen compounds follows the first-order reaction with respect to the total nitrogen concentration.[2]

When the feed is present as a liquid and the content in the reactor consists of gas, liquid and solid phases, the reactor is referred to as a trickle-bed reactor. The catalytic activity decreases gradually due to coke formation and reaction temperature is increased to maintain a constant activity.[1,3,104] Coke can be removed by burning the catalyst. The regenerated catalysts can be used repeatedly.

5.2.2 Hydrotreating and Hydrocracking of Residues

Although the direct desulfurization process was originally developed to produce low sulfur fuel oil, the demand for fuel oil is decreasing. In recent years, therefore, this process has become more important as a pretreatment process before RFCC, which is one of most effective processes to obtain lighter fractions from residues.

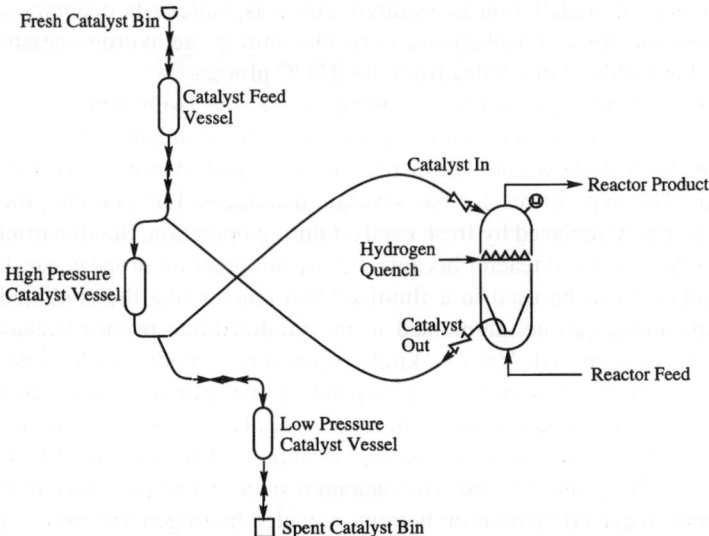

Fig.5.11 OCR reactor system.
(From B. E. Reynolds and R. W. Bachtel, *Petrotech*, **17**, 659 (1994))

Down-flow fixed-bed reactors are still used in the HDS of heavy petroleum feeds such as atmospheric residue. Since the HDS reactivity of heavy feeds is low, the reaction condition is more severe, as shown in Table 5.11. Further, residues include large amounts of heavy metals and asphaltenes which bring about coke and metal depositions and subsequently the deactivation of desulfurization catalysts during operation. To avoid this, a demetallation reactor is generally set up upstream of an HDS reactor. Metal accumulation on the catalyst varies from the top to the bottom of the reactor. Therefore, the choice of catalyst is important in the use of the fixed-bed reactor for heavy feed processing. When the vanadium content in feeds are higher, the fixed bed reactor needs the exchange of HDM catalysts which stop the operation of the reactor for a long time. Hydrocracking processes using the fixed bed reactor is similar to HDS processes using the fixed bed reactor. Although the conversion increases with severer reaction conditions, dry sludge is formed at higher conversions (40–50%).

To solve the problems in the fixed-bed reactor, the OCR (Onstream Catalyst Replacement)[105] and bunker flow processes[106] adopting a moving-bed reactor, and the H-Oil and L-C fining processes[107–111] adopting an ebullated-bed reactor, which enable the exchange of catalysts during the hydrotreating operation, have been developed. In the OCR and bunker flow processes the moving-bed reactor is used as a guard reactor which removes solids and heavy metals included in feeds, protects the fixed-bed reactor downstream, where exchange of catalysts is impossible, from plugging of the catalyst bed, and prolongs the cycle of the catalyst exchange. In the OCR process shown in Fig. 5.11,[105] the reactor feed is introduced from the bottom of the reactor and flows upward. Catalysts are introduced from the top of the reactor and waste catalysts are withdrawn from the bottom of the reactor. Since feeds and catalysts come into contact with countercurrent flow, the catalysts work effectively and the metal content in withdrawn waste catalysts is twice as much as the average metal content of the catalysts inside the reactor. 2–5% of catalysts are intermittently exchanged once or twice a week depending on the metal content in the feeds. The reactor product comes out from the top of the reactor and is introduced into subsequent HDM and HDS reactors. In the HDM reactor where deep demetallation is required, some asphaltene is decomposed to remove metal. The decomposition of asphaltenes in residue during the hydroprocessing contributes to increase in the yields of distillates from the RFCC process.

In the ebullated-bed reactor used in, for example, the H-Oil process[110] or LC-Fining process,[111] the catalyst particles are suspended by the upward stream of the liquid reactant and hydrogen. Fig. 5.12(a) shows the H-Oil reactor. The catalyst bed is expanded because of upflow stream. This type of reactor has several advantages. For example, the deactivating catalysts can be easily replaced by fresh catalyst during operation. Small particles of catalyst are avoided in the fixed-bed reactor because significant pressure drops occur. However, fine particles of catalyst can be used in a fluidized-bed reactor like the ebullated-bed reactor. Uniform temperatures can be maintained in the fluidized-bed reactor because the reactor contents can be fully mixed. Various kinds of operation modes such as desulfurization, demetallation and thermolysis can be used depending on the demand of products and 90% of vacuum residue can be converted. This process does not have the restriction by metals, sulfur, nitrogen, CCR, etc. because of the intrinsic advantage of the ebullated-bed. Fig. 5.12(b) shows the process flow where 55–90% of vacuum residue can be converted to distillates. The mixture of feeds, recycled distillation bottom, recycled hydrogen and make-up hydrogen is introduced from the bottom of the reactor.

When the conversion of residues increases in the moving-bed and ebullated-bed reactor

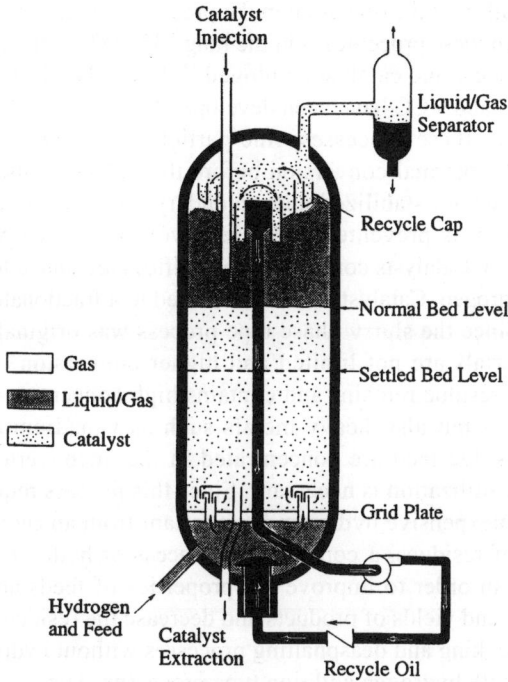

(a) H-Oil reactor.
(From T. Takahashi, *Petrotech*, **14**, 863 (1991))

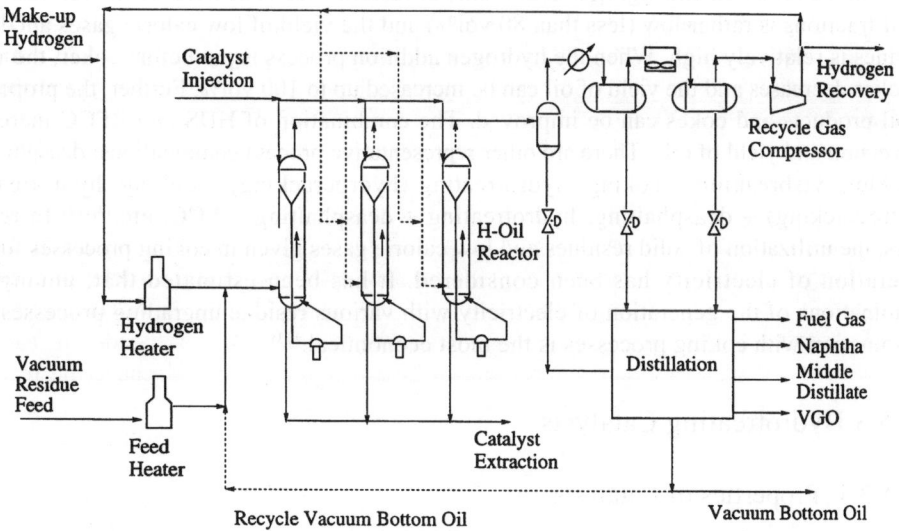

(b) H-Oil process flow sheet.
(From T. Takahashi, *Petrotech*, **14**, 864 (1991))
Fig.5.12

as well as in the fixed-bed reactor, dry sludge formed causes contamination or plugging of instruments and the stability of the operation in the processes is impaired. In general, the upper limit of the conversion in these processes is in the range 45–75%, although it differs depending on the kind of feed, process and catalyst employed.[13] Recently, slurry phase type processes such as CANMET, SOC and VCC have been developed to achieve higher conversion and long continuous operation. In these processes, fine particle catalysts supply hydrogen to free radicals generated in the thermal conversion before the polycondensation of these radicals occur. Since free radicals are stabilized and secondary conversion and polymerization are inhibited, coke formation is prevented and a relatively higher conversion of 90–95% is achieved by once-through. Catalysts containing iron sulfides are added to feeds and introduced into the reactor with hydrogen. Catalysts are concentrated in a fractionator and discharged with unconverted residue. Since the slurry phase type process was originally developed for coal liquefaction, raw materials are not limited and higher conversion is obtained. However, 5–10% of unconverted residue remains and contains high concentrations of sulfur. Further, not only waste catalysts but also heavy metals such as vanadium and nickel which are included in the raw residue feed are concentrated in the unconverted residue. Therefore, development of residue utilization is necessary. Since this process requires large amounts of hydrogen, a supply of inexpensive hydrogen is important from an economical point of view.

In the upgrading of residues, a combination process of hydrogen addition and carbon rejection is often used in order to improve the properties of feeds and oil products and to increase the conversion and yields of products and decrease the residue.[14,112] Carbon rejection type processes such as coking and deasphalting processes without hydrogen have economical advantages compared with hydrogen addition type processes. The coking process has higher vacuum residue conversion, high tolerance for various kinds of feeds, and low construction cost. Further, most heavy metals and CCR, and a significant amount of sulfur and nitrogen in feeds can be removed and significant amounts of residues are converted. However, the yield of oil fractions is rather low (less than 80 vol%) and the yield of low calorie gases and solid residues is relatively high. When the hydrogen addition process is put before coker, the yield of cokes decreases and the yield of oil can be increased up to 100 vol%. Further, the properties of oil products and cokes can be improved. The combination of HDS and RFCC increases conversion and yield of oils. There are other representative process combinations: deasphalting + coking; visbreaking + coking; hydrotreating (hydrocracking) + coking; hydrotreating (hydrocracking) + deasphalting; hydrotreating + deasphalting + FCC, etc.[14,112] In recent years, the utilization of solid residues and low calorie gases given in coking processes for the generation of electricity has been considered. It has been estimated that, among the combinations of the generation of electricity with various residue upgrading processes, the combination with coking processes is the most economical.[113]

5.3 Hydrotreating Catalysts

5.3.1 Properties of Catalysts

Large amounts of hydrotreating catalysts are used extensively for HDS, hydrocracking, and hydrogenation reactions in refineries. Consumption of fresh catalyst amounts to 70000 tons per year worldwide, with 10000 tons per year in Japan, following FCC catalyst consumption. Hydrotreating catalysts were originally developed from these for the hydrogenation of coal and coal-derived liquid. In hydrotreating catalysts, metal components with the activities of hydrotreating reactions such as HDS and hydrogenation are dispersed on an inorganic oxide support with high porosity. Generally, cobalt and molybdenum are used for the metal components and alumina is used for the support. A report for industrial application of Co-Mo/Al$_2$O$_3$ catalyst appeared as early as 1943.[114] In practical catalysts, about 3–5 wt% of CoO and 12–20 wt% of MoO$_3$ are supported on alumina. Ni and W are used instead of Co and Mo, respectively. Molybdena-alumina and tungsten-alumina are basic catalysts and Co and Ni are promoters. Various catalysts have been made by the combination of a basic catalyst with a promoter. Co-Mo, Ni-Mo and NiW catalysts are the most common combinations. The composition of these active components varies extensively depending on the designed products or selectivity for reactions in the hydrotreating process. The products and selectivity for the reactions are affected by both catalysts and reaction conditions. Excessive hydrotreatment consumes excess hydrogen. In contrast to supported Pt catalysts for reforming, these hydrotreating catalysts have hydrogenation activity even at higher concentrations of sulfur compounds.

Generally, hydrotreating catalysts are sulfided to obtain the active phase. This sulfiding procedure appears to affect the catalytic activity and stability significantly. Therefore, a number of attempts have been made to improve the sulfiding procedure. The presulfiding is typically performed at the beginning of hydrotreating process by introducing sulfur-containing feed into the catalyst.[115] Although H$_2$S is also used for presulfiding by adding it to the recirculating hydrogen stream, this method is being replaced by other improved methods. For example, carbon disulfide, dimethylsulfide (DMS), dimethyldisulfide (DMDS) are added to the recycle gas instead of H$_2$S. These reagents as well as H$_2$S are still toxic and need considerable time to achieve a sufficient sulfidation state. Another method is the use of an organic polysulfide (SULFICAT process),[116,117] which is added to the catalyst before the hydrotreating reaction. In the subsequent reaction, presulfiding initially occurs by sulfur or hydrogen sulfide released by hydrogenolysis of polysulfide. This off-site presulfiding seems to have many advantages: Polysulfides are stable and not so toxic as other reagents,e.g. carbon disulfide, DMS, DMDS and H$_2$S. However, the applicability of this method to all hydrotreating reactions has not been clarified yet, and further investigation is needed.

In hydrotreating reactions which mainly aim at removing heteroatoms, the cracking of C-C bonds is not needed so the acidity of the catalysts is weak. When heavy feeds containing large amounts of aromatics are treated, high catalytic activity of hydrogenation is required. In hydrotreating of gas oil or residues, the fixed-bed reactor is in the trickle-bed mode as mentioned above and the catalyst pore is filled by liquid. Reactants and hydrogen diffuse in this liquid phase and the reaction occurs on the inside surface of the catalyst. When a surface reaction is the rate-determining step, the effect of pore diffusion can be ignored. When the surface reaction is very rapid at higher temperatures, however, the concentrations of reactants in the pore inside become lower compared with those at the entrance of the pore and the pore diffusion significantly affects the reaction rate. Therefore, the design of the catalyst pore is very important to control the reaction and to achieve high catalytic activity. In general, the catalytic activity increases by increasing the surface area of the catalyst. However, when the pore diameter is made small to increase the specific surface area, the diffusion of the reactants to the pore inside the catalyst is inhibited and the overall reaction rate decreases. On the other hand, active surface area decreases more with larger pores than with smaller pores. The

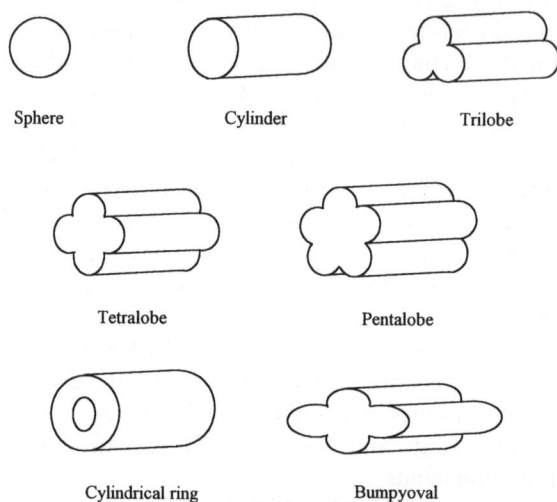

Fig.5.13 Catalyst shapes.

optimum pore size increases with increasing particle size since the diffusivity becomes lower with the use of larger particle size. Regarding the effect of pore size on HDS, optimum pore sizes have been observed.[118–122] The effect of catalyst pore size is significantly different for different reactions.

There are many kinds of hydrotreating catalysts for industrial use (Fig. 5.13). Typically, supports having dimensions of 0.8–5.0 mm are used. The most common shape of extrudates of alumina is the cylinder. As mentioned above, the rates of hydrotreating reactions are often affected by diffusion in the catalyst pores. Thus, the selection of catalyst particle size and shape as well as pore structure is important, especially for the hydrotreating of heavier feeds. For this purpose, small catalyst particles which have larger external surface area per unit volume are often used. Larger pores and shapes with larger external surface area are desirable: for example, tri-, tetra- or pentalobe, cylindrical ring, bumpy oval, etc. When the diameter is the same these catalyst shapes minimize the pressure drop. However, the activity per unit volume is less than that of catalysts with the cylinder shape.

During operation, deposition of metals and coke at the pore entrance occurs, limiting the diffusion of reactants and products. In hydrotreating of heavy feeds, bed plugging also occurs due to deposits on the exterior of the catalyst particles, especially at the inlet of fixed-bed reactors. These deposits consist of V and Ni sulfides, coke, iron compounds, etc. When the particle size of catalysts is very small, bed plugging at the top of the reactor is remarkable because of the action of fine catalyst particles as a filter. Therefore, the pressure drops and destruction of catalysts or increase in operating cost occurs. This problem of pressure drop can be reduced by packing larger particles of catalysts at the top of the reactor and smaller particles towards the bottom.

Another problem due to the use of small particles is related to the breaking strength of catalysts. Since larger particles of catalysts have larger mechanical strength, destruction of catalyst by pressure drop can be avoided. However, the pressure drop in a catalyst bed packed with smaller particles of catalysts is more significant than that with larger particles of catalysts.

When the difference in pressure becomes too large and exceeds the mechanical strength of the support, the catalyst is destroyed.

In hydrocracking, bifunctional catalysts having two kinds of active sites for hydrogenation and cracking are used. The use of acidic porous supports (SiO_2-Al_2O_3, zeolites, etc.) give rise to the active sites for cracking. The selectivity for cracking and hydrogenation in a catalyst must be controlled to achieve a particular purpose.[123] When zeolite supports are used, the hydrocracking reaction is preceded by a hydrotreating reaction to remove nitrogen and sulfur.

5.3.2 Catalyst Deactivation

Catalyst deactivation is caused mainly by coking and metal sulfide deposition. Many reviews describe deactivation studies of hydrotreating catalysts.[120,124–136] In many studies, the initial deactivation is attributed to coking[124,137,138] and long-term aging is attributed to metal deposition.[120,125,126,129,130,139] On the other hand, it has been shown that coking and metal deposition causing deactivation of catalysts cannot be separated.[134] A deactivation mechanism of the residue HDS catalysts in commercial reactors has been investigated.[136] Both coke and metals were responsible for decreasing the intrinsic reaction rate and effective diffusivity. It was proposed that, in the fast deactivation regime, metals poisoned the virgin active sites as well as the active site which coke covered initially. It was shown that deactivation by coke could occur throughout the whole run in residual oil hydrotreating, and that, in one instance, coke was responsible for approximately 80% of the deactivation.[133] It was also reported that both carbon and vanadium have a deactivating effect on toluene hydrogenation, cyclohexane isomerization, and thiophene HDS performed on used Ni-Mo/Al_2O_3 catalysts under 6 MPa hydrogen pressure.[140] However, a small amount of vanadium well dispersed inside the catalyst grain has been found to be more deactivating than a large amount of carbon. Less studied but no less important are changes of chemical state[141] and agglomeration of the active phase.[142–144] Eijbouts and Inoue studied the dispersion and homogeneity of a typical Ni-Mo/Al_2O_3 VGO hydrotreating catalyst by TEM. After extended use at high severity or after exposure to high temperatures, MoS_2 dispersion decreased, a separate Ni_3S_2 phase was formed, the samples became less homogeneous and contained more crystalline material. The HDN activity correlated with the MoS_2 dispersion. Reaction conditions had a major impact on the extent of the MoS_2 dispersion loss.[144]

Model compound activity measurements using Co-Mo/Al_2O_3 catalysts showed that the HDS activity was less sensitive to the effects of coke deposition than the HDN and hydrogenation activities.[145] Further, it was shown that the initial metals deposits had a stronger effect on the loss of activity for HDS than for HDN and hydrogenation, indicating that different sites were involved in these hydrotreating reactions. In another study using a Co-Mo/Al_2O_3 catalyst and DBT, naphthalene, and nickel and vanadium etioporphyrins as model compounds, the catalytic activity for nickel etioporphyrin HDM was maintained at least up to 50% metal loading.[146] HDS activity (hydrogenolysis of C-S bond) decreased with increasing nickel and vanadium on the catalyst but the hydrogenation activity increased. The hydrogenation activity of a Ni-sulfide phase was higher than that of a Co-Mo-sulfide phase, which caused an increase in the catalyst hydrogenation activity with increasing nickel loading on the catalyst. Effects of nickel sulfide and vanadium sulfide deposits on naphthalene hydrogenation and DBT hydroprocessing catalyzed by Ni-Mo/Al_2O_3 have been reported.[147] Nickel sulfide deposits only slightly affected the rate constants for hydrogenation in either network, but the vanadium sulfide deposits led to a decrease of at most 50% in the rate constants for hydrogenation of naphthalene and to doubling of the rate constants for hydrogenation of DBT. Nickel sulfide deposits did not affect hydrogenolysis of DBT to give BP but vanadium sulfide deposits decreased the rate of this reaction significantly. Further, nickel sulfide deposits have little activity for cracking to give low molecular weight products, but vanadium sulfide deposits have a relatively high activity for cracking. On the other hand, it has been shown that addition of specific promoters may lead to better resistance to deactivation. It was reported that fluorination of Co-Mo/Al$_2$O$_3$ catalysts had an inhibiting effect on deactivation by nickel and vanadium naphthenates.[148] XPS intensity data suggested that dispersion of deposited metals decreased with increasing fluoride concentration and that deposition occurs to a higher extent on the exterior of the particles than on the interior.

Coking is caused by high molecular weight polynuclear aromatics (PNA).[149,150] When the amount of high molecular weight PNA in feeds is larger, a larger amount of coke is observed on the catalyst.[149] Further, it was shown that coking depends on the acidic properties of the catalyst rather than on its pore size distribution.[151] More than one type of coke has been reported.[152–154] The deactivation of catalysts caused by coking is related to plugging of catalyst pore-mouths and covering of active sites.[120,155,156] In another case where either a light cycle oil or a coal residuum was hydrotreated on a commercial Ni-Mo/Al$_2$O$_3$, it was proposed that the loss in catalytic activity resulted from blockage of the MoS$_2$ crystallite edges, which was regarded as active sites, by coke formed on the alumina support.[157] Muegge and Massoth prepared coked Ni-Mo-P/Al$_2$O$_3$ catalysts containing 2–12% C by coking reaction with anthracene.[150] In this case, the coke was evenly distributed throughout the catalyst particles, blocked only a small fraction of the pores and constricted internal pore mouths. The activities of HDS, hydrogenation (HYD) and HDN decreased with increasing coke content, and the order of deactivation was HDS > HYD > HDN. With a Co-Mo/Al$_2$O$_3$ catalyst previously used for the processing of a heavy oil fraction (VGO), it was shown by XPS that the catalyst surface was almost completely (96%) covered by coke.[158] Mo/Al$_2$O$_3$ and Mo/SiO$_2$ catalysts showed a much lower degree of coverage of the catalysts with coke (about 50%), corresponding to a relatively low degree of catalyst deactivation. With respect to the location of the coke, for Co-Mo/Al$_2$O$_3$ catalyst, the coke was randomly distributed over the entire catalyst surface, while with the Mo catalysts the coke tended to cover the active Mo phase. The above results suggest that the extent of coke formation and the location may change depending on the kind of catalyst used and feed processed. In general, the drop in the catalytic activity by coking can be minimized by operation under higher pressure. After the initial activity drop, which is still present, the rate of activity drop become constant and stable.

Larger amounts of nitrogen are observed in spent catalysts.[159–161] A significant enhancement in nitrogen content of coked catalysts over that in feed has been found.[137] Nitrogen is present as inorganic as well as organic nitrogen adsorbed on coked catalysts.[162–166] It is believed that adsorbed nitrogen compounds can play a significant role in catalyst deactivation and act as coke precursors.[160,167–170] Zeuthen et al. have proposed that sulfur in the coke is associated with the upper layers of coke while nitrogen may adsorb preferentially during initial coke laydown.[145] However, details are not well known.

Yoshimura et al. investigated the influence of oxygen-containing compounds on the changes in chemical states of supported metals, and on the deactivation in HDN activity for Ni-Mo catalysts in hydrotreating coal-derived oils.[141] The effect of oxygen-containing compounds on the deactivation in HDN was in the order O$_2$ > H$_2$O$_2$ > benzofuran > phenol >

Pore-mouth
plugging

Catalyst metal

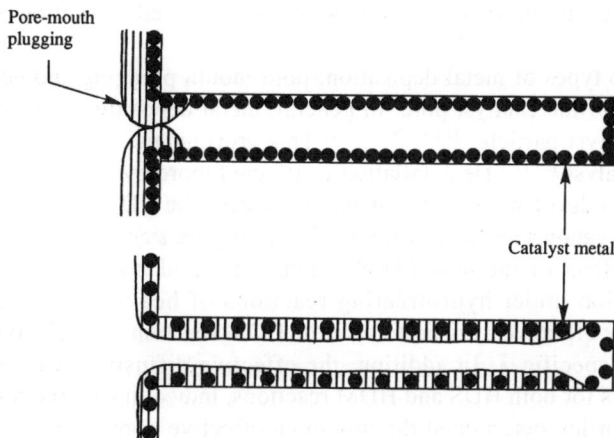

Fig.5.14 Effect of catalyst metal amount on heavy metals deposition.
(From H. Ozaki, *J. Jpn. Petrol. Inst.*, **36**, 180 (1993))

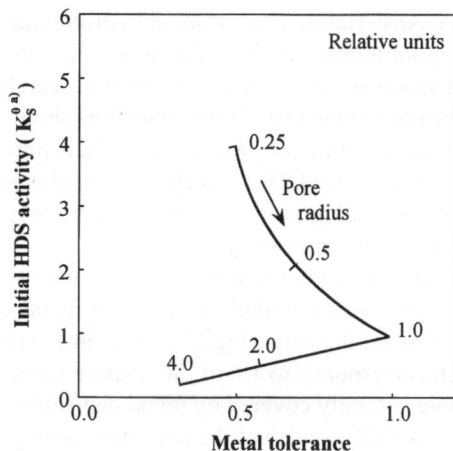

Relative units

0.25

Pore
radius

0.5

1.0

4.0

2.0

Initial HDS activity ($K_S^{0\,a)}$)

Metal tolerance

Fig.5.15 Activity metal tolerance.
a) K_S^0: Initial overall first-order rate constant for HDS.
(From H. Ozaki, *J. Jpn. Petrol. Inst.*, **36**, 180 (1993))

H_2O. XPS analyses of the heavily deactivated Ni-Mo catalysts showed that the amounts of Mo^{4+} decreased significantly, and almost all of the nickel sulfide and sulfidic sulfur was converted into Ni^{2+} and sulfate sulfur, respectively. When the degree of oxidation of the sulfided Ni-Mo catalyst via air was higher, the HDN activity was lower. They proposed that some of the metal sulfide was converted into the lower reactive oxides/sulfates via oxygen in the oxygen-containing compounds even under hydrotreating conditions.

Silicon occurs naturally in relatively small amounts in crude oils. Recent analysis of used catalyst from a commercial unit showed silicon deposits of 1% to 6% inside catalyst particles.[171] When the effect of silicon (methyltriethoxysilane) on hydrotreating of petroleum residue was investigated, it was shown that the deposition of silicon is relatively rapid compared to HDM of nickel and vanadium, and that silicon deposition reduced both catalyst

HDS and HDM activity. Further, silicon deposition influenced deposition profiles of metals, sulfur, and carbon.

There are two types of metal deposition: pore-mouth plugging and poisoning of active sites in the inside of the catalyst pore. In general, metal deposition occurs near the exterior surface of the catalyst particle.[125,126,138,172–174] This prevents the diffusion of reactants to the interior of the catalyst.[173,175] Demetallation is affected more by diffusion than HDS. Model reactions have been developed by several authors to describe HDM.[120,125,126,176,177] The activity and stability of the catalyst are significantly affected by pore size.[106,178,179] Chen and coworkers investigated the effect of the ratio (λ) of reactant molecular size to catalyst pore size on restrictive diffusion under hydrotreating reactions of heavy residue oils over Co-Mo catalysts.[180,181] The effective diffusivities for the HDM reaction were always smaller than for the HDS for any specific λ. In addition, the effective diffusivity values decreased with increasing λ values for both HDS and HDM reactions, indicating a large restrictive diffusion effect. Dai and Bartley determined the minimum effective pore diameters (MEPD) for the HDS and HDM reactions in residue hydroprocessing using data on catalyst pellet size, metal content, and intrinsic HDS activity of Co-Mo and Mo alumina catalysts.[182] For Arabian Light and medium vacuum residue feedstock, MEPD was *ca.* 4 nm for the HDS reaction and *ca.* 5 nm for HDM reaction. Absi-Halabi *et al.* studied hydrotreating of vacuum residue over four catalysts with different unimodal and bimodal pore size distributions having different proportions of meso- and macropores.[183] A unimodal pore catalyst with maximum pore volume in the medium mesopore range (10–25 nm diameter) showed the highest activity for HDS. Large pore catalysts having a major proportion of their pore volume in 100–300 nm diameter pores were more effective for HDM and HDN. Bimodal pore catalysts having large amounts of narrow pores showed a higher rate of deactivation than unimodal pore catalysts with a maximum amount of medium mesopores.

Figure. 5.14 shows the effect of pore size on the metal deposition. Fig. 5.15 shows the effects of relative value of pore size on HDS activity and metal capacity.[184] A wide pore enables deep penetration of metal deposits (Fig. 5.14 bottom) and can give a typical HDM catalyst. In this case, which corresponds to 1.0 of the relative value of pore size in Fig. 5.15, the inside of the catalyst pore is totally covered by metal deposition, and approximately first-order HDS reaction, where the diffusion is not the rate limiting step, is significantly retarded. Therefore, HDS activity of the catalyst is relatively lower while its metal capacity is relatively high. In contrast, in the case of a narrow pore which gives a typical HDS catalyst and corresponds to 0.25 of the relative value of pore size in Fig. 5.15, relatively high initial catalytic activity is maintained by inhibiting the penetration of metals to the interior of the catalyst. However, since its metal capacity is relatively low, pore-mouth plugging is caused by metal deposition (Fig. 5.14 top), resulting in rapid deactivation of not only HDM but also other reactions such as HDS. In Fig. 5.15, when a catalyst has one larger pore size than 1.0 of the relative value of pore size, the metal capacity drops rapidly and the HDS activity gradually decreases. In this case, poisoning of the inside of the pore is predominant.

The above phenomenon indicates that an optimal hydrotreating reactor which can treat various kinds of raw feeds containing different amounts of sulfur, nitrogen or metals can be designed by controlling the combination of the activity and the metal capacity of catalysts. When different kinds of catalysts, e.g. a HDM catalyst and a HDS catalyst, are used simultaneously in the same catalyst bed, better run length and activity are obtained than when either of the catalysts is used alone.[185] This composite catalyst filling is important to obtain relatively longer life and higher activity of catalyst.[134] When the HDS activity is important and a HDS catalyst is mainly packed into a reactor, the ability of HDM, that is, the capacity for metal accumulation, is relatively low. In contrast, when a HDM catalyst is predominantly packed into the reactor, the HDS activity is minimal. The optimum condition to attain an acceptably long life and relatively high activity of the catalyst can be obtained from an appropriate combination of HDS and HDM catalysts. In a typical composite catalyst filling, HDM catalyst is enriched in the inlet and HDS catalyst is enriched in the outlet of the reactor.

Based on the phenomena of the catalyst deactivation mentioned above, the direction for the design of the catalyst system has been described according to the metal content in feeds.[106]

1) To treat feeds having a metal content of less than about 25 ppm:

One kind of catalyst which has small pore size, high HDS activity and low metal capacity is enough.

2) To treat feeds having a metal content of 25 to 50 ppm:

The use of two kinds of catalysts is effective. The catalyst having a higher metal capacity than the catalyst downstream is put upstream, corresponding to the distribution of metal concentration in feeds. The HDS activity drops in this system.

3) To treat feeds having a metal content 50 to 100 ppm:

Besides 2), the use of the HDM catalyst, which is specially prepared to have the high metal capacity, upstream of the catalysts in 2) has economical advantages.

4) To treat feeds having a metal content more than 100 ppm:

Use of the moving multistage trickle-bed reactor, in which the catalysts can be compensated during the operation, at the upstream is economically attractive. To treat feeds having metal content more than several hundred ppm, the regeneration of the HDM catalyst has economical advantages. For this purpose, the DCR (Demetallation Catalyst Regeneration) process to compensate the resid bunker HDM process has been developed.[106]

5.3.3 Regeneration of Hydrotreating Catalysts

Regeneration of the spent catalyst is an attractive process to recover the original activity and stability, and a number of attempts to regenerate catalyst have been made.[163,164,186–198] Until the first half of the 1970's, regeneration process was performed *in situ* using a mixture of air

Table 5.12 Ratio for Recycle of Spent Hydrotreating Catalysts in Japan

Classification of Catalyst Use	Recycle Methods	Amount (*t*)	Ratio (%)
Residue Upgrading	Regeneration	100	1
	Metal Recovery	8600	96
	Waste	300	3
Others	Regeneration	1400	47
	Metal Recovery	1500	50
	Waste	100	3
Total	Regeneration	1500	13
	Metal Recovery	10100	84
	Waste	400	3

(From T. Suzuki, *J. Jpn. Inst. Energy*, **77**, 305 (1998))

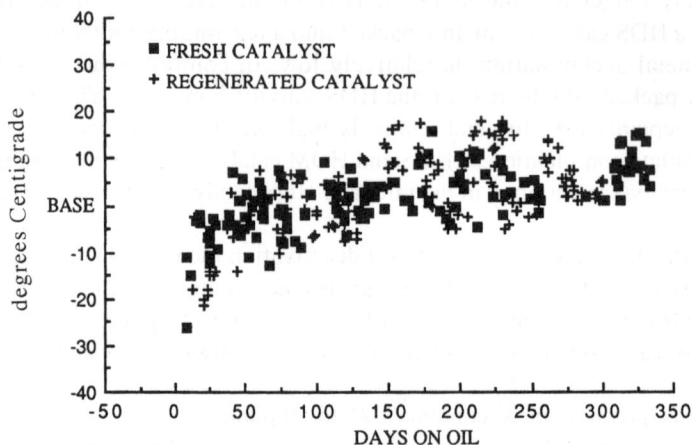

Fig.5.16 Normalized temperature for CFI.
(From T. Takahana, H. W. Homan Free and J. Mertens, *Nippon Keijen Seminar 1995*, **H-3**, 13 (1995))

and steam or oxygen and nitrogen. However, nowadays regeneration for most spent catalysts is performed in an exclusively *ex-situ* process. *Ex-situ* regeneration offers better recovery because of better temperature control.[117] The spent catalyst to which oil still adheres is extracted from the reactor and is followed by regenerative calcination or metal recovery. Table 5.12 shows the ratio of treatment of spent catalysts.[199] 75% of the catalyst is consumed in residue processing. In processing other than residue processing, the ratio of regeneration to metal recovery is almost same while most treatment of the spent catalyst in the residue processing is metal recovery because of large vanadium and nickel deposits cannot be removed by regeneration. 3% of the spent catalyst is disposed while the ratio of catalyst disposed is decreasing, approaching zero.

The coke on the catalyst is calcined in the presence of oxygen. This is called carbon burn-off regeneration, and used extensively in the industry. In the generation procedure, the initial evaporation of oil attached to spent catalysts (up to 250°C) and the subsequent oxidation of metal sulfide (300°C) occur at lower temperatures, and finally carbonaceous materials are burned at over 350°C. Therefore, it is important to consider the exothermic oxidation and to avoid overheating of catalysts. In general, burn-off temperature at regeneration is kept below the calcination temperature of fresh catalysts or the temperature at which reactions between active metal species and supports occur. To find the upper limit of burn-off temperature, the relationship between regeneration temperature and surface area is measured, although surface area does not mean directly catalytic activity. Agglomerated MoS$_2$ structures can be dispersed again to the original level after the regenerative calcination.[144] However, this does not mean complete recovery of the original activity. The ratio of activity recovery in regeneration changes depending on the severity of hydrotreating. When there is no metal deposition and the catalyst is not used under severe conditions, however, the activity of spent catalysts can be recovered up to the same activity level as that of a fresh catalyst by regeneration. For example, the activities of fresh catalyst and regenerated catalyst were compared for about one year in the Akzo-Fina CFI process for hydrotreating of distillates.[200] As shown in Fig. 5.16, there is no difference between the two catalysts.

Since deactivation occurs when cobalt and nickel sulfides are transformed into sulfates, these sulfides must be oxidized into oxides. Yoshimura *et al.* studied the oxidation behavior of metal sulfides by using TGA, XPS, EXAFS, and thermodynamics and showed that cobalt and nickel sulfates easily formed at lower regeneration temperature and higher oxygen partial pressure.[196] In actual operation, the formation of sulfates is inhibited by decreasing SO_2 partial pressure. Oh *et al.* studied oxidative regeneration of HDS catalysts which were compulsively deactivated by coking from 1,5-hexadiene.[198] During regeneration, physicochemical properties such as surface area, crystallinity, reducibility, and metal disribution changed significantly with the regeneration temperature. Increase in the dispersion of promotor species was observed in the catalysts regenerated at low temperatures and this gave rise to the enhancement of activity in comparison with the fresh catalyst. On the other hand, promoters migrated into the sublayer of alumina support at higher temperatures and thus resulted in the formation of $CoAl_2O_4$ or $NiAl_2O_4$ phases.

Metal sulfide deposits in spent catalysts are transformed to the oxide form by regenerative oxidation, but cannot be removed from catalyst surface and inhibit the complete recovery of catalytic activity. When vanadium deposits are present in spent catalysts, original surface area cannot be recovered even after regeneration.[199] The reduced surface area may bring about a decrease in catalytic activity and difficulty of recovery of the activity. As described in the previous section, metal sulfide deposits occur on the exterior surface of the catalyst particle and significantly in the reactor inlet. A large amount of deposit causes pore mouth plugging. The activity is recovered to a small extent in the regeneration of spent catalysts with metal deposition, especially in the reactor inlet. When the spent catalysts are crushed, however, the activity is remarkably recovered independent of the location of the catalysts in the reactor. The above indicates that metal deposition is physical poisoning rather than chemical poisoning of active sites and prevents sulfur compound from diffusing into the pore. The permissible upper limit of the amount of metal sulfide deposits for the regeneration of spent catalysts is about 2%.

Rejuvenation is a technology where spent catalysts are regenerated by leaching of foulant metals such as vanadium with a reagent. As described above, such metal deposits as vanadium and nickel sulfides cannot be removed by regeneration (regenerative calcination). In rejuvenation, these metal deposits are removed by reagents such as organic acids. Although approaches to recover catalytic activity by rejuvenation have been ongoing since the 1960's, none has yet been industrialized. In recent studies, sulfuric acid[201] and mixtures of oxalic acid with iron nitrate,[202] aluminum nitrate[203] and hydrogen peroxide[204] have been reported as reagents for rejuvenation.

Metal recovery is an alternative recycle method to regeneration when the activity of regenerated catalysts cannot be recovered to a level appropriate for reuse. Since the metal content in the used hydrotreating catalysts is higher than that in ore, these catalysts are attractive as mineral resources.[205] However, because of the complex composition and the variety of catalysts, it is not easy to recover metal economically. There exist industrialized metal recovery processes, i.e. a selective recovery process for molybdenum and vanadium[206] and a recovery process for all metals.[199,207] In the former process, some degree of molybdenum and vanadium is recovered by leaching with hot water after roasting together with sodium carbonate at temperature above 650°C and then precipitating vanadium as ammonium vanadate by adding ammonium chloride and subsequently molybdenum as molybdenum hydroxide after pH adjustment. According to this method, all nickel and cobalt in addition to a small amount of molybdenum and vanadium remained in alumina. In contrast, in the latter process, roasted spent catalysts are successfully dissolved in sulfuric acid by adding aluminum and all metals containing aluminum can be recovered. Molybdenum and vanadium are recovered as oxides after separation by solvent extraction, cobalt and nickel as hydroxides by neutralization and aluminum as aluminum sulfate. The completion of the latter process for all metal recovery has established the complete recycle of hydrotreating catalysts.[199] In recent studies, recovery of molybdenum and vanadium,[208,209] or cobalt and nickel[209,210] from spent HDS catalysts by means of liquid-liquid extraction has also been reported. Silva *et al.* proposed catalyzed electrochemical dissolution for spent catalyst recovery which offers advantages over conventional spent catalyst reclamation processes.[211] Raisoni and Dixit reported that molybdenum could be selectively leached from a spent Co-Mo/Al$_2$O$_3$ HDS catalyst using a solution of SO$_2$ in DMSO.[212] Cobalt remained unleached under the condition where molybdenum was leached.

References

1. H. Topsøe, B. S. Clausen and F. E. Massoth, *Hydrotreating Catalysis*, p.4, Springer-Verlag, Berlin Heidelberg (1996).
2. A. G. Bridge and E. M. Blue, *Kagaku Hannou To Hannouki Sekkei* (H. Tominaga and M. Tamaoki eds.), p.265, Maruzen, Tokyo (1996) [in Japanese].
3. B. C. Gates, J. R. Katzer and G. C. A. Schuit, *Chemistry of Catalytic Processes*, p.394, McGraw-Hill: New York (1979).
4. J. K. Minderhoud and J. A. R. van Veen, *Fuel Proc. Tech.*, **35**, 87 (1993).
5. J. W. Ward, *Fuel Proc. Tech.*, **35**, 55 (1993).
6. *Sekiyu Seisei Process* (Sekiyu Gakkai ed.), p.41, Kodansha Scientific Co. Ltd. (1998) [in Japanese].
7. P. Dufresne, P. H. Bigeard and A. Billon, *Catal. Today*, **1**, 367 (1987).
8. A. Corma, A. Martínez, V. Martínez-Soria and J. B. Monton, *J. Catal*, **153**, 25 (1995).
9. J. A. Anabtawi and S. A. Ali, *Ind. Eng. Chem. Res.*, **30**, 2592 (1991).
10. T. Takatsuka, Y. Wada, H. Suzuki, S. Komatsu and Y. Morimura, *J. Jpn. Petrol. Inst.*, **35** (2), 179 (1992).
11. X. Ma, K. Sakanishi and I. Mochida, *Fuel*, **73** (10), 1667 (1994).
12. X. Ma, K. Sakanishi, T. Isoda and I. Mochida, *Ind. Eng. Chem. Res.*, **34** (3), 748 (1995).
13. Y. Hori and T. Takatsuka, *Petrotech*, **18** (4), 289 (1995) [in Japanese].
14. M. Inomata, *Manufacturing Processes of Highly Purified Products, Handbook of High Purification Technology*, 887, Fuji Technosystem, 877 (1998) [in Japanese].
15. J. G. Speight, *The Desulfurization of Heavy Oils and Residua*, Marcel Dekker, New York (1981).
16. G. D. Hobson, *Modern Petroleum Technology*, John Wiley and Sons, New York, 1984.
17. J. P. Dickie and T. F. Yen, *Anal. Chem.*, **39**, 1847 (1967).
18. J. G. Speight and S. E. Moshopedis, *Chemistry of Asphaltenes* (J. W. Bunger and N. C. Li eds.), **195**, 1, Advances in Chemistry Series (1981).
19. S. Sato, Y. Morimoto, T. Takatsuka and H. Hashimoto, *J. Jpn. Petrol. Inst.*, **29** (1), 60 (1986).
20. C. Takeuchi, Y. Fukui, M. Nakamura and Y. Shiroto, *Ind. Eng. Chem., Proc. Des. Dev.*, **22** (2), 236 (1983).
21. S. Asaoka, S. Nakata, Y. Shiroto and C. Takeuchi, *Ind. Eng. Chem., Proc. Des. Dev.*, **22** (2), 242 (1983).
22. S. Peurex, S. Bonnamy, B. Fixari, F. Lambert, P. Le Perchec, B. Pepin-Donat and M. Vrinat, *Bull. Soc. Chim. Belg.*, **104** (4-5), 359 (1995).
23. D. K. Lee, S. K. Park, W. L. Yoon, I. C. Lee and S. I. Woo, *Energy & Fuels*, **9**, 2 (1995).
24. Z.-S. Ying, B. Gevert, J.-E. Otterstedt and J. Sterte, *Ind. Eng. Chem. Res.*, **34**, 1566 (1995).
25. Q. Guohe, L. Peipei and L. Wenjie, *Prepr.-Am. Chem. Soc., Div. Petrol. Chem.*, **38** (3), 703 (1993).
26. Y. W. Chen, M. C. Tsai and C. Li, *Ind. Eng. Chem. Res.*, **33**, 2040 (1994).
27. Y. W. Chen, W. C. Hsu, C. S. Lin, B. C. Kang, S. T. Wu, L. J. Leu and J. C. Wu, *Ind. Eng. Chem. Res.*, **29**, 1830 (1990).
28. D.-S. Yang, R. Dureau, J.-P. Charland and M. Ternan, *Fuel*, **75** (10), 1199 (1996).
29. N. Papayannakos, V. Kaloidas and S. Megalofonos, *Fuel Process. Technol.*, **28**, 167 (1991).
30. S. Kushiyama, R. Aizawa, S. Kobayashi, Y. Koinuma, I. Uemasu and H. Ohuchi, *Ind. Eng. Chem. Res.*, **30**, 107 (1991).
31. S. Kushiyama, R. Aizawa, S. Kobayashi, Y. Koinuma, I. Uemasu and H. Ohuchi, *Appl. Catal.*, **63**, 279 (1990).
32. T. Yasuhara, *Petrotech*, **9** (11), 1006 (1986) [in Japanese].
33. H. Tominaga, *Chenistry and Education*, **43** (3), 141 (1998) [in Japanese].

34. K. Ukegawa and S. Sato, *J. Jpn. Inst. Energy*, **71** (3), 202 (1992) [in Japanese].
35. H. Ozaki, *Kagaku Kogyo*, **48** (7), 532 (1997) [in Japanese].
36. S. Sato, M. Enomoto and S. Takahashi, *J. Jpn. Petrol. Inst.*, **32** (5), 268 (1989).
37. R. F. Sullivan, *Proc. 43rd Midyear Meeting, Refining Dep., Am. Petrol Inst.*, May 10 (1978).
38. K. M. Jeong and J. C. Montagna, *Am. Chem. Soc., Div. Petrol. Chem.*, **Vol. 29** (3), p.307 (1984).
39. T. Hisamitsu, K. Gomyo, F. Maruyama and H. Ozaki, *J. Jpn. Petrol. Inst.*, **30** (6), 404 (1987).
40. D. J. Curtin, U. S. Pat. US 4 029 571.
41. P. L. Cottingham and H. C. Carpentar, *Ind. Eng. Chem., Proc. Des. Dev.*, **6**, 212 (1967).
42. D. A. Young, U.S. Pat. US 4 046 674.
43. T. Hisamitsu, K. Gomyo and F. Maruyama, *J. Jpn. Petrol. Inst.*, **36** (6), 479 (1993).
44. T. Hisamitsu, K. Gomyo and F. Maruyama, *J. Jpn. Petrol. Inst.*, **36** (6), 485 (1993).
45. T. Hisamitsu, K. Gomyo and F. Maruyama, *J. Jpn. Petrol. Inst.*, **37** (2), 155 (1994).
46. S.-Y. Jeong, J. W. Bunger and C. P. Russell, *Energy & Fuels*, **8**, 1143 (1994).
47. S. Sato, S. Takahashi, A. Matsumura and M. Enomoto, *J. Jpn. Petrol. Inst.*, **37** (2), 123 (1994).
48. K. Murakawa, *Petrotech*, **15**, 529 (1992) [in Japanese].
49. R. K. Sharma and E. S. Olson, *Prepr. Pap.-Am. Chem. Soc., Div. Fuel Chem.*, **40** (3), 604 (1995).
50. K. J. Smith, L. Lewkowicz, M. C. Oballa and A. Krzywicki, *Can. J. Chem.* Eng., **72**, 637 (1994).
51. M. R. Gray, A. R. Ayasse, E. W. Chan and M. Veljkovic, *Energy & Fuels*, **9**, 500 (1995).
52. E. C. Sanford, *Energy & Fuels*, **9**, 549 (1995).
53. E. C. Sanford, *Energy & Fuels*, **8**, 1276 (1994).
54. D. C. Longstaff, M. D. Deo and F. V. Hanson, *Fuel*, **73** (9), 1523 (1994).
55. S. Kwak, D. C. Longstaff, M. D. Deo and F. V. Hanson, *Fuel*, **73** (9), 1531 (1994).
56. P. L. Jokuty and M. R. Gray, *Ind. Eng. Chem. Res.*, **31**, 1445 (1992).
57. J.-W. Kim, D. C. Longstaff and F. V. Hanson, *Fuel*, **76** (12), 1143 (1997).
58. S. M. Ricardson and M. R. Gray, *Energy & Fuels*, **11** (6), 1119 (1997).
59. S. M. Yui and S. H. Ng, *Energy & Fuels*, **9**, 665 (1995).
60. D. C. Longstaff, M. D. Deo, F. V. Hanson, A. G. Oblad and C. H. Tsai, *Fuel*, **71**, 1407 (1992).
61. L. C. Trytten, M. R. Gray and E. C. Sanford, *Ind. Eng. Chem. Res.*, **29**, 725 (1990).
62. S. Nakajima, *J. Jpn. Inst. Eng.*, **73**, 683 (1994).
63. T. Kotanigawa, *Petrotech*, **19**, 375 (1996) [in Japanese].
64. E. C. Moroni, *Am. Chem. Soc., Div. Fuel Chem., Prepr.* **31** (4), 294 (1986).
65. J. B. McLean, A. G. Comolli and T. O. Smith, *Am. Chem. Soc., Div. Fuel Chem., Prepr.* **31** (4), 268 (1986).
66. D. Gray, G. Tomlinson, A. El Sawy and A. Talib, *Am. Chem. Soc., Div. Fuel Chem., Prepr.* **31** (4), 300 (1986).
67. M. T. Martínez, I. Fernandez, A. M. Benito, V. Cebolla, J. L. Miranda and H. H. Oelert, *Fuel Processing Technol.*, **33**, 159 (1993).
68. F. Bergius, British Patent 148436 (1913).
69. O. Weisser and S. Landa, *Sulphide Catalysts: Their Propeties and Application*, Pergamon Press, London (1973).
70. Fine Particle Catalyst Testing US DOE Advanced Research Liquefaction. Symposium-Iron-based Catalysts for Coal Liquefaction", *Prep. Pap.-Am. Chem. Soc., Div. Fuel Chem.*, **38** (1), 1 (1993).
71. T. Suzuki, *Energy & Fuels*, **8**, 341 (1994).
72. T. Kabe, K. Yamamoto, K. Ueda and T. Horimatsu, *Fuel Process. Technol.*, **25**, 45 (1990).
73. T. Kabe, K. Kimura, H. Kameyama, A. Ishihara and K. Yamamoto, *Energy & Fuels*, **4**, 201 (1990).
74. T. Kabe, T. Horimatsu, A. Ishihara, H. Kameyama and K. Yamamoto, *Energy & Fuels*, **5**, 459 (1991).
75. T. Kabe, A. Ishihara and Y. Daita, *Ind. & Eng. Chem. Res.*, **30**, 1755 (1991).
76. A. Ishihara, H. Takaoka, E. Nakajima, Y. Imai and T. Kabe, *Energy & Fuels*, **7**, 362 (1993).
77. A. Ishihara, S. Morita and T. Kabe, *Fuel*, **74**, 1, 63 (1995).
78. M. Godo, M. Saito, J. Sasahara, A. Ishihara and T. Kabe, *Energy & Fuels*, **11**, 2, 470 (1997).
79. M. Godo, A. Ishihara and T. Kabe, *Energy & Fuels*, **11** (3), 724 (1997).
80. M. Godo, M. Umemura, A. Ishihara and T. Kabe, *AIChE J.*, **43** (11), 3105 (1997).
81. M. Godo, M. Saito, A. Ishihara and T. Kabe, *Fuel*, 1998, in press.
82. W. Qian, A. Ishihara, H. Fujimura, M. Saito, M. Godo and T. Kabe, *Energy & Fuels*, **11** (6), 1288 (1997).
83. S.-J. Liaw, R. A. Keogh, G. A. Thomas and B. H. Davis, *Energy & Fuels*, **8**, 581 (1994).
84. M. Machida, S. Ono and H. Hattori, *J. Jpn. Petrol.Inst.*, **40** (5), 393 (1997).
85. Y. Sato, *Catal. Today*, **39** (1-2), 89 (1997).
86. E. S. Olson and R. K. Sharma, *Prepr. Pap.-Am. Chem. Soc., Div. Fuel Chem.*, **39** (3), 706-9 (1994).
87. A. P. Raje, S.-J. Liaw, R. Srinivasan and B. H. Davis, *Appl. Catal.*, A, **150** (2), 297 (1997).
88. R. G. Dosch, F. V. Stohl and J. T. Richardson, ACS, Symp. Ser., 437, *Novel Mater. Heterog. Catal.*, **279** (1990).
89. A. B. Garcia and H. H. Schobert, *Fuel Process. Technol.*, **26**, 99 (1990).
90. H. Shimada, T. Sato, Y. Yoshimura, A. Hinata, S. Yoshitomi, A. C. Mares and A. Nishijima, *Fuel Processing*

Technology, **25**, 153 (1990).

91. H. Shimada, T. Kameoka, H. Yanase, M. Watanabe, A. Kinoshita, T. Sato, Y. Yoshimura, N. Matsubayashi and A. Nishijima, *Stud. Surf. Sci. Catal.*, **75**, New Frontiers in Catalysis, Pt.C, 1915 (1993).
92. A. B. Garcia and H. H. Schobert, *Coal Prep.* (Gordon & Breach), **9** (3-4), 185 (1991).
93. C. W. Curtis, K. J. Tsai and J. A. Guin, *Ind. Eng. Chem. Res.*, **26**, 12 (1987).
94. S. A. Fouda, J. F. Kelly and P. M. Rahimi, *Energy & Fuels*, **3**, 154 (1989).
95. P. M. Rahimi, S. A. Fouda, J. F. Kelly, R. Malhotra and D. F. Mcmillen, *Fuel*, **68**, 422 (1989).
96. Z. M. George, *Can. J. Chem. Eng.*, **68**, 519 (1990).
97. H. A. Rangwala, Z. M. George and A. H. Hardin, *Energy & Fuels*, **5**, 835 (1991).
98. T. Sato, T. Kameoka, Y. Yoshimura, H. Shimada, N. Matsubayashi, M. Imamura and A. Nishijima, *Sekiyu Gakkaishi*, **37** (3), 285 (1994) [in Japanese].
99. M. Ushio and M. Hatayama, *Petrotech*, **17** (8), 701 (1994) [in Japanese].
100. US Patent 3,841,995.
101. Tokkai hei 5-78670.
102. A. J. Suchanek & E. L. Granniss: 1995 NPRA Annual Meeting, p.AM-95-40 (1995).
103. M. Ishii, *Petrotech*, **17** (2), 174 (1994) [in Japanese].
104. A. V. Sapre and B. C. Gates, *Ind. Eng. Chem., Proc. Res. Dev.*, **20**, 68 (1981).
105. B. E. Reynolds and R. W. Bachtel, *Petrotech*, **17** (8), 659 (1994) [in Japanese].
106. P. B. Kwant, *Petrotech*, **8** (6), 543 (1985) [in Japanese].
107. R. E. Boening, N. K. McDaniel, R. D. Petersen and R. P. Van Driesen, *Hydrocarbon Proc.*, 59-62, September, (1987).
108. K. G. Tasker and L. I. Wisdom, *Hydrocarbon Tech. Intl.*, 73 (1988).
109. *Heavy Oil Processing Handbook* (H. Kashiwara *et al.*, eds.), p.48, The Chemical Daily Co. Ltd., Tokyo (1982).
110. T. Takahashi, *Petrotech*, **14** (9), 863 (1991) [in Japanese].
111. H. Yamakoshi, *Petrotech*, **14** (9), 865 (1991) [in Japanese].
112. T. Hisamitsu, *Manufacturing Processes of Highly Purified Products, Handbook of High Purification Technology*, 901, Fuji Technosystem, Tokyo (1998) [in Japanese].
113. M. Inomata, T. Takatsuka and N. Kawata, *Petrotech*, **19** (6), 499 (1996) [in Japanese].
114. A. C. Byrns, W. E. Bradley and M. W. Less, *Ind. Eng. Chem.*, **35**, 1160 (1943).
115. Y. Morimura, S. Nakata, T. Takatsuka and M. Nakamura, *J. Jpn. Petrol. Inst.*, **38** (3), 192 (1995).
116. S. R. Murff, E. A. Carlisle, P. Dufresne and H. Rabehasaina, *Prepr.-Am. Chem. Soc., Div. Pet. Chem.*, **38** (1), 81 (1993).
117. T. Suzuki and P. Dufresne, *Stud. Surf. Sci. Catal.*, **92**, Science and Technology in Catalysis 1994, 215 (1995).
118. A. L. Dicks, R. L. Ensell, T. R. Phillips, A. K. Szczepura, M. Thorley, A. Williams and R. D. Wraag, *J. Catal.*, **72**, 266 (1981).
119. A. Nielsen, B. H. Cooper and A. C. Jacobsen, *Symp. on Residuum Upgrading and Coking, Petroleum Chemistry, Inc.*, ACS, Atlanta Meeting **26** (2), p.440 (1981).
120. P. N. Hannerup and A. C. Jacobsen, *Am. Chem. Soc., Pet. Div. Preprs.*, **28**, 576 (1983).
121. B. M. Moyse, B. H. Cooper and A. Albjerg, Am-84-59, NPRA Annual Meeting, National Petroleum Refiners Association, Texas, (1984).
122. P.-S. E. Dai, D. E. Sherwood and B. R. Martin, *Chem. Eng. Sci.*, **45** (8), 2625 (1990).
123. R. Bezman, *Catal. Today*, 13, 143 (1992).
124. H. Beuther and B. K. Schmid, *Proc. 6th World Petrol. Congr.*, **3**, 297 (1963).
125. F. M. Dautzenberg, J. van Klinken, K. M. A. Pronk, S. T. Sie and J.-B. Wuffels, *Am. Chem. Soc. Symp. Ser.*, **65**, 254 (1978).
126. P. W. Tamm, H. F. Harnsberger and A. G. Bridge, *Ind. Eng. Chem., Proc. Des. Dev.*, **20**, 262 (1981).
127. B. Delmon, *Appl. Catal.*, 15, 1 (1984).
128. H. Topsøe, R. Candia, N.-Y. Topsøe and B. S. Clausen, *Bull. Soc. Chim. Belg.*, **93**, 783 (1984).
129. J. Laine, J. Brito and F. Severino, *Appl. Catal.*, **15**, 333 (1985).
130. D. S. Thakur and M. G. Thomas, *Appl. Catal.*, **15**, 197 (1985).
131. A. C. Jacobsen, B. H. Cooper and P. N. Hannerup, 12th World Petrol. Congr., Houston, Paper 3, p.97 (1987).
132. L. Kellberg, P. Zeuthen and H. J. Jakobsen, *J. Catal.*, **143**, 45 (1993).
133. J. Bartholdy and B. H. Cooper, *Prepr.-Am. Chem. Soc., Div. Pet. Chem.*, **38** (2), 386 (1993).
134. J. Bartholdy, P. Zeuthen and B. H. Cooper, AIChE Spring Nat. Meeting, Paper 56c (1994).
135. M. Ushio, M. Hatayama, T. Waku, M. Akiyama, Y. Okamoto and A. Inoue, *Prepr.-Am. Chem. Soc., Div. Pet. Chem.*, **40** (3), 537 (1995).
136. H. Koyama, E. Nagai and H. Kumagai, *Prepr.-Am. Chem. Soc., Div. Pet. Chem.*, **40** (3), 446 (1995).
137. K.-S. Chu, F. V. Hanson and F. E. Massoth, *Fuel Processing Technol.*, **40**, 79 (1994).
138. Z. Sarbak and S. L. T. Andersson, *Appl. Catal.*, A: Gen., **79**, 191 (1991).
139. E. Newson, *Ind. Eng. Chem., Proc. Des. Dev.*, **14**, 27 (1975).

140. G. Gualda and S. Kasztelan, *J. Catal.,* **161**, 319 (1996).
141. Y. Yshimura, T. Sato, H. Shimada, N. Matsubayashi and A. Nishijima, *Appl. Catal.*, **73**, 55 (1991).
142. Y. Yshimura, S. Endo, S. Yoshitomi, T. Sato, H. Shimada, N. Matsubayashi and A. Nishijima, *Fuel*, **70**, 733 (1991).
143. S. Eijbouts, J. J. L. Heinerman and H. J. W. Elzerman, *Appl. Catal.*, *A Gen.*, **105**, 53 (1993).
144. S. Eijsbouts and Y. Inoue, *Stud. Surf. Sci. Catal.,* **92**, Science and Technology in Catalysis 1994, 429 (1995).
145. P. Zeuthen, B. H. Cooper, F. T. Clark and D. Arters, *Ind. Eng. Chem. Res.*, **34**, 755 (1995).
146. E. P. H. Rautiainen and J. Wei, *Chem. Eng. Comm.*, **98**, 113 (1990).
147. M. Yumoto, S. G. Kukes, M. T. Klein and B. C. Gates, *Catal. Lett.*, **26**, 1 (1994).
148. Z. Sarbak and S. L. T. Andersson, *Appl. Catal.*, **69**, 235 (1991).
149. P. Wiwel, P. Zeuthen and A. C. Jacobsen, *Stud. Surf. Sci. Catal.*, **68**, 257 (1991).
150. B. D. Muegge and F. E. Massoth, *Fuel Processing Technol.*, **29**, 19 (1991).
151. L. W. Brunn, A. A. Montagna and J. A. Paraskos, *Am. Chem. Soc., Div. Petrol. Chem. Prepr.*, **21**, 173 (1976).
152. D. Alvarez, R. Galiasso and P. Andreu, *Oil Sands*, p. 206 (1977).
153. M. Ternan, E. Furimsky and B. I. Parsons, *Fuel Processing Tech.*, **2**, 45 (1979).
154. J. van Doorn, H. A. A. Barbolina and J. A. Moulijn, *Ind. Eng. Chem. Res.*, **31**, 101 (1992).
155. C. C. Hughes and R. Mann, *Am. Chem. Soc., Symp. Ser.*, **65**, 201 (1978).
156. A. G. Bridge, Paper presented at *Advances in Catalytic Chemistry II*, Symp. Salt Lake City, May (1982).
157. F. Diez, B. C. Gates, J. T. Miller, D. J. Sajkowski and S. G. Kukes, *Ind. Eng. Chem. Res.*, **29**, 1999 (1990).
158. K. P. De Jong, H. P. C. E. Kuipers and J. A. R. van Veen, *Stud. Surf. Sci. Catal.*, **68**, Catal. Deact., 289 (1991).
159. E. Furimsky, *Ind. Eng. Chem., Proc. Des. Dev.*, **17**, 329 (1978).
160. J. G. Speight, *Symp. on Advances in Resid. Upgrading, Div. Pet. Chem., ACS*, **32**, 413 (1987).
161. P. Zeuthen and A. C. Jacobsen, Tenth North American Meeting of the Catalysis Sciety, San Diego (1987).
162. F. E. Massoth, *Fuel Processing Technology*, **4**, 63 (1981).
163. Y. Yoshimura and E. Furimsky, *Appl. Catal.*, **23**, 157 (1986).
164. J. van Doorn, J. L. Bosch, R. J. Bakkum and J. A. Moulijn, Stud. *Surf. Sci.*, 391 (1987).
165. P. Zeuthen, P. Blom, B. Muegge and F. E. Massoth, *Appl. Catal.*, **68**, 117 (1991).
166. P. Zeuthen, P. Blom and F. E. Massoth, *Appl. Catal.*, **78**, 265 (1991).
167. P. Zeuthen, J. Bartholdy, P. Wiwel and B. H. Cooper, *Stud. Surf. Sci. Catal.*, **88**, 199 (1994).
168. E. Fitzer, K. Mueller and W. Schaefer, *Chem. Phys. Carbon*, **7**, 237 (1971).
169. J. van Doorn, J. A. Moulijn and J. J. Boon, *J. Anal. Appl. Pyrol.*, **15**, 333 (1989).
170. P. M. Boorman, K. Chong and R. A. Kydd, *Stud. Surf. Sci. Catal.*, **73**, Prog. Catal., 11 (1992).
171. M. D. Phillips and E. L. Sughrue, *Fuel Sci. Technol. Int'l.*, **9** (3), 305 (1991).
172. J. J. Stanulonis, B. C. Gates and J. H. Olson, *AIChE. J.*, **22**, 576 (1976).
173. T. L. Cable, F. E. Massoth and M. G. Thomas, *Fuel Processing Technology*, **4**, 265 (1981).
174. A. C. Jacobsen, Proc. of the NATO Advanced Study Institute on Surface Properties and Catalysis by Non-Metals: Oxides, Sulfides and Other Transition Metal Compounds (J. P. Bonnelle B. Delmon and E. Derouane eds.), p.305, D. Reidel Publishing Company, Dordrecht (1983).
175. B. G. Johnson, F. E. Massoth and J. Bartholdy, *AIChE. J.*, **32**, 1980 (1986).
176. B. J. Smith and J. Wei, *J. Catal.,* **132**, 41 (1991).
177. J. Bartholdy and P. N. Hannerup, *Stud. Surf. Sci. Catal.*, **68**, 273 (1991).
178. R. J. Quann, R. A. Ware, C.-W. Hung and J. Wei, *Adv. Chem. Eng.*, **14**, 95 (1988).
179. A. H. Hardin, M. Ternan and R. H. Packwood, CANMET report81-4E, Energy, Mines and Resources, Canada, 1981.
180. C. Li, Y.-W. Chen and M.-C. Tsai, *Ind. Eng. Chem. Res.*, **34**, 898 (1995).
181. M.-C. Tsai, Y.-W. Chen and C. Li, *Ind. Eng. Chem. Res.*, **32**, 1603 (1993).
182. P.-S. E. Dai and B. H. Bartley, *Stud. Surf. Sci. Catal.*, **75**, New Frontiers in Catalysis, Pt.C, 2543 (1993).
183. M. Absi-Halabi, A. Stanislaus, T. Al-Mughni, S. Khan and A. Qamra, *Fuel* **74** (8), 1211 (1995).
184. H. Ozaki, *J. Jpn. Petrol. Inst.*, **36** (3), 169 (1993).
185. A. C. Jacobsen, P. N. Hannerup, B. Cooper, J. Bartholdy and A. Nielsen, American Institute of Chemical Engineers, Spring Nat Meeting, Houston, 1983.
186. D. L. Trimm, *Stud. Surf. Sci. Catal.*, **53**, 41 (1990).
187. E. Furimsky and F. E. Massoth, *Catal. Today*, **17**, 537 (1993).
188. A. V. Ramaswamy, L. D. Sharma, A. Singh, M. L. Singhal and S. Sivasanker, *Appl. Catal.*, **13**, 311 (1985).
189. A. Arteaga, J. L. G. Fierro, F. Delannay and B. Delmon, *Appl. Catal.*, **26**, 227 (1986).
190. J. M. Bogdanor and H. F. Rase, *Ind. Eng. Chem., Prod. Res. Dev.*, **25**, 220 (1986).
191. E. Furimsky, J. Houle and Y. Yoshimura, *Appl. Catal.*, **33**, 97 (1987).
192. Y. Yoshimura, E. Furimsky, T. Sato, H. Shimada, N. Matsubayashi and A. Nishijima, *Proc. 9th Int. Congr. Catal.* (M. J. Phillips, M. Ternan eds.), p.136, Chem. Inst. Can., Ottawa (1988).
193. K. W. Babcock, L. Hiltzik, W. R. Ernst and J. D. Carruthers, *Appl. Catal.*, **51**, 295 (1989).
194. J. Walendziewski, *Appl. Catal.*, **52**, 181 (1989).

195. F. T. Clark, A. L. Hensley, J. Z. Shyu, J. A. Kaduk and G. J. Ray, *Stud. Surf. Sci. Catal.*, **68**, 417 (1991).
196. Y. Yoshimura, N. Matsubayashi, H. Yokokawa, T. Sato, H. Shimada and A. Nishijima, *Ind. Eng. Chem. Res.*, 30, 1092 (1991).
197. C.-S. Kim, F. E. Massoth and E. Furimsky, *Fuel Process. Technol.*, **32**, 39 (1992).
198. E.-S. Oh, Y.-C. Park, I.-C. Lee and H.-K. Rhee, *J. Catal.,* **172** (2), 314 (1997).
199. T. Suzuki, *J. Jpn. Inst. Eng.*, **77**, 305 (1998).
200. T. Takahama, H. W. Homan Free and J. Mertens, Prepr. Nippon Ketjen Seminar 1995, **H-3** (1995).
201. S. J. Hildebrandt, R. O. Koseoblu and J. E. Duddy, *Prepr.-Am. Chem. Soc., Div. Pet. Chem.*, **38** (1), 40 (1993).
202. M. Marafi, A. Stanislaus and M. Absi-Halabi, *Appl. Catal. B Environ.*, **4**, 19 (1994).
203. M. Marafi, A. Stanislaus and C. J. Mumford, *Catal. Lett.*, **18**, 141 (1993).
204. A. Stanislaus, M. Marafi and M. Absi-Halabi, *Appl. Catal. A Gen.*, **105**, 195 (1993).
205. M. E. Godoy, M. del C. Ruiz, M. W. Ojeda and J. B. Rivarola, *Lat. Am. Appl. Res.*, **25** (1), 47 (1995).
206. K. Sudo, *J. Jpn. Inst. Eng.*, **69** (11), 1042 (1990) [in Japanese].
207. P. Zhang, K. Inoue, K. Yoshizuka and H. Tsuyama, *Nippon Kagaku Kaishi*, 407 (1995) [in Japanese].
208. P. Zhang, K. Inoue and H. Tsuyama, *Kagaku Kogaku Ronbunshu*, **21** (3), 451 (1995) [in Japanese].
209. P. Zhang, K. Inoue and H. Tsuyama, *Energy & Fuels*, **9**, 231 (1995).
210. P. Zhang, K. Inoue and H. Tsuyama, *Kagaku Kogaku Ronbunshu*, **21** (3), 457 (1995) [in Japanese].
211. L. J. Silva, L. A. Bray and D. W. Matson, *Ind. Eng. Chem. Res.*, **32**, 2485 (1993).
212. P. R. Raisoni and S. G. Dixit, *Ind. Eng. Chem. Res.*, **29**, 14 (1990).

Index